KENYA
A Natural History

KENYA
A Natural History

Stephen Spawls and Glenn Mathews

T & AD POYSER
London

First published in 2012 by T & AD Poyser, an imprint of
Bloomsbury Publishing

Copyright © 2012 text by Stephen Spawls and Glenn Mathews
Copyright © 2012 chapter opener illustrations by Martin Woodcock;
other illustrations by Tim Spawls
Copyright © 2012 photographs as individually credited

The right of Stephen Spawls and Glenn Mathews to be identified as the
authors of this work has been asserted by them in accordance with the
Copyright, Designs and Patents Act 1988.

All rights reserved. No part of this publication may be reproduced
or used in any form or by any means – photographic, electronic or
mechanical, including photocopying, recording, taping or information
storage or retrieval systems – without permission of the publishers.

Bloomsbury Publishing Plc, 50 Bedford Square, London WC1B 3DP
Bloomsbury USA, 175 Fifth Avenue, New York, NY 10010

www.bloomsbury.com
www.bloomsburyusa.com

Bloomsbury Publishing, London, New Delhi, New York and Sydney

A CIP catalogue record for this book is available from the British Library
Library of Congress Cataloging-in-Publication Data has been applied for

Commissioning editor: Nigel Redman
Project editor: Jasmine Parker
Design and typesetting by Susan McIntyre

UK ISBN (print) 978-1-4081-3471-9

Printed in China by C&C Offset Printing Co., Ltd.

This book is produced using paper that is made from wood grown in
managed sustainable forests. It is natural, renewable and recyclable. The
logging and manufacturing processes conform to the environmental
regulation of the country of origin.

10 9 8 7 6 5 4 3 2 1

Page 2: *Vieillot's Black Weaver* Ploceus nigrerrimus. *(Stephen Spawls)*

Contents

Acknowledgements		6
Preface		7
Chapter 1	Geology	11
Chapter 2	Fossils and Hominins	40
Chapter 3	The Peopling of Kenya	65
Chapter 4	The Landscape, Climate and Weather	97
Chapter 5	The Vegetation and Habitats	135
Chapter 6	The Mammals	172
Chapter 7	The Birds	212
Chapter 8	The Reptiles	255
Chapter 9	The Amphibians	291
Chapter 10	The Freshwater Fishes	315
Chapter 11	The Arthropods	338
Chapter 12	The Marine Environment	368
Chapter 13	Conservation	398
References and Bibliography		431
Index		438

Acknowledgements

The production of a book like this invariably involves a lot of people. We are indebted to them all. We would like to start by thanking the hard-bitten professionals who looked at our chapters and kindly and painstakingly corrected our mistakes; in particular, we thank Sanda Ashe, Bob Drewes, Brian Finch, Kim Howell, Paula Kahumbu, Quentin Luke, Dino Martins, Ian Parker, Jos Snoeks and Alan Walker. Quentin Luke and Jos Snoeks also kindly identified pictures for us. David Anderson and Jo Darlington kindly assisted us with technical queries. Those errors that remain are ours. We would also like to thank those who lent us their superb photographic material: Bob Drewes, Peter Gorvett, Dino Martins, Tim Spawls, Paolo Torchio, Laura Spawls and two old friends and relatives who are sadly no longer with us, Jim Ashe and Harry Spawls. Over the years Gordon Boy has assisted us with his encyclopaedic knowledge of all things Kenyan. At Bloomsbury, our thanks are due to our editor, Nigel Redman, who suggested the concept and offered support and encouragement throughout, and Jasmine Parker for her kindly expertise. Barry Hughes assiduously hunted out museum literature for us, and the staff at Kew Herbarium and the Natural History Museum were extremely helpful. Professor James Newman and Yale University Press kindly gave us permission to quote from Newman's excellent book *The Peopling of Africa*. We would also like to thank those generous scientists who sent us their papers on request and also those who make copies of their publications free via the Internet. In these days of rapacious scientific publishing, where absurd amounts are often charged for a single paper, it is heartening to find that many scientists do not believe that scientific knowledge should be expensive.

We are grateful to Tim Spawls, who painstakingly did all the line drawings, as well as providing stimulating discussion and Martin Woodcock, whose lovely illustrations enliven the chapter openings. Steve extends his thanks to Ian and Joy MacKay, Andrew and Zoe Nightingale, Andrew and Margaret Botta, and Sanda Ashe, for their friendship and hospitality, to Jolly and Naaz Esmail, of Central-Rent-A-Car, for service beyond the call of duty, and his friends at the department of herpetology at the National Museum, especially Patrick Malonza, Victor Wasonga and Beryl Bwong. Steve also thanks Dick Palmer, his college principal, James Barr (sadly no longer with us) and Ian Cummings for intellectual discussions. Steve is also grateful to his students over 30 years for their stimulating repartee. Glenn thanks his co-author, for the writing, and companionship over 45 years, most recently on our safari to western Kenya. And finally, we would also like to thank our families, in particular our long-suffering wives, Laura and Karen, for forbearance and encouragement (and computer skills in the case of Karen, as Glenn is computer illiterate). It is their efforts that enabled us to find the time to do the work.

Preface

This book is about Kenya, which is a remarkably diverse country in terms of its topography and biodiversity. Sited astride the equator, in East Africa, Kenya has a landscape of great beauty, that varies from unspoilt palm-fringed beaches beside a sapphire sea to snow-capped peaks rising to more than 5,000m, from game-filled grassy plains to papyrus-fringed lakeshores, and from humid rainforest to near-desert. Its people, and its fauna and flora, are drawn from the four quarters of the continent of Africa and beyond. Kenya is a country of stunning scenery, with its lakes strung like pearls along the Great Rift Valley, mountain moorlands and tree-fringed rivers that twist through dry scrubland. It is a country that lies in the heart of the African savanna, where our ancestors came down from the trees and took those first tentative steps on the road to becoming us, modern humanity. The remains of those ancestors may be seen in Kenya's museums and the sites where they were found can be visited.

Kenya is also a showcase – we are tempted to say the showcase – of African wildlife. In Kenya, within half a day's drive of the capital city, on largely good roads, there are more than ten world-class conservation areas and reserves in which the visitor may see fifteen or more types of large mammal, including the 'big five' – elephant, Lion, Leopard, Buffalo and rhino – in a single day. A week's guided safari to a selection of national parks in Kenya should yield sightings of more than 40 large mammal species; for the enthusiastic birdwatcher a 'bag' of 500 species or more can be seen in a couple of weeks. For those who are interested and have the skills or the support, a wide range of reptiles, amphibians, insects or plants may also be found. Nowhere else in the world, outside of Africa, can such a variety of wildlife be easily seen; nowhere else in Africa is there such a variety of accessible wildlife and conservation areas in such a range of habitats. You can drive for six hours from Nairobi and be on a cold, bright moorland that resembles Scotland, on the banks of a shallow river running through desert, watching elephants drink, on the crater of a grumbling volcano, in a warm sea on a coral reef, or in golden grassland with a relaxed pride of Lions asleep a few feet from you.

Kenya is also a modern and thriving country, with a functioning infrastructure. Businesses work. You can buy a ticket for a plane, a train, a coach or bus, or hire a chauffeur-driven or self-drive car, without problems. Petrol stations are open and they have fuel. The telephones, post offices, computers and satellite communications function; the IT sector and mobile/cell phone services are vibrant. Banks are open regular hours, ATMs are common and functional, and money can be changed easily. The shopping is good. Hotels are open and contactable; you will be quoted an honest tariff and your reservations will function.

Kenya is not without its problems, however. For a country where most people and goods travel by motor vehicle, the state of the roads can be astonishingly variable, as can the standard of driving. Pernicious and long-term corruption, fuelled by greed, amongst public servants has resulted in the disappearance of, *inter alia*, much of the money intended to create a countrywide serviceable road system. There are still many desperately poor people in Kenya, and the occasional high-profile, violent crime occurs. Outside of the cities no emergency services functions (with the shining exception of the flying doctor service). There is still some tendency within Kenya to initially identify the other person by his or her tribe, rather than as a fellow Kenyan. Occasional acts of terrorism, by foreigners using Kenya as a base, have taken place. A three-tier charging system exists for wildlife reserves, and even for hotels; visitors may be asked to pay over fifteen times as much to visit a sanctuary as a citizen, which often causes resentment – not all visitors to Kenya are comfortably affluent. There are occasional shortages of essentials like water, gas and electricity. Kenya also has a number of animals that can bite or otherwise hurt you, and there are nastier diseases than those in temperate countries. Kenya is also possessed of a starkly beautiful but agriculturally unrewarding landscape; more than 60 per cent of the country receives less than 250mm annual rainfall; in those arid lands, crops cannot be relied upon. Such areas may be ravaged by drought and famine. This has pushed much of the population into the fertile areas of the south and west, creating a lot of pressure on the land.

But nevertheless, Kenya is a stable and pleasant land, with a track record of steady improvement. It has managed to avoid the military madness and the ill-conceived social experiments that have blighted so many of Africa's countries. Unusually for Africa, the traveller in Kenya can largely proceed without harassment, or demands for bribes, by men in uniform, with or without guns. Your identity papers are rarely asked for. You do not get stopped at roadblocks and shaken down. Kenya has a free press, and in most Kenyan towns you can buy a morning paper, printed the night before, that honestly reports, largely without fear or favour, what the government is up to.

Kenya has a lot going for it and is a popular destination for visitors. It is also home to an increasing number of interested naturalists and scientists. Many guidebooks and travel guides are available. These are strong on practical information but they usually lack detailed background. One book we recently consulted has only a single page on the natural history of Tsavo East National Park – one of the largest and most fascinating reserves in the country – but has four pages on the hotels of that park and how to get there. This sort of information is useful if you want to compare prices, but little more. On the other hand, the scientific papers on Tsavo – for example Lack, Leuthold and Smeenk's 1985 paper on the birds of Tsavo East, Greenway's 1969 checklist of Tsavo plants, Pohl and Horkel's work on the geology of Tsavo's stony hills, or the Bowkers' work on Ukambani amphibians – are often hard to obtain, and often highly technical. And so we have written this book, where we intend to cover the middle ground, to bridge the gap between the traveller's guide and the technical books and papers. We cover the natural history of the country – its geology, climate, flora, fauna and people, among other things. This book is for the reader who wants to know 'what' and 'why'. We have tried to summarise what we feel is the essential background information, and we have described the relevant literature, for those who want to find out more. This book will help you get to know Kenya. To use the jargon of business, this book should 'get you up to speed' on most aspects of Kenya's natural history. As far as we know, no other book quite like this exists. We hope it is useful, and points the way for those who may wish to find out even more. As eminent palaeontologist and scientific author Richard Fortey has said, the more you know about nature, the more you see, and this is enriching.

We have taken a few liberties. We have not dealt with the very tiny organisms, with the non-arthropod invertebrates, or fungi. Our chapter on the peopling of Kenya takes the view that *Homo sapiens*, us, is just another large and successful inhabitant of the Kenyan landscape. We have given relatively little space to things like mosses and invertebrates, and much space to vertebrates and big landscape. This is a concession to popularity; most visitors and naturalists have come to see the large flora and fauna.

We are also aware that we may be charged with over-simplification and/or superficiality. For example, we have dealt with the peopling of Kenya in a single chapter of 30-odd pages, yet Africa's historians have written many thousands of pages, in erudite books and scholarly journals, on the migrations and tribulations of the people who today make up the thriving population of Kenya. Likewise, we have dealt with birds in a chapter of 43 pages, and there are a number of books on Kenya's birds alone; there are at least two major works on the bird faunas of Africa and their significance; and a vast number of scientific papers. However, our compression of the data has certain benefits: we are able to give an overview and insights into the big picture, benefits that cannot always be gained from wading through the primary literature.

This isn't a book that you have to work through from beginning to end; you can dip into it. Our chapters are self-contained; you can read them independently, or even read the subsections within each chapter independently. In order not to interrupt the flow, we have tried to keep referencing to a minimum. If you want the background information, read the sections on the literature at the end of each chapter, which explain where the bulk of our data have come from, or review the references listed at the back of the book. Most of the important literature can be located through the Internet. In order to make each section self-contained, we have repeated a few bits of information here and there. We make no apology for this; repetition of information is an accepted teaching technique, and once you have come across something five times, as educational philosophy shows, you remember it forever. Some of our material may seem both basic and obvious, but we have followed the advice of a wise ex-soldier, who once said, 'always state the obvious, so that everybody knows the same obvious'.

We want to make a somewhat weighty final point. With the rise of the personal computer, the Internet, satellite and multichannel television, there has been a depressing increase in the number of people – particularly young people – who find their entertainment solely indoors. It is a development that we view with alarm. Computers and the Internet are useful both in terms of learning and entertainment; so is television. But neither involves any real interaction or physical activity. No high-powered car chase in a computer game, or any other second-hand thrill obtained from staring at a screen can compare with the excitement of making discoveries outdoors. A day spent in front of a computer is not the same as a day spent on Crescent Island. A day in a forest is better than a day spent on YouTube. Passion for the natural world should come from the heart, not the hard drive. We both benefitted from our upbringing in Kenya; it is a country where it is a sheer pleasure to be out in the open air. This book is designed to make accessible the data on Kenya's natural history. It is designed to enhance the experiences made from actually being there. Get out there and get to know Kenya, and encourage others. Get out of your car. If you have travelled to Kenya's wild places, open the door and step into the wilderness, walk where you can. The wild places are not meant to be viewed solely through the windows of a car (although it is sometimes necessary to remain inside for safety). Visit Kenya's museums, both indoor and outdoor. If you are a parent, take your children. Experience the thrill of Kenya's wide-open spaces. Our young people will not want to preserve a landscape or creatures that they do not know. The words of Robert Michael Pyle, one of America's foremost conservationists, come to mind: 'What is the extinction of a condor or an albatross to a child who has never known a wren?'

Chapter 1

Geology

… the resultant series of earth movements have kept the region [Kenya] … in a condition of disorder and unrest. One region has been raised and another depressed … the evidence of these changes is apparent on every hand – J. W. Gregory, *The Great Rift Valley*, 1896

Why start a book on natural history with a long chapter on geology? Geology is not natural history, but we are not being perverse, and nor is it because one of the authors took a degree in geology and has a liking for it. There is a logical reason. Kenya has a distinctive landscape: it has a sea coast, a huge central plateau and a couple of big mountains; a rift valley with a chain of lakes bisects the land north to south; the north of the country is mostly near-desert, the eastern part is flat and arid, there is rainforest in the west. The types of vegetation, and the assemblages of animals that live in these places, are consequences of the environment and the climate, and these themselves are a consequence of the landscape and its underlying geology. And the landscape has changed: our earth is a dynamic place, continents move. In starting with the geology, we are explaining how, and why, Kenya has got the landscape that it has. If you find this chapter daunting, however, there is no need to read right through it now – have a glance at what's in here, try another chapter and come back to this one later.

We start by discussing the formation of Kenya and its ancient landscape in broad detail, and then, in a section entitled 'Fire and forces' we will look at what has happened in the last 65 million years. Other sections describe Kenya's economic minerals and the story of geological exploration in Kenya right up to the present day. An overview of the literature of Kenya's geology concludes the chapter. Each section is self-contained and can be read separately. If you find yourself struggling with the dates and names, like Miocene, or Cretaceous, or Cenozoic, our simplified geological timescale in Figure 1(1) should be a help.

THE FORMATION OF KENYA AND ITS LANDSCAPE

Geology and geography are useful disciplines; they give us clues about the landscape. The movement of plates and flows of molten rock creates landforms, weathering then wears them down. Some large rock masses move up, down or sideways, a process known as faulting. Landscapes give visual clues. In this chapter, we are going to talk about the geological processes that formed Kenya's landscape. Later, in Chapter 4, we describe the physical processes that have altered it.

Geology is the study of the earth and its origins; you are looking into the past. Kenya's oldest rocks are more than 3,000 million years old. There are two main kinds of geologists: economic geologists, who look for valuable and useful minerals (of which Kenya has distressingly small quantities), and academic geologists, who try to work out what happened in the past, what shaped it and what lived there. Although a lot of geological work has been done in Kenya, there is often disagreement between geologists as to what took place. This is the way science works: a theory or idea is put forward, it may be criticised, another idea is offered. After a while, a consensus emerges … or not.

Geology is often a difficult science. Much of the evidence is underground and inaccessible; the events occurred in the past, over huge time intervals. We can sometimes use the present for clues. But the further back we look, the more slender the evidence becomes. In addition, large areas of Kenya, for example much of eastern Kenya and the lower Tana River, have never been surveyed by a professional geologist. There is work to be done. Hypotheses are not proven. For example (of which more later), there is yet no scientific agreement as to how the Rift Valley, arguably Kenya's most magnificent landscape feature, was formed. But there are several plausible theories.

Mineral wealth: Tsavo green zoisite, Kajiado orange marble and Wamba amethyst. (Stephen Spawls)

SOME SMALLER DIVISIONS	
Holocene epoch:	11,700 years ago to the present
Pleistocene epoch:	2.6 million to 11,700 years ago (the term Quaternary includes both the Pleistocene and Holocene)
Miocene epoch:	23 million to 2.6 million years ago
Oligocene epoch:	35 million to 23 million years ago
Eocene epoch:	56 million to 35 million years ago
Palaeocene epoch:	65 million to 56 million years ago
Cretaceous period:	145 million to 65 million years ago
Jurassic period:	200 million to 145 million years ago
Triassic period:	250 million to 200 million years ago
Permian period:	300 million to 250 million years ago
Carboniferous period:	360 million to 300 million years ago
Devonian period:	420 million to 360 million years ago
Silurian period:	445 million to 420 million years ago
Ordovician period:	490 million to 445 million years ago
Cambrian period:	540 million to 490 million years ago

SOME LARGER DIVISIONS	
Cenozoic era:	65 million years ago to the present (the term Tertiary may be used for the period from 1.8 million years ago back to 65 million years ago although some geologists dislike it)
Mesozoic era:	250 million to 65 million years ago
Palaeozoic era:	540 million to 250 million years ago

SOME EVEN LARGER DIVISIONS	
Phanerozoic eon:	540 million years ago to the present
Proterozoic eon:	2,500 million to 540 million years ago
Archean eon:	3,800 million to 2,500 million years ago
Precambrian supereon:	from the start of the earth 4.6 billion years ago to 540 million years ago (this includes the Archean and Proterozoic eons)

Figure 1(1) Simplified geological timescale. In our text we have used a very simplified mixture of epochs and periods, which may upset geological purists, but should enable readers to get to grips with the timescale. You can find out more by putting 'geological timescale' into any search engine.

The Kenyan landscape, in its present form, has been developing over the last 3,000 million years. There are rocks of that age that can be dated in western Kenya, around Lake Victoria; they are the oldest rocks in the country. But before summarising Kenya's geological history (and then going into detail), it is worth saying something about how Kenya actually came to be where it is today, strange though this may sound.

Our present universe began about 14 billion years ago, with a massive explosion, the so-called 'big bang', which scattered matter outwards. From the light that reaches us from distant galaxies we can tell that they are travelling away from us, so we know the universe is still expanding. Our earth is just over four and half billion years old. Our solar system was formed when debris from a supernova, or exploding star, was pulled together by gravity. Most of the debris went into the sun, the remainder formed the planets. The earth was originally so hot that the material that makes up solid rock today was liquid then. The interior of the earth remains hot; in the deep gold mines of southern Africa the miners wear refrigerated suits. This temperature increase inwards is called the geothermal gradient.

After its formation the earth cooled most rapidly on the outside, like a pie removed from the oven. As it cooled, metals such as iron and nickel sank to the centre, while the lighter silicon-based minerals floated up. At a crucial point, more than 3,000 million years ago, the liquid rock, or magma had cooled enough to become solid at the surface, forming the first continents. Once the land masses appeared, life started shortly after that.

We don't know exactly how life started; it is an improbable event, although the American biochemist Stanley Miller's experiments give some clues. Miller circulated a number of simple compounds and elements, the sort that might be present in a young earth, in a glass container, and subjected them to heating and cooling and electric shock, simulating the effects of the sun and lightning. After a time, to Miller's delight, the mix was found to contain some complex biochemicals, although how life went from there to producing a replicating organism is not known … and may never be known. Life did start, just once, in water. This hypothesis is supported by several pieces of evidence. All living things make proteins in the same way, using the same small range of amino acids; all rely on a chemical called ATP (adenosine tri-phosphate) to power energy-using functions in cells; virtually all have similar DNA and every living organism (viruses excepted) relies on water for its transport medium.

Once the surface rock began solidifying, geologists believe that several 'supercontinents' (very large undivided landmasses) formed in succession. The last really big one was Pangaea, created about 250 million years ago, which almost immediately split into two: Laurasia (north) and Gondwana (south). Africa as we know it did not come into existence until fairly recently (in geological terms).

These landmasses ride on large chunks of the earth's crust, called plates, and they move slowly. This is the theory of continental drift, and the study of the features involved is called plate tectonics. Until surprisingly recently, the supporters of the theory were often ridiculed, even by fellow geologists. What forces, they asked mockingly, could cause a continent to move? The man who worked it out was a brilliant pioneering geologist called Arthur Holmes. Holmes went to Mozambique in 1911 and nearly died there (from blackwater fever) but his observations on the rocks of the Mozambique belt (part of which underlies much of eastern Kenya) lead him to his first big idea, that the earth was older than the 20 million years suggested by the physicist Lord Kelvin. This lead to Holmes' first book, *The Age of the Earth* (1913), published when he was only 23.

Holmes' second big idea concerned plate tectonics. He realised that these plate movements are powered by two processes: the cooling and sinking of the plates and the upwelling of hot magma from the interior. Holmes saw that the earth's plates move relative to one another. They go up, down and sideways, powered from below, by a process called convection (the same process as causes hot air balloons to ascend). The heated fluid, the magma, down inside the earth, expands, thus gets less dense and rises up. In the centre of the Atlantic, rising hot rock at the mid-Atlantic ridge comes up and spreads out both ways; pushing both east towards Africa and west towards the Americas; the process is called sea-floor spreading. It's like a gigantic conveyor belt, but rolling both ways (albeit very slowly), moving the continents apart. Holmes laid out these mechanisms very clearly in a classic textbook entitled *Principles of Physical Geology* (1944). Not everyone accepted these new ideas. One hostile reviewer fretted that Holmes' explanations were so lucid that students might actually believe them!

The supercontinent Gondwana split up over a period of 150 million years, pushed apart by these stunning and slow forces. The huge land became a group of smaller plates on which rode South America, Antarctica, Africa, Madagascar, India and Australia, plus a few other bits. These lands are widely separated, but if you draw their shapes (including their continental shelves) you will find that, like a jigsaw, they fit very precisely together. Africa, as we know it today, became visible about 120 million years ago, when South America split off and began drifting

Geothermal heat from the hot rock: steam jet, Hell's Gate. (Stephen Spawls)

away westward. Thirty million years ago, Africa's drift northward caused it to bump into Eurasia, it pulled away again about 23 million years ago, and then rejoined 17 million years ago. A huge sea out to the east of the two landmasses, the Tethys Ocean, was closed off to become the Mediterranean Sea and – as explained in Chapter 6 – a number of Eurasian mammals took the opportunity to nip down into Africa.

Three further processes shaped – and continue shaping – Kenya into the landscape we see today. One process is the eruption of the molten rock from the mantle; that portion of the earth's interior that lies below the crust. When this hot liquid cools, it forms what are called igneous rocks. The type and structure of such rocks depends on the conditions when the magma starts to cool, and the minerals within them.

If the magma does not emerge onto the surface, it cools slowly. (Imagine turning off the oven and leaving the pie inside. The heat will escape from it very gradually.) When this happens, as the magma solidifies rock crystals grow slowly and the grain or crystal size is large; such rocks have a visible crystalline structure of differing minerals and often differing colours. Typical igneous rocks include granite (a rock rich in quartz) and basalt (which has no quartz at all). Basalt forms the magnificent cliffs around Hell's Gate. Hell's Gate gorge itself was formed when water from a superlake where Lake Naivasha stands today flowed outwards. A thousand metres under the gorge are the hot rocks that power the Olkaria geothermal project there, Africa's first geothermal electricity-generating station.

If hot magma moves rapidly upwards, under pressure, it may erupt onto the surface, forming lava flows and volcanoes. Once outside, it loses heat very quickly. If it forms rocks under these conditions, the rocks look homogeneous and the grain size or crystals cannot be seen. Much of the present landscape of central Kenya is due to a mixture of rock movement and such volcanic activity. If the magma is liquid, it will flow, and it may flow huge distances before it cools. The Yatta Plateau is one such lava flow.

The second process involves the huge forces and heat flows caused by moving plates or eruptions of magma. Liquids cannot be compressed. The forces created by a huge rising mass of molten rock are large. The physical stress involved can cause existing rocks to bend or break; the heat involved can chemically change the nearby

Lava flow, Kibwezi, volcanic Chyulu Hills in background. (Stephen Spawls)

rocks. Rocks affected by heat and pressure are called metamorphic rocks; examples include marble, schist and gneiss. Most of the hills in Tsavo are metamorphic, and the hot rock that baked and pressurised them also injected streams of molten, mineral-rich fluid into the original rocks along lines of weakness. These reacted and cooled to produce semi-precious and precious minerals like ruby and tsavorite, but in relatively small quantities.

The third process is erosion and weathering – the effects of water, air and temperature on exposed rock. We have expanded on this in Chapter 4. Weathering can transform a landscape; it can grind down a mountain in time. The debris produced is transported downwards by water, or air, and deposited in suitable basins. Eventually, such sediments become compacted, and form sedimentary rocks. Their names give a clue to the particles that formed them: sandstones, mudstones, siltstones. Limestone is a sedimentary rock, formed when calcium carbonate (often from the remains of dead sea creatures) is deposited in shallow water. Fossils are usually found in sedimentary rocks: living things died and when their remains were covered by falling sediment, some became fossils. Dead animals and plants cannot get into igneous rock, although sometimes ash from volcanic explosions (which is thus igneous) will cover up or encase a dead organism, turning it into a fossil; this is a significant process in the formation of hominin and associated animal fossils. Very rarely fossils are found in sedimentary rocks that have been metamorphosed; usually metamorphosis cooks and crushes them.

These processes take place over very long periods. The time spans involved are really gigantic and often difficult for us, who only live on average 60–80 years, to comprehend. A helpful analogy for grasping the vastness of time is to imagine that the entire history of the earth, 4,600 million years, compressed into a 24-hour day, i.e. the history of the world starts at midnight and ends the following midnight. The continents start to form between 05:00 and 06:00 hrs. The first living things appear just after 6 o'clock in the morning. All life is confined to water between 06:00 and 22:00 hrs. Plants make it onto land around 22:00, dinosaurs arrive at 22:45, the first mammals appear at 23:30 hrs. Africa is formed at 37 minutes to midnight. Virtually all dinosaurs become extinct at 20 minutes to midnight. The first hominins appear around 2 minutes to midnight and Kenya becomes independent less than 1/1,000th of a second before midnight. We humans haven't been around for long in terms of the earth's history.

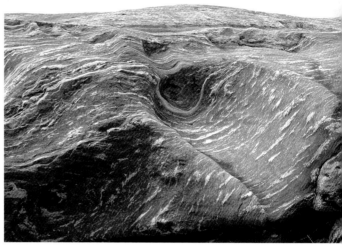

Convoluted metamorphic rocks, schists at Lugards Falls, Tsavo East. (Stephen Spawls)

Left: Lugard's Falls, Tsavo East, the 13-million year old Yatta Plateau on the horizon. (Stephen Spawls)

We have two main processes that tell us how old rocks are: radioactive dating and geochronology; they are often used together. Geochronology consists of looking at a succession of rock layers and trying to work out how long they took to form. Radiometric dating consists of calculating how old a rock is by finding out how much of the original radioisotopes within the rock has disappeared since it formed.

The earliest East African geological event that we have any clear idea about was the formation of the Tanzanian craton, more than 3,000 million years ago, in an eon known as the Achaean. A gigantic mass of liquid rock emerged from the mantle, spread out, cooled, solidified and sank. The Tanzanian craton extends southwest from Tanzania to Angola, and west to Gabon. It is a largely flat, crystalline plain; its northern edges just touch Kenya. This huge mass of rock probably formed the major part of an ancient continent, but it was not Africa. When the Tanzanian craton was formed, Africa did not exist.

At the same time, large rivers of volcanic lava erupted in what is now western Kenya, forming mountains. These mountains were then eroded by fast-flowing rivers, and the rock debris that formed was deposited in layers. Plate movement and associated forces caused these rocks to become folded and injected with fresh hot magma. Rocks from this time can be seen along the road from Kisumu to Kakamega. Then, about 1,000 million years ago, more floods of lava poured out, also in western Kenya; forming a complex known as the Nyanza shield, on the north and south sides of the Winam Gulf. In geological terms, a complex means a series of strata of various different origins, but it is an appropriate name because the Nyanza Complex is complex, consisting of a mixture of lavas, granite and sediments, and its relationships with the nearby rocks and relative ages is poorly known.

Kenya's geological record then goes quiet for a while, until the formation of the Mozambique mobile belt, a huge belt of disturbed rocks that underlies most of central Kenya and is exposed in eastern Kenya. Evidence indicates that two huge blocks of land – East and West Gondwanaland – collided, twice along a north–south line, roughly 800 and 580 million years ago. The resulting heat and pressure associated with this cataclysmic collision created a great band of changed rocks – they were almost certainly mountains – which geologists now know as the Mozambique belt (they were originally recognised in Mozambique). This belt consists mostly of gneisses, typical metamorphic rocks that have been formed by heat and pressure. Evidence for this collision is the heavily folded strata and rocks called ophiolites ('snake stones', they look like snake skin) in north-western Kenya, in West Pokot. Ophiolites are associated with all the great collision mountains of the world. The heavily folded strata between Baragoi and Barsaloi are indicative of two huge plates meeting and causing buckling.

At the time, climates were different: much of the Sahara was covered with ice and Africa was, oddly, close to the South Pole (as indicated by the horizontal and vertical orientation of the minerals within the rocks). The massive plate collision closed off a huge body of water, the 'Mozambiquan Ocean', and formed a massive mountain range. The original rock – what geologists call the country rock – was so changed by heat and pressure that its initial identity was obliterated, although much of it was probably sedimentary. Evidence for this is in the remaining rocks of the Mozambique belt that are accessible, largely in the crystalline hills of eastern and north-central Kenya. These hills consist of gneisses, schists, quartzite and marble, all of which are metamorphosed sedimentary rocks (marble is metamorphosed limestone).

Gneisses are absolutely typical metamorphic rocks, banded and changed by great heat (temperatures in excess of 350°C) and great pressure (more than 1,000 megapascals), meaning that they were formed at great depth, a condition called high-grade metamorphism. These gneisses and schists are often dramatically folded and layered, and they can be easily observed in the rocky outcrops of eastern Kenya. The landscape there has been eroded. Only the harder rocks remain and the hills in it are relatively small – although anyone who has climbed, or tried to climb, Ithumba Hill, or Ngulia, or Warges in the Mathews Range might object to the term relatively small! Look at the photograph on page 28.

Some valuable minerals occur in the Mozambique belt, including mica, fluorite, asbestos (not as popular now as it was) and graphite, although none in large enough quantities to bring a big smile to the face of Kenya's ministers of finance. But the massive plate movements of the Mozambique belt also deposited, in south-eastern Kenya and northern Tanzania, some of the world's most spectacular gems – unfortunately in relatively small quantities. Two of these gems have received names that indicate their places of origin, tanzanite and tsavorite. Other beauties laid down in this dry and largely featureless, elephant-ridden and harsh hinterland include rubies, green tourmaline, sapphire and garnet. They were placed there at random, in suitable cracks and gaps in the savagely twisted strata of eastern Kenya. As the huge masses of rock in the continental plates were ground into one another, bent, burnt and crushed, a lot of heat was generated. It was enough to melt, or partially melt, the very minerals that the rocks were made up of. As the temperature fell and big chunks of rock solidified again, a mass of superheated water circulated around the freezing rock. Out there on the edge of the Mozambique belt, south and east of where the Taita Hills are today, in the border country of Kenya and Tanzania, as these pockets of superheated fluid froze and melted and froze again, they deposited in cracks and bubbles, relatively small quantities of some astonishingly beautiful minerals.

After this great collision, comparable in power to the formation of the present-day mile-high Tibetan plateau as the Indian plate continues to crunch up against the Asian plate, Kenya's geological history again goes quiet for an amazingly long while. At this time (roughly 580 million years ago), all the present continents were grouped together to form the supercontinent Pangaea. A huge, and probably continuous mountain range came down from the area of the present-day Red Sea, right through Kenya, looped around westwards across Angola and went back up through West Africa to Morocco.

Over the following 300 million years, things are rather blank in Kenya. During the time, the periods known as the Cambrian, Ordovician, Silurian, Devonian and the Carboniferous, an explosive radiation of life occurred. Many arthropods arrived on the scene, and then invaded the land. The first vertebrates appeared in the oceans and then made it onto land, some fleeing a deadly sea predator, the shark, which appeared more than 350 million years ago. The first vascular land plants appeared, reptiles and insects appeared and massive 'coal forests' and extensive swamps dominated the earth. Two mass extinction events took place, the Ordovician and Devonian extinctions; some authorities estimate that roughly 80 per cent of all living species at the time were wiped out. All the trilobites became extinct. And yet, during this time in Kenya, things were quiet. From 580 million years ago until after

Kibwezi, folded and mineralised metamorphic schists. (Stephen Spawls)

300 million years ago, no rock strata at all– as far as we know – were laid down in East Africa. Thus there is no fossil record. Compared with the cornucopia of superb fossils known from Kenya from the Cenozoic (the last 65 million years), there is virtually nothing prior to 65 million years ago.

Life existed during this time of course; it just has not left any trace in Kenya. Presumably, in East Africa there was no volcanic activity, and no basins for sediments to be laid down in; there was just weathering. Out in eastern Kenya (and presumably in central Kenya, although that was subsequently covered by Cenozoic volcanic material, hiding the evidence), the huge hills and mountains that resulted from the collision were worn down, leaving the subdued landscape that we see today in eastern and north-central Kenya. And then, just under 300 million years ago, some huge rivers began meandering across north-eastern and south-eastern Kenya, presumably running eastwards from the remaining high hills of the Mozambique belt, and laid down sedimentary rocks that are named after the localities where they outcrop: the Mombasa Basin and the Mandera Basin.

These rocks belong to a system of distinctive sedimentary rock formations called the Karoo System, named for a semi-desert area of South Africa. Rocks of this system occur all over the southern half of Africa. They range in age from about 300 million years old (Upper Carboniferous period) to about 200 million years ago (the Triassic/Jurassic boundary). The Karoo beds in South Africa contain fossils of some very large mammal-like reptiles, the synapsids, animals that are important in mammalian evolution. One group of synapsids, the cynodonts, are – or so we believe – the ancestors of mammals. Many of the Karoo synapsids were found and named by the eccentric Scottish palaeontologist Robert Broom, a man who enjoyed doing his African fieldwork naked; we will meet him again. The Karoo beds of East Africa are very interesting. They are up to 7,000m thick in places. They all lie on rocks of Precambrian age (i.e. older than about 530 million years) and yet none of them are older than about 300 million years: such a gap in succession is called an unconformity.

The Tanzanian Karoo strata contain some remarkable fossil material, including ancient and now extinct plants known as seed ferns, of the genera *Glossopteris*. These plants are often regarded as typical indicators of Gondwana as they were widespread throughout that ancient continent when it was intact. The only usable coal deposits in East Africa are in Karoo strata. Long known from southern Tanzania, good quantities of coal have just been located in Karoo strata near Kitui, at depths between 100 and 500m.

Kenya's Karoo rocks contain the country's oldest fossils, mostly small water animals, some fossil wood and the skeleton of an early crocodile-like reptile. A dominant feature of the Karoo rocks in Kenya is the Shimba Range, south-west of Mombasa, that rises up some 300m above the coastal plain. The actual rocks of the Karoo succession, mostly sandstones and conglomerates, underlie much of eastern Kenya, but cannot be precisely dated. Some Karoo rocks are exposed in the Mombasa basin. They form the basis of the Nyika, that arid raised plain that caused so many problems to the early explorers and the builders of the railway. These rocks are sometimes called the Duruma Group, and are split into several divisions. Going eastwards and getting younger, their names tell you where they were first found: the Taru formation, the Maji ya Chumvi formation, the Mariakani formation, the Matolani formation and the Mazeras formation. In the Lali Hills on the Galana River, rocks of the Taru formation, 300 million years old, contain rocks formed in glacial meltwater. Three hundred million years ago it was cold in Tsavo.

Boreholes in the Maji ya Chumvi beds have yielded fossil arthropods and an Eosuchian reptile. Eosuchian means 'Dawn Crocodile', the animal itself is named *Kenyasaurus mariakaniensis*, revealing where it comes from. It is roughly 240 million years old. It has a massive sternum, or breastbone, which indicates that it used its forelimbs as paddles. Birds also have a sternum like this, as you will have observed if you have taken apart a roast chicken, and the huge wing muscles attach to it (look at the top left photograph on page 217).

The Karoo succession came to an end about 200 million years ago, in the Mesozoic era. The Mesozoic ('time of middle life') lasted from 250 to 65 million years ago and the Cenozoic ('new life', also called the Tertiary era), from 65 million years ago until now. Kenyan rocks from this time, from 200 million years ago to the present, are of three distinctive types: sedimentary rocks found in the east and laid down in the sea; more recent, largely unconsolidated sediments – sands, gravels and so on – of various origins, also mostly in the east; and volcanic rocks that have emerged in vast quantities throughout most of central Kenya, associated with the East African Rift Valley. Also, during this time, Gondwana split into East and West Gondwana, and Africa split from South America.

The period between 200 and 145 million years ago is known as the Jurassic, named for the Jura Mountains in France, which are of this age. East Africa has its own 'Jurassic Park'. In the early twentieth century a massive hoard of dinosaur bones was found in sandstone beds around Tendaguru Hill in south-east Tanzania. They were spotted in 1906 by a German mining engineer, Bernhard Sattler (those looking for interesting coincidences will recall that the name of Dr Alan Grant's fellow scientist in the film *Jurassic Park* was Ellie Sattler). Sattler saw a very large bone weathering out near a path at the base of the hill. The Berlin Museum raised funds to the tune of 200,000 Marks and an expedition excavated at the site from 1909 until 1912. The expedition leader was Dr Hans Reck, whom we shall meet again. The scientists shipped back 250 tonnes of fossil bones of nine species of dinosaur, including a massive dinosaur (probably the largest ever known), originally called *Brachiosaurus brancai* and now known as *Giraffatitan brancai*. The mounted skeleton, 26m high, may be seen at the natural history museum in Berlin. Another of the dinosaurs was named *Dryosaurus lettowvorbecki*, after the redoubtable German general Paul Von Lettow Vorbeck, commander of the East African campaign in the First World War.

After the war Tanzania changed hands (although Von Lettow was never defeated) and in 1923 the British Museum sent an expedition to their country's new colony to get some dinosaur bones for themselves. The expedition was led by a Dr William Cutler, a Canadian palaeontologist. Cutler had no African experience, and his field assistant, hired for his African expertise, was none other than the young Louis Leakey. Leakey was taking a break from his Cambridge studies to recover from post-traumatic epilepsy, brought on by being twice kicked in the head during a rugby match. Field conditions were tough, food was short, Cutler and Leakey found few bones (they were not actually looking in the right place) and did not get on; both were often ill. Cutler was a bizarrely cruel man, which considerably upset Leakey, who liked animals. In one of his journals Cutler describes torturing a

Wajir, limestone outcroppings and a line of Second World War tank traps. (Stephen Spawls)

live hornbill, for no obvious reason, by hanging rocks from its feet and stabbing it with a spike. After Leakey left, to return to his studies, Cutler died of blackwater fever.

Kenya's Jurassic rocks indicate that during Jurassic times a shallow sea covered large areas of eastern Kenya and extended as far east as the base of Mt Kilimanjaro, Mt Kenya and Mt Marsabit (although those mountains did not exist at that time, all were formed more recently). Virtually all Kenya's Jurassic rocks are of marine origin and they contain a small number of invertebrate fossils. An economically significant outcrop of Jurassic limestone extends along the coast from Kwale through Mombasa to north of the Galana River. Johann Ludwig Krapf, the explorer, discovered a spiral ammonite from this limestone at his mission in Rabai. This limestone is used by the factory in Bamburi, just north of Mombasa, to make cement for Kenya's construction industry. It produces well over a million tons of cement each year, and there is a remarkable conservation story involved in the reclamation of the huge opencast limestone pits. But the spread of the Jurassic sea across central and eastern Kenya did not seem to have lasted very long.

At the same time, some major structural faulting off the Kenya Coast caused huge areas of land to slump into shallow water. In addition, East Gondwana (Madagascar, India, Australia and Antarctica) separated from West Gondwana (Africa and South America) roughly 160 million years ago, although this date and the mechanisms involved are controversial. That Madagascar and eastern Africa were attached is conclusively proved by the existence of identical evaporite beds (beds of various salts deposited in a shallow sea) in north-western Madagascar and on the coastal strip between Lamu and Kismayu. One theory holds that when Madagascar moved away from Africa it was still attached to India, another indicates that India detached first and then Africa split from Madagascar.

The Jurassic is followed by the last period of the Mesozoic era, the Cretaceous; after the Mesozoic is the Cenozoic or Tertiary era (see Figure 1(1) for a simplified timeline). The Cretaceous lasted from 145 million years ago until 65 million years ago, and the end of the Cretaceous period is signified by a remarkable event –

Jimba Caves, Gede, formed by acidic rainwater erosion of coral rag. (Stephen Spawls)

the disappearance of the dinosaurs. In the rocks that mark the Cretaceous–Tertiary boundary there is layer of red clay that contains a lot of iridium. Iridium is an element that is much more common in space, in dust and meteoritic material, than on earth. It seems that – although not everyone agrees – around 65 million years ago a large meteorite struck the earth at a place called Chicxulub, in Mexico's Yucatan Peninsula. The explosion threw up huge clouds of dust into space. The global fall in temperatures killed off virtually all the dinosaurs, saved those that were to become birds.

The Cretaceous in Kenya was marked by another invasion by the sea. Kenya's Cretaceous rocks also contain the remains of its only dinosaurs. Just northwest of Lake Turkana, in the remote Lapurr Range north of Lokitaung, is an outcrop of Cretaceous rocks: the Lubur sandstone. In 1968, a fragment of a dinosaur bone was found there. In 2004 a joint expedition to the Lapurr Range, by Kenya's National Museum and the University of Utah, led by Meave Leakey and Joseph Sertich, found more than 150 fossil bones dated to the late Cretaceous, between 94 and 65 million years ago. The bone cache includes crocodiles, turtles, two large theropod dinosaurs (which are bipedal carnivores, *Tyrannosaurus rex* was a theropod) and two sauropods, which are mostly large herbivorous dinosaurs.

The Mesozoic thus ended 65 million years ago, to be followed by the Tertiary. Offshore faulting, connected with the final break up of Gondwana, took place, and this is why the island of Pemba is separated from the mainland by a 700m deep channel. A shallow sea covered parts of south-eastern Kenya, its depth fluctuating. Ancient shorelines in the Miocene (22 million to 5 million years ago) were up to 100m higher than they are today. In recent times (i.e. in the last 5 million years) the sea extended inland at least as far as 6km in many places. The sea left behind the reef complex of limestone, rubble and sandstone known as coral rag, much used for building nowadays and found all along Kenya's shore, and up to 2km offshore in places. Turn off the main road heading from Mombasa to Lunga Lunga at Kanana and drive to Shimoni and you are crossing a peninsula entirely composed of ancient coral, of thickness nearly 100m, laid down relatively recently.

FIRE AND FORCES

In the last 65 million years nothing very exciting geologically has occurred in eastern Kenya. But up in the centre of the country, it is a very different picture. Here, Kenya has been cracked apart, from north to south, creating one of the world's most astonishing landscapes; as anyone who stands on the eastern edge of the escarpment on a clear day and looks north-west, past Suswa and Mt Longonot into the dim blue distance, to the huge mass of Eburru and the western wall, will agree. This is the East African Rift and up here tectonic activity has been taking place for more than 30 million years; huge masses of rock have moved. Associated with the rift are a host of mountains, volcanoes, faults and lava flows; volcanics of this period – the Cenozoic – cover about 30 per cent of Kenya. The appearance of this dramatic landscape drastically altered the climate and natural history of central Kenya; indeed it seems likely that it was the formation of the rift valley that led to the evolution of humanity. And the astonishing thing is, no one is absolutely certain how the rift valley was formed, although there are several plausible theories. Another odd thing is, few of the early adventurers who crossed the rift saw fit to remark upon it. In the early twentieth century, an expatriate German resident of Lamu, Clemens Denhardt (after whom a curious burrowing caecilian was named), constructed a map of the interior of Kenya even though he had never visited it. It was based on the narratives of 33 Arab and Swahili traders who had travelled deep into the interior. His map is accurate in many respects: it has Lake Bogoria, Ol Donyo Lesatima, the Lorian swamp and the Tana River, and yet the rift valley does not appear on it anywhere!

The East African Rift System (known jocularly by many geologists as the EARS) forms part of the gigantic Afro-Arabian Rift System (which perhaps fortunately is not known by its acronym) and extends from Mozambique to Turkey, some 6,500km. The EARS extends from the Afar Triangle in Ethiopia south to the Zambezi River, but discontinuously, as shown in Figure 1(2). It cuts clearly through the heart of Ethiopia. When it drops down to the northern Kenya border, it jumps westwards, and extends from Lake Turkana down to Lake Baringo (although

Mt Longonot, a young volcano, active within the last 200 years. (Stephen Spawls)

between these two lakes its walls are poorly defined). From there it goes through Lakes Bogoria, Nakuru, Elmenteita, Naivasha and Magadi, into Tanzania to Lake Natron, where it splits in two and then disappears. There is also a Kenyan side branch, the Kavirondo or Nyanza Rift, which extends from the Winam Gulf up towards Londiani.

Another major branch, the western branch, appears north of Lake Albert in Uganda, and then heads south through Lakes Edward, Kivu, Tanganyika and Malawi. This western branch is known both as the Western Rift Valley or the Albertine Rift (after Lake Albert). The eastern branch is sometimes known as the Kenya Rift Valley (for the section in Kenya) or the Gregory Rift, in honour of John Walter Gregory who explored the rift in the 1890s. Gregory introduced the term 'rift valley', which he defined as a 'linear valley with parallel and almost vertical sides, which has fallen owing to a series of parallel faults'. Gregory wrote a classic book on Kenya and the valley, of which more later.

The Albertine Rift is generally about 40–50km wide, its sides are steeper and lakes deeper than the

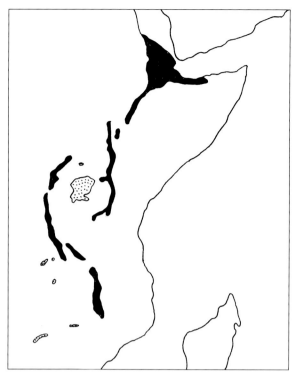

Figure 1(2) East African Rift System

The Gregory Rift Valley, with downthrown fault block and central volcanoes. (Stephen Spawls)

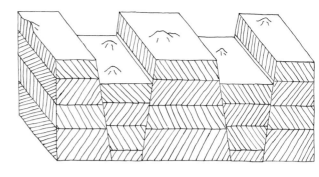

Figure 1(3) Horsts (uplifted blocks) and grabens (downthrown blocks)

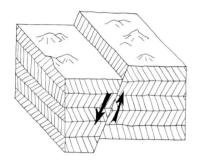

Figure 1(4a) A normal fault, one block slides down

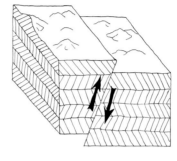

Figure 1(4b) Reverse fault, one block slides up

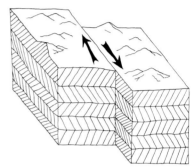

Figure 1(4c) Tear fault, the two blocks move sideways

Gregory Rift, but the Gregory Rift is a bit broader, at 50–80km wide. In Kenya, the altitude of the valley floor varies from 250m above sea level in the Suguta Valley south of Lake Turkana (often said to be Kenya's hottest place) to over 2,000m in the Naivasha area. In places along the rift the altitude difference between the floor and the top of the escarpment is huge, notably in the Albertine Rift where the brooding hulk of the Rwenzoris looms over the flood plains; Margherita Peak is more than 4,000m above the Semliki River. The differences are not so dramatic in Kenya but nevertheless the Nguruman escarpment is 1,500m higher than the valley floor and the Aberdares tower up nearly 2,000m above Lake Naivasha.

Most valleys are formed by flowing water, or occasionally ice. But the rift valley was formed by large-scale rock movement. Both branches of the rift valley consist mostly of grabens (see Figure 1(3)), which are fallen blocks of rock. Imagine holding three cubes of wood, one in each hand and the third held between them. If you move your hands apart, the central one will fall. This is how a graben forms and this is what has happened in the rift valley; the floor consists of downthrown blocks of rock. The downthrown blocks contain the same strata, or rock layers, as the ones on the walls (as shown in the diagram), so we know where they came from. A mass of rock forced up in similar circumstances, above the surrounding country, is known as a horst (also in Figure 1(3)). Both terms are German: *graben* means 'ditch', *horst*, rather delightfully, means 'eyrie', an eagle's nest.

When a force causes rocks to break and the resulting blocks of rock to move against each other, it is called a fault; fault movement is associated with earthquakes. If a block falls, it is called a normal fault, and happens when the blocks move apart. If the rocks are pushed together and one block slides up, it is a reverse fault. If one block moves sideways with respect to the other, you get a tear fault. All three are shown in Figure 1(4.) (There are other types of fault but they are not relevant here.) Fault movements sound innocuous, but the forces and masses involved can be gigantic, causing huge earthquakes, tsunamis if under water, destruction and loss of life.

Gregory recognised that the floor of the East African rift valley was a downthrown block or graben. Something had sent the floor downwards and the walls upwards; this much is agreed upon. At first, there was debate about whether this was caused by compression or tension. Were the rock layers being squeezed together and the valley floor was suddenly forced down, or were the layers were being pulled apart and the valley floor fell? One crackpot early theory suggested that the rift valley was formed as the earth cooled and shrunk like an apple removed from an oven, and the 'skin' formed ridges on the surface.

The tension theory was suggested by the Austrian geomorphologist Eduard Suess (a geomorphologist studies landforms) and taken up by Gregory. Suess was probably the first earth scientist to recognise the significance of the fractured zones between the Red Sea and Lake Turkana. He never actually visited Africa but wrote a remarkable memoir entitled 'The Bridge of East Africa' from the comfort of his office in Vienna. Gregory was an admirer of Suess, and gave the name Lake Suess to the huge ancient lake that existed between Hell's Gate and Gilgil. Unfortunately – or perhaps fortunately – the name never stuck.

The compression theory was advocated by E. J. (Jim) Wayland and A. W. Groves. Wayland was chief government geologist in Uganda for a number of years. An interesting man, he was one of the first earth scientists working in East Africa to recognise the extent of African prehistory. However, Wayland made some mistakes; one of them was over the Gregory Rift. He located some genuine reverse faults (i.e. blocks being shoved up rather than falling down) and thought they were associated with the formation of the rift valley; they turned out to be more than 500 million years old.

Nowadays it is accepted that tension faulting formed the rift valley. The rocks shifted apart and the bit in the middle fell down. The angles on the slip faces confirm this. The big picture is that two large blocks of rock are moving away from one another; a process of spreading called crustal extension. And large areas of East Africa have been uplifted, a process called doming. So far, so good; geologists agree on this. What causes these movements and the consequent fall of rocks, however, is not agreed upon.

The first point of contention is whether the East African Rift is evolving into a mid-oceanic rift system. Does it lie across the junction of two major plates that are moving apart, spreading in the same manner as the sea floor in the mid-Atlantic? This is a popular point of view, leading to speculation that in the distant future the eastern half of East Africa will sail off into the ocean, to become another Madagascar; perhaps the Island of Greater Somalia. Another theory is that that the EARS is a 'failed arm'; a minor crack associated with much larger cracks, in this case the Red Sea. This theory indicates that in Kenya the plates are indeed moving apart but the spreading is small and decreasing, so that eventually it will just grind to a halt.

Associated with the tectonic movements of the rift is a lot of volcanic activity, or magmatics – hot rocks upwelling and sometimes coming out onto the surface. The second main point of contention is whether the hot rock down below is driving the extension, or the extension is allowing the hot rock to come up. Another theory – not popular, however – is that the rift valley has nothing to do with spreading or any sort of forces below, but is due to the fact that Africa was at one time near the South Pole, where the earth is more flattened. As the continent has drifted north to the equator, where the earth's curvature is more pronounced, it has simply cracked; as a huge rigid mass must do if its centre is elevated. It is worth having a look at these theories, and whether or not the evidence supports them.

The spreading rates of the East African Rift are relatively small; this is significant. The opposite sides are still moving apart. If you were to stand on the escarpment near Kijabe and gaze across at the Mau for long enough, it would eventually move out of sight – although you would have to wait a very long time! The actual rate of spread in the southern Kenyan rift is less than 2mm per year (but is greater in northern Kenya). The Red Sea is spreading faster, at 10–20mm per year; and mature ocean ridges, like those in the mid-Atlantic, are spreading at 20–100mm per year. This means that what is happening in the East African Rift is unusual, and unlikely to lead to the breaking-up of Africa. It is approximately 10 million years since the rift valley began to form in southern Kenya, and if it was spreading at the rate that mature ocean ridges spread (i.e. about 20mm per year), the gaps between the original rocks, filled by recent sediments or upwellings, should be around 200km. In fact, the gaps are less, often much less, than 10km, indicating a rate of spread of about 1mm per year. This indicates that what is happening under Kenya is not that similar to the great rifts being formed under the Atlantic,

Ol Njorowa Gorge, Hells Gate, formed by a huge river flowing out of Lake MegaNaivasha. (Stephen Spawls)

although geologists agree that extension is taking place. The plates really are moving; the gaps are not due to crustal doming only.

The second important bit of evidence is that deep under the Kenyan rift are low-density sialic rocks; which are rocks composed on silica and aluminium. Such rocks are usually indicators of stable continental conditions; they were the ones that rose up and cooled. The rocks under areas of moving plates and in the vicinity of sea floor spreading are usually high-density basalts, belonging to a class of rock called sima, made of silica and magnesium.

These two facts indicate that the East African Rift cannot be compared to the oceanic rifts. So what did happen – and what is happening now – under central Kenya? Evidence indicates that, starting about 70 million years ago, a mass of hot rock, a mantle plume, began to well up under north-east Africa. Mantle plumes originate from deep within the earth, in the mantle, below the crust, as shown in Figure 1(5). Driven by convection (the hot liquid magma is less dense than surrounding rock), the hot material rose up like a hot air balloon. It hit the base of the lithosphere, which is the strong, rigid, cool continental crust. Here it spread out, forcing a large area of land above it upwards. This is a process called doming and what it formed is the Kenya dome, the high land of central Kenya. And the fracturing of the rigid lithosphere meant that cracks formed, and the hot magma from below went zooming up these cracks, often spreading out, like mushrooms. Such intrusions are called diapirs. In some cases the hot rock came out into the open air. And this fracturing also led to faulting, with the central blocks plunging downwards.

This then leads to a question of succession. The concept of the mantle plume is widely accepted. Something absolutely gigantic booted central Kenya up into the air. This scuppers the theory of Africa cracking just because it drifted near the equator, and also the baked apple theory. But was this mantle plume just part of a long-existing region of mantle activity, with magma emerging relatively fast from deep within the mantle under Kenya, a so-called 'active' rifting mechanism, or did some sort of external forces cause the split that is the East African rift and allow hot magma to simply well up slowly towards the surface, cooling as it went – a 'passive' rifting mechanism? And here the geologists disagree.

The 'active' theory is supported by a fair amount of evidence. It's hot down there – we have mentioned the geothermal gradient. A German geologist called Ulrich Achauer, working on the Kenya Rift International Seismic Project (known, naturally, by the snappy acronym KRISP) has calculated that the temperature 65km deep below the Kenyan Rift is around 1,400°C, and this is about 200 degrees hotter than it should be if magma was welling up slowly from lower down in the crust. This is obviously a theoretical calculation as no one can get down there; the deepest any drill has gone down into the earth is 12km. However, several experts agree with Achauer's prediction.

The 'passive' theory, however, has its supporters. This theory suggests that the African plate, in East Africa, is riven by cracks from ancient stresses. As the plate slowly moves, it passes over reservoirs of hot magma in the mantle, and the magma just wells up the cracks. Supporters of this theory point to evidence like the slow rate of plate movement and repeated and yet variable volcanic activity over long periods of time, evidence consistent with an erratic supply of lava making use of the same cracks over time. And these escape channels act as centres that focus heat, rather as a hot tap warms up when hot water flows through it. This may lead to local melting, where the heat causes deeply buried slabs of the original rock to partially melt and get carried to the surface. This is a process known as metasomatism. The moving plate effectively generates its own supply of magma, but it comes up out of the same cracks. A problem with this theory, however, is that there is lots of evidence of ancient rifting prior to the formation of the rift valley, but no magma associated with it; so why did hot rock suddenly become available recently?

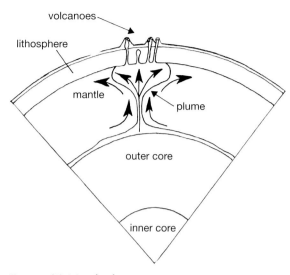

Figure 1(5) Mantle plume

Ngulia Hills from the air, hard metamorphic outcroppings on a long-eroded landscape. (Stephen Spawls)

Nairobi National Park, situated on a huge flat lava flow. (Stephen Spawls)

Another unusual theory of how the rift valley was formed, using evidence from earthquakes in eastern and south-eastern Africa, suggests that a gigantic mantle plume lies under Gabon and Cameroon, out there in Africa's western armpit. This rising plume puts the whole continent under radial tension. If you put your hand under a stretched sheet of plastic, in the middle, and push up, it creates rings of force around the point where you are pushing. This theoretically would explain the rifting in East Africa, although it does not explain why this would generate the rising of hot magma in East Africa.

So the jury really is still out on how the East African Rift was created. Most geologists accept the existence of a mantle plume under East Africa, and some suggest that a mixture of active and passive mechanisms may be operating; i.e. rising plume and moving ruptured continental plate. But there is a spectre that haunts all this cheerful theorising, and that spectre is the Rwenzori. We presume that the mechanisms that created the eastern rift also created the western one. But the Rwenzori is not igneous, nor is there igneous material below it. It is a massive block of metamorphic rock, a huge horst, hanging there, 4km above the surrounding plains. What got it up there? Jim Wayland was aware of the problem, and his prescient comment, made back in 1923, still holds: 'no theory of the rift will pass muster if it leaves Ruwenzori hanging in the air'. At the moment the Rwenzori is still hanging.

What is certain, however, is that the rifting and associated volcanic activity have produced a spectacular landscape. All you have to do is drive from Nairobi to Narok, and then back across the Aberdares and around Mt Kenya to Meru on a clear day to see the best of it. This is where you can see how the fire below irrevocably cracked and changed Kenya's heart. And, as a basin with little outflow, it forms a huge sediment trap. It is for this reason that so many important fossils have been found in the rift valley.

Kenya's two big mountains, Mt Kenya and Mt Elgon, are both volcanic, as is Mt Kilimanjaro, although none of the three actually lies in the rift. Mt Meru and a host of smaller ranges and hills, such as Suswa, Longonot, the Nyambeni Hills, Menengai, the Tugen Hills, the Ngongs, the Aberdares, the Chyulu Hills, Mt Marsabit and Eburru are all volcanic. All were formed in the last 30 million years. Also within this time, massive floods of volcanic material that give central Kenya its identity – the Rumuruti, Uasin Gishu and Kapiti plateaux – poured out and settled; similarly large amounts of lava were deposited all over northcentral; and north-west Kenya. Stand at the bottom of the Kerio Valley where the road between Iten and Kabarnet crosses the river and look westwards. What you see, lying on the top of the escarpment, giving it its flat appearance, is a lava plateau.

It looks as though, about 70 million years ago, the previously mentioned mantle plume began elevating north-eastern Africa, and this gave rise to two domes, pushed up by upwelling hot rock: the Ethiopian and Kenyan domes, separated by the lower ground around Lake Turkana, which sank downwards as the domes rose. Then came a time of activity, rock rising, breaking and falling, accompanied by earthquakes and lava squirting out. Action on the Gregory Rift began around Lake Turkana some 33 million years ago, with volcanics erupting on the Lotikipi plain, west of the top end of Lake Turkana, followed by rifting some 5 million years later. The rifting then moved southwards over the next 15–20 million years, reaching Nakuru 20 million years ago, with major faults forming in the central Gregory Rift about 16 million years ago; the Elgeyo escarpment was formed around this time. The Albertine Rift got going much later, with initial volcanism some 12 million years ago. In Kenya, Mt Elgon was formed about 22 million years ago (although some authorities put Elgon at nearer 18 million years old) and at the same time huge floods of basalt were erupted along the northern rift valley between Lakes Turkana and Bogoria.

Twelve million years ago a series of volcanic eruptions filled the central and Nyanza rift with an alkaline lava called phonolite; and roughly 13 million years ago phonolite lava also flowed out to create the huge Kapiti Plains east of Nairobi, and then overflowed from a point just east of Thika and ran south-east, down almost to the coast, creating one of the world's longest lava flows, 300km long, the Yatta Plateau. Originally encased within a valley (or so it is believed), subsequent erosion of the surrounding rock (presumably less hard than the actual lava) has left the black volcanic flow standing proud of the surrounding country. Often 150–200m higher than its surroundings, preventing the Galana River shifting to the north-east, the Yatta forms a continuous ridge visible on the left to anyone driving down the Mombasa road between Mtito Andei and Voi. Although it is nowhere more than 10km wide, there is only one major gap in this natural barrier, where the Kibwezi–Kitui road slices through at the appropriately named Yatta Gap. The Yatta lava flow finally puddles out in a mess of lava north of Aruba, on the far bank where the Galana becomes the Sabaki.

The Aberdares, those dramatic mountains on the eastern shoulder of the rift, formed between 6.5 and about 5 million years ago. Essentially, the Aberdares are a basalt volcano, a consequence of the mantle plume. Quietly edging up skywards, as south-central Kenya bulged, the Aberdares were rising nicely when the extension of the crust had its inevitable effect. To the west, a gigantic crack appeared, the Sattima Fault; it extends from South Kinangop to El Joro Orok. The shoulders of the mountains plunged, falling away from the central Aberdares, crunching down 500m; forming the Kinangop Plateau. If you drive up towards the Aberdares National Park from the Naivasha side and look to the north as you approach the final steep slope towards Mutubio Gate, near the fire tower, you are looking along the line of the Sattima Fault. The isolated volcano Kipipiri was downthrown at the same time. The same fate befell the Ngong Hills (also a volcano) at around the same time; a crack appeared, the Ngong Fault, and down went the western slope. You can see the downthrown block clearly if you drive down the Magadi road; it is straight in front of you as you cross the low pass over the southern end of these charismatic hills. And across the rift to the south-west of the Ngong Hills, the Nguruman Fault formed, and the southern sector of the rift valley dropped. It created a trough for not only Lake Magadi to form, but also Lake Olorgesailie; on the shores of this now dry lake, Gregory and C. W. Hobley found what is arguably the world's most important site for the manufacture of stone tools. And after the fall of the Kinangop, the centre of the rift – where Longonot and Suswa are today – sank further, forming in effect a series of steps. As you drive off the Aberdares down to Lake Naivasha, you are coming down these steps.

The last 3 million years have seen a lot of volcanic activity, including the formation of Mt Kilimanjaro, Mt Kenya and many smaller volcanoes, including Suswa, Longonot and Menengai. These last two still have clear-cut craters at the summit; Mt Menengai is 12km wide. Suswa is unusual in that it has two craters, or calderas, and several phases of growth, with an isolated graben in the centre; in the middle of this is an island block,

Waterfall, Kerio Valley, the Uasin Gishu lava plateau clearly visible at the top. (Stephen Spawls)

romantically called the 'lost land' as it is difficult to reach. Suswa is discussed in some detail by Celia Nyamweru (1980) in her excellent book *Rifts and Volcanoes*.

As the rifting moved south, volcanism has gradually drifted to the east of the main rift, as the figures indicate; Mt Elgon and the Tinderet volcanoes were active more than 20 million years ago; lava filled the central rift 12 million years ago, the Aberdares rose up 5–6 million years ago and Mt Kenya, under 3.5 (possibly as little as 2) million years ago. This eastward movement is probably associated with continued fracturing as the Kenya dome rises. Mount Marsabit formed between 1.7 and 0.7 million years ago and the Nyambeni Hills, 2 million years ago. Mount Kenya was originally a cone-shaped stratovolcano, comprised of layers of lava and pyroclasts, which are volcanic materials that were blown into the air and then fell in layers. According to Celia Nyamweru, Mt Kenya was probably more than 6,000m high, but has been intensely eroded by ice (it is now just over 5,000m). The high peaks of Batian and Nelion are a massive volcanic plug, more than 2km wide, of hard lava called nepheline syenite (a type of phonolite). Forced up from below into the vent and jammed in the crater, it cooled slowly, and hence grew big feldspar crystals, up to 30mm long. The original crater is long since gone, but the lava in the volcano's neck, the 'plug', has resisted erosion, giving the mountain its characteristic spiky appearance, unlike the smooth profile of Mt Kilimanjaro. However, the erosion on Mt Kenya pales besides that of Kisingiri. Just west of the present town of Homa Bay, Kisingiri started blasting out about 25 million years ago, creating a volcano several thousand metres high. Subsequent erosion has reduced it to a small hill, Rangwe, only 500m higher than the lake. Out to the east, the Chyulu Hills first erupted 1 million years ago. But most volcanic activity in these spectacular hills took place in the last 10,000–15,000 years and has created the world's longest lava tunnels, exciting places to explore. Mount Kilimanjaro, also well to the east of the rift, is around 1.2 million years old; the oldest peak, Shira, is now heavily eroded and is only just over 4,000m high, although it may have been originally as high as 4,900m. Mawenzi is just over 5,100m and Kibo, the youngest peak and thus least eroded, is, at 5,895m, the highest point in Africa. Mount Kilimanjaro is now dormant, but the fire remains; the taciturn British climber Bill Tilman discovered both sulphur deposits and fumarole activity in the Kibo Crater in 1933.

Volcanic activity continues in East Africa. Teleki's volcano, at the southern tip of Lake Turkana, was erupting when Count Teleki arrived there in 1888. In the past 200 years Longonot has erupted, as have some of the cones in the Suguta Valley and near Kibwezi. Central Island, in Lake Turkana, blasted out ash clouds in the late 1960s. At present, however there is little volcanic activity in Kenya, although there are three centres of active volcanism in eastern Africa: on the Albertine Rift where Uganda, the Democratic Republic of the Congo (the former Zaire) and Rwanda meet, in the Afar Depression in Ethiopia and in northern Tanzania, where the volcano Ol Donyo Lengai (meaning 'the mountain of God' in Maa) is gently erupting, spewing out low-temperature carbonatite lava. The occasional earthquake occurs in Kenya. In 1969 three seismologists from Columbia University in New York carted some portable seismographs down into the southern Kenyan rift and monitored earthquakes; at the bottom of the escarpment just west of Kajiado, where there are a lot of small faults, they measured 47 minor earthquakes in one day! Kenya's strongest earthquake took place on 6 January 1928. Popularly called the Subukia earthquake, the epicentre was actually on the Laikipia escarpment just east of Lake Bogoria. It measured 7.1 on the Richter scale, but – as far as is known – no one was killed, although a lot of buildings fell around Solai and Subukia, Muranga and even Kabete, several fissures opened and rivers changed direction.

MINERALS AND MONEY

Certain minerals have been of great importance in Africa's history. One thinks of salt and the Sahara caravans, the gold of the pharaohs, and the copper used in the Katanga crosses that were currency in central Africa for hundreds of years. Modern Africa has vast mineral riches: oilfields, precious metals and diamonds. The existence of these minerals have made countries rich; but often not without controversy. Money from diamonds and gold in South Africa financed the savage insanity of the apartheid regime. So-called 'blood diamonds' have financed hideous small-scale wars and atrocities in West Africa. Demand for coltan in 2002, for mobile phones, brought lawlessness and poaching to the eastern DR Congo.

About 1 per cent of Kenya's gross domestic product comes from minerals, including 500,000 tonnes a year of soda ash (trona, a mixture of sodium carbonate and sodium hydrogen carbonate, a versatile chemical) from Lake Magadi, and this will continue for the foreseeable future. Other profitable minerals include fluorspar from the Kerio Valley mine, a little gold and useful building minerals such as slate, limestone for cement, granite, sand and gravel. Kenya also has, although unfortunately in rather small quantities, some beautiful precious and semi-precious gemstones. The story of their finding and exploitation is one of skulduggery. Following the discovery of the fiery bluish crystals of tanzanite in Tanzania in 1967, two American geologists discovered a deposit of high-quality, blood-red rubies near the huge block mountain Kasigau, at the edge of the Mangeri Swamp. An attempt to get Kenyan politicians on their side backfired and both men were deported.

Then there's the story of tsavorite, a vivid, radiant green garnet, worth as much as comparably-sized emeralds. A Scottish geologist named Campbell Bridges located a strike of the fiery gems in the dry country between Kasigau and Tsavo. Some gemologists believe that tsavorite should be considered the fifth genuinely precious stone (the other four are diamond, ruby, emerald and sapphire). The story ends in tragedy: in August 2009 Bridges was attacked by a mob of miners near Voi and stabbed to death in a dispute over mining rights. Sadly, Kenya has few diamonds, although Africa has some of the world's most important diamond mines. Diamonds may be found in 'pipes' of the rock named kimberlite, or, if a pipe has emerged and exploded, scattered across country. Big diamond strikes were made in Tanzania, by the persistence of John Thoburn Williamson, a Canadian geologist who nearly died from hunger and thirst and got into massive debt as he doggedly hunted for diamonds in the dry savanna south of the Speke Gulf in northern Tanzania. There are a few kimberlite pipes in Kenya, mostly in the ancient rockscape north of the Winam Gulf, but no diamonds have been found there. But just to the south of Simba town, near Hunter's Lodge, in eastern Kenya, a small volcano, Losogori, has a kimberlite-type rock and contains sporadic diamonds.

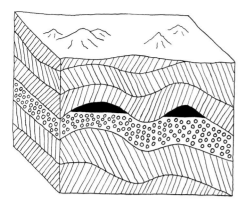 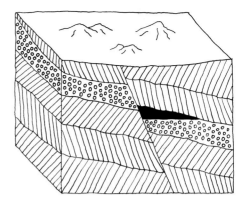

Figure 1(6a) Anticline: the oil is in the sandstone, the rock above is impermeable

Figure 1(6b) Fault trap: the oil in the sandstone has been sealed off by the fault

The story of oil and gas in Kenya looked to follow a similar pattern, with only small uneconomic discoveries located, although some 30 exploratory wells had been drilled into the suitable folded sedimentary rocks that form traps (as shown in Figure 1(6)). Such rocks, usually older than 20 million years, exist near Lamu and in parts of northern Kenya. But as this chapter was finalised, exciting news emerged. Tullow Oil, an Anglo-Irish oil prospecting company, have found a reservoir of potentially 2 billion barrels of oil at Lokichar, some 80km south of Lodwar. Although the size of the reservoir, trapped in Miocene sediments above the ancient crystalline basement, has yet to be fully proven, prosperity may beckon.

Gold was located in the ancient crystalline rocks east of Lake Victoria in the late nineteenth century, and the reef continued north and outcropped in the Kakamega area. The big Kakamega gold strike started in 1930. Kakamega became a boom town (briefly!), with two hotels, appropriately called 'The Golden Hope' and 'El Dorado'. The miners

Gold miners working the tailing dumps, Macalders Mine. (Stephen Spawls)

'Face' in ancient conglomerate, near Macalders Mine, western Kenya. (Stephen Spawls)

panned river gravel and worked upstream to find where the gold-bearing strata outcropped in the river. One of Kakamega's more famous miners was George Adamson. To guard his gold, Adamson recruited a fat Rhinoceros Viper *Bitis nasicornis*, which he named 'Cuthbert Gandhi'. The story goes that the surname Ghandi signified that the viper appeared to want to starve itself to death. When Cuthbert was offered a live rat, however, he apparently awoke and struck like lightning! After two years the Kakamega deposits became unreachable. Evidence of the mines remains today. If you enter the National Park at Kakamega and drive a few kilometres eastwards, Lirhanda Hill looms up on your right. If you climb the hill, with its spectacular views over Kakamega Forest and wonderful birds, like Great Blue Turacos *Corythaeola cristata* and White-headed Saw-wing Swallows *Psalidoprocne albiceps*, you will find, halfway up, in a clump of trees, an old adit, or horizontal mine shaft; gold was mined here. Inside, amongst the spectacularly coloured yellow and red rocks, a colony of bats roost, and in the trees outside the shaft you may find a Blanding's Tree Snake, *Boiga blandingii* which is partial to bats. Other gold strikes were made up here in the west, including the North Mara Mine and Macalder's mine, near the Tanzania border; which produced 950kg of gold between 1935 and 1966, in which year the mine closed. At present, gold mining in western Kenya consists largely of a little gold panning, digging shallow shafts by hand and reworking of the old mine dumps, or tailings. This is mostly carried out by individuals and is a risky business, with the miners having a short life expectancy. Cave-ins occur and the miners handle mercury, which has deadly fumes, carelessly. The mine tailings contain a lot of heavy metals, which are poisonous, and the miners extract them with pressure burners.

Kenya's most recent mineral debacle took place near the town of Kwale, on the coast, on the land south-east of the Shimba Hills. The Magarini Sand Dunes proved to contain large commercial quantities of heavy minerals, in particular ilmenite and rutile (titanium ores) and zircon, an ore of zirconium. Titanium is a super-strong metal, and zirconium resists corrosion better than almost any other known material. So the Kwale and Kilifi sands seemed to have great potential. However, environmental and economic concerns stalled the development. It looked as though the lone and level sands would remain unexploited, but an Australian company, Base, is now in the process of commencing a mining operation there.

GEOLOGICAL EXPLORATION IN KENYA

In the nineteenth century, even as Western explorers were really beginning to penetrate Africa, the general view was that African geology was unexciting. In 1852, Sir Roderick Murchison (after whom the great waterfall on the Nile in Uganda was named) stated in his presidential address to the Royal Geographical Society that Africa was a land of 'great antiquity and simplicity'. As late as 1891 Professor Henry Drummond, pontificating on African geology, said 'throughout this vast area, opening up to science, there is nothing new – no unknown force at work, no rock strange to the petrographer, no pause in denudation, no formation texture or structure to put the law of continuity to confusion'. He might as well have added that there was no unknown topic upon which someone eminent wasn't capable of making a fool of himself.

Kenya's people had, of course, been making use of its geological resources for a long while, using its pigments for decoration and painting, its stone for building and its hard rocks for tools. Tool technology started in

Kenya; our ancestors skinned and cut up game with wickedly sharp stone knives made from obsidian. But Western geological exploration in Kenya effectively starts with Joseph Thomson. A Scot, who took a degree at Edinburgh, he is the man after whom Thomson's Gazelle *Gazella rufifrons thomsoni* was named. Before he made his famous journeys across Kenya, Thomson had twice been adventuring in Africa. His third safari, the most exciting, and the one that nearly killed him (and probably led to his death, in the final analysis) was to Kenya, from Mombasa to Baringo and Lake Victoria. It is described in his best-selling 1885 book *Through Masai Land*. Thomson started twice. Initially he turned back near Mt Kilimanjaro as he thought his caravan was about to be attacked by the Maasai. Thomson was a thoroughly decent man, who believed in negotiation and being nice to all he met, unlike many of those who explored before him (and even after), who travelled in huge, well-armed caravans and whose response to any trouble with the local people was to reach for their firearms. Thomson's motto was 'he who goes gently goes safely, he who goes safely goes far', although he was not above trying to fool local people into thinking he was a magician.

Thomson's second start was more successful, although he was lucky not to bump into the Maasai as he made his way past Mt Kilimanjaro. He made his way up the southern Kenyan Rift Valley, crossing the Ngong Hills, climbing Mt Longonot and pottering around Lake Naivasha. He named the Aberdares, after the then-president of the Royal Geographical Society, the First Baron Aberdare, Henry Bruce, who had funded his safari. Thomson came across the beautiful waterfalls that bore his name for a long time, Thomson's Falls, although this has now been changed to Nyahururu. He then made his way to Baringo, and across to Lake Victoria. His geological observations were accurate. Thomson realised that the Uasin Gishu Plateau was a layer of lava overlying metamorphic rocks, and what he saw in the central rift caused him to suggest that a chain of volcanism extended all the way from South Africa to Ethiopia. Thomson's description of the setting of Lake Baringo from his viewpoint on the Laikipia escarpment is the first account of the geomorphology of the rift valley.

Returning to England, after a hazardous journey in which he was gored by an African Buffalo *Syncerus caffer*, walked within a few miles of Lake Bogoria and managed to completely miss it (if he had found it, it might have been called Lake Thomson, at least for a while) and suffered a frightful bout of amoebic dysentery. Thomson recovered enough to deliver a sensational lecture on his adventures to the Royal Geographic Society, but his health had been wrecked by his bout of dysentery and he also had schistosomiasis. He worked briefly in West Africa and was employed by Cecil Rhodes in southern Africa, but he was rarely free from pain. He returned to England and died in 1892; he was only 37.

At around the same time Thomson was exploring in Kenya, a German naturalist, Gustav Fischer, was exploring in northern Tanzania and southern Kenya. Fischer, an army surgeon, had come to East Africa in 1876 with the Denhardt brothers, who founded the German enclave of Wituland, Deutsch-Witu, around the town of Witu just north of the Tana delta. Fischer initially practised as a doctor in Zanzibar, and then explored along the Kenya coast and inland up the Tana River. As well as collecting geological specimens and researching on the landscape, Fischer was an all-round naturalist, and has several animals named after him, including a spectacular Tanzanian chameleon, a beautiful lovebird and a turaco. Fischer climbed on Mt Kilimanjaro, but not to the top. Fischer's explorations took him up the rift valley to as far north as Lake Naivasha. His monument can be seen by all today; not far inside the entrance of Hell's Gate National Park stands Fischer's Tower, a beautifully tapered rock spike and an enjoyable and reasonably testing climb. The tower is actually a plug of rhyolite lava that was left standing when a superlake centred on Lake Naivasha overflowed, a prosaic explanation. A Maasai legend about Fischer's Tower is much more charming but improbable; the storytellers say that the monolith is the petrified body of a Maasai maiden, who was told not to look back at her home village as she left to marry a warrior, but she did look back, and was turned into stone.

Fischer was actually forced to retreat at Naivasha by the Maasai, three months before Joseph Thomson got there, which in a way was fortunate for Thomson, who travelled north of Naivasha into country not seen by Europeans. After his return, Fischer was involved in an abortive attempt to rescue the German naturalist Emin Pasha (whose real name was Eduard Schnitzer); out on the eastern shores of Lake Victoria he nearly starved to death. He returned to Germany and died shortly afterwards at the age of 38, another young victim of Africa's harsh regime.

In late 1886, Count Samuel Teleki de Szek (Teleki for short), a Hungarian aristocrat, and Lieutenant Ludwig Von Hohnel, an Austrian, met up in Zanzibar, their mission to explore northern Kenya. Teleki was not an altogether happy man at the time. He had intended to go hunting in Tanzania, but was persuaded by his friend, Crown Prince Rudolf, to take Von Hohnel, who then persuaded Teleki they should be doing scientific exploration, not hunting. And yet the expedition was not only successful, but each man came to like and respect the other; there was rarely a cross word between them. They set forth from Zanzibar, ascended part of the way up Mt Kilimanjaro, and then made their way north right up through the Kenyan Rift Valley, during which they noted old lake levels and recorded the geology of their route. North of Lake Baringo, looking east they spotted the prominent range of hills the Samburu knew as Ol Donyo Lengeyo, and named them the 'General Mathews Range', after General Lloyd Mathews, the British agent in Zanzibar who had given them much help and advice. Today this beautiful chain is still known as the Mathews Range.

The resolute explorers made their way across to Lake Turkana and then right up the eastern side, to the Omo Delta, inside what is now Ethiopia. They also went east to the great salt lake of Chew Bahir. Teleki named the long lake after his patron, Crown Prince Rudolf, and he named the salt lake after Rudolf's Belgian wife, Princess Stephanie. Von Hohnel had a remarkable geological insight here; he realised that the depression in which Lake Rudolf lay must be linked with what Thomson called the 'Meridional trough' in southern Kenya. On their way back, Von Hohnel called the smoking mountain on the extreme southern end of the lake Teleki's volcano. Not to be outdone, Teleki named the bay that forms the southern end of the lake Von Hohnel's bay. These were clearly men who respected each other! Both names are still used today, unlike Rudolf and Stephanie, which were dropped in favour of Lake Turkana and Chew Bahir (which means, literally, salt lake, and was in all fairness the lake's original name). (The lives of Rudolf and Stephanie for whom the lakes were named ended in scandal and tragedy: Rudolf abandoned Stephanie, began a love affair with the attractive 17 year-old Marie Vetsera, and then in 1889 shot himself and his lover in the hunting lodge at Mayerling.) Von Hohnel, however, despite being savagely gored by a rhinoceros on a second trip to Kenya, with William Chanler, survived to the age of 84; he died in 1942 and is commemorated by both the bay and a beautiful little high-altitude chameleon that also bears his name.

The man who put Kenya on the map where geological research was concerned was the magnificently moustachioed John Walter Gregory, after whom the Gregory Rift is named. Born in 1864, Gregory was a Londoner, who had a long and illustrious career in the geological field, not just in Africa. He made two trips to Kenya, between 1892 and 1893, and again between 1918 and 1919. His adventures in Kenya had a hideously inauspicious start. As part of the 'Great Lake Rudolf Expedition', Gregory and his companions landed at Lamu. Everything went wrong. Recruits deserted, the rain came down, many of the porters caught malaria, and others mutinied as the expedition made its way to Witu and was attacked by hostiles. Gregory suffered badly from ulcerated legs following a severe case of sunburn. He then had severe malaria and lost his sight for a while. He describes a harrowing episode where, sightless, he was trying to find his quinine but the only way he could work out which packet it was in was by licking each and seeing which one tasted bitter. Many would have quit, but not Gregory. He made his way to Mombasa, and re-equipped. Tellingly, he prefaces the description of this debacle in his book with the appropriate Swahili proverb '*Kulekeza si kufuma*', i.e. to aim is not to hit!

The second expedition was much more successful. Gregory marched inland, and noted that the Kapiti Plains were a lava plateau. Within five weeks of leaving Mombasa he was in sight of the rift valley, opposite Mt Longonot. As he descended, he noted the parallel sides, so very different from a river valley. He also observed the platforms that mark old shore lines, and worked out that lake levels had fluctuated. He then made his way north to Lake Baringo and crossed the north end of the Aberdares to Mt Kenya, which he climbed up as far as the glaciers. While his team were cutting their way up through the bamboo belt, a hideous episode that lasted four soaking days (worth remembering these days when you pass through it in a 4x4 on a well-marked track or road), they found a block of lava that showed distinctive striping and grooving. This proved it was transported by ice, revealing that the glaciers had extended much further down the mountain in the past. In the alpine zone, Gregory named the features and many names remain in use: the Teleki Valley, Point Piggott, the Lewis glacier and the Darwin Glacier

Tsavo West, primary colours, metamorphic hills and red laterite soil. (Stephen Spawls)

are all his. In a kindly gesture, Gregory invited his favourite porter, a stalwart and philosophical man called Fundi Mabruk (who had ascended with Teleki on a previous expedition), to be the first man to step onto a Kenyan glacier. Fundi knelt and prayed, and then stepped onto the glacier, Gregory urged him to go further, but Fundi shook his head: 'No, master, it is too white.'

Gregory had a tough time, particularly with the Maasai. Those resolute warriors, to whom Kenya owes so much, for they kept the slavers out, were not above holding back the foreign explorers as well. In addition, Gregory's route overlapped that of Dr Karl Peters, a trigger-happy German colonist who was trying to annexe parts of eastern Africa for the German empire. Peters' response when crossed or held back was to shoot those in his way; consequently those who had encountered him were very wary of white foreigners. But Gregory was much liked by his men, and it was them who kept the expedition going. They called him *mpokwa*, meaning 'pockets', due to his habit of filling his pockets with specimens.

Gregory's book *The Great Rift Valley*, published in 1896, sold well. He was the first geologist to correctly define a rift valley, in the sense of a narrow valley caused by subsidence between fractures. On his return to Kenya in 1918 he not only inspired the then 15-year-old Louis Leakey, who was at one of Gregory's lectures (some say it was this lecture that galvanised Leakey to go hunting for stone tools), but while exploring in the southern rift valley with C. W. Hobley, found the now-famous prehistoric site of Olorgesailie. After covering himself with glory in Australia, Gregory accepted the chair of geology at Glasgow University. In 1932, at the age of 68, Gregory drowned in Peru, on an expedition; the journalist Anthony Smith speculates that it might have been his ever-loaded pockets that weighed Gregory down, a geologist to the very end.

In 1890 C. W. Hobley arrived at Mombasa to work for the Imperial British East Africa Company as a geologist. Hobley was a something of a polymath; he soon became an administrator, but continued to research and publish on the people, the geology and the natural history of Kenya (indeed, in 1910 Hobley and Sir Fredrick Jackson founded the East Africa and Uganda Natural History Society). As mentioned, Hobley located the fabulous prehistoric site of Olorgesailie with Gregory, he also discovered significant fossil-bearing beds of Miocene age on the shores of Lake Victoria. The fossil of a curious shovel-toothed ancient elephant found in these beds was named after Hobley, *Deinotherium* (nowadays *Prodeinotherium*) *hobleyi*. Hobley was the man who organised for Louis Leakey to go to Tendaguru. He published an enjoyable book, *Kenya, from Chartered Company to Crown Colony* (Hobley 1929), with some perspicacious observations on the country and its inhabitants, and a bit of geology. Hobley is commemorated by a small volcano in Hell's Gate.

In 1919, with the sympathetic support of Sir Robert Coryndon, the governor of Uganda, the first geological survey in East Africa was set up at Entebbe by E. J. ('Jim') Wayland, East Africa's first professional geologist. Wayland remained its director for 20 years. When a publisher once asked Wayland to write a book on his Africa adventures, Wayland is said to have replied, 'I had none, only incompetent people have adventures.' Trained at Queen's University in Belfast, he came to East Africa with field experience in Sri Lanka under his belt and insisted on walking 900 miles around Uganda. He also travelled 300 miles by canoe, in order to familiarise himself with the territory.

Wayland discovered the cache of fossils on Kenya's Rusinga Island; in Uganda he found numerous stone tools. He recognised that East Africa has a long prehistory; in 1920 he wrote in an annual report that 'Uganda has been inhabited by people … for a very lengthy period, indeed probably quite as long as Europe has.' He did some good work. Unfortunately he is remembered by his mistakes. He described various Stone Age cultures in Uganda (Kafuan, Sangoan, etc.), some of which have had to be rejected, as what Wayland thought were stone tools turned out to have been shaped by natural forces, not humans; others were described from the wrong strata. As mentioned earlier, Wayland championed the compression theory to explain how the rift valley was formed, in opposition to Gregory's (correct) theory of tension. Wayland also suggested that the European ice ages could be correlated with periods of intense rainfall in Eastern Africa, so called 'pluvial periods', an alluring but ultimately inaccurate theory that bedevilled Kenya's geologists and palaeontologists for years.

Not many of the diligent geologists who have worked in Kenya in the latter half of the twentieth century have had the impact of men like Gregory, Fischer or Thomson, but then those adventurers were the pioneers. We might mention Robert Shackleton; a cousin once-removed of the famous explorer Sir Ernest Shackleton. Robert Shackleton spent six years in Kenya during the Second World War and then returned frequently thereafter; he developed some major ideas on the tectonics of the East African Rift. An indefatigable field geologist, Shackleton explored large areas of Kenya on foot or by camel. Apart from documenting the geology of such diverse areas as Laikipia and Migori, he also located some interesting Neogene fossils near Maralal. On one occasion Shackleton was tossed by a Black Rhino *Diceros bicornis* he had disturbed, and was almost immediately bitten by a Puff Adder *Bitis arietans*.

THE LITERATURE

The classic work on the geology of our region, without doubt, is Dr Thomas Schluter's *Geology of East Africa*, published in 1997 by Gebruder Borntraeger in Berlin. This is a 484-page scholarly description of the geology of Kenya, Tanzanian and Uganda, with more than 180 figures and maps. It starts with a historical review and ends with a definitive summary of East Africa's biostratigraphy and palaeontology. We have drawn extensively on it for this chapter. It is still in print. No serious student of Kenya's geology should be without it; it will get you up to speed on our geology quickly. The bibliography lists the bulk of significant papers published on East African geology up to 1997. However, the potential reader (and purchaser) should be aware that it is an expensive book, costing over £100, and it is also written by a professional geologist for other professional geologists; don't attempt to read it without a geological dictionary to hand.

Dr Celia Nyamweru's book *Rifts and Volcanoes* gives an excellent summary of the geology of the East African Rift System and includes 'big picture' stuff on plate tectonics and earthquakes. Intended as a school text, this super little book, with its clear text, maps, diagrams and photographs, serves the purpose admirably. Published inexpensively by Nelson Africa in 1980, it is now out of print but can be found via the Internet. Another useful text aimed at the layperson is a handy little softback published in 1984 by the National Museums of Kenya, entitled *Kenya's Place in Geology*; copies are sometimes available at the National Museum shop. No author is directly named, but John Thackeray and Martin Pickford wrote the text. It starts with an introduction to geology, moves on to the history of the universe and covers basic geology. The latter half of the book summarises Kenya's geology, palaeontology, development of fossil faunas and provides geological notes for two major trans-Kenya journeys – and all this in a total of 39 pages!

Gregory's book *The Great Rift Valley* was published in 1896 by John Murray in London; a facsimile was published in 1968 by Frank Cass. Students of Kenya's natural history should read it. Of its 422 pages, the first 210 are an entertaining travelogue, covering Gregory's initial false start and then his trip to Baringo and Mt Kenya. This is followed by nearly 200 pages on the landscape, flora, fauna and peoples of East Africa; mostly covering Kenya. There is a 24-page chapter on the geography and geology of East Africa. Gregory's surface geology is generally sound, and he correctly surmised that tension had caused the rift faulting, although his ideas on the actual mechanisms are outdated.

Kenya's oldest rocks: 3 billion year-old greenstones near Migori, western Kenya. (Stephen Spawls)

Sonia Cole's book *Leakey's Luck*, published by Collins in 1975, although mostly concerned with Louis Leakey and East African palaeontology, gives background details on geology (Cole had two years of geological training) and many East African geologists, in particular those who were contemporaneous with Leakey Senior.

A lot of good material has been published on Kenya's geology since the first pioneering papers of the early twentieth century. Thomas Schluter reckons there are some 7,000 papers published on East African geology. His reference list mentions nearly 1,500 of them. But Kenya's geology has not been fully documented. As Schluter points out, when considering the total number of publications, this only actually amounts to one publication per 250 square kilometres. Large areas have never been fully investigated or mapped. Who knows what riches – both intellectual and commercial – still lie beneath Kenya's soil?

Chapter 2

Fossils and Hominins

It is somewhat more probable that our early progenitors lived on the African continent than elsewhere – Charles Darwin, *The Descent of Man*, 1871

The bulk of this chapter is devoted to the story of human fossils and to paleoanthropology, which is the name given to the study of human origins. During the last 160 years, paleoanthropologists have unearthed fossils of about 25 extinct species that are more closely related to modern humans than to chimpanzees, our nearest living relatives. These species are called hominins, although the older synonym hominids is still sometimes used. This chapter looks at the emerging picture of human evolution and Kenya's contribution to the study of ancient and progressively modern humanity. Some scientists feel that too much emphasis is given to the study of human fossils, as opposed to other fossils, but many disagree. Paleoanthropology is about us and how we became what we are. What could be more interesting?

The chapter is split into three sections, looking first at fossils and how they can be dated, then at hominins and the evolution of humanity. There is a concluding review of the literature, and again, each section is self-contained.

FOSSILS, DATING AND KENYA'S FOSSILS

Fossils are traces of past life, our link with the world in an earlier time, and the study of fossils is called palaeontology (from the Greek word *palaios*, meaning 'old'). Fossils indicate that there have been great changes in the kinds of organisms that have dominated life on earth; many forms have disappeared, although in general life has become more complex. More than 95 per cent of all organisms that have ever lived are now extinct.

Fossils form in several ways, but usually an organism dies and is covered up by something that stops it decomposing to some extent. The whole organism may be preserved, or just some bits, especially hard parts like teeth, or bones. Sometimes a solution of dissolved minerals (solutes) flows past the body, and replaces it. This is petrification, and the copy may be perfect. Not all fossils are formed from living things; a fossil can be a footprint, or a wormcast.

It is often possible to work out how old a fossil is. Geochronology is the science of rock dating, and is used with sedimentary rocks. One technique that can be used is known as stratigraphy, where the time taken for modern sedimentary beds to form is used to age similar ancient beds, and the characteristics and orders of such beds is used to correlate them to similar beds elsewhere. Absolute dates can be calculated by radioactive dating, although there is a degree of uncertainty about this, especially if the sample is very old. In simple terms, radioactive isotopes such as carbon-14 for recent fossils, or potassium-40 for older material, are deposited in the material. After the death of the organism or solidification of the rock, the radioactive isotope decays away. If you know its decay rate (the half-life), how much was in it when it started and how much is left, you can calculate its age.

The principles of palaeontology are fairly straightforward. You find suitable strata (usually sedimentary rocks but not always) and look for fossils. Luck may enter into it. Hobley and Gregory happened on Olorgesailie while on a camping safari. A German entomologist, Wilhelm Kattwinkel, almost fell into Olduvai Gorge while pursuing a butterfly, and thus stumbled on that classic prehistoric site. Richard Leakey spotted the sediments around Koobi Fora from his plane.

In looking for suitable rocks for finding fossils you first need rocks that appear to be layered and are therefore likely to be sedimentary. Then you examine the rocks. If you are looking for hominins, then the strata need to be fairly young, since no hominins have been found in rocks older than 7 million years. If you want dinosaurs then the rocks need to be older than 65 million years. Despite the movie industry's best efforts, aided by Raquel Welch in her fur bikini in *One Million Years B.C.*, dinosaurs and humans were never on the earth at the same time. The best places for fossils are where gradual erosion is taking place on a (preferably) gentle slope, so that some fossils or fossil-bearing rocks have been washed out and are exposed. You survey, dig trenches and pits, sieve soil. If things look promising, you hopefully get some funding and bring in experts, like geologists. You make a 3-D map and mark precisely where fossils are found. You might have to do a lot of work; the team at Nariokotome, on the north-western shores of Lake Turkana, moved one-and-a-half thousand tons of debris by hand! Great care is needed as you remove valuable material and you may need witnesses. Kenya teams often excavate using 'Olduvai Picks', a six-inch nail set in a short wood handle. You may need to fix your fossil with a preservative like butvar or bedacryl resins and then put in a plaster jacket for transport to the laboratory. And you compare, describe and publish, and see what your colleagues think.

Most fossil hominins have been found in Africa, the bulk in four countries; Kenya, Tanzania, South Africa and Ethiopia. A few specimens are also known from Sudan, Chad, Malawi and Zambia. Western Kenya and the rift valley are rich in fossil sites; the carbonatite volcanoes of western Kenya preserve bones well and the rift valley is a natural sediment trap. These fossil hominins give a picture of how life has developed over the millennia, although there are some noticeable gaps. In the previous chapter we mentioned some smaller and earlier fossils. They have come from a wide variety of sites. Olduvai Gorge, probably the most famous fossil location in the world, is actually in northern Tanzania, as is Laetoli – although one book, *The Official Field Guide to the Cradle of Humankind* (2002), by Lee Berger, an American palaeontologist based in South Africa, and Brett Hilton-Barber, a radio presenter, managed to place Olduvai Gorge in Kenya. This did nothing for Berger's credibility; a review of this book by Tim White, a top palaeontologist, is one of the most devastating scientific demolitions of a shoddy piece of work you are ever likely to come across (White, 2002). But the list of Kenya's major fossil sites is a long roll of honour: Olorgesailie, Gamble's Cave, Rusinga Island, Fort Ternan, Maboko Island, Koobi Fora, Lothagam, Kanam, Kipsaramon, Lemudong'o and Koru, to name a few.

Richard Dawkins, in his book *The Blind Watchmaker*, (1986) has pointed out how lucky we are to have fossils. Kenya has more than its fair share of recent fossil material (by recent, meaning within the last 60 million years, which is recent in geological terms). So many fossil mammals are known from Kenya that, as Thomas Schluter explains in his classic work on East African geology, biostratigraphic sequences in East Africa are largely based on mammal faunal assemblages – meaning that you can identify the age and nature of particular rocks by the fossil mammals in it. Elsewhere such marker organisms are usually fossil pollen or tiny invertebrates.

There is plenty of material, other than mammals, of course. Non-hominin fossils are mentioned in the relevant chapters; Kenya has its share of fossil plants and animals. But a huge number of fossil mammals are known from Kenya. More than 10,000 specimens of nearly 250 species have been found in the northern Lake Turkana basin alone, and many more are known from the sites around Lake Victoria. Too little attention has been paid to this mammalian fossil fauna, frequently eclipsed by the often infrequent hominin remains with which there is such an obsession. Many – most – of these mammals lived in the last 30 million years. The discoveries around Lakes Turkana and Victoria paint a remarkable picture of a changing landscape and a developing fauna.

HOMININS AND THE EVOLUTION OF HUMANITY

'I have found a hominin fossil, but it isn't particularly unusual or significant' – What paleoanthropologists don't say

Fossil and molecular evidence indicates that humanity emerged in Africa, and spread out from there to populate the world. It is one of the most exciting scientific stories of our time; it is a popular topic and it is also astonishingly controversial. This chapter cannot do full justice to it but we hope to provide a decent summary and point interested readers in the direction of some good literature (there are very many books on the subject). We have expanded a little more on the evolution of primates and on the development of bipedalism, as these topics flesh out the story of the emergence of the species we call 'man'.

Certain facts are indisputable. We are here, a single species, *Homo sapiens*, modern humanity. Every human being on earth is capable of breeding with another human being (of the right sex, age and persuasion, of course) and producing fertile offspring; and this is one definition of a species. All human beings have very similar genes (a gene is a sequence of hereditary molecular information that enables a cell to build a protein).

If you look back at the fossil record, there is nothing that looks much like us older than 10 million years. There are no fossil apes older than about 35 million years. There are no true mammal fossils older than 140 million years. There are no land vertebrates older than 360 million years and no fishes older than 470 million years. Before that, no vertebrate fossils are known, but fossils of multicellular eukaryotic organisms (eukaryotic means they had organised nuclei; the nucleus is the 'brain' of the cell) 2,000 million years old have been found. Even older, more simple single-celled organisms from 3,500 million years ago are known from the fossil record.

This all points to a gradual evolution: modifications take place from generation to generation, with organisms changing, becoming more and more complex and advanced. We, modern humanity, are one of the most complex. Living things did not all come into existence at the same time. There has been a gradual change, from very simple, small, single-celled organisms, without organised nuclei, to more and more complex and larger living things, with huge numbers of cells that contain organised membrane-bound nuclei. Occasional opponents put forward the argument that evolution cannot be disproved. But it can. A fossil mammal from Cambrian rocks, a fossil human, or flowering plant, from the Jurassic, a fossil trilobite from the Permian or a fossil vertebrate from the Proterozoic, 1,000 million years ago, would prove that evolution as we see it has not happened.

Hominin palaeontology is both fascinating and exciting stuff. It's about us, where we came from and how we became so advanced – which we are. The occasional scientist will talk as though dolphins, or cockroaches, or termites, or some other organism 'is more advanced' than us, but they are not. We are studying other organisms, they are not studying us. Forget the whimsy. We have developed a complex, abstract method of communication, and some astonishingly sophisticated tools. Only humans can communicate the meaning of 'tomorrow', or 'wait over there'. Only humans can design and operate a computer, or kill from a distance. Only humans can carry out large-scale, purposeful modifications of their environment. We are the most advanced species to inhabit this earth, to date, even if we did get here by some large slices of luck.

Hominin palaeontology is also cutting-edge science. Ideas are constantly changing. New fossils are constantly being found. The data are regularly reworked. Someone gathers certain evidence and puts forward a theory. Human evolution is a scientific battleground, littered with injured, dead and still living theories about the nature of the path that led to us. Bipedalism; was it important; what prompted it? Which is more significant: having a large brain or being able to stand upright? Who was ancestral to whom? Can it be proved? Does it matter? Is one distinctive fossil lineage the ancestor of another, or a side branch? Are we naturally aggressive and can fossils prove it? Is the ability to kill from a distance significant? How many species of hominin were around at a certain time? Are the dates certain? Could a fossil hominin have been buried at a level below where they should be, which will give a false age? These things are vigorously debated. We will try to briefly address some of them here. And these contentious points lead onto another, violently polarised, field of biology (although beyond the scope of this book): the part that our genes and ancestry play in our behaviour – nature or nurture. If you want a good, solid popular exposition of this, read *The Blank Slate* by Steven Pinker (2002).

Hominin palaeontology is the lifeblood of good science, of active debate moving towards a consensus. And – we accept that this is a shameless plug for Kenya in particular and East Africa in general – virtually all the important fossil evidence for the development of modern humanity comes from the eastern side of Africa. Most of the really significant fossils from the period from 6 million years ago up to one million years ago were found in Africa (probably because our ancestors were confined to Africa, of which more later) and most actually in East Africa, in a small number of important sites that lie between Olduvai Gorge in northern Tanzania, through Olorgesailie and the Lake Turkana sites to the rift valley north and east of Metahara in Ethiopia. This area really is, to use a hackneyed phrase, the 'Cradle of Humanity'.

And the fossils themselves have often been found and popularised by some of the most larger-than-life, strongly opinionated characters that any field of scientific endeavour has ever produced. Many have made startling mistakes, others have documented these. Some have misled, deliberately or by accident, some have turned savagely on their scientific colleagues. Some have been ridiculed by their peers and the general public. Many are startlingly egotistical and believe that their personal discoveries are the most important; it is a rare hominin palaeontologist who announces he or she has found a hominin fossil but it isn't particularly exciting. Many discoveries have been announced in the media as 'breakthroughs'. Almost every new hominin discovery, it is said, will lead to the 'textbooks having to be rewritten'. Underlying it all seems to be a desire for the fossil hunter's discovery to be regarded as a 'true ancestor of humanity' (to the exclusion of all others), or the 'earliest member of the genus'. If this doesn't work, then there seems to be a desire to describe your specimen as a new species, or erect a new genus specifically for your fossil. And yet all have, in their own ways, done science and humanity a service, in heightening awareness and stimulating debate. It is exciting stuff. Palaeoanthropologists have featured on the covers of popular magazines and newspapers, and television series have covered their activities. Many of us want to know where we have come from, and why we behave the way we do; the study of our ancestors – or at least those bits of them that remain – may provide us with vital clues.

What do we know for certain? During the last 150 years, a number of large and advanced primate fossils that closely resemble us have been found (primates are the order of mammals that includes humans, lemurs, monkeys, apes and a few other climbing creatures). Many of these fossils have been found in Africa. Most are not the remains of any living organism; they represent extinct forms. The ones that start to look like us and are not the ancestor of any living ape are all less than 7 million years old. And molecular and anatomical data indicate that the last common ancestor of humans and chimps lived between 8 million and 6 million years ago; try to get more accurate than that and you invite controversy. In general, the closer you get to the present, the more 'modern' some of these fossils become (by modern, meaning that they look like us). The oldest fossil remains – usually skulls – that look just like us, and thus can be said to be of our own species, *Homo sapiens*, are about 160,000 years old, perhaps slightly older, and virtually all are from Kenya or Ethiopia.

It is worth mentioning a bit of terminology here. Our fossilised ancestors used to be called hominids, but research in the 1960s indicated that great apes are so closely related that they should be included in the same family, also hominids. So the more modern term hominin came into use to cover those fossil remains that are related to us and excludes apes.

During the last 6 (or possibly 7) million years hominins evolved at what is, in geological terms, quite a startling speed. Our ancestors stood upright and began to walk on two legs (as opposed to four – apes walk on their feet and their knuckles); this is called bipedalism. Their brains increased dramatically in size. In the last 2 million years average brain volume has tripled, from 500 to 1,500 cubic centimetres. Their jaws got smaller, but their number of teeth didn't change. This may be why modern humans often have trouble with wisdom teeth: it's because our jaws are too small. Alternative theories are that our teeth have shrunk but our slushy diet means our jaws don't grow properly. The ancestors began to make and use tools, including weapons. They developed a complex means of communication, the date of which is still much speculated upon. And all these advancements took place in Africa, except possibly the refinements of communication. Thus they became us and, finally, they set off from Africa, in at least two waves. Those of the second wave populated the known world, and here we are. But this is very much the broad-brush picture. Let's now look at the developments and personalities involved in more detail.

The science of evolution has taken off since the mid-nineteenth century. Although earlier scientists put forward ideas about evolution, it was the work of Alfred Russel Wallace and Charles Darwin that guided and formalised the field. The fact that humans have evolved was not initially well received, but evidence began to mount up in the nineteenth century. In 1856 some workmen in the Neander Valley in Germany found part of an old and unusual skeleton in a cave in a limestone quarry; the biologist Thomas Huxley named it – in honour of its locality – Neanderthal Man, *Homo neanderthalis*. In 1887, a young Dutch doctor, Eugene Dubois, arrived in Sumatra to do what no one had done before, look for the remains of our ancestors. His diggers were excavating near a village called Trinil, and found a tooth and a smooth skullcap of a hominin; the following year a thighbone was found. Dubois decided that all three bits of fossil came from the same creature. He described the remains of a new species of upright chimpanzee, which he called *Anthropopithecus erectus* (now *Homo erectus*), literally the 'upright, man-like ape'. Some scientists were sceptical, and Dubois essentially faded from the scene, embittered and upset. But his discoveries had enlivened the debate about human ancestors. Then followed the spectacular hoax of Piltdown Man, which fooled some of science's brightest brains, including the staff at London's Natural History Museum. Fortunately, it was left to a fine museum man, Kenneth Oakley, to expose the hoax. A witty letter to *The Times* newspaper asked: 'Sir, may we now regard the Piltdown Man as the first human being to have false teeth?' But in the meantime, real hominin discoveries were being made in Africa.

The story of man's evolution in Africa story starts in northern Tanzania – German East Africa, as it was then – in 1911, with a German entomologist, Professor Wilhelm Kattwinkel, stumbling upon and nearly falling into a massive eroded valley, nearly 100m deep, while pursuing a butterfly. Kattwinkel then spent some time collecting in the gorge, which was called Olduvai, a misspelling of the Maasai words Ol-Tupai, meaning 'the place of the wild sisal'; a spiky plant of the genus *Sansevieria*, known colloquially as 'mother-in-law's tongue'. In the past the gorge did not exist; the area was a huge lake filled by rivers flowing down from the volcanic highlands to the south. Animals lived beside it, and our ancestors had camped on its shores and made stone tools.

Kattwinkel found a number of fossils that he took back to Germany. Scientists there were interested and an official expedition was launched under the leadership of a Berlin geologist, Professor Hans Reck; the man who excavated the dinosaurs at Tendaguru. Accompanied by more than 100 porters, Reck made his way to Olduvai in 1913 and spent three months in the gorge. He worked out the basic geology, naming five beds, one to five; one was the oldest, at the bottom. Reck also found plenty of fossils (although he did not find any stone tools). But his big moment came one evening, when Manjonga, his right-hand man, came to his tent and told him, 'Bwana, we have found something which we have not seen before. I believe it is an Arab.' Mad with excitement, Reck hurried to the spot. Manjonga had found a hominin skull and part of a skeleton, on its side with its knees drawn up in a fetal position. It was in bed two. It had to be old, or so Reck thought.

The professor took his skeleton to Berlin. It caused both excitement and controversy. At the same level of the skeleton, which seemed to be rather similar to those of modern humans, were fossil mammals older than one million years. This was puzzling; an older type of human was expected. Reck's fellow scientists said that the body must be that of a recent human (they suggested a Maasai), buried in a grave and thus artificially lower in the strata than expected. Reck was unhappy, and a proposed second expedition which was scuppered by the outbreak of war. Reck did not return to Olduvai until 1931, with Louis Leakey.

In the meantime, there were developments in Southern Africa. In 1924, Raymond Dart, Australian professor of anatomy at Witwatersrand University, received a skull of a fossil baboon from a lime-quarrying operation at Taung (literally ' the place of the lion' in Setswana), south-west of Johannesburg. Dart realised this was important; virtually no fossil primates were known from sub-Saharan Africa at the time. He asked for more specimens from the quarry. Of those he received, one was a fossilised skull of a young hominin – the first definite such find from Africa. It was a momentous discovery.

Accompanying the skull was an endocast – a rock replica of the brain – and it was surprisingly large, about 410 cubic centimetres, which is about the same volume as the brain of a chimpanzee (a gorilla has a slightly bigger brain, around 500 cubic centimetres). In addition, the position of the foramen magnum, where the spine enters the skull, seemed to indicate that the creature walked upright. Dart could not be certain of the skull's age – it came from a limestone breccia, a pile of ancient rock rubble – but he estimated it was more than a million years old. He described the skull as *Australopithecus africanus*, meaning literally 'southern ape from Africa' (Dart 1925). (The Latin *australis* has nothing to do with Australia; it means 'southern', although occasional critics suggested that Dart chose it as he was an Australian.) Nowadays the skull is often called the 'Taung Child'. The name, the species, is still valid today. Dart was right: he had found Africa's first genuine hominin. Early man was in Africa; that much was clear.

But in 1925, although the popular press were excited, science didn't like it. Dart did receive strong support as to his claims that this fossil represented a link between apes and modern humanity – the 'missing link' – from Robert Broom, a well-respected but highly eccentric palaeontologist. However, many other scientists did not agree. For a start, it confounded accepted theory at the time, which was that humans had developed a big brain first and then their jaws had shrunk. Piltdown Man, which at the time was as yet unmasked, seemed to confirm this erroneous theory. Dart's Taung Child had a small brain and small, human-like jaws. No one could be certain of the skull's provenance; where exactly it came from, and thus its age. No supporting fossils had been found. Other people disliked the idea that this supposed 'ancestor' came from Africa. The world was not ready for the idea that humans emerged in Africa (some people still are not). Dart could not find a journal prepared to publish further papers, and he faded into obscurity. For years, his valuable fossil was used as a paperweight.

It is time to usher in Louis Leakey, the grand old man of East African palaeontology. Christened Louis Seymour Bazett Leakey (his first name was pronounced 'Lewis' by the family), he was born on the 7 August 1903, just north of Nairobi, at Kabete, the son of Canon Harry Leakey, a British missionary, and his wife, Mary. The young Louis Leakey was brought up in highland Kenya, speaking fluent Kikuyu and English. He got a Double First at Cambridge University – where, incidentally, he was the first man to wear shorts to play tennis, and was ordered off court for indecency! He devoted the rest of his life to hunting fossils, mostly in East Africa. A European born in Africa, something pretty unusual at the start of the twentieth century, Leakey had no problems with identity: his first volume of autobiography (Leakey 1937) is called *White African*. In Sonia Cole's excellent biography, *Leakey's Luck* (1975), the frontispiece is a photograph of Leakey taken at Lodwar in the 1950s. It encapsulates the man: tall, bronzed, with a gunfighter's moustache and a devil-may-care smile, dressed in a filthy bush jacket and khaki slacks, a cigarette dangles from his lips, a muscular arm casually clasps the barrels of a huge hammer-lock double-barrelled gun. This, you feel, is someone you would trust in the bush. In later years, Leakey became a good bit stouter, and took to wearing odd-looking khaki boiler suits, looking like an elderly garage mechanic. A bronze in the National Museum in Nairobi shows him dressed thus. But he was a field man: he not only knew his subject, he got out there and actually found the things he was talking about. And when he did, he let the public know about them; his charisma, enthusiasm and flamboyant style ensured he got good publicity and raised the profile of his topic. East African palaeontology is forever in his debt.

Bronze of Louis Leakey in Kenya's National Museum. (Stephen Spawls)

Leakey did make a few mistakes, but in a life of field activity, academic publication and reaching out to the public, no one can avoid errors. He had a tendency to spread himself too thin and work quickly, often without carefully documenting his localities, a shortcoming which – as we shall see – nearly blighted his career. He moved quickly from one discovery to the next; dropping one idea and latching onto another. He was also possessive of both his academic field and his sites, and he did not forgive others who disagreed with him. If you didn't get on with him, you didn't get access to the prime fossil areas. If he didn't approve of your interpretation of what he had found, you didn't get to see his specimens. He believed all his major ape and hominin finds were the direct ancestors of modern humanity; the finds of other were all side shoots. He upset many museum-and university-based academics, who disliked his never-ending stream of discoveries and the special specific status he gave to them. A cartoon in the British magazine *New Scientist* in the 1960s shows a thoughtful gorilla saying to itself, 'I wonder if Leakey is a separate species …' But nevertheless, he was the man who brought East African palaeontology to the world. For many years he hunted fossils with hardly a cent to his name, keeping on and on, in the hope of the big strike, which he eventually got. He also fathered four sons, three of whom, Richard, Jonathan and Philip, have become almost as major a part of the Kenyan legend as Leakey Senior.

In the 1920s, Leakey had carried out a number of investigations at sites in Kenya's central rift valley, between Eburru, Gilgil and Nakuru, sites with names such as Willey's Kopje, Kariandusi, Bromhead's Site, the Makalia Burial Site and Gamble's Cave (on a farm belonging to a farmer called Gamble). Some of these sites are still open to the public. Based on excavations at these sites, Leakey published his first paper in *Nature*, entitled 'Stone Age Man in Kenya Colony' (Leakey 1927). At Gamble's Cave a lot of stone tools and several skeletons were found; the positioning of the skeletons seemed to indicate that they had actually been buried. Leakey estimated their age at 20,000 years, although many years later radioactive dating indicated they were more like 6,000 years old. The Leakeys' work here, and later at Hyrax Hill and the Njoro River Cave near Nakuru, shed light on life in Kenya in the last 3,000 years. Too little attention has been paid to their discoveries there, although there is some good material in *The Prehistory of East Africa* by Sonia Cole (1954). But the Gamble's Cave skeletons were to provide clues to the next piece of scientific investigation that Leakey became involved in: the provenance of Hans Reck's Olduvai skeleton.

Leakey visited Munich in 1925 and 1927, to examine the skeleton, and was struck by its resemblance to his Gamble's Cave specimens. He thought it might be of similar age. Reck, you recall, was convinced that it was over a million years old. In 1931 Leakey organised an expedition to Olduvai, and invited Reck, and several other scientists, including Sir Vivian Fuchs, the explorer, an old college friend Donald MacInnes, and the British Museum palaeontologist Arthur Hopwood. Leakey had a £10 bet with Reck (quite a lot of money in those days: Leakey reckoned that £10 would cover most of his research costs for the year) that he would find stone tools within 24 hours of arrival. Reck was convinced he would find none. Leakey won his bet, finding a handaxe shortly after dawn on the first full day there. It turned out that Reck had not found stone tools because he was looking for flint implements, as all the tools known from Europe were made from flint. But Leakey knew African stone implements when he saw them; he was looking for tools made of quartzite, chert or lava, and he found them. Olduvai is a treasure house of stone tools, in fact, and has given its name to a particular stone tool industry, the Oldowan, although both Oldowan (simple bashers and cutters) and Acheulean (more complex, with two worked faces) tools are found there.

But to return to Olduvai in 1931. On the hominin front, things didn't go so well. The team managed to locate the spot where Reck's skeleton had come from; that wise geologist had marked it with wooden pegs. Reck also managed to persuade Leakey – one of the few known occasions, according to Sonia Cole, that someone managed to get Louis Leakey to change his mind – that the skeleton was in its natural position, in bed two, meaning that others had not buried it at a lower level. While they were still at the gorge, Leakey drove to Nairobi and he, Hopwood and Reck stuck their necks out and suggested in both a letter to *Nature* and an article in *The Times* that 'Oldoway man', as Reck's skeleton was known, was the most ancient hominin skeleton in Africa. It was a rash move; they had made a mistake. The geologist John Solomon found that the rock debris around the skeleton contained minerals that were only present in the beds above. The only way they could have got there would be if the skeleton had been buried. So Oldoway man was not old at all. He was a fairly modern chap who had been buried by his friends in a deep grave; later analysis indicated he was only about 17,000 years old, not millions.

Leakey was furious. Solomon had debunked the best hominin find he was associated with to date. He still had no unequivocal evidence that man had emerged in Africa. He stated that Solomon's test was 'rubbish'. And he announced that he would find another Oldoway skeleton that would put his critics' doubts to rest. In retrospect, Leakey should not have felt he needed to shoulder the blame. He had not actually found the skeleton, and he originally thought that it might be relatively young. It was just unlucky that he had put his name first on the paper. Unfortunately, worse was to come.

The area Leakey next chose to go fossil hunting was in western Kenya, on the shores of Lake Victoria, on the Homa peninsula, which sticks out into the lake just north of Homa Bay. At Kanam, just north-west of Homa Mountain, is a series of light-coloured sedimentary beds, and the runoff from the higher slopes has eroded the

White sedimentary beds below Homa Mountain, site of one of Louis Leakey's greatest mistakes. (Stephen Spawls)

sediments, exposing them as rich fossil-bearing strata. And here they found the bit of hominin that nearly wrecked Leakey's nascent career; it was a bit of fossilised jaw with two teeth. The man who actually found it was one of Leakey's team, Juma Gitau bin Shariffu, and Leakey did not accurately mark the spot, or photograph it. (The full details of the story can be found in *Leakey's Luck* (Cole 1975).) The fossil itself is not very prepossessing – just a chunk of lower jaw with two stained molars. But it was a hominin, small and un-apelike. Leakey gave it the name *Homo kanamensis*, Kanam Man, and said it was 2.5 million years old. He exhibited the jaw at a Royal Society meeting in March 1933, and later at the Natural History Museum. But Percy Boswell, Professor of Geology at Imperial College in London, having examined the site, pointed out in the journal *Nature* – a devastating place to attack another scientist – that Leakey could not prove the provenance of the specimen, and had misidentified the rock it was in, claiming it to be a clay when it was actually an agglomerate (Boswell 1935). The implication here was that Louis Leakey, palaeontologist, did not know the difference between a clay and an agglomerate. It was a devastating attack on the Kenyan's professionalism. No wonder that Leakey wrote darkly in his diary, 'Boswell's findings may ruin my career'. They did not, but he was in the wilderness for quite a long time afterwards.

But fortunately nothing it seemed stopped Louis Leakey. In 1935 he published a book, *The Stone Age Races of Kenya*, divorced Frida, his first wife and mother of his children Colin and Priscilla, and married Mary Nicol, a talented artist and self-taught expert on both archaeology and stone tools. As Mary Leakey she rose to become as great a palaeontologist as Louis. After a brief period in Britain, they returned to Kenya, where Louis wrote a major work on Kikuyu customs, and assisted the Kenyan police with several cases, including the Lord Errol murder. Louis and Mary excavated at Hyrax Hill and the Njoro River cave near Nakuru, both localities that were 3,000 years old. Eighty bodies had been cremated at Njoro. During this time they were almost penniless and were camped on Nellie Grant's farm. Nellie (the mother of author Elspeth Huxley and immortalised in her books as 'Tilly'), noted they 'lived on the smell of an oil rag'. During the war, Louis was active in intelligence operations, building up a network of informers and broadcasting in Kikuyu, Kamba and Swahili; he was fluent in three languages. He also managed to relocate the prehistoric site of Olorgesailie, and after the war he was appointed the curator of what was then the Coryndon Museum, now the National Museum. In 1948 Mary found not only the fossil of a giant hyrax, as big as a calf, but a superb skull of *Proconsul*, a Miocene ape, on Rusinga Island in Lake Victoria. This story was widely reported in the British Press.

Leakey was also the court interpreter at the trial of Jomo Kenyatta, at Kapenguria in 1952. At the trial, Leakey clashed on several occasions with Denis Pritt, QC, who was defending Kenyatta; Pritt thought Leakey was getting beyond his brief and interpolating as well as interpreting. After five such clashes, Louis withdrew from the trial, but his services were retained as a translator. Leakey wrote two books about the Mau-Mau during the time of the emergency. He also published a major paper on fossil African pigs, naming eleven genera and 27 species; many of which were later rejected, unfortunately. And he and Mary returned to work at Olduvai Gorge, and made some momentous discoveries.

In the later 1930s and 1940s, some more interesting hominins had been found in South Africa by Dr Robert Broom, including two species now known as *Australopithecus africanus* and *Paranthropus robustus*. Both between 2 and 3 million years old, these discoveries are very significant in terms of human evolution. Firstly, they represent two different evolutionary lines, sometimes called 'gracile' (slender) and 'robust' (stocky) Australopithecine. There is much debate about their relationships with other hominins and us. It was hypothesised that they occupied different niches, i.e. they behaved differently. Secondly, following Broom's publication of his new specimens, most paleoanthropologists came round to the idea that *Australopithecus* might be the ancestors of *Homo*, our own genus. And lastly, the pelvis and spine from a partial skeleton from Sterkfontein indicated that *Australopithecus* walked upright. This was in the late 1940s. Science was gradually accepting the idea that (a) humanity evolved in Africa and (b) bipedalism was significant. But not every worker in the hominid field accepted that the Australopithecines were ancestors of *Homo*, and one who did not was Louis Leakey, who was at Olduvai Gorge.

On 17 July 1959 Louis wasn't feeling well and stayed in bed. Mary went out for a walk, and spotted a skull eroding out at the top of a gully they had named Frida Leakey Korongo (Frida was Louis's first wife, *korongo* is Swahili and means 'valley', as well as 'stork' and 'Roan Antelope'!). She called Louis, who was initially excited but when he saw the teeth was disappointed; he was hoping for *Homo*, not *Australopithecus*. Rarely can have gloom been more inappropriate; this was the hominin that was to shoot the Leakeys to fame and establish their reputation. As luck would have it, the prominent natural history filmmaker Des Bartlett was on hand and filmed the excavation of the skull.

This remarkable skull put Leakey in a dilemma. He didn't believe that *Australopithecus* was ancestral to *Homo*, and in this belief he was in a minority. But he also didn't believe that *Australopithecus* was capable of making tools, and in this he was among a majority. Man was the toolmaker, not some possibly ancestral ape-man. Now, not only had his wife found the ancestor that he didn't believe in, but the specimen in question lay among a huge range of tools that Leakey didn't believe it could have made. Even now, there is vigorous debate amongst experts as to who made what tools, and this debate is complicated by the fact that, many times in the past, more than one species of hominin was alive at the same time. The rock tools don't talk, they can't tell us who made them. Recent evidence, from eastern Ethiopia, indicates that stone tools were being made up to 3.4 million years ago, and they were being made by *Australopithecus afarensis*,

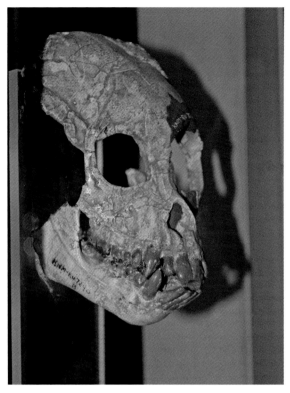

Skull of Proconsul, early ape, from Rusinga Island, as shown by the museum acronym RU7290. (Stephen Spawls)

which would seem to settle the matter. If the Australopithecines were making tools that long ago, they are unlikely to have lost the skill … or so we presume.

But at the time the idea was up for debate. Leakey did not want to call his new 'dear boy' (the Leakey family nickname for the skull) *Australopithecus*, because he was – as always – convinced that he had found a human ancestor and the Australopithecines were not. So Louis described it, in a letter to *Nature*, as *Zinjanthropus boisei*; Zinj being an ancient Persian word for East Africa, and the specific name was for Charles Boise, a philanthropic London businessman who had financially supported the Leakeys for twelve years. New genus, new species. That was forever Leakey's way … and the way of a great many other hominin hunters, before and since. Nowadays the species is placed in the genus *Paranthropus* by most palaeontologists (but not all, some believe it belongs in *Australopithecus*). It is quite an unusual skull in that it has a very broad face. It also has a sagittal crest; this is a ridge on top of the skull that increased the surface area for attachment of jaw muscles, suggesting that its owner had very powerful crushing jaws. And thus the public, who picked up on this, got to know *Zinjanthropus* as 'Nutcracker Man'. Modern humans have lost this crest. Radiometric dating in 1961, using the potassium–argon decay series, put Nutcracker Man at 1.8 million years old. This was astonishing as few geologists believed the Olduvai beds were that old. The discovery brought funds to Louis and Mary to enable them to continue their excavations. The skull itself is now housed in the strong room of the Tanzanian National Museum in Dar es Salaam.

The next big discovery at Olduvai was made by Jonathan Leakey, Louis and Mary's eldest son, who later made a name for himself as a herpetologist. Consequently this hominin is sometimes called 'Johnny's Child'. It is a partial skull and some hand and foot bones. Leakey brought in two experts to analyse this hominin – the doctor and anatomist John Napier, and another anatomist, Michael Day. Napier found that the bones came from two hands, one adult, one a juvenile, and this included an opposable thumb. This was a very significant discovery. The opposable thumb is an important part of our anatomy, it enables us to grip and control something very precisely. Try writing, or threading a needle without using your thumb – it's difficult. This hominin possessed fine control and walked erect, and had a brain capacity of 680 cubic centimetres. This is bigger than any ape but smaller than modern humans.

In 1964 Leakey, the eminent South African paleoanthropologist Philip Tobias and Napier described Jonathan's new man in the journal *Nature*. Out of respect for the probable tool-manufacturing abilities of their man, they named him *Homo habilis*. *Habilis* is Latin for 'handy', so this was the handy man. But this paper was controversial. Sir Arthur Keith had defined *Homo* –somewhat arbitrarily – as having a brain size larger than 750 cubic centimetres. *Homo habilis* was just under this value, so the authors moved the goalposts. They also suggested that *Homo habilis* made the tools at Olduvai, reversing the previous suggestion that Nutcracker Man had made them. And then a specimen of *Homo erectus* was found by Louis, incidentally the only hominin he actually discovered by himself. Olduvai Gorge had now turned up three major hominins: *Paranthropus boisei*, 1.8 million years old, with a brain size of 530 cubic centimetres, *Homo habilis*, also 1.8 million years old, but with a brain size of 680 cubic centimetres, and *Homo erectus*, 1.5 million years old, with a brain size of over 1,000 cubic centimetres (although Margaret Cropper, the first wife of Louis's son Richard, found a *Homo erectus* in 1962 with an estimated brain size of 750 cubic centimetres, which received the inevitable but unfortunate nickname of 'pin head'). Evolution was happening, brains were getting bigger; as time progressed our ancestors were starting to look more and more like us.

The scene now shifts to Lake Turkana, in northern Kenya, and a skull found in 1972 by Bernard Ngeneo, who at the time was a cook, but his find got him promoted to a fossil hunter. There are a number of superb fossil sites in the vicinity of this long, narrow and remote lake, including Koobi Fora, Nariokotome, Lothagam and the lower Omo River. This last is in Ethiopia, the others in Kenya. It is a harsh environment: the lake is at an altitude of under 400m, and in the driest area of Kenya, with less than 250mm of rain per year (and less than 250 can mean zero). Daytime temperatures can reach 40°C. A violent wind blows strongly at night. The green waters of the lake have led to its being named 'The Jade Sea'. The surrounding country is the haunt of bandits, scorpions and Carpet Vipers *Echis pyramidum*, dangerous little snakes; the lake contains some huge crocodiles. It is no country for the faint-hearted. No one was initially aware that the area might contain rich fossil deposits; it was spotted by accident, and was a breakthrough for Richard Leakey.

Born in 1944 in Nairobi, Louis and Mary's second son, Richard Leakey is an extremely interesting man, with a remarkable, varied and controversial career. He left school at the age of 16, and has been a safari guide, palaeontologist, museum director, director of Kenya's wildlife service, politician and finally conservationist. During his adventurous life he has lost both his original kidneys, both his lower legs and when he was fifteen nearly died from anaphylactic shock after being bitten by a Puff Adder. In 1964, at the age of nineteen, he organised a major expedition, including air support from the RAF, to the Peninj River in extreme northern Tanzania, on the north-western shore of Lake Natron. The expedition found a lot of fossils, including a lower jaw of a *Paranthropus boisei* (at the time, it was described as *Australopithecus*). He was then asked by his father to organise an expedition to the Omo River, which flows into Lake Turkana in southern Ethiopia. This was at the behest of the then Emperor of Ethiopia, Haile Sellassie, who had noted the fossils that the Kenyans were finding and hoped some might be found in his country.

It was not a happy expedition. There were three teams – French, American and Kenyan – and there was rivalry between them. Some of the Western scientists were racially prejudiced and superior, and they sniped at Richard on account of his lack of a university degree. They had some successful fossil finds, including a skull of a modern human, *Homo sapiens*, dated to 130,000 years old, which pushed the age of modern humans back 70,000 years (it is now reckoned to be 190,000 years). But Richard had a lucky break while on this expedition. The pilot diverted from the western shore of the lake to the east to avoid a thunderstorm while flying back to Nairobi. Richard noticed that, although the geological maps of the lake showed the shoreline as being volcanic rock, north of Allia

The birthplace of industry with a million years of production: the handaxe factory at Olorgesaile. (Stephen Spawls)

Bay there were large areas of sediment. Exposed recent sediment in dry country is rapidly weathered, and this can wash out fossils, giving rapid clues as to the potential of the site with a simple surface survey. On his return, Richard flew down to Allia Bay by helicopter. A quick ground search revealed both tools and fossils. It was a new site. Richard had been feuding with his father, and he wanted to break away and do his own thing. Here was the place to do it. He mounted a small, motorised expedition to the area. It was successful, and a number of fossils including some hominin material, were found.

Richard then attended a National Geographic funding meeting in the United States with Louis, but surprised his father by asking for money to run his own excavation in East Turkana. Richard was given the money, on the proviso that if he had no luck, no further funds would be available. But the team was lucky. They found several hominin fossils, and in 1969 established a base camp at Koobi Fora, the jutting peninsula north of Allia Bay, where an intact volcanic tuff with stone tools embedded in it was found. This was important: a tuff is a rock made of consolidated volcanic ash and it can be dated; this tuff was dated to 2.4 million years old. Richard's team, the so-called 'hominid gang', began hunting and more hominins were found, including a skull of *Australopithecus boisei*, found by Richard, the only one he actually found himself. The hominid gang grumbled that it was such a big skull that even a *mzungu* (white person) could have found it. Numbered KNM-ER 406, it was to prove very important some years later, in what was called the 'single species hypothesis'. (KNM is the Kenya National Museum, ER stands for East Rudolf – the lake's name was not changed to Turkana until 1974.)

The year 1972 was to be a significant one in the life of the Leakey family. In July 1972, shortly after Richard and Meave Leakey's first daughter, Louise, was born (Richard having divorced his first wife in 1969 and married anthropologist and palaeontologist Maeve Epps in 1970), Bernard Ngeneo found the hominin skull that was to cause so much controversy. Ngeneo had spotted the fragments of the skull in a small pile of broken fossil material; other team members had seen the pile but had observed only bits of antelope bones. Ngeneo was not so sure, and took a closer look. He thought it was hominin, and called over top fossil hunter Kamoya Kimeu. Kamoya agreed, and called in Richard Leakey. The team began excavating and sieving, and finished up with over 150 fragments. They pieced the skull together, and measured the approximate size of the brain by sealing the cracks with plasticine and pouring in sand. It turned out to have a volume of about 800 cubic centimetres, which was not far away from the brain capacity of *Homo erectus*, and larger than the Australopithecines. It was possibly an example of *Homo*. And the tuff that it had come from, the KBS tuff (KBS standing for Kay Behrensmeyer Site, Kay being one of the team) had been provisionally dated at 2.6 million years old. If the data were correct, this was, by a long margin, the oldest member of our genus ever found.

The palaeontologists were tremendously excited. Richard took the skull, which has the museum number KNM-ER 1470, down to Nairobi to show Louis. Father and son had been at loggerheads for a while but this remarkable find reconciled them. But it was to be their last meeting. Louis then flew to London where he had a heart attack on 1 October. He was in Vanne Goodall's flat at the time; he was admitted to hospital and died. Vanne was the mother of Jane Goodall, the chimpanzee expert. Louis had met Jane in Nairobi, and arranged for her to study chimpanzees; the rest is history. Sadly, Mary Leakey was particularly upset by the fact that Louis had suffered his final heart attack in the home of another woman. Louis was always something of a womaniser, trading on his fame and charisma. Mary speaks harshly of this in her own autobiography, *Disclosing the Past* (Leakey 1984).

But to return to skull 1470 and Lake Turkana. There are, or were, two controversial things about this skull. Firstly, the shape, secondly the age. Let's take the shape first. This is a story that has appeared in a number of books about palaeontology, with many of the authors putting their own slant on the significance. Essentially; Richard assembled the skull with a particular orientation, with the face at right-angles to the top of the skull. This made it look like a member of the genus *Homo*. During a conference on hominins at the National Museum in Nairobi, the eminent British palaeontologist Alan Walker, who had been working in East Africa for a number of years, observed that the bones would fit properly only if the face were angled outwards from the skull, which made it look like an Australopithecine, more primitive. Walker also suggested that it should be described as such, whereas Leakey wanted to call it *Homo habilis*. Walker left the room, but returned and compromised. Richard Leakey told Walker that if he had not returned he would have been removed from the team. The implication here is that Richard wanted things his way and, like his father, if you didn't agree with him, you no longer got to work in what he considered his domain.

The second controversy surrounding skull 1470 was the age. Two expert geochronologists, Jack Miller and Frank Fitch, both dated the skull and initially got wildly varying dates but settled on 2.6 million years old. If correct, this meant that this specimen of *Homo habilis* would be the earliest representative of our genus and a very important fossil indeed. But other experts thought the dates were wrong. The fossil pig expert Basil Cooke had found that a nearby assemblage of fossil pigs, from a horizon dated 2.4 million years old, were actually indicative of a fauna known to be 2 million years old. Fossils from the Omo, 80km north, inside Ethiopia, indicated a horizon 2 million years old. A near identical assemblage of fossils from Koobi Fora was being claimed to be 2.6 million years old.

There was a lot at stake here. Richard and his team had said that their *Homo habilis*, 1470, was 2.6 million years old. This meant that our genus *Homo* extended back at least that long, and since Australopithecines were known that were younger than that, they could not have been ancestral to *Homo*, just a side-branch that did not make it through. Now this suggestion was under threat, on two fronts: firstly that the fossil was not a *Homo* at all, and secondly that it was not as old as had been claimed.

The response from the scientists on the Kenyan side took two forms. They claimed that evolution might have moved faster on the Kenyan side and hence produced certain forms earlier. At the time, this was not as daft a suggestion as it seems. The evolutionary zoologists Stephen Jay Gould and Niles Eldredge had just put forward their theory of punctuated equilibrium, which suggested that species remain unchanged for a long time and then undergo sudden genetic (and consequently morphological, or shape) changes. It is a theory that has since been comprehensively discredited by Richard Dawkins (Dawkins 1986), but at the time had a certain appeal when trying to explain gaps in the fossil record. Some sort of isolating mechanism kept the Ethiopian and Kenya

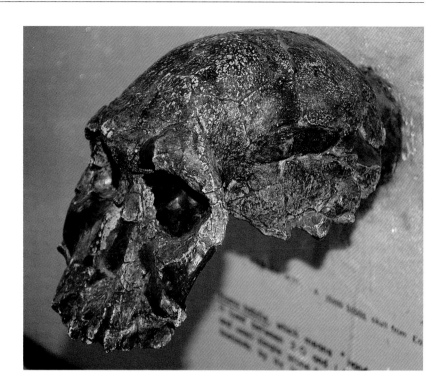

Cast of KNM–ER 1470, highly controversial skull from East Turkana. (Stephen Spawls)

fauna apart, and they evolved at different rates. However, the workers in Ethiopia (and most other disinterested scientists) felt that this was special pleading by the Kenyan team. The Koobi Fora scientists' next move, however, was definitely beyond the pale. They decided not to allow any other scientists to have or date a sample of the Koobi Fora tuff where 1470 had been found. This was a mistake. The entire ethos of good science is that of truth and transparency. In refusing access, they implied they had something to hide. Eventually, under pressure from top geochronologist Garniss Curtis and others, a sample was given for dating. It turned out to be 1.6–1.8 million years old. Our early ancestor was not so early.

Many more hominins have been collected since then. Dramatic discoveries were made at Hadar, on the northernmost loop of the Awash River in the Afar region of Ethiopia by Dr Donald Johanson between 1973 and 1978, none more memorable than the partial skeleton that Johanson and colleague Tom Gray found on 30 November 1974. According to Johanson, the shape of the pelvis indicates she was a female; officially she is known as A.L. (Afar locality) 288-1 Partial skeleton. But that night in camp, as the team celebrated, the Beatles song 'Lucy in the sky with diamonds' was playing, and the skeleton is now known popularly as Lucy, although the Ethiopian workers called her 'Dinkqnesh', meaning 'you are wonderful'. More skeletons were found, including thirteen individuals close together, possibly killed by a flash flood, and now known by palaeontologists as 'the first family'. In 1978 Johanson (with Tim White and Yves Coppens) named this species *Australopithecus afarensis*; they are now known to have had a brain size of about 550 cubic centimetres, lived between 3.7 and 2.8 million years ago and walked upright. General opinion – if such things can be known – is that Lucy's species is ancestral to us all.

Lucy was also, at the time, the only hominin find big enough to be called a skeleton; all earlier finds had been fragments, mostly skulls. This is a significant point. The description of a new species of living animal usually covers the entire creature in great detail, from external appearance, colour, shape and size through to bones, teeth and soft anatomy where relevant. Many measurements are taken; the organism has to be fully described so further specimens can be identified. But when a new fossil is described, often the authors have only a few bits of bone and teeth to go on. Many hominins have been described on the basis of their skulls alone, some on just a small bit of the skull or other bits of bone. This causes problems. Suppose a hominin is known from your area and was described from its skull and maybe a leg bone. If you find fossil hand bones, how can you be sure they are of the same species? You can't. You may not have a whole skull, but several bits. How can you be sure they are from the same individual – or even the same species? This is a particularly pertinent question in Kenya, where sites have often yielded several species. And the situation is complicated by the fact that the organism may be changing through time. You've got one species in an area, then you find a younger fossil, and it's a little bit different. Should it be called the same species? Or a new species? Or maybe a new genus? And this partially explains why there is much vigorous debate about hominin fossils, and what they mean.

More remarkable discoveries were on their way. Back in 1935 in northern Tanzania, Louis and Mary had visited a river called Laetolil (more correctly called Laetoli) south-west of Olduvai, guided by a farmer, Sanimu, of Kikuyu-Maasai ancestry, who knew the area and had brought along some specimens. The Leakeys found a few fossils. Louis's account of his trip there (in *By the Evidence*, his second volume of autobiography, published in 1974) is dramatically enlivened by his description of how he treated a Maasai elder who had been speared in the skull years earlier, an injury that left the brain exposed. The elder complained only of headaches. Louis's description of how he flushed the man's wound with hydrogen peroxide, and pus and dead flies came out from the brain cavity, can be read only by those with a strong stomach. But the old chap survived, thanks to Leakey's help. Mary Leakey returned to Laetoli in 1975, and in 1976 a treasure was revealed. The palaeontologist Andrew Hill was walking down the river bed with David Western (top Kenyan conservationist, his friends know him as 'Jonah') when Western threw a ball of elephant dung at Hill, and as he ducked and fell, Hill spotted a pattern of tiny markings on the rocky river bed. They were fossil raindrop prints in a layer of ancient volcanic ash.

Further investigations over several years yielded over 9,000 footprints of dozens of different animals, and although most were guineafowl and hares there were also elephant, Giraffe, rhino and even insect trails. They could be dated precisely, at 3.6 million years, as they were made in a layer of volcanic ash, chucked out by the nearby volcano Sadiman. Rain had then fallen, turning the ash into mud, creatures walked across it and the ash then set like concrete, leaving a simply fabulous hoard of fossil evidence for us all. The prize exhibit, without doubt, is a set of footsteps extending over nearly 50m made by hominins. The spacing of the prints shows that the hominins who made them were walking upright, unequivocal proof that they were bipedal organisms. It is also reasonably certain that three individuals were involved, a tall one who was walking very close to a slightly shorter individual. A diorama of this event at the American Museum of Natural History in New York shows the taller male with his arm around the female; a third individual, possibly a child, may have played follow-my-leader and stepped in the footprints of the two in front.

The hard science here is unequivocal and momentous. Australopithecines were walking upright 3.6 million years ago. But it is the speculative side of this tableau that has captivated the world, particularly the little human touches; a family, with a protective male, an apprehensive female and a child having fun. Some American feminists castigated the diorama as 'paternalistic'. But Mary Leakey considered the Laetoli discoveries her finest work. *Nature* devoted six pages to the discovery. The hominin fossils found at Laetoli were largely *Australopithecus afarensis*, the same species (supposedly) as Don Johanson found in Ethiopia, and it is presumed that they made the footprints.

But not everything about the Laetoli discoveries pleased Mary Leakey. She fell out with the palaeontologist Tim White, to the extent that White refused permission for his hominin work to be included in the official report. Alan Root, the famous Kenyan film-maker, and the man who actually suggested the explanation of how the third trail came to be made, filmed the excavation of the footprints, but Anglia Television in Norwich lost his film and it has never come to light. And Don Johanson and Tim White, in naming their new Ethiopian species with Yves Coppens, in 1978, said that the Laetoli fossils were also members of this species, a move that angered Mary Leakey. She was originally listed as an author, and she insisted that Johanson remove her name; the issue of the journal had to be reprinted.

More groundbreaking fossils keep turning up in East Africa. In August of 1984, Kamoya Kimeu found the first bits of a remarkable skeleton at Nariokotome, southeast of Lokitaung on the north-western shores of Lake Turkana. Kamoya is a professional fossil hunter, a Kamba from eastern Kenya who has worked with the Leakeys for over 40 years. His skill at finding hominin fossils is legendary; he was presented with the John Oliver LaGorce Medal of the Royal Geographic Society by President Ronald Reagan at the White House in 1985. The hominin fossil that Kamoya located that day in August turned out to be the most complete skeleton ever found of *Homo erectus*. The team actually shovelled out and sieved more than 1,500 tonnes (that is one-and-a-half million kilograms) of sediment to get most of the fossil bones. They even found ribs and a scapula, bones that are virtually never found, as the ribcage is the first bit that carnivores crunch into.

Nicknamed Turkana Boy, this hominin is believed to have been about 11 or 12 years old when he died, 1.6 million years ago. It is speculated he might have grown to over 6 foot. Turkana Boy is the oldest known *Homo erectus*. He had a very narrow neural canal (which the spinal cord runs within), about half the cross-section area of a modern human. This indicates that his spinal cord lacked the nerves that command the muscles for the fine control of speech and breathing, so he might not have been able to talk. This is significant in terms of language development and one of the things unique to us humans is our use of complex language.

Cast of KNM–ER 3733, Homo erectus, *the skull that destroyed the 'single-species' hypothesis. (Stephen Spawls)*

Some time before Turkana Boy was discovered, Kamoya had found another *Homo erectus* skeleton, in 1973, at Koobi Fora. This female was given the museum number KNM-ER 1808, and gave a unique insight onto human behaviour. This skeleton proved initially very frustrating for Alan Walker and Meave Leakey, because it was scattered over a wide area, mixed up with hundreds of other fossils, such as crocodiles, Giraffes and antelopes. In addition, none of the hominin fragments was the right shape, save the skull; the female in question had suffered from some bone disease, which coated the normal bone with a layer of fibrous bone of varying thicknesses. In discussion with a team of doctors, it was realised that the hominin had probably suffered from hypervitaminosis A, a disease caused by eating too much vitamin A. Although it is impossible to know, the source could well have been a carnivore liver. What was particularly fascinating about this skeleton, however, is the implications of the disease. Hypervitaminosis is a debilitating condition; victims cannot help themselves and normally die. But the growths on the bones of this female indicated she had lived for a fair while after developing the disease. So it seems likely that someone else took care of her. This is something very human, caring for the suffering; animals generally don't do it. It was a glimpse of the real life of our ancestors, something that bones rarely show.

Another fascinating skull from Koobi Fora was KNM-ER 3733, *Homo erectus* (some authorities call it *Homo ergaster*). The braincase was full of a stony matrix, so the brain size could not be calculated. In a bold move, Walker plastered over the skull with cigarette papers stuck on with glue, so he would know what bits went where, and then cracked it with a sledgehammer and cold chisel. Richard Leakey had driven out of Nairobi, as he could not bear to watch this potentially destructive manoeuvre, but fortunately, instead of shattering into tiny bits, the skull split perfectly, and when Walker had removed the rock using his air scribe (a small air-powered hammer), the brain size was found to be 850 cubic centimetres.

This important skull destroyed a curious theory called the 'single-species hypothesis', originally put forward by the ornithologist Ernst Mayr, one of our finest evolutionary biologists. Mayr (1950) suggested that because man occupies more different ecological niches available for a *Homo*-like creature than any other animal, he cannot speciate. This theory was then given a cultural dimension by C. Loring Brace and Milford Wolpoff, (Wolpoff, 1971) anthropologists at the University of Michigan, who said that since only one cultural species exists at present, us, there could never be two species of hominin existing at the same time. The logic behind this was that hominid species would have developed culture, or learned behaviour, and that all hominin species would occupy that same niche, i.e. they all would take on the existing culture. Now a basic tenet of biology is that two similar species cannot occupy the same niche, and that if they do find themselves there, one always outcompetes the other. So theoretically, if two hominin species find themselves in the same area, since they have the same culture, one wipes the other one out.

There are two problems with this supposition. One is that two species often do occupy the same niche, and yet manage to rub along quite well. They do this by developing very slight differences in behaviour, or microhabitat, although a purist might say that in doing so they have moved into different niches. For example, two relatively large dangerous snakes of the same genus, the Black Mamba, *Dendroaspis polylepis*, and the Green Mamba, *Dendroaspis angusticeps*, live in much the same habitat along the northern Kenya coast, and if you live in Watamu you may find both in your garden. A careful study shows that they occupy very slightly different microhabitats in the same area, with the Black Mamba tending to be in larger trees and slightly more open situations, the Green in smaller trees and thickets and deeper within the bush. And a recent study of doves in West Africa indicates that three different species occupy exactly the same niche and all three survive. The second problem with the single-species hypothesis is that the concept of culture defining part of a niche is not one that applies to any other organism.

Why Brace and Wolpoff never thought of these points is hard to say, but one thing is for sure, the work by Walker and his team destroyed their hypothesis. *Homo erectus* – or *H. ergaster* – had been found in strata of exactly the same age as the earlier skull, *Australopithecus boisei*, KNM-R 406. They were contemporaneous. Out went the single-species hypothesis, dreamed up without evidence. It is thus that science is practised. If the facts don't suit the theory, then change the theory; the facts are sacred.

In 2002 a team led by Martin Pickford and Brigitte Senut found some leg bones, bits of jaw and teeth of a hominin in strata that were about 6 million years old, at four sites including Kapsomin, in the Tugen Hills, just west of Lake Baringo in Kenya. Although they did not have much to go on – there was no skull – Pickford and Senut believed that the leg bones of the fossil suggest that it was (a) definitely a hominin, and (b) bipedal, it walked upright, although not everyone accepts this. They called their fossil *Orrorin tugenensis* (Orrorin is a Tugen word meaning 'original man'); the media dubbed it 'Millennium Man'. Pickford and his co-authors suggest that *Orrorin* was not only the first Miocene hominin but was ancestral to *Homo* (and thus the Australopithecines were not), a claim that made many hominin researchers uneasy. If *Orrorin* is a hominin, then for a short time it held the title of being the oldest known one, and indicated that the human–chimpanzee line split more than 6 million years ago. Fossils found with the Millennium Man included both ancient Impala *Aepyceros melampus* and colobus monkeys, indicating that he may have lived in woodland, not grassland. This may be significant in light of the debate about bipedalism, as one theory is that the hominins stood upright because they had left the forest. However, *Orrorin* is a notorious fossil for all the wrong reasons. Not only did it lead to a great deal of theorising with very little data, but also politics and skulduggery were involved. Pickford was thrown into jail shortly after its discovery, after Richard Leakey suggested to the National Museum that Pickford was working illegally. Pickford was released eventually; no charges were brought. The sorry affair made headlines in the British newspaper *The Independent*.

Since the discovery of Millennium Man an even older skull has been found in northern Chad – a fossil named *Sahelanthropus tchadensis* ('Sahara Man from Chad'), and nicknamed 'Toumai', meaning 'hope of life'. Reliably dated at 7 million years old, the Chad skull might be a human ancestor, or have lived before the human–chimpanzee split; reconstructions of the skull look rather ape-like. In addition, a multidisciplinary team has been working on the middle Awash River between Metahara and Tendaho in eastern Ethiopia during the past 25 years, under the stern discipline of American palaeontologist Tim White, a man who has not only revolutionised hominin palaeontology but also produced the definitive study of cannibalism among native Americans of the

American southwest. White clashed with Kenya-based paleoanthropologists in the 1970s and hence has ensured that his team embodies the professionalism and commitment that he felt was lacking from the Kenyan teams, especially in the early days. (White has called some Kenyan collecting expeditions 'hominid treasure hunts'.) His team never rushes to publish, he has ensured that Ethiopian scientists are both trained up to post-doctoral level and are full members of the multinational team, the surveys are long-term, methodical and painstaking. White has insisted that Ethiopian scientists are lead and major authors on most papers. Their careful approach has resulted in a slew of hominin fossils, including *Ardipithecus ramidus* and *Ardipithecus kadabba*, from a genus even older than *Australopithecus*, and *Australopithecus garhi*, a gracile Australopithecine that many professionals think may link *Homo* and *Australopithecus*. The epithet *garhi*, delightfully, means 'surprise' in the Afar language. South Africa is also adding to the hominin database, with two new species recently described, *Australopithecus sediba* and *Homo gautengensis*, although these are both controversial.

So, if we try to look at the big picture from all these discoveries, we see that over the last 120 years many species of hominin have been described. As we have seen, even top professionals disagree about their specific names, the genera they are placed in, their ages and their significance. Early diagrams depicting the relationships usually showed, without qualification, one species as being directly ancestral to another and thus evolving into it; a process called anagenesis. Sometimes a species would split into two, a process called cladogenesis. Early evolutionary texts are full of such diagrams. The researchers themselves were very fond of showing 'their' species as ancestral and the discoveries of their rivals as side-shoots and blind alleys. Strange though it may seem, virtually all these diagrams were largely educated guesses. There is no unequivocal evidence that any one species evolved into another, diagrams showing them doing so are what the researcher Henry Gee calls 'Voodoo palaeontology' (Gee 2000).

Exasperated with such guesswork, in the 1970s many hominin researchers began to make use of a classification system devised by an East German entomologist called Willi Hennig. Hennig called his system 'phylogenetic systematics'; it is now called 'cladistics', derived from the Greek for 'branch'. The fundamental concept of this system is that organisms are grouped together on the basis of their shared, common heritage. No assumption is made that any species is ancestral to another. Such a system was introduced at the Natural History Museum in London in the 1980s, for an exhibition entitled 'Man's Place in Evolution'. It enraged a number of workers; the brilliant and controversial palaeontologist Bev Halstead, in a letter to *Nature*, accused the museum of 'introducing a fundamentally Marxist view of the history of life'.

Strong stuff indeed. But all the cladists were saying was that they were not prepared to speculate on ancestry without unequivocal evidence, and this was lacking. A perfectly sound viewpoint, even if rather tame. As Gareth Nelson, a bright young American palaeontologist and populariser of Hennig's scheme, said, 'Do the rocks speak?' To some scientists, too, the names are not important. Many would disagree; wisdom begins with the naming of names, as the Chinese proverb says. You have got to know what you are talking about. But the point is this; whether you call your fossil *Homo* or *Australopithecus* is not as important as where it fits in the scheme of things. Evolution is not static and nor does it progress in jumps; it moves steadily and we should expect to see fossil forms gradually changing as time passes. And evidence does show that. In his book *Deep Time*, Henry Gee (2000) takes the view that we can never be certain if any particular fossil species is ancestral to another. This is a rather extreme view, although it has some merit. But Ian Tattersall, curator of anthropology at the American Museum of Natural History in New York, has stuck his neck out and drawn up a provisional chain of inheritance, that indicates that *Australopithecus afarensis* is one of our ancestors, as is *Homo ergaster* and *Homo antecessor*. All members of *Paranthropus*, robust Australopithecines, are on a side branch, as is *Homo erectus*. And the ancestor of all Australopiths – and us – is *Australopithecus anamensis*, which lived over 4 million years ago on the shores of Lake Turkana, and was named by Alan Walker and Meave Leakey after the Turkana name for the lake, Anam.

So what information can we take away from the wonderful crop of ancient hominins that have been found in eastern and southern Africa? What is known for certain? The changing nature of science cautions us that the answer is; not very much. However, it appears that the branch of primates that leads to us – the hominins –

originated in Africa; no member of the genera *Australopithecus* or *Ardipithecus* has been found outside of Africa, and the fact that most come from southern and eastern Africa is probably important.

A few brief points about primate evolution may be useful here. Primates evolved in Eurasia, and became arboreal, meaning that they lived in trees. Adaptations to an arboreal life include eyes on the front of the face, giving binocular vision to judge distances when jumping; and grasping hands and feet, to hold onto the branches. These adaptations became significant later, as we shall see. Living in trees has a big advantage: you can get away from carnivores, unless they can climb trees, and even if they can, you may be able to use your smaller weight, grasping and jumping ability to stay safe. Thus you don't need to be able to run fast. Diurnal primates also lived in groups, and group behaviour is a good defence against predators. Primates moved into Africa more than 30 million years ago.

Evidence also indicates that our ancestors came down from the trees and moved into the savanna in Africa, that they stood upright first, and then their brains got bigger. Their canine teeth shrunk. They also started making tools, around the time that the increase in brain size occurred. There remains the question of why they came to the ground. In South America, which has over 80 species of monkey, none are terrestrial; all are tree dwellers, although they may descend briefly. But a number of African monkeys spend much of their time on the ground, including the Patas Monkey *Cercopithecus patas* and the baboons *Papio* sp. The suggestion is not that our primate ancestors left the forest, but the forest left them. Binocular vision and grasping hands are evolutionary characters that are favoured by life in the trees. During the Miocene, between 25 and 6 million years ago, earth movements and volcanism forced up the high ground of the rift valley. The high ground captured rain moving in from the west, creating a dryer zone, a rain shadow. Or so one theory goes. It has also been suggested that the massive increase of the ice sheets in the northern hemisphere was more significant in creating an arid environment in north-eastern Africa. But one thing is sure; the African rainforest decreased in size, and was replaced progressively by woodland and then savanna. A new habitat formed; and some primates exploited it, foraging on the ground.

So primates were using the ground *and* the trees. Expert opinion is divided about when they stood up and then started walking upright. Pickford and Senut think Millennium Man stood upright. The Laetoli footprints show that *Australopithecus* was bipedal 3.6 million years ago. And evidence that hominins were bipedal comes from their bones; walking on two legs is connected with several major changes in the skeleton, three of which are very significant, and can be observed in some hominin fossils if we have a skull, leg bones and some of the pelvis. Firstly, the pelvis becomes broader and stronger, as it must bear some of the weight of the abdominal organs – the viscera – such as the alimentary canal, although the abdominal muscles also help. Apes are not bipedal; they walk on their feet and their knuckles. Their ribcage and anterior abdominal muscles bear the weight of the viscera. Secondly, the angle of the femur (the big bone in your upper leg) changes relative to the ground. In an ape, the femur points straight down, and when an ape stands upright and walks, it must swing the body to balance it over each leg as it picks up the other. You can try this out yourself. Try walking slowly while keeping your feet about 50cm apart, and look at the ground as you do so. You are waddling, your body weight shifting about dramatically. An ape can walk like this, but not with ease, or far.

In humans and some hominins, particularly the later ones, the femur does not point straight down, but is angled inwards. This means that our feet are almost directly below the centre of gravity of the body; when we walk the central positioning of the feet means our torso does not sway about. Try it, walk with your legs positioned normally, no great distance apart, and look down. Your body does not sway. We are the only primates that can walk like this. And thus we can easily stand on one leg, without much weight-shifting. It is tempting to theorise that our easy, bipedal walk has contributed to our development.

The third big skeletal difference between us and apes involves the position of the foramen magnum, the hole at the base of the skull where the spinal cord enters, in an ape it is near the back of the skull, in humans it is near the centre of the skull (see the photograph on the right). This is important to us: if your spine is vertical it is easier to look forward if the bone support enters at the centre, and it also means that the turning forces at the back and front of the skull are almost balanced, and thus when we stand erect it is not tiring to look forward.

Early hominins on display in the National Museum in Nairobi. (Stephen Spawls)

Right: Position of the foramen magnum, the orang-utan's (left) is much nearer the back of the skull than the human's (right). (Stephen Spawls)

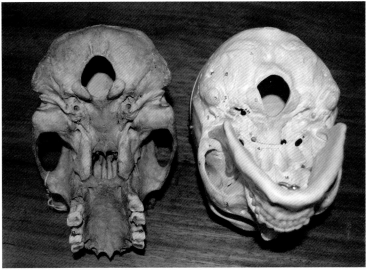

So we definitely became bipedal more than 3.6 million years ago, possibly a lot more. A dramatic increase in brain size came after this. The next question is, why did we stand and walk upright? In evolutionary terms, it must have been an advantage – but what? There is no definitive answer. Some of our finest scientific minds, including Jonathan Kingdon, John Napier, Charles Darwin and Richard Dawkins, have put forward their own theories on bipedalism. None is exactly the same. Darwin suggested it frees the hands, in defence of a terrestrial way of life. You can hold a weapon. Raymond Dart suggested bipedalism protects us from predators. Stand up and you can see more. Dawkins thinks it is a gimmick, spread by imitation and then favoured by selection. Jonathan Kingdon suggests that we developed an upright posture while 'squat feeding', sitting on the ground and reaching for food. Gelada Baboons *Theropithecus gelada* do this. Other theories (some more bizarre than others) suggest that it is a phallic display for showing off the penis, a way of foraging in water, a technique for intimidating others (looking taller and hence more dangerous), a thermoregulatory response (it's cooler away from the hot ground), a means of carrying babies and food back to base, a way of walking down thick tree branches. One obvious point is that when you stand up, you can reach more food. John Napier, the anatomist, suggested simply that bipedalism might result from a mixture of all or some of these potential advantages. With so many great scientists disagreeing, we can't be certain. As the sideshow barker might have said, 'you pays yer money and you takes yer choice'.

But one thing is for sure. When our ancestors stood upright, it conveyed, perhaps purely fortuitously, one tremendous advantage. It freed up the upper limbs and hands. And they were grasping hands, with an opposable thumb. They could hold things firmly, with fingers wrapped around it – the power grip, the way you hold a hammer – and also precisely – the precision grip, the way you hold a pen; both these terms were coined by John Napier. No carnivore can hold something in this manner. And suddenly, those two forward-facing, close-focusing eyes became very important: you could see what you were holding. The way was open for our ancestors to grip something and manipulate it, such as tools. At the very least, you could pick up a rock and chuck it at an enemy, a predator or potential prey. And if you stand up you can throw something a lot further, and more easily than when you are on four legs. You can tower over any adversary, and in a fight gravity assists you in the downstroke.

And after bipedalism came an increase in brain size. Since we cannot be sure which (if any) of the known Australopithecines, or members of the genus *Homo*, are our ancestors, we cannot draw a graph of increasing brain size as compared with body size, but we can make certain generalisations. Tourmai, *Sahelanthropus*, 7 million to

Model of the skeleton of Turkana Boy, a triumph for Kamoya Kimeu and Alan Walker. (Stephen Spawls)

Below: Chimpanzees walk and run quickly on their knuckles, they can only waddle on two feet. (Stephen Spawls)

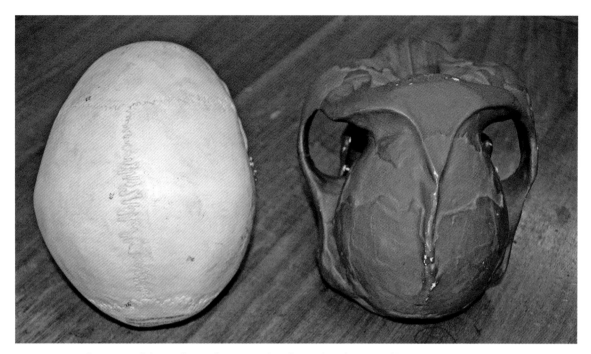

Brain size, we have a much bigger brain than Australopithecus. (Stephen Spawls)

6 million years ago, had a brain volume somewhere between 300 and 400 cubic centimetres. Most Australopithecines, living between 5 and 2 million years ago, had brain sizes between 400 and 600 cubic centimetres, which even at the top value is not much more than a gorilla. *Homo habilis*, between 2.5 million and 1.5 million years ago, had a brain size of 500–700 cubic centimetres. *Homo erectus*, living between 2 million and 0.4 million years ago had a brain volume between 750 and 1,000 cubic centimetres. Next comes *Homo neanderthalensis*, who lived in Eurasia (none have been found in Africa) and had a brain size around 1,300 cubic centimetres (or possibly even larger), and then comes us, *Homo sapiens*, with a big brain, 1,200–1,600 cubic centimetres. And the remains of the oldest humans that were just like us were found at Kibish, in Ethiopia, and are just under 200,000 years old. So as we get near the present, with hominins starting to look more and more like us, brain size increases rapidly, with roughly a tripling in volume in the last 3 million years. What led to this? Variation existed; something favoured the survival of individuals with bigger brains. Was it our use of tools? At Dikika, in Ethiopia, evidence has been found of the use of stone tools 3.4 million years ago. The use of tools thus starts around the time that our brains started getting bigger. It is tempting to speculate that those with the larger brains were able to make and use better tools, which aided survivability. But in 1995, Leslie Aiello and Peter Wheeler put forward the theory that an increase in meat-eating led to an increase in brain size. They suggest that a meat-rich diet means your gut can become smaller; and this is logical, carnivores have shorter guts than herbivores, most breakdown of the protein-rich diet takes place in the highly acidic and protease-loaded stomach. Guts are expensive tissues, constantly shedding short-lived cells, producing secretions and carrying nerve cells. So the shorter gut frees up energy for an increase in brain size. It is a beguiling hypothesis.

There is also the vexed question of language. Did this contribute to the booming brain size? All human beings speak highly complex languages, and the written languages that evolved 5,000 years ago were also complex. Some think language evolved suddenly within the last 100,000 years or so, some believe it has been doing so gradually over the last 5 million. We cannot be sure. Alan Walker spotted the narrowed spinal canal of Turkana Boy; Anne MacLarnon then hypothesised that he could not speak, and that was 1.6 million years ago. But he might have been able to make a range of sounds. Changes to the skull, especially the positioning of the foramen magnum, which developed in *Australopithecus* over 3 million years ago, may have made it easier to produce a range of sounds, as the load on the larynx changed. You can test this, try speaking in a clear loud voice and then, while still talking, tip your head forward and press your chin hard against your neck. The compression reduces the range of sounds you can make, your voice becomes slurred. We also have an unanchored bone in our throats, the hyoid bone, which braces, among other things, the mouth muscles, and aids us with clear speech. The fossil hyoid bone of a Neanderthal found in Israel might suggest that Neanderthals could speak ... although there is no single bit of anatomy that can unequivocally prove that its owner had speech.

So what is the big picture? Primates entered Africa, and in eastern Africa they split from chimpanzees, left the trees, stood upright, starting making tools and their brains increased dramatically in size. And then some of them departed from Africa, in at least two waves. And Kenya, and Kenya's fossil hunters, have provided, and will continue to provide, some of the most dramatic and useful fossil evidence that fleshes out the continuing story. Go and see the sites for yourself, and wonder.

THE LITERATURE

A huge amount of literature is available on hominins; reflecting our strong interest, and it is still being busily produced. The original descriptions and discussions are, of course, in the scientific literature. A great number appeared in what is arguably the world's most prestigious scientific journal, *Nature*, and continue to do so; Raymond Dart described his discovery of *Australopithecus africanus* in *Nature* in 1925. Groundbreaking papers were also published in *Science*, *Nature*'s main rival, and in such journals as the *Proceedings of the National Academy of Sciences of the United States of America*, *Evolutionary Anthropology*, the *Journal of Human Evolution*, *Anthropological Science* etc. Most of the big sites are well documented. The Clarendon Press in Oxford has had a long association with the Leakeys, and published the Njoro River Cave Excavations, the 500-page Laetoli report and the voluminous Koobi Fora reports. *Lothagam: The Dawn of Humanity in Eastern Africa*, a remarkable 600+ page work with some super reconstructions of the extinct mammal fauna, was produced by the Columbia University Press in 2002.

The popular literature is voluminous, with many books, some strongly opinionated. It is a constantly changing field and of great interest. The books tend to be split into three major categories: those written by the palaeontologists themselves, describing their discoveries and coupling these with their take on human evolution, those written by museum scientists, giving an overview, and those written by journalists and popularisers of science. There are also a handful of books, and chapters within books, written by sceptics and debunkers, which range from those who doubt the accuracy of skull reconstructions and dates to outright crackpots who question the whole business of human (and any other) evolution. There are few that are not worth reading; even the crackpots often have valid points to make, although some may set your teeth on edge. We will deal with a few of them here; but bear in mind there are many, this is a personal choice.

A good, short and accurate summary of hominin evolution, and some of the controversies within it, is provided by chapters 28 ('The mysterious biped') and 29 ('The restless ape') in Bill Bryson's popular work on science, *A Short History of Nearly Everything* (Doubleday 2003). Bryson's style is enjoyable and erudite; he mentions many of the notable characters and controversies, and is skilled at quickly grasping the essential scientific points. A nice book covering the whole of human evolution is Robin McKie's *Ape Man* (BBC Books 2000).

An excellent, solid, fact-filled book that deals with the whole history of the search for human origins, from 1857 up to 1980, accompanied by what are arguably the best photographs in existence of the fossils and their discoverers, is John Reader's *Missing Links*, published by Collins in 1981. Obviously, it does not cover the many momentous discoveries made since 1980. Reader, a photojournalist, has written some good books on Africa, and spent time at many of the crucial sites; he has also written a fine biography of Africa. Two more recent books on hominid evolution are Jon Kalb's *Adventures in the Bone Trade* (Copernicus Publisher 2001) and Ann Gibbons' *The First Human: The Race to Discover Our Earliest Ancestors* (Doubleday 2006), both books cover not only the discoveries but some of the savage rivalries that, as we have seen, characterise much of African hominin palaeontology.

Books by the fossil hunters themselves include Louis Leakey's two volumes of autobiography, *White African* (Hodder and Stoughton 1937) and *By the Evidence* (Harcourt Brace Jovanovich 1974). For any Kenyaphile these books are delightful and essential reading; adventures in East Africa, written in a friendly and open style by a man with a keen eye who knew the land and its people. *White African* has some fascinating pictures of the people and places. The natural history observations in both books are as interesting; Leakey must be one of the few if not the only human to have observed a Giant Pangolin *Smutsia gigantea* catching water beetles in Lake Victoria by the light of the moon. Both books are characterised by Leakey's prescient view of Kenya's future development and his singular regard for ordinary Kenyans; a view which brought him into conflict on several occasions with the white settler community. A touch of sadness attaches to *By the Evidence*, which covers the years from 1931 only up to 1951. Leaky's sudden death in 1972 came the day after he completed it and he never got the chance to write his own account of what were undoubtedly his most productive years.

A biography of Louis, *Leakey's Luck*, by Sonia Cole, was published by Collins in 1975. Although she did not complete her degree, Cole studied geology for two years and it shows; this book is well written, accurate, extensively referenced and excellent on detail. Although Cole was both a family friend and worked in the field with Louis, she does not gloss over his peccadilloes. Cole is also the author of *The Prehistory of East Africa*, published by Penguin in 1954 (and revised in 1965). Many of the theoretical principles underlying Cole's work are outdated (there is no mention of the Lake Turkana sites, which had not been found) or wrong (the pluvial theory, Kafuan stone culture, use of bolas etc.) and the book is somewhat paternalistic and colonial in outlook. However, the actual descriptions of the excavations, the stone tools and the skulls, as well as the then state of the science, are accurate and useful; it provides a reasonably sound (and as yet unsurpassed) synthesis of the Stone Age in East Africa. The history of the whole Leakey family, from Harry's arrival in Kenya in 1902, up to 1995, is covered in Virginia Morell's book *Ancestral Passions* (Simon and Schuster 1995). This is an unswerving, thorough, 638-page chronicle, and although Morell is a friend of the family, both the clan's shortcomings and their good points are covered in equally meticulous detail; you read this and you get a pretty good picture of the people, the country and the science. It also has a nine-page bibliography, three pages of which list papers and books published by the family themselves. Richard Leakey has written several books on palaeontology and two volumes of autobiography, *One Life* (Michael Joseph 1983), about his early life and paleoanthropological work, and *Wildlife Wars* (described in Chapter 13).

Alan Walker and Pat Shipman's *The Wisdom of Bones* is a superb book by a field team (Hodder and Stoughton 1996). The book is Walker's story but Shipman, his wife, did the writing. This is essentially the story of the Nariokotome *Homo erectus* skeleton and what it meant. However, Walker manages to fit in a good summary of virtually all the significant events in paleoanthropology; he conveys the thrills and spills of a fossil hunt and also explains clearly a good deal of the biology of human evolution. The entire book is thoughtful and attacks nobody's reputation; it is arguably one of the best books on African palaeontology. Another good read is *Lucy* by Don Johanson, the discoverer of the famous hominin, and Maitland Edey, the journalist, published by Simon and Schuster in 1981. Johanson is a controversial palaeontologist and has upset a lot of his fellow scientists, but this is a nice book, easily readable and with a lot of good science; Johanson is one of the few people who attempts to explain how the mass spectrometer works.

We might end with a few opposing views. *Fantastic Invasion* by British journalist Patrick Marnham (Jonathan Cape 1980) took a long, caustic look at Africa and the activities of foreigners there, and this book includes a short blast at Louis and Richard Leakey and Kamoya Kimeu (misspelt as Kimoya). But Marnham is a cynic, unable to allow that palaeontology has any benefit, and his attacks on the Olduvai skulls and Laetoli footprints are wide of the mark. David Lamb, an American reporter, spent four years in Africa, and in his book *The Africans*, published by Methuen in 1983, has a shot at Mary Leakey and her strange lifestyle. After inviting him to Olduvai for an interview, Mary disappeared shortly after his arrival and did not return for some hours. Virginia Morell also deals with this antisocial behaviour in *Ancestral Passions* (Simon and Schuster 1995). Mary didn't like talking with visitors to the gorge, and she didn't socialise at all with her black workers, who were forbidden to sing or talk while excavating in her presence. Her dogs were fed meat daily, but her African team only once a week. Mary's autobiography, *Disclosing the Past* (Doubleday 1984), is dedicated not to her band of stalwart assistants or to any of the scientists who worked with her and gave generously of their expertise, nor to her husband or her parents or her children – it is dedicated to her dogs.

For those who weary of reading about Africa and its hominins purely from the point of view of foreigners, a refreshing alternative is provided by Professor Bethwell Ogot's book *My Footprints on the Sands of Time*, published in 2003 by Anyange Press. Ogot, a distinguished historian from western Kenya, was responsible for recommending the appointment of Richard Leakey as administrative director of the National Museum. Later Ogot himself was appointed first director of a research institute know by the unwieldy acronym of TILLMIAP, which stands for The International Louis Leakey Memorial Institute for African Prehistory. Ogot then left his post under considerable stress, and his book contains an 80-page chapter on the vicissitudes of the institute, as seen and painstakingly documented by the eminent historian; both expatriate experts and prominent Kenyan politicians come in for a lot of criticism in his account.

The British geologist Martin Pickford, who was raised in Kenya, has written two books that are given over solely to attacking Richard and Louis Leakey. Pickford is a highly professional geologist who has contributed a lot to East African palaeontology and geology; he is, after all, the man who was involved in revealing *Orrorin tugenensis* to the world. Pickford's scholarly publications are both weighty and erudite. And yet, both his anti-Leakey books are largely without merit, being mere indicators of a fruitless academic vendetta. Pickford fell out with Richard Leakey over an incident involving National Museum notebooks, which Leakey thought Pickford was attempting to steal. Pickford's first book, *Richard Leakey, Master of Deceit*, published in Nairobi by White Elephant Publishers in 1995 and written jointly with Eustace Gitonga, a Kikuyu sculptor who had also crossed swords with Leakey, is a dense and enraged flow of non-stop invective, loaded with scurrilous, unsubstantiated and often unbelievable stories, including some bizarre sexual ones. There is no attempt at even-handedness, and Leakey's conversations, reported in bold type, make him sound like a action-movie villain. This book, which is poorly produced and bound, tending to fall apart on opening, was produced, not coincidentally, at the time of the 1995 election when Leakey was standing for parliament with a new political party and thus represented a major setback for him. Pickford's second shot at the Leakey clan, *Louis Leakey: Beyond the Evidence* (Janus Publishing 1997) is an unswerving relentless attack on Leakey Senior, documenting the various mistakes that Louis made in his early days over Kanam Man and the pluvial theory, among others. This is an unkind book, the title itself being a sniggering play on Leakey's own book. Other authors have dealt more sympathetically with Leakey's mistakes, which are well documented, even by Louis himself, and they have been weighed in the generally favourable verdict that the scientific community has bestowed on him. However, both Pickford's books contain some material of interest. 'Beyond the evidence' conveys vividly the excitement of fossil hunting in the country around the Winam Gulf, while 'Master of deceit' contains, reproduced verbatim (often scanned, in a higgledy-piggledy manner), a good amount of correspondence between Richard Leakey and such scientific luminaries as Nicholas Toth, John Onyango-Abuje, Tim White and Don Johanson, as well as Pickford's own correspondence. These, being genuine, present a fascinating view of the violent spats that characterise this field, and of the rapidly hardening attitudes between men who hated each other and each other's standpoint. So much for scientific cooperation – but nevertheless fascinating.

Chapter 3

The Peopling of Kenya

In the interior of eastern Africa, at various times and in various places Khoisan, Cushitic, Nilotic and Bantu peoples met one another to produce what is arguably the most complex ethnolinguistic region in the continent – James L. Newman, The Peopling of Africa, 1995

This chapter will discuss the story of humanity in Kenya, looking at archaeology, migrations and settlement patterns. After a brief look at the rationale and methods of such study, the rest of the chapter is chronological, following the dispersal out of Africa and the Stone Age, settlement and development of food production in the Iron Age, the movement of peoples through Africa and into Kenya, and the era of exploitation, explorers, colonials and independence. A review of the literature concludes the chapter. Although presented chronologically, the sections could be read independently.

RATIONALE, METHODS AND BACKGROUND

In this chapter, we want to look at the peopling of Kenya, but from the standpoint of detached natural historians, documenting the developments and movements of that significant species *Homo sapiens*, modern humanity. For, as Jared Diamond, eminent physio-ecologist has said, we are just another species of large mammal, although admittedly a rather exploitative one.

The archaeology and history of humanity in Kenya can be conveniently divided into five phases: (1) the activities of our earliest ancestors; (2) migration into and settlement in East Africa; (3) exploration by adventurers from the west; (4) colonial history; and (5) post-colonial. Many competent historians have covered the last three topics. It has been said, somewhat caustically, that many history books about Kenya are really just the history of foreigners in Kenya. Thus we shall only refer to the exploration, colonial and later eras briefly, towards the end of the chapter. We have provided some pertinent references in the literature section for those who wish to find out more. The activities of our earliest ancestors, the hominins, were covered in Chapter 2.

So this chapter largely concerns the archaeology, the migrations and settlement; what we know about humanity from the time we became *Homo sapiens* and the peopling of Kenya from about 5,000 years ago up to the age of exploration. We start with early migrations, some material on the movements of our ancestors, and a section on the Stone Age that predates the later migration. We are going to try to answer the questions: what happened in Kenya between the dawn of modern humanity (roughly 200,000 years ago) and the time of the great migrations, where did today's Kenyans come from, and how did they fit into the landscape?

In the last 100 years, due to the rapidity of transport and an increasing open-mindedness, Kenya's human population has undergone a fairly considerable homogenisation, although there has been the occasional xenophobic blip. But at the beginning of the twentieth century, most of Kenya's people had well-established homelands, and to some extent they remain there today. How did they get there? What happened between 5,000 years ago and the twentieth century that put Kenya's peoples where they are? Why are the Kikuyu in the high country around Mt Kenya and the Aberdares, the Luo around Lake Victoria, the Giriama at the coast, the Somali around Wajir and the Turkana west of Lake Turkana? We shall try to address these points.

We want to start with a few observations on history itself. It can be an emotive subject, often written by those who were not there. Kenya's history – before the time of the European explorers, anyway – is complicated by the fact that little of it was recorded in writing. Historical texts are not always accurate, of course, as shown by conflicting accounts of battles involving the Pharaoh Rameses the Second. History really does depend upon those who write it. However, written records are at least one source of evidence. A second source of evidence is tangible structures; buildings like churches, temples and monuments. However, prior to about 200 years ago, Kenya's interior has no written history and relatively few structures, although the coastal strip has both buildings and written records going back 1,200 years. Thus documentation of the history of the interior of Africa has traditionally relied upon four types of evidence: present clues, languages, archaeology and oral history.

There are conflicting views of the benefits of oral history. Kenya has produced a great pioneer of that discipline, Professor Bethwell Ogot. But Niall Ferguson, Scottish historian, has famously said that oral history is a recipe for misrepresentation, because few tell the truth, even if they intend to. In lieu of anything written down, there is always going to be speculation, and nowhere is this more evident than in the case of a magnificent African historical site, Great Zimbabwe, probably sub-Saharan Africa's most spectacular monument. Built between the eleventh and sixteenth centuries, it covers an area of seven square kilometres, and is made of stacked dressed granite blocks, rock-solid stable yet without any mortar. The walls around the outer enclosure are close to 11m high and over 250m long; within the main complex is a solid stone conical tower nearly 10m high.

Excavations at Great Zimbabwe were puzzling, yielding large amounts of cattle bones, big pots for storing and brewing beer, iron hoes and gold wire. Other findings include ivory carvings, a glazed Persian bowl, some Chinese pottery, falconry bells from the Middle East, thousands of glass beads, and, most famously, the eight beautiful Zimbabwe birds and the monoliths they stand on, carved from soapstone. Nowhere in the complex did the archaeologists find any pottery of the sort found in an ordinary African household, or the remains of houses, or toilets, or fabric, or the sort of food eaten by the common people, such as plant seeds and products, chicken, sheep or goat bones. Nobody was buried there. So whatever happened at Great Zimbabwe, the common people did no living, eating, drinking or dying there. And several widely separated communities claim that their ancestors built it. They cannot all be right.

Great Zimbabwe is still largely a mystery. The strange artefacts excavated there and the absence of any written evidence have led to much speculation. Cecil Rhodes decided that foreigners must have built it, as no black Africans could have had the skills needed to construct such a beautiful monument. This racist nonsense was part of a crackpot theory that foreigners from the north (Phoenicians, Arabs or some other non-Africans) had an

Great Zimbabwe, mysterious monument. (Stephen Spawls)

empire in southern Africa. Other theories about Great Zimbabwe have suggested it was a centre for circumcision, a cultural centre to display the power of a cattle-herding society, a compound for holding slaves, an astronomical observatory, a replica of the Queen of Sheba's palace, a royal palace with different sections allocated to kings, priests and wives, a religious centre or just a folly to celebrate the building skills of the makers. The lovely soapstone birds and the enigmatic round tower suggest a religious purpose; the Zimbabwean historian Peter Garlake believes that trade was also important. The variety and provenance of the artefacts found there support this role. But no one really knows. As with Stonehenge in Britain, the true function is a mystery. And if this is the situation at a staggeringly large, spectacular, elaborate and fairly well documented monument, how is it possible to be sure about the accuracy of historical accounts when all there is to go on are the words of ancestors?

But oral history is better than no history. As natural historians, we claim no great expertise; we shall try to present both sides of the evidence where there is clear-cut conflict. But it is worth bearing in mind that many African historical events are in dispute, even among eminent experts. Consider iron smelting. In a paper discussing the debate, the historian Stanley Alpern (2005) marshals twelve historians who believe the technology arose independently in Africa, and nine who believe it was imported from the north. Whose side do you take?

The arrival of a powerful biological technique has recently revolutionised the history of the last five millennia, that technique being the analysis of biochemical molecules. For humanity, the important thing about biological molecules is that they say something about (a) relationships and ancestors; who is related to whom and who is descended from whom, and (b) time; who split from whom and how long ago. This information is intensely valuable to historians. But the evidence the molecules offer is not unequivocal and there is often angry debate about what they mean. In recent African biochemical research, where humanity is concerned, quite often, the molecules have said things that are violently disputed or simply refuted, for various reasons, including culture and tradition, as we shall see.

Historical evidence is often fragile, and can be interpreted in more than one way. Even in a subject that is academically rigorous, there are a lot of loaded, emotive or pejorative terms. Many historical texts (and even some modern ones) used the term 'Hamite' (the 'son of Ham', Ham being one of Noah's sons) to describe northerners in Africa, and this is often coupled with the pervasive myth that pale-skinned foreigners had some sort of major influence in sub-Saharan Africa's history. Consequently, most modern African historians do not use the term Hamite, or the adjective Hamitic. An equally angry debate rages about the identity of the Swahili, the people of the East African coast, and how African they are, fuelled by bitter comments from the Nigerian Nobel Laureate Wole Soyinka on one side and the great Kenyan historian Ali Mazrui on the other. And much of Africa's historical material (especially archaeological artefacts) has been plundered and removed from the continent. Many early expeditions were more interested in looting riches than furthering understanding, and those African countries who have tried to get their heritage back have all too frequently faced a stonewall of bureaucracy where ethics were the very last thing on the holders' minds. Zimbabwe's attempts to get their beautiful stone birds returned involved in the end an astonishing piece of blackmail, which saw the Bulawayo Museum's entire insect collection (which was the best in Africa; its herpetological collection still is) handed over to what was then apartheid South Africa.

DISPERSAL OUT OF AFRICA AND THE STONE AGE

Present evidence indicates that our ancestors, genus *Homo*, made at least two group movements out of Africa. The first was unsuccessful, the second spectacularly successful. In the first, hominins of the species *Homo erectus*, 'Upright Man', left the continent some time between 2 million and 1.7 million years ago and made their way into Asia and Europe, spreading east to China and Indonesia, north to Spain, Great Britain and Georgia. They had a range of stone tools, but nevertheless by 300,000 years ago they were extinct. We don't know what happened to them. There was a suggestion at one time that they in fact evolved into modern humans, *Homo sapiens*, a theory pushed in particular by some Chinese scientists, but now shown to be incorrect. Modern Chinese people all have a DNA sequence indicative of modern *Homo sapiens*.

The oldest indisputably modern humans of our species are known from skulls from Ethiopia and are at least 190,000 years old. So modern humanity, on present evidence, originated in Africa. Where exactly is not certain; for a long time, it was presumed to be Ethiopia. More recent work indicates that the hugely varied DNA of southern-African Khoisan speakers suggests they are the oldest humans. Be that as it may, the second migration, this time by us, modern humans, *Homo sapiens*, left Africa some time ago. There is a lot of debate about the date of their departure; several dates between 50,000 and 125,000 years ago have been proposed. Nor is it certain whether those pioneers left via the Nile Valley, or across the mouth of the Red Sea. It is not known exactly when they reached places like Australia, or China. But what is certain is that the adventurers who left spread throughout the world; by foot (or possibly by boat), the last place they reached was southern South America, about 12,000 years ago. This great migration is sometimes called the 'out of Africa' movement or theory. Those early travellers, and their compatriots who chose to remain in Africa, became us, all of us; molecular evidence proves this.

One problem, which we will address briefly as it is not particularly relevant to Kenya, is that several well-recognised species do not seem to fit in, and their disappearance is a mystery. *Homo heidelbergensis*, lived between 300,000 and 600,000 years ago and is known from several places in Europe, including Germany, Greece and Italy. *Homo neanderthalensis*, Neanderthal Man, widely known from fossils in Europe, Asia and possibly north-east Africa (but not sub-Saharan Africa) first appeared 230,000 years ago (some authorities say 130,000, some say over 350,000); but by 35,000 years ago the Neanderthals were extinct. Some Neanderthals had bigger brains than us, and prominent brow ridges. They were well adapted to the cold or northern Eurasia. When modern humans, *Homo sapiens*, reached Eurasia, the Neanderthals disappeared. Where did they go? Did they die out? Did *Homo sapiens* eat them, or give them some deadly disease? There are no signs of slaughter; the Neanderthals just quietly

Khoisan hunter, Khutse, Botswana, with his skin bag, ancient receptacle. (Stephen Spawls)

fade away, like *Homo erectus*. Some scientists think that Neanderthals interbred with modern humans, other say they didn't; recent work with Neanderthal DNA does indicate an overlap of a few percent. However, a number of eminent palaeontologists believe that *Homo heidelbergensis* was the direct ancestor of both *Homo sapiens* and *Homo neanderthalensis*, others believe that a Spanish fossil, *Homo antecessor*, was ancestral to both.

Thus the Neanderthals and *Homo heidelbergensis* pose a problem. We are reasonably certain that all existing humans are descended from a group of African ancestors, *Homo sapiens*, and all non-Africans are descendants of those who left Africa in that final migration. This means that if *Homo heidelbergensis* is an ancestor of us, then he must have been in Africa. However, very few unequivocal African fossils of the species are known. Another possibility is that modern *Homo sapiens* arose in Eurasia, or travelled there; migrated back to Africa, then left again, which is improbable. It's a mystery. An easy explanation for the Neanderthals would be that they are descended from *Homo erectus*, which is logical in terms of time and place, but is not accepted by the scientific community. For the time being, the best explanation is probably that *Homo heidelbergensis* is the ancestor of both us and Neanderthal Man, in Africa. The Neanderthals (or their ancestors) left Africa before us, flourished for a while in Europe and Asia and then gradually disappeared as we spread into their territory.

One big ancient event in African history that we have a fair amount of evidence for is the Stone Age. This is a controversial term; it has often been used as an insult, rather like the terms 'primitive' and 'native'. This is a pity, because the Stone Age is a rather important phase in the history of humanity. It was from stones that our ancestors fashioned the first durable and developed tools. It is possible that they may have made earlier use of wood, or animal bones, but wood tends not to last. Stone tools do and what we see in eastern Africa is a gradual (sometimes very gradual) development of tool technology, and with it comes some engaging artefacts. Many scientists feel that it is what we did with stones that led to us becoming the advanced creatures we are. The African Stone Age is divided into three sections; the Palaeolithic (Old Stone Age, from 3 million years ago to about 100,000 years ago), the Mesolithic (Middle Stone Age, 100,000 to 35,000 years ago) and Neolithic (New Stone Age 35,000 years ago until fairly recently). During this time, various stone industries developed; a stone industry is characterised

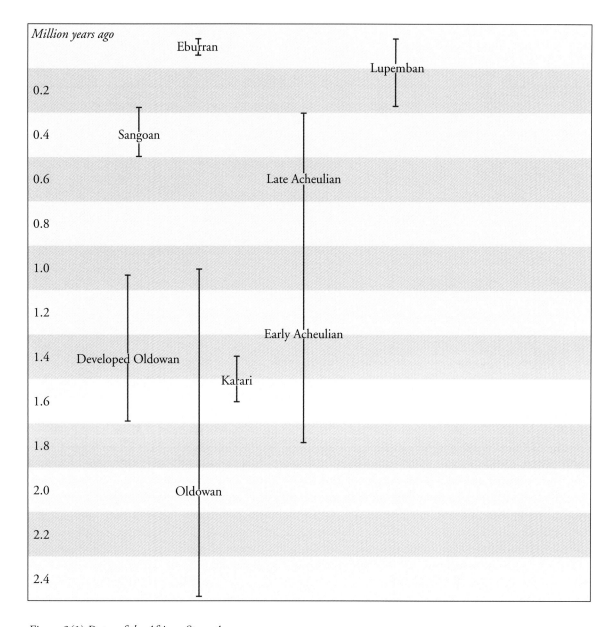

Figure 3(1) Dates of the African Stone Ages

by a distinctive set of tools. The sub-Saharan African stone industries include the Oldowan, Acheulian, Sangoan, Lupemban and Eburran. Figure 3(1) shows the dates of the African stone industries; most cultures are based in the Palaeolithic. There is controversy regarding both the dates and the industries.

Stone tools have some shortcomings. They cannot be directly dated – they are as old as the parent rock – and it is not possible to know exactly who made them. We can speculate, of course, especially if we can date the layer that the stone tools are in. If there are hominin remains at the same level, you can speculate that the hominins made the tools. Unfortunately, hominins have not, so far, done us the favour of dying with the tool they were using in their hands. And, as Alan Walker and Pat Shipman drily remark, if the most common fossils in a geological horizon are used to indicate who made the stone tools found there, you conclude that stone tools are made by antelopes.

The earliest evidence of stone tool use by our ancestors supposedly dates from 3.39 million years ago. *Australopithecus afarensis*, at a place called Dikika on the Awash River in eastern Ethiopia, used tools to make cuts on fossil animal bones. The tools themselves were not found, the hominins might have just used sharpened bits of rock that came to hand. It was the oldest example of a technology that many would say leads to us; terrestrial primates using something from their environment to improve their life. It represents a momentous leap; using something from your surroundings for useful purposes. Although it has been suggested that, in the absence of any actual tools of that age, the cuts *might* actually be the teeth marks of crocodiles.

Our ancestors used stone for tools for obvious reasons: it's available and it's tough, the hardest natural material known to man. The right material, hard fine-grained igneous or strongly metamorphosed rocks, will break to give sharp edges, which can be used to cut. Stone can be worked and shaped. The use of tools is an unusual skill, and many would argue that it is our tool-making skills that have made us the unusual species that we are. It gives us – and this is a very important point – abilities that we do not naturally possess. Think about that when you next use a panga or a hammer … or a computer. Imagine trying to do whatever you are doing with just your bare hands. But you cannot use a tool unless you can hold it, and hence only a few animals use tools. Birds, of course, can grip with both their beak and their claws, and there are a few tool-using birds, like the Egyptian Vulture *Neophron percnopterus* and the New Caledonian Crow *Corvus moneduloides*. Likewise, a few mammals: Jane Goodall (one of Louis Leakey's protégés), discovered Chimpanzees *Pan troglodytes* using grass stems and twigs as fishing tools, to catch termites. The twig is stripped of leaves and stuck into a termite hill. The insects bite onto the twig and are then extracted and eaten. The stripping off of leaves means the tool is being developed, as opposed to just being something handy that is used as it is found. Humans are the only other animals that develop tools. Over the last 3 million years hominins have used tools, and they have become increasingly sophisticated.

The study of stone tools has had its pitfalls. Jim Wayland thought he had found evidence of stone tools in Uganda but the rocks he found had been broken by natural processes. Louis Leakey found a rock, fossils of a 14-million year old Miocene ape called *Kenyapithecus wickeri* and some smashed animal bones at Fort Ternan. He said the rock came from miles away, and concluded the ape used it to break the bones. Alas, the rock turned out to be local, and not hand-crafted.

The African Stone Age occupies 3 million years, almost up to the present. The oldest beds where stone tools have been found are at Gona, in the Afar Triangle in Ethiopia, and are dated 2.6 million years old. Humanity then continued to use stone tools until almost the present day; small exquisite stone tools from the Eburran tool industry (named after Eburru, the big forested volcano just north-west of Lake Naivasha) have been dated to 8,000 years ago. Stone has served us well.

Stone tool manufacturing in Africa has been divided into several distinct phases or industries. The first, named after Olduvai Gorge, is the Oldowan Industry and in Africa the Oldowan seems to have extended from 2.6 to about 1 million years ago, or even nearer the present. Oldowan tools are fairly simple: big tools, consisting of a rounded original or core stone that has been chipped or flaked in two dimensions, or even just one, to make simple bashers, short-bladed cutters and sharp flakes. Their shape varies. David Phillipson, one of Africa's most eminent archaeologists, points out that Oldowan tools have no standardisation. Some authorities think that the flakes that were chipped off the bigger rocks were used, rather than the parent rock. But nevertheless, Oldowan tools represent the start of the stone ages. One of our ancestors sat down and

Oldowan handaxe (quartzite) and scraper (obsidian), functional but cruder and older than Acheulian tools. (Stephen Spawls)

Above: Beautifully fashioned hammer stone, Kariandusi. (Stephen Spawls)

Left: Acheulian handaxe, made of obsidian, Kariandusi. (Stephen Spawls)

made an artefact that enabled them to do something they could not do before. Many Oldowan tool sites are known from the edges of lakes and rivers, other tools have been found in caves or just lying on the ground. Known Oldowan tool sites include the eponymous Olduvai Gorge, Koobi Fora, the lower Omo River, the Awash Valley and Sterkfontein in South Africa. And although no one can be certain who was the last to use Oldowan tools, it is fairly certain that *Australopithecus* was the first.

The Oldowan tool industry was followed by the Developed Oldowan tool industry, which started about 1.5 million years ago, and is characterised by tools with two faces (bifacial) and some more sophisticated flakes; a version of Developed Oldowan, called Karari, has been identified at Koobi Fora, dated between 1.6 and 1.4 million years ago. The Karari industry is distinguished by distinctively shaped, rounded and delicately edged scrapers and heavy-duty cores with both faces sharpened. The Karari tools show something quite unusual. They are fairly standardised and often the finished tool is quite different from the raw material, which indicates that the manufacturer could visualise their finished product, this is a very human skill. Phillipson, as a consequence of this, sees the Karari as representing the start of the Acheulian, rather than the end of the Oldowan. Interestingly, both the Oldowan and Developed Oldowan are uniquely African industries, and are not found outside Africa.

The Developed Oldowan tool industry was followed by the Acheulian (sometimes spelt Acheulean) industry. Although the industry developed in Africa, it is named after St Acheul in France, where these tools were first found. Acheulian tools are more sophisticated than Oldowan tools; they have been shaped in three dimensions, and include hand axes and cleavers, often of remarkable strength, beauty and even delicacy. They are excellent for butchering animals; you grip them at the curved top end and chop or slice downwards. The Acheulian industry lasted for a long time; from about 1.65 million years ago to 250,000 years ago, or possibly even to as recently as 100,000 years ago. Acheulian sites occur all over Africa outside of the forest; most sites are near water sources. Known Acheulian localities in Kenya include Lewa Downs, Olorgesailie, Kariandusi, Nariokotome, Koobi Fora, Mt Kilombe and the Kinangop; the Kinangop site is one of the youngest, at an age of 400,000 years. Acheulian tools are remarkably standardised; handaxes from Olorgesailie, Montagu Cave in South Africa, the Jos Plateau in Nigeria and Sidi Zin in Tunisia are very similar in shape and construction. Some researchers suggest that the time of the Acheulian industry was an important period in our history, with exploitation of a wider range of areas and altitude. Known Acheulian sites at relatively high altitude, such as the Kinangop, Konso in Ethiopia and the highlands of Lesotho, may be an indication that clothing was coming into use; those places are chilly at night.

Some of Kenya's best examples of Acheulian tools, especially handaxes, are found at Olorgesailie and Kariandusi. Olorgesailie was a tool 'factory' for close on a million years. This lovely but baking hot monument is one of Kenya's most important historical sites; everyone should visit it. Many of the artefacts there are made from quartz and obsidian, rocks that are not found at that location. It seems that the Acheulian masons camped by the shores of a shallow lake, their factory was there, but the tools they worked on were carried in from elsewhere. David Phillipson argues that the workers probably made skin bags for transport of the stones, similar to those used by the Khoisan hunters of southern Africa. This is an important cultural point; humans are the only

Peter Ole Tunai explains the use and variety of handaxes, Kariandusi. (Stephen Spawls)

creatures that carry things using receptacles. The humble Kenya kikapu, or the goatskin bag, is a descendant of an implement that our ancestors first used hundreds of thousands of years ago.

Olorgesailie gives us some unique insights into humanity in the early days; the wide variety of tools found there, often at the same horizon, indicates that the masons' skills varied a lot, and this may mean that tool-making was a taught skill. This may indicate that the workers used language. It was also a site – possibly the birthplace – of industrial organisation; in some areas tools were merely sharpened, at others they were made. The only question that remains is, who were the makers? Mary Leakey was convinced the *Homo erectus* was the manufacturer, and the recent discovery of a bit of *Homo erectus* skull there supports this, but no one is certain. A few other curious facts attach to the Acheulian. Some tools appear to have been made purely for ornamental purposes; for example huge handaxes weighing over 10kg are known from Tanzania. They cannot have had any practical use, and this suggests they were made for pleasure. In addition, some handaxes are, according to Phillipson, unnecessarily elaborate, being very finely finished and never used. This may suggest the beginning of culture; the manufacture of things for aesthetic reasons. The Acheulian tool kit was exported to Eurasia.

For most of its tenure, the Acheulian stone tool kit did not change very much, a fact that led Mary Leakey to call *Homo erectus* 'dim-witted'. This seems harsh; they were simply sticking with the technology they knew and trusted. One may draw modern parallels. But even Phillipson calls the Acheulian one of the least understood periods in archaeological history; an industry that did not change very much and occupied a gigantic distribution, both in time and space. And in defining the Acheulian as a time of big handaxes and cleavers only, we run the risk that smaller, finer tools might have been made in that time and because they don't fit in with our concept of the Acheulian, we don't classify them as such. In general, the first appearance in Kenya of small very fine tools, or microliths, post Acheulian, is dated about 50,000 years ago, although it has been suggested that some fine stone blades found at Kapthurin, near Baringo, are over 500,000 years old.

At a number of sites across Africa, some 500,000 years ago, another stone industry began to emerge, the Sangoan. It persisted until about 250,000 years ago. It was named by Jim Wayland in 1920, for Sango Bay on the Ugandan shore of Lake Victoria. Sangoan tools include some rather heavy, crude, triangular-sectioned core tools often called picks (which may have been used for digging), up to 300mm long, thick handaxes or core axes and a variety of small retouched flakes, with sharp cutting edges. Many Sangoan stone tools are known from in and around Ghana and Botswana. In Kenya, the industry is associated entirely with the west and the environs of Lake Victoria. Localities that have yielded tools of this culture include the Yala River valley, near Kakamega, Sotik and Muguruk near Kisumu.

It has been suggested that some Sangoan tools were not used for butchering animals but for woodworking – cutting trees and making wooden tools (which have not survived). Some show signs that they might have been hafted; i.e. mounted on a presumably wooden shaft, although these shafts have not survived. The Sangoan industry might have developed as a response to a wetter climate and the consequent increase in forest cover, outwards from the core of central African forest. It's a plausible theory; there are few Sangoan sites in dry northeast Africa. One would only expect the fine Acheulian tools to get heavier and cruder if they were being developed for a different job, cutting wood. But the theory is entirely speculative. Stone tools do not tell us what they were used for, and no Sangoan site has yielded a decent crop of animal fossils, or any associated hominin remains. Some scientists treat the Sangoan industry as an advanced or late stage of the Acheulian.

The Middle Stone Age is usually regarded as extending from 100,000 (some say 200,000) to 35,000 years ago. The stone tools of this time, especially the small ones, known as microliths (literally 'tiny stones') were generally smaller and more finely crafted than the Acheulian and Sangoan tools. Some hafted tools are known from this period: stone blades set into the shafts of arrows or spears, although none are known from Kenya.

Modern humanity was on the scene by this time: the Herto skulls from Awash in Ethiopia are 160,000 years old *Homo sapiens*, the skulls from the Kibish River, just northwest of Lake Turkana, are believed to be about 190,000 years old. No Australopithecines were around then and there are few *Homo erectus* fossils from this time; the Middle Stone Age in East Africa was a time of modern humans. So we can speculate that the stone industries in operation then were the work of our own species. The Sangoan had given way, on the eastern side of central Africa, to another industry, the Lupemban, named after a stream in the southern DR Congo by the Abbé Henri Breuil, a French Catholic priest whose hobby – some would say obsession – was archaeology; Breuil was jocularly known as the 'pope of Palaeolithic history'.

The Lupemban was originally thought to date from between 30,000 and 12,000 years ago, but Lupemban tools dating back to 300,000 years ago have been found, notably from Muguruk, near Kisumu in Kenya. Lupemban stone artefacts are largely known from south-central Africa. The Lupemban is characterised by small, well-made tools such as adzes, chisels, planes (i.e. woodworking tools) and beautiful long lanceolate blades that may have been used as spearheads. As well as Muguruk, Kenyan sites with Lupemban artefacts include Yala Alego and the Awach River valley (not to be confused with the Awash in Ethiopia). Most sites, in the country north of the Winam Gulf, are in unconsolidated sedimentary beds and thus cannot be dated accurately.

However, by 20,000 years ago, in the Neolithic, or New Stone Age, tiny backed blades (sharp on one side, blunt on the other) were being made in some quantities in the highlands of southern Kenya. Backed blades can be used as a knife or, if faceted, they could be set into a handle or shaft. Two classic sites in Kenya are on the Kinangop and at Lukenya Hill, south-east of Nairobi. The blades at the Lukenya site were associated with fragments of a skull that seems very similar to the skulls of modern Kenyans. A beautiful suite of microliths are known from a number of sites near the western wall of the rift valley, such as Gamble's Cave, Nderit Drift and the gorge of the Malewa River, plus a few other sites, such as Kabete near Nairobi. This period of tool manufacture is now called the Eburran, after the big volcanic massif Eburru, just east of the Mau Escarpment, north-west of Lake Naivasha. The Eburran is dated between 13,000 and 9,000 years ago; most Eburran artefacts are less than 60 or 70mm long and made of obsidian, that hard black volcanic glass so typical of the high central rift. And in the same area, but from before the Eburran, the Enkapune Ya Muto rock shelter on the slopes of the Mau Escarpment yielded some astonishing artefacts. Discovered by the eminent archaeologist Stanley Ambrose, the debris on the floor of the shelter yielded a set of disc-shaped ostrich-eggshell beads, not dissimilar to the beads worn as necklaces by modern Turkana women. Made more than 40,000 years ago, it is East Africa's oldest jewellery. Similar discs have also been found at Loyengalani, near Lake Turkana; those cannot be dated so accurately but might be even older.

What was life like for our Stone Age ancestors? Some lived in, or at least made use of, caves. Others may have made shelters for themselves using stone and branches; Mary Leakey found a stone circle at Olduvai that might represent a shelter made of branches supported by rocks. It seems likely that our ancestors, considering their tool-making abilities, would have quickly caught on to the idea of making shelters to protect against predators and the weather. They lived in a hot and harsh environment. But the essence of life as a hunter–gatherer is mobility: you have to go to where the food can be found. The material we have shows little snapshots of their life. We can speculate, and base our speculation on the lives of present hunter–gatherer societies, some of which still exist in Kenya today (although some historians would profoundly disagree with this technique, arguing that the present may not be the key to the past). Our early ancestors probably had lives little different from a band of savanna chimpanzees. They would have been concerned with

Elegant and relatively recent Lupemban lanceolate blades. (Stephen Spawls)

survival. As primates, they needed water regularly, so unless they had means of carrying or storing water, they would have had to stay within half a day's walk of a permanent water source. Much of the daylight hours would have been spent in procuring food, which leads us back to the theory that meat-eating might have made a big difference in terms of brain size, and that meat-eating took off about 2 million years ago, but not cooking. But an increase in meat eating argues that our ancestors either began to hunt, or scavenge, or drive other carnivores off their kills. Since chimpanzees hunt cooperatively, it seems reasonable to suppose our ancestors did so as well. As well as hunting, and looking for vegetable food, they would also have been on the lookout for predators and danger, which might have included bands of other hominins. Like chimps, our ancestors probably rested during the heat of the day. This may have been the time for some social activity, possibly involving grooming and some sort of communication.

Until about 250,000 years ago, they would not have had the use of fire, and since gregarious primates are diurnal, the night would have been a frightening time. Leopards *Panthera pardus* and Lions *Panthera leo* were out there. Nothing could be seen. Did they shelter in trees, or caves, or make some sort of defensive perimeter? Presumably as the sun began to set they would have made their preparations for sleep. Did they roam widely, maybe following the Wildebeest, or the rain, or did they stick to the area they knew?

SETTLEMENT, FOOD PRODUCTION AND THE IRON AGE

Some time in the last 20,000 years, possibly earlier, a major, momentous change took place: people began to settle down. A transition took place from hunting and gathering to the formation of settlements. Some of those early settlements were based on fishing. There is logic to this: a settlement requires a nearby steady source of food. If hunter–gatherers try to settle they are liable to run out of local food fairly quickly. But on the shores of a lake, fish will always be available, if not overexploited. Handy protein packages will be there for the taking throughout the year, and the cycle of rainy and dry seasons will not affect the lake unduly, whereas river levels and clarity can fluctuate widely in tropical Africa. In addition, water is always available and the lakeshore harbours Hippos *Hippopotamus amphibious* and attracts birds and mammals wanting water, all potential food sources. There are disadvantages, like mosquitoes, crocodiles and dangerous predators coming to drink, but the advantages must outweigh the disadvantages.

About 10,000 years ago, the waters of Lake Turkana rose some 80m above its present level and a river ran through it. Civilisations sprang up in the area. Around the lake, particularly in the Omo Valley, at Lowasera, on the south-east side of the lake, and at Lothagam there were fishing communities; remains of bone harpoons and fish bones have been found. Finely developed stone tools from Lowasera include backed microliths, flakes and hammerstones. Bits of broken pottery (potsherds) with 'wavy line' decorations were also found. Sonia Cole suggests these lines were made using catfish spines. A fishing community at Lothagam, a remarkably productive site discovered in 1965 by the veteran archaeologist Lawrence Robbins near the Kerio delta on the south-western shore of Lake Turkana, has yielded a number of skeletons of modern humans, as well as more than 200 barbed bone harpoon or spear points, microliths, fish bones and pottery.

The wavy line pottery found at Lothagam has been dated to 8,000 years ago; and is Kenya's oldest, a little younger than the wavy line pottery known from the Sahara west of the Nile, which is believed to be about 9,500 years old. Similar pottery has been found along the Nile near Khartoum, leading Robbins to speculate that the communities may have been connected. It would appear that he was correct. Robbins also suggested that travel along the linking rivers might have been by boat. There was no actual evidence of boat use in eastern Africa at the time, but a bone harpoon at Lothagam was found embedded in a deepwater bed; it can only have been launched from a boat. The remains of a dugout canoe dated at 8,000 years old was subsequently found at Lake Chad; so it looks as though some boating was enjoyed by those early fishermen on the shores of the desert lake. There is also some evidence of fishermen on the shores of a then much deeper Lake Nakuru.

An astonishing object was dug out of volcanic debris at the site of Ishango, a long-established Neolithic fishing community on Lake Edward's Congo shore in the Western Rift Valley. Found by a Belgian geologist with the magnificent name of Jean de Heinzelin de Braucourt, it is a tool made from the leg bone of a baboon and dated at over 20,000 years old. It has a piece of quartz embedded at one end, possibly for engraving, but what is particularly interesting are a series of marks cut into the shaft of the tool. Originally thought to be tally marks, the notches, lines cut at right-angles to the length of the bone, are grouped in a way that suggests complex mathematical understanding. One column consists of descending odd numbers, the other even numbers; these may indicate knowledge of multiplication and division. Some archaeologists think it is a lunar calendar, and may have been drawn up by a woman in connection with the menstrual cycle; others argue that it is concerned with prime numbers. Whatever the explanation, it is obviously more than just a tally stick; it is believed to be the world's oldest mathematical artefact. Mathematics started in Africa.

It is around this time, the last 10,000 years in the late Neolithic, that evidence of humanity's cultivation of wild plants and domestication of certain wild animals begins to appear. As always, the dates are not certain. Some archaeologists believe that the presence of sickles and grinding stones at a site indicates that the inhabitants cultivated plants, others point out that they might just as easily have been used to process wild plants that had been gathered. The dates of appearance and significance of domesticated cattle in Kenya is also a point of debate. Cattle, those sources of meat, milk and wealth, have performed remarkable service to humanity, particularly in Africa, since they were domesticated. We are reasonably certain, from molecular evidence, that all domestic cattle are originally descended from the Aurochs *Bos primigenius*. This big broad-horned black ox was widespread across Eurasia, from the Atlantic to the Pacific, ranged north from India to Siberia and extended into North Africa … and is now extinct, the last one died in 1627.

Nowadays, there are two main 'species' of cattle, taurine or regular (humpless) cattle (*Bos taurus*), and zebuine (humped) cattle (*Bos indicus*). They are not true species in the biological sense, as they can interbreed and produce fertile offspring; subspecies might be a better definition. The zebu is a characteristic animal of the dry country; it is drought-and disease-resistant. Studies of modern cattle DNA indicate that there was a single domestication 'event' of cattle, somewhere in the Near East, between 9,000 and 11,000 years ago, although the Kenyan-born zooarchaeologist Fiona Marshall suggests that the domestication may have occurred in northern Africa. A molecular study of Indian, African and European breeds of both zebu and taurine cattle clumps all the Indian breeds (all zebu) together in one lineage and all the African and European breeds (both zebu and taurine) together in a separate lineage. This is a simple and probably correct scenario; cattle were domesticated once, then the zebu and taurine breeds were produced by selective breeding, the zebu in India or Pakistan, the taurine further west. Africa got the taurine breed first, and the zebu was later brought in by traders via the East Coast. In some places it cross-bred with the taurine.

Bones from Lake Turkana provide the first evidence of cattle in Kenya. The earliest records of cattle in Africa are bones from the eastern desert of Egypt, 9,000 years old. The cattle herders must have moved slowly south. It has been suggested that the prime mover was the drying out of the Sahara 4,500 years ago, forcing the cattle herders to move as the Sahara savanna became the Sahara Desert. This aridity would have opened corridors free from tsetse flies. Tsetse are cattle killers, and you cannot keep cattle where these hard-biting flies exist. Tsetse carry trypanosomiasis, but they cannot tolerate aridity and they need woody plants for shelter and breeding. In 1983, John Barthelme excavated cattle bones at a site called Dongodien, in the Koobi Fora area, and these are dated to roughly 4,000 years ago; goat bones were also found (goats were domesticated in Asia roughly 10,000 years ago). At this time Lake Turkana had dropped by 40 or so metres, from 80m above its present level to 40m. It is possible that the drying climate created tsetse-free areas around the lake. Cattle and goat remains from this time are also known from the Ileret area, near the Ethiopian border.

This evidence is tied in with some rather intriguing language-based evidence. This is detailed in the next section, but, briefly, the linguist Christopher Ehret's research indicates that Cushitic-speaking people (Oromo, Somali and Rendille) moved into the Lake Turkana area from Ethiopia some 5,000 years ago. Did they bring their cattle with them? Were they the fisherfolk? If so, they must not have yet acquired the customary Cushitic taboo against eating fish.

Well-argued work by Fiona Marshall, at the University of Washington, in a 2002 paper entitled 'Cattle Before Crops' in the *Journal of World Prehistory*, indicates that cattle herding spread south of the Sahara before the spread of cultivation. The archaeological sites around Lake Turkana support this intriguing theory, at present anyway, because no evidence of domestic crops have been found there, although the stone bowls indicate that food plants were used. It is presumed that the Lake Turkana cattle were taurines. This is because the earliest evidence for zebuine cattle in eastern Africa are humped figures from Axum, in Ethiopia, that are 1,800 years old, and cattle figurines from sites on the Sudanese Nile change from humpless to humped about a thousand years ago. Bones of zebuine cattle 800 years old were also excavated at Hyrax Hill. From Lake Turkana, cattle herding then spread southwards into Kenya, and some major farming complexes developed. Pottery, stone bowls and microliths were recovered from an extensive settlement in the Chalbi Desert.

Sites from further south in Kenya that show evidence of cattle herding include Narosura on the Mau Escarpment, Crescent Island in Lake Naivasha, the Enkapune Ya Muto rockshelter on the Mau Escarpment and around Lemek, just north of the Maasai-Mara Game Reserve. Peter Robertshaw, the prominent archaeologist, investigated literally dozens of sites in the Lemek Valley. Those early pastoralists must have moved south down the rift valley, which was obviously then, as today, good cattle country. The settlers in the Lemek area arrived there about 2,700 years ago; they had large herds of cattle and goats. They may also have cultivated some crops such as Sorghum and Finger Millet. They penetrated into the northern Serengeti, but, significantly, they did not go further south for a long time. The reason for this may have been disease; the thick bush of the better-watered savannas of southeast Africa sheltered that deadly pest of cattle, the tsetse fly. Cattle herders did not reach southern Africa until about 2,000 years ago, and there is some debate who took cattle to there because the arrival date of Bantu herders in the south, 1,500 years ago, does not match the arrival date of the cattle. A possible clue lies in the fact that people of the Botswana/Namibia border country have a genetic marker that is also possessed by speakers of Nilotic languages in Tanzania.

The complete Acheulian tool kit, on display at Hyrax Hill. (Stephen Spawls)

Camels, sheep and goats seem to have reached Kenya around 5,000 years ago, so possibly earlier than cattle. Evidence of this comes from bones from the Ele Bor rockshelter in the north near that lonely road that wanders from Mt Marsabit up to Moyale on the Ethiopian border. Sheep and goats then spread throughout Kenya; but the camel remains largely confined to the north and northeast. No one is quite sure when the camel was domesticated.

Louis Leakey named a number of 'stone age cultures' from this period, such as the Kenya Capsian, and Kenya Aurignacian, later most of these were scrapped as they were poorly defined, but two that have been rehabilitated today are the Oldishian and the Elmenteitan cultures, the latter named after the lake. The people of these industries lived in Kenya. They herded cattle, used stone tools, including long two-edged blades and backed blades, and manufactured basketwear, jewellery, stone bowls and pottery. Some of the pottery had a pointed base, indicating it stood between three stones over a fire. Many of the potsherds show a distinctive pattern created by jabbing the surface with a wedge-shaped tool; this pottery is called Nderit ware, after the drift on the river of that name just south-east of the south-east corner of Lake Nakuru. Nderit pottery has been found at many of the sites around Lake Turkana; in East Africa Nderit ware extends from close to 5,000 years ago up to 1,000 years ago. The word *nderit* is Maasai, and means 'dust'. The Elmenteitan period lasted at least 2,000 years, from 3,000 to 1,000 years ago, classic sites include the Njoro River Cave, Gamble's Cave, Bromhead's Site, Lion Hill Cave and Hyrax Hill, all situated in the Western Rift Valley between Lakes Elmenteita and Nakuru. The Elmenteitan herders were cultured people; they buried or cremated their dead with ceremony. The 80 individuals cremated in the Njoro River Cave, buried with goods (a stone bowl, pestle and mortar), were drenched after death with red ochre, possibly because its blood-red colour was symbolic. Twenty-eight people were buried at Bromhead's site. Sonia Cole (1954), describes the skeletal appearance of the people buried there: many were tall with long faces. Other practices at that time included erecting stone cairns over the burial site.

It was originally believed that in East Africa only the highlands around the rift valley were settled by herdsmen who used stone tools, but this is now known not to be the case. Overlapping the Elmenteitan, but further to the west, was a group of hunter–gatherers known after their own distinctive style of pottery, the Kansyore. The Kansyore were hunter–gatherers, they were around from 8,000 to 3,000 years ago, and their communities centred on Lake Victoria.

The Peopling of Kenya

Rock work, ancient cupules dug for the game of mbao, Hyrax Hill. (Stephen Spawls)

One the most remarkable late Neolithic sites in Kenya is the Jarigole Pillar complex, located on the 80m beach near Allia Bay on Lake Turkana, just north of the barren Jarigole lava hills. About 4,000 years old, it consists of an oval-shaped platform and some 28 finely worked basalt pillars, most less than 2m high, with petroglyphs (engravings) on a number of them. The anthropologist Charles Nelson believes it is a mortuary site, the earliest known from eastern Africa, with the pillars raised to commemorate the dead. Jarigole is one of a number of 'Namoratunga' sites in northern Kenya, others are at the Losidok Hills near Kalokol and in the Kerio Valley near Lokori. These pillars are believed by the Turkana to represent dancers who were turned to stone after jeering at a malevolent spirit; there is a frightening Turkana chant about the Namoratunga which ends 'the hell fires to come will be more horrible than the first'. The Jarigole site contains a number of burial pits, where the dead were interred with grave goods.

Nderit pottery with its distinctive pattern, Hyrax Hill. (Stephen Spawls)

Important discoveries were made in a grave at Jarigole. The chamber contained not only Nderit pottery, stone tools, ivory artefacts and ostrich eggshell beds but also amazonite pendants from Ethiopia and a number of beads of an Indo-Pacific marine Mitre Shell, *Mitra paupercula*. These discoveries mean only one thing: Jarigole was on a trading – or at least travelling – route that linked Ethiopia to the Lake Turkana area and the sea, 4,000 years ago. Subsequently a Money Cowrie (*Cypraea moneta*) shell 1,400 years old was found at Kathuva on the Galana River in Tsavo East, more evidence of an ancient trade route into the interior. The Jarigole site also contained a number of pottery figurines, the oldest from East Africa. They include sheep, cattle and some game, including an elephant.

The American archaeologist Mark Lynch suggested, controversially, that the positions of the pillars at the Namoratunga site at Kalokol, near Ferguson's Gulf, meant it was an astronomical observatory. The alignment of the stone pillars was matched with the Cushitic calendar to a date 2,300 years ago. The pillars indicated the rising of seven stars and constellations, including the Pleiades, Orion and Sirius; these are used by the Oromo (Borana) to produce a complex 2–year calendar of 12 months and 354 days. Some said Lynch's measurements were inaccurate. Lynch was about to publish a rebuttal when he was tragically killed in a hit and run accident in California. Later work indicated Lynch was right, although a number of the petroglyphs (a word of Greek origin: *petra* 'rock' and *glyphs* 'carvings') on the pillars were recognised by the Turkana as being stock brand symbols.

As we saw earlier, cattle seem to have arrived in Kenya earlier than domesticated crops, material revealed by Fiona Marshall's assiduous research. In most other places, crops came first or animals and plants were domesticated simultaneously. But Marshall suggests that the situation in north-eastern and eastern Africa is a consequence of the high mobility of herders and the risk associated with trying to grow crops in arid environments. In other words, it's safer to have stock and move them if drought threatens, rather than settle down and try to grow crops (this is still the situation in much of dry northern and eastern Kenya). The other complication is that most of East Africa's food crops – at present, anyway – came here from elsewhere, at a time when transport and communications were slow, although this is not the case in West Africa and Ethiopia, where a number of the food crops are indigenous. The first domestic crops in Africa, in Egypt, were wheat and barley from the Fertile Crescent. This is where cities and civilisation began, and more than 9,000 years ago those crops were cultivated. Elsewhere, domestic plants seem to have spread out from one centre, but in Africa they appear to have been initially cultivated in a variety of places, dependent on the conditions.

The earliest archaeological record of a domestic crop in Kenya is Finger Millet, *Eleusine coracana*, known in Swahili as *wimbi*. Finger Millet was first domesticated in Ethiopia, its country of origin, around 7,000 years ago Remains of the seeds were found by the historian Stanley Ambrose, at Lord Francis Scott's old farm, Deloraine, near Rongai, dated to nearly 1,200 years ago. Finger Millet might also have been domesticated on the White Nile, near to the present-day border of Uganda and South Sudan. Seeds of Sorghum, *Sorghum vulgare* (*mtama* in Swahili), 2,500 years old, are known from the Sudan, it is believed that Sorghum may have been cultivated close to 5,000 years ago, in the Sahel. A plant also originating in the Sahel, Pearl Millet, *Pennisetum typhoideum*, known as *mawele* in Swahili, was cultivated in West Africa at least 2,000 years ago, but it is not known when it reached Kenya. These three crops are particularly important in dryland cultivation; they were the basis of agriculture in Kenya until introduced crops spread inland. Even now they are widely grown around Lake Victoria and eastern Kenya. Maize, Kenya's most important food crop, is nowadays grown almost throughout the country where there is over 600mm annual rainfall (at altitudes below 2,400m). However, maize originates from Central America, and reached Kenya via Spain and West Africa, four to five hundred years ago. Bananas and plantain, another staple food (Tanzania produces 800,000 tonnes of bananas per year) originated in Southeast Asia and likely reached Kenya via Indian Ocean trade routes, maybe as much as 1,000 years ago.

We seem to have jumped forward here, and must briefly go back to an important era of Kenya's history, the Iron Age. It was mastery of metals that advanced our species, for they were the first hard materials that could be shaped for a purpose. The first metal tools were made of copper and bronze, 6,000 years ago in the Fertile Crescent. Copper artefacts are known from Egypt, nearly 6,000 years old; 700 years ago copper crosses were used in Katanga, in the modern-day DR Congo, as a form of currency, but there was no age of copper in East Africa. The Stone Age turned smoothly into the Iron Age.

Iron changed the world. With it, one can make relatively large, strong tools that can be sharpened. And it can be mended and re-used, whereas a stone tool, once broken, is largely useless. Iron artefacts 5,000 years old, and even earlier, are known from Egypt and the Middle East. However, these are believed to have been made with iron derived from meteorites, and produced by smithing, which is hammering hot iron into shape, rather than by

Africa's famous and controversial 'White Lady' rock painting, Brandberg, Namibia. (Stephen Spawls)

forging, which is producing crude iron from the ore in a furnace, and bringing it to shape by repeated hammering. Smithing can be done at 600–800°C, and forging can be done at temperatures just above 1,000°C. The first blacksmiths in Africa do not seem to have achieved smelting, which is the melting into liquid and casting of iron from the native ore; for this you need temperatures of 1,500°C.

Iron smelting began about 4,500 years ago in Anatolia and the skill spread across the known world. It revolutionised warfare and agriculture. With iron tools and weapons you can clear forest, plough, cut stone and kill enemies. You can make long-lasting artefacts and their size is limited only by your technology; remains of great iron sheet gongs have been found scattered across Africa, from Nigeria to Zimbabwe, symbols of kingship. Those who worked with iron often had a mystique attached to them; in parts of Africa ironworkers were feared, in others they might even be despised as being slightly unclean. But they were also important members of the community (Smith is the most common surname in the English-speaking world). The name Angola comes from *ngola*, which means ironsmith. The Wata people of Tsavo are sometimes called Wasanye, which means 'those who work iron', i.e. smith.

In Africa, there is vigorous debate about whether or not iron-smelting technology was developed independently, or came in from the north, and this debate has been polarised on racial grounds. The material is lacking; in the warm, wet soils of sub-Saharan Africa, iron artefacts rapidly rust away to nothing. Thus evidence of iron working often depends upon finding ancient furnaces or slag, the debris that floats off the top when iron is smelted. To smelt iron, you need a high temperature: pure iron melts at 1,500°C. You have to find the ore, and break it up into fist-sized chunks. A furnace must be made, usually of clay, the iron is heated with charcoal and air is forced through the heating bed using bellows. The more carbon (charcoal) you can get to dissolve in the iron, the lower the melting point. The proportion of fuel to ore and the rate of air supply determines whether you get cast iron, steel, wrought iron or a useless lump of mixed material.

The earliest African iron workings, 4,500 years old, are at Termit and Egaro, in the southern Sahara in Niger. They are as old as the iron workings of the Middle East, which, in the view of the historian Gerard Quechon, proves that iron technology arose independently in Africa, and did not necessarily come from the Middle East via Carthage and/or Egypt. David Phillipson favours the latter view. Historian Stanley Alpern summarised the debate in a paper entitled 'Did they or didn't they invent it; iron in sub-Saharan Africa' (2005). Twelve of Alpern's experts say it was invented in Africa, nine do not. Some historians believe that since the use of iron was preceded by copper elsewhere, and copper smelting technology is useful when you come to work with iron, the sudden appearance of iron smelting in Africa lends credence to argument that iron-working expertise was introduced to Africa from the north.

East Africa's oldest iron workings are in Rwanda and Burundi, and are over 2,500 years old. So the skills took a while to drift eastwards across Africa from Niger. The term 'African Iron Age', as used by archaeologists, is usually taken to mean the period between 2,000 and 1,000 years ago, when iron smelting was widely practised in sub-Saharan Africa. The first use of iron in Kenya was, as far as we know, in the Lake Victoria area, and is attributed to a culture called the Urewe, named after a site of that name in the Yala River Valley in western Kenya. The Urewe made a distinctive type of pottery that Louis Leakey called 'dimple-based' as the potsherds he found all had a characteristic hollow in the base. Urewe pots are stylish, with a mixture of ribboned geometric lines on their necks and shoulders. The Urewe smelted iron with a furnace consisting of a basin and cone-shaped chimney.

In eastern Africa, the early Iron Age is associated with a period called the Chifumbaze, named by David Phillipson from a rockshelter of this name in Mozambique. Some archaeologists call this period the Early East African Iron Age. Around 1,800 years ago, a rapid dispersal of the iron-using, Bantu-speaking farmers of the Chifumbaze complex took place. Spreading out from an area centred on northern Mozambique, a group of such farmers moved into coastal Kenya, and along the hill ranges of the Kenya–Tanzania border country. The remains of their villages can be identified by the bits of pottery turned up there, known as Kwale ware after the town where the first samples were found. Located in brown sands, the original hoard included pottery, a corroded arrowhead, some carbonised wood and iron slag. Carbon-14 dating indicates an age of 1,700–1,800 years. These Iron Age Bantu farmers spread up to the north-west, to the Chyulu Hills, the Pare Mountains and the forests around Mt Kilimanjaro but do not seem to have entered the rift valley in Kenya. The Kilimanjaro sites are dated to 500 years ago. Who were they? It is tempting to speculate they were the ancestors of the Kamba peoples.

And at this point, having moved into the last 2,000 years, we are going to pause slightly and discuss what is known about the migrations of people through Africa and into Kenya, and its significance.

MOVEMENT THROUGH AFRICA AND INTO KENYA

There are five main sources of evidence regarding the movement of humanity through Africa: those sources are molecules, archaeology, oral history, languages and the present. So quite a decent crop of data exists, but nevertheless it is not unequivocal. And the big picture is extremely complex. At this point we need to move back in time, to the emergence of modern humanity, *Homo sapiens*. And here the molecules, in particular DNA, start to become important.

DNA (deoxyribonucleic acid) is a chemical found in the nuclei and mitochondria of cells. It looks like a long, twisted ladder, the rungs of which consist of pairs of four chemicals called bases, symbolised by the letters A, T, C and G, which always match up in pairs, C with G and A with T. If you can identify corresponding bits of DNA from two organisms, the closeness of resemblance between the sequences of bases tells you how closely related the organisms are.

Work on human DNA has revealed some interesting material. Firstly, there is far more variation in the DNA (and other molecules) of Africans than of all the people of the rest of the world put together. This indicates that humanity started in Africa, diversity increased, and then a small band of fairly closely related people left Africa, and populated the rest of the world. The size of this variation (geneticists call it a 'mismatch') also indicates that the

African population underwent a rapid expansion in size (and presumably in distribution) at least 80,000 years ago, DNA from Asian and European populations also underwent such an expansion, but 60,000 and 40,000 years ago respectively. This means that the pioneers left Africa before 60,000 years ago, but the exact date is not known. As has been sometimes said, there are more differences between the DNA of a Kenyan Maasai and a Kenyan Kikuyu, living 20 kilometres apart, than there is between the DNA of a Chinese from Beijing and a Spaniard from Madrid. Native Venezuelans show virtually no variation in their mitochondrial DNA. In genetic terms, non-Africans are a pretty homogenous group.

Modern-day Tanzanians seem to have the largest genetic variation of any country, so far. This indicates that Tanzania may have been – or been near to – the ancestral 'dispersion point' from which our species spread out through Africa. What the molecules indicate is that between 80,000 and 60,000 years ago, in Africa, people dispersed from a 'homeland' (probably in eastern Africa) southwards and westwards. And as they went, and after they arrived, they began to accumulate genetic changes. Those changes involved, among other things, height, facial appearance, limb length and skin colour. The people who made their way right down into the dry country of southern Africa became (or possibly started out as) small, with light brown or golden skins, whilst those who made their way across to the west became dark-skinned and some were very tall, others short. As for those who left Africa, some became very pale skinned and pale haired indeed, and some acquired short limbs and blue eyes.

One may hazard guesses at the evolutionary significance of these changes. There are risks with this; as the noted geneticist Steve Jones points out in *The Language of the Genes* (1993), it's easy to make up stories about how selection may favour certain genes, but none can be taken seriously without experiments. But a degree of speculation is essential in science. Among other things, people with light skin synthesise vitamin D (which keeps bones tough) six times faster than those with dark skin. So if you move to a climate where the light is less intense than equatorial Africa, it is probably an advantage to have pale skin. Likewise, a dark object loses heat more rapidly than a white one; in hot Africa a black skin keeps you cool and neutralises dangerous rays, in cold northern Europe a white skin slows your heat loss. Likewise, if you are rounded with short limbs, you have a small surface area to volume ratio, heat escapes slowly; if you are tall and thin you have a large surface area to volume ratio and lose heat quickly. Evolution is a passive process, but when those pioneers moved north, into the cold, those who variation favoured with lighter skin were more likely to survive.

Most of Africa's original human population have brown or black skin. Between roughly 5,000 and 4,000 years ago there were three different 'ethnic' groups of people in mainland Africa. Much of the north was occupied by people with brown skin, straightish hair and aquiline faces. According to James Newman, in *The Peopling of Africa*, they were the ancestors of modern-day Berbers. Black Africans occupied Western and Central Africa, and Khoisan people occupied southern Africa. And around this time, a most significant migration began that was to effectively populate much of Kenya. Some really useful evidence for this migration comes from studies of the language families of Africa.

Our complex language is one of the things that makes us human. It is also generally accepted that such language evolved over 100,000 years ago. And yet, according to the historian John Sutton, the four main language groups of Africa seem to have no relationship to each other. The four families are Nilotic, Niger–Congo (also called Niger–Kordofian), Khoisan and Afro-Asiatic. The standard work, entitled *The Languages of Africa*, was published by Stanford University's great linguist Joseph Greenberg in 1963. There is still general agreement with his conclusions. Afro-Asiatic languages are spoken by most of the people of the northern half of Africa. Khoisan languages contain clicks and are centred in the dryer regions of Southern Africa. Nilo-Saharan languages are spoken in several widely separated areas, including the East African Rift Valley, the upper White Nile, parts of north-central Africa and the northernmost bend of the Niger River. And Niger–Congo languages are spoken everywhere else – essentially all of Africa south of the line connecting the mouth of the Senegal River with Lamu, with the exception of the areas of Khoisan languages. The Niger–Congo languages are split by some linguists into Bantu and Non-Bantu languages; the Bantu languages being spoken roughly everywhere south of the equator. The Afro-Asiatic languages include the subgroups Cushitic and Semitic; terms originally used for the languages of north-eastern Africa (Somali and Oromo are Cushitic) and the languages of North Africa and the Middle East. Semitic languages include Hebrew, Arabic and Aramaic, although the Semitic language subfamily is mainly African, with twelve of the nineteen surviving languages found only in Ethiopia.

It may seem odd that the four major groups of African languages seem to have no relationship to one another. But languages, like plants and animals, are capable of rapid change, particularly in isolation, and the huge differences between Africa's language families may suggest that their speakers were physically separated. And the fact that the Bantu languages are so similar lends support to the idea that the Bantu expansion through Africa was very rapid, as we shall see shortly.

All four African language groups are represented in Kenya, some more so than others. The Khoisan languages are characterised by clicks that do not involve the vocal cords. The tongue or palate is used. Several of them are easy to make, such as the kiss, the cluck of the tongue against the roof of the mouth and the sound made by sucking quickly at the cheek, although fitting several into the same word is a remarkably skilled operation. In Tanzania, the Sandawe and the Hazda speak Khoisan-type languages. In Kenya the Dahalo, who live north of the Tana Delta, also speak a click language, although some linguists suggest it isn't, and the actual click words are loan words from the south. The Hazda and the San have very different DNA, however, (the Hazda cluster with nearby Tanzanian non-click speakers), indicating their common ancestors lived 40,000 years ago, and this suggests to some linguists that the Khoisan click languages are very old, and may have been the first languages of humanity. Fascinatingly, some language experts believe that it is a language developed for hunting. The sounds of normal human speech scare animals, especially antelope, but they are not spooked by click words. Khoisan hunters also have a range of hand signals that are designed for hunting and are quite astonishingly expressive – the differing curl of the index and second fingers, for example, indicating the difference between a Roan Antelope *Hippotragus equinus*, a Hartebeest *Alcelaphus buselaphus* and an Eland *Taurotragus oryx*; over 30 species of potential food animals can be shown. Digressing slightly, one wonders if elaborate sign language might have predated sound language and whether it was hunting that was the creative impetus behind the development of such languages, the clicks and signs enabling hunters to approach their quarry and plan the kill without having to use words.

Kenyan speakers of Bantu languages include the Swahili, Mijikenda, Kikuyu, Meru, Kamba, Taita and Luhya; Nilo-Saharan speakers include the Luo, Kalenjin, Turkana, Maasai, Samburu, Pokot and Njemps; speakers of Afro-Asiatic languages include the Somali, Mukogodo, Rendille and Oromo. The language that you speak tells you something about whom you are related to, your identity and where and when your ancestors entered the country. The relationships of the languages and the families were worked out by Greenberg. He used a method called mass or multilateral comparison, looking for a wide range of similarities in certain crucial words and constructions, and based on the idea that mass borrowing by one language of another's basic vocabulary is unknown. Similarity of certain basic words is a useful clue. For example, in most Bantu languages the word for Lion is *simba*, or some variation of it (*shumba* in Shona), whereas in Cushitic languages it is different, in Somali it is *libaax*, and in Amharic it is *anbassa*. However, the word for a francolin, *kwale*, is the same in Swahili, Setswana and Shona (and gave its name to the coastal town). The word for a crocodile in the Nilotic languages shows their relationship: in Luo it is *nyang*, in Turkana *aginyang* and in Maasai, *Ikinyang* (but in the Bantu languages spoken all around Mt Kenya it is *king'angi*). In most Bantu languages the word for a snake is *nyoka*, or *njoka*, or some variation on this, but in Oromo it is *bof*, or *boffa*, and in Somali it is *mas*.

There is no contemporary written record of human movement into and through Kenya, so there is room for speculation about exactly who arrived when and where. And archaeological artefacts do not necessarily identify any particular language speakers, as the distinguished American linguo-historian Christopher Ehret has pointed out. It's tempting, for example, to believe that the cattle-herding pastoralists who held sway in the Lemek area 2,500 years ago were Maasai, or their direct ancestors, but there is no unequivocal evidence that links the present Maasai inhabitants with those who were there two-and-a-half millennia ago, other than circumstantial evidence, for example that both herded cows and goats. Oral evidence indicates that the Maasai arrived at Lemek much later. The big picture, where Kenya is concerned, seems to be that Afro-Asiatic (specifically Cushitic) speakers who herded cattle and goats entered from the north around 4,000 years ago, the Nilotic language speakers entered Kenya from the north-west some 2,500 years ago, and finally, during the last 2,000 years, Bantu people entered Kenya from both the west and the south; and all the time there was relative movement, some peoples were assimilated or engulfed by others and some struck out on their own. We will expand on these movements below, but some speculation is involved. A quotation from David Cohen's essay on the Nilotes in *Zamani* (1968) sums it up: 'historians … of Africa find no more difficult problem than the reconstruction of the great population movements of the past'.

Those who initially entered Kenya from the north, 4,000 years ago, were speakers of a southern Cushitic language, and they made their way southwards into western Kenya and north-central Tanzania. Those in western Kenya then gradually disappeared; no southern Cushitic language is spoken there today. They were presumably assimilated by later arrivals. Southern Cushitic languages still survive in Tanzania, but not in Kenya, unless the Dahalo click language is regarded as southern Cushitic, which some authorities claim it is.

Between 4,000 and 6,000 years ago, Nilotic speakers and Cushites mingled in the border country between the Sudan, South Sudan, Ethiopia and Kenya. The Nilotes were cattle husbandrists who bled their stock and are significant in Kenya's history. Some time in the past the Nilotic ancestors of the Luo lived in south-west Ethiopia, and then moved into a 'cradleland' in south-east Sudan, south of the junction of the Bahr-El-Gazal and the Nile – the name Luo is said to mean 'swamp'. They did not stop there, they spread out, to the north and then the east, and those of significance to Kenya moved south, initially to Pubungu, where the modern Ugandan town of Pakwach is, on the Albert Nile. Luo oral history tells of a massive military camp there. Some pioneers went south-west from here, and gained remarkable power, or so the tradition tells, taking charge of the fabled empire of Kitara, originally ruled by the shadowy Bachwezi dynasty. The explorer John Hanning Speke took this story literally, but it is now thought to be largely legendary. Two Luo brothers, Kimera and Rukidi, are said to have founded the dynasties in Buganda and Bunyoro, massive empires that dominated present-day Uganda until white westerners arrived. The Luo in Kenya did not achieve that degree of political centralisation. However, some 500 years ago another group of Luo pioneers made their way past Lake Kyoga and the Mt Elgon foothills, were involved in battles with the Maasai, and by 300 years ago were in western Kenya, having displaced some Bantu peoples. They were successful stockmen, but they struggled to keep their cattle on the lakeshore plains, which harbour tsetse flies. However, they occupied some of the high land north and east of the lake, and between 1750 and 1800 they crossed the Winam Gulf, and spread out in the Homa Bay area; and this remains Luo heartland today. A prominent Luo of this time was the dreaded magician Gor Mahia, who lived on the Kanyamwa escarpment and dominated South Nyanza and beyond. Legend states that he could change from a man to a dog or a bull. A famous Kenyan football team are named after him and occasionally make the pilgrimage to his hill, looking for assistance with some wizardry in the opponents' penalty box.

The Luo are sometimes known as lake or river Nilotes, to distinguish them from the highland or southern Nilotes, and the plains Nilotes. The highland Nilotes comprised three main peoples: the Dadog, the Kenya–Kadam and the Kalenjin. They moved into Kenya from Ethiopia between 1,000 and 2,000 years ago. The Dadog then migrated into Tanzania, where they are today, and the Kenya–Kadam peoples were absorbed by the plains Nilotes. Those highland Nilotes who remained in Kenya assimilated a number of Cushitic speakers, and over some 1,000 years became the Kalenjin, and occupy the high country east of the lake, centred on the towns of Kericho, Kapsabet and west of Kitale. They were not known as the Kalenjin then; this is a name that the modern Kalenjin (principally the Kipsigis, Nandi and Sabaot) took from a radio programme for Nilotic speakers. The broadcaster used to begin his programme with the word *Kalenjin*, meaning 'I tell you this'. The Nandi were originally called the Chemwal; the story is that they were called WaNandi, meaning 'raiders', by porters on the explorers' caravans, liked the name and kept it.

Five hundred or so years ago, most of Kenya's western highlands, between Sotik and the slopes of Mt Elgon, and east as far as Lake Elmenteita, were the territory of a people called the Sirikwa. They have disappeared, but not without trace. They may have been ancestors of the Kalenjin, or been absorbed into the Luhya or the Maasai, but they have left interesting evidence of their presence. Scattered over the western highlands and western rift are a number of so-called 'Sirikwa Holes'; saucer-shaped depressions or hollows 10 or 15m wide and several deep. A nice example can be seen at Hyrax Hill Prehistoric Site, at Nakuru. These depressions occur on hillsides, in groups of five to 50, in some places there are over a hundred. Kalenjin elders knew what they were and told archaeologists so, evidence of the putative relationship between the Kalenjin and the Sirikwa. Excavation proved the old men right; the holes are simple stock pens that had been dug out, designed to protect the cattle overnight, with a narrow downhill entrance. They had been surrounded with a thorny fence, long since gone, of course, although on the Uasin Gishu plateau some fences were replaced by stone walling. In John Sutton's book *A Thousand Years of East Africa* (1990) there is an elegant essay on Sirikwa holes. Around this time, too, a little further to the north, a remarkable system of irrigation canals fed the farmers' fields on the slopes of the Kerio Valley. These may be of the same age as the irrigation canals and stone-walled terraces of the massive farming complex that flourished between 300 and 600 years ago at Engaruka, just east of the Crater Highlands in northern Tanzania. The builders are unknown.

Famous people of the plains, or eastern, Nilotes are the Maasai, named after their language group, Maa, other eastern Maa-speaking Nilotes include the Samburu and the Njemps. No one is certain how the Maasai reached Kenya, although it is speculated that they came from north of Lake Turkana; there is a curious legend that they represent a lost Roman legion. However, their biological molecules say otherwise: studies of the DNA of Nilo-Saharan speakers in East Africa show a close relationship with people from the western Sahel, and suggest that some 10,000 years ago the Nilo-Saharan speakers lived in a heartland in the southern half of the Sudan. Beyond that, it is impossible to say. But between 1,000 and 500 years ago the Maasai came down the Kerio Valley, and went onto the Uasin Gishu plateau, where they split the Kalenjin people there into two sections; Nandi and Kipsigis to the west, Tugen and Marakwet to the east, by the mid-eighteenth century they were in central Tanzania. As cattle herders, the Maasai need open country. You cannot keep cattle in forest or woodland, so the high country of Kenya's central and southern rift valley suited them perfectly. Most of Kenya's southern border is Maasai country. Their military competence stems from their age-set organisation, all young men of the same age passing through fixed age grades. In the seventeenth and eighteenth century, the Maasai also started a tradition of eating only products from livestock raising. One group spread across the beautiful Laikipia grasslands north of Mt Kenya and the Aberdares, the Laikipiak Maasai. Oral legend tells that they raided far to the north, beyond Marsabit; some claim they went as far as the Daua River. But their herds were devastated by castle disease and the people were affected by smallpox. To add insult to injury, they were then evicted from the plateau by the colonial government in the early twentieth century, a move that still causes tension and resentment today.

The Samburu (who actually call themselves the Loikop, or Lokob, meaning 'people of the world') were part of the Maasai expansion, occupying parts of north-western Kenya. They are believed to have been pushed east into the country surrounding Maralal and the Mathews Range by the Turkana (also eastern Nilotes). who came down from the long escarpment along the Uganda border. The Turkana broke away from the Karamajong, also cattle-herding plains Nilotes of north-eastern Uganda; the story is that a group of Turkana chased a runaway ox down the escarpment, and liked the look of the country they were in. This part of Kenya, the north-west, is cattle-raiding country *par excellence*; your stock in this area represents your wealth. It is a huge area, and at one time the largest branch of the Kenya police was the stock theft unit that dealt with the north-west. This is a land of proud nomadic people, with the usual macho culture, and it is thus a land of instability. Nomads always trouble governments, because all their wealth is concentrated in their stock; a single raid can transform a rich stockowner into a beggar, and vice versa. The Turkana acquired their camels, so the story goes, by raiding the Rendille.

Which brings us to the Cushitic people, speakers of a sub-branch of the Afro-Asiatic group of languages. Speakers of eastern Cushitic languages in Kenya include the Rendille, Oromo, the El Molo, the Orma and the Somali. In Kenya, the eastern Cushitic homelands are situated to the north, north-east and east of the highlands; it is believed that the eastern Cushites originally came from the Ethiopian southern rift valley, south of Lake Zwai. Before that is anyone's guess. Most of south-central Ethiopia is dominated by Oromo-speakers, an interesting people. In Ethiopia, where there are over 15 million of them, the Oromo speak a language called Oromiffa; these names have their approval. In Kenya (and Ethiopia), Oromiffa speakers were known for a long time as Galla, but

Sirikwa hole, ancient stock pen at Hyrax Hill. (Stephen Spawls)

Ali Adan Hussein, water well driller, with his camels, Wajir. (Stephen Spawls)

they neither like nor chose this name, which means simply 'a wanderer', whereas Oromo means 'nation'. Kenya's largest group of Oromiffa speakers are the Borana, (which means 'free people'). Their homeland stretches from Ethiopian across the border west of Moyale, south through Marsabit and east of there, down to Garba Tula and the northernmost bend of the Tana River. The Oromo heartlands are the southern Ethiopian highlands; the Oromo pioneers entered Kenya from Ethiopia over 400 years ago and gradually spread southwards, taking up camel herding rather than cattle herding, as a response to the harsh climate of northern Kenya. An interesting group of Oromiffa speakers, the Wata or Sanye, inhabit the lower Tana, the Taru desert and parts of Tsavo East National Park. The famous elephant hunter Arthur Neumann called them the Wasanya; they have also been called (among other names) the Waata, Wasi and WaLiangulu. This last name means 'eaters of meat' in Giriama and that they are – these are the people who hunt elephants using bows and arrows. They are slight people, with muscles and tendons of steel, with the strength and skill needed to draw a 5-foot bow far enough to send an arrow deep into an elephant. Brave, resourceful and courteous, they have earned the respect of many of the hard-bitten professionals of the wild country of eastern Kenya, men like Terry Mathews, Ian Parker and Bill Woodley, the charismatic warden of Tsavo. Dennis Holman (1967) wrote a book about the Sanye, which he entitled *The Elephant People*.

Three related Cushitic languages, Somali, Rendille and Boni, are known collectively as Sam, meaning 'nose'. The Boni, who live in the north-east coastal hinterland, inland from Lamu, are believed to have migrated south from Ethiopia, along the Tana River and then north-east along the coast. The Rendille inhabit the country south and east from Loyengalani, on the shores of Lake Turkana, to the Merille River. Somali people inhabit most of north-eastern Kenya. Somalis are believed to have moved into Kenya about a thousand years ago, possibly earlier. There is debate about their origins; many sources suggest they originated in Arabia, others in Ethiopia. There is a political element to this argument. Somali DNA contains certain sequences that are found in Asian peoples, and are rare in sub-Saharan Africa. Some workers suggest that the peoples of north-east Africa, in particular Somalis, Eritreans and eastern Ethiopians, became genetically isolated from sub-Saharan African some time ago. The question is; did those who left Africa for Asia have this sequence and take it with them? This would mean that those who left had, among other things, Somali ancestry. Or did they go to Asia and then migrate back to Africa? Nobody is sure.

Languages have proved of considerable importance in reconstructing the next part of Kenya's history, the Bantu migration. Bantu-speakers make up more than 50 per cent of Kenya's population. There are in excess of 1,000 Niger–Congo languages, and about 500 of these are Bantu. The Bantu languages are very similar; it has been facetiously said that there is only one Bantu language and 500 dialects. Although they are spoken widely through sub-Saharan Africa, the majority of the different Niger–Congo languages, some 180 subfamilies, are all spoken in a small area in eastern Cameroon and adjacent eastern Nigeria. This is thus the heartland, proven by its diversity (a similar logic pinpoints the origins of English in coastal north-western Europe, one of a cluster of closely-related languages). It has to be said that not all historians agree with this argument, but most mainstream researchers support the idea that the similarity between the Bantu languages in Africa south of the equator indicates that a group of related Bantu speakers left their heartland. The first departure was some 5,000 years ago, and they spread relatively rapidly throughout central, eastern and southern Africa. With them they took wet climate crops like yams, and they picked up iron technology as they moved into central and eastern Africa. Essentially, they conquered sub-Saharan Africa, freed by their lifestyle from having to rely on open areas to graze stock. On the way they assimilated the small peoples (often called pygmies, although this is a somewhat pejorative term) who lived around and in the central Africa forests; and who today have no language of their own, although they may have some unique words. They also displaced or engulfed the Khoisan speakers from a number of areas, notably the fertile lands of present day Angola, Zambia, Zimbabwe and eastern South Africa. The Bantu did not bother going into the dry areas of north-eastern Africa, nor did they move across the lands of the Nilotic peoples (there are no Bantu in Ethiopia) but they entered Kenya. Where and when, however, are matters of debate.

Essentially, there are three clusters of Bantu-speaking peoples in Kenya: they are the Luhya, Kisii and Kuria in the west, the Kikuyu, Meru and Kamba in the central highlands and the Mijikenda, Swahili and Pokomo on the coast. Bantu peoples in Kenya are lucky: they occupy fertile country because they have the lands with the consistent rainfall. Bantu pioneers had reached lakes Kivu and Edward some 3,000 years ago. It seems possible that the makers of Urewe pottery were Bantu peoples; they were on Lake Victoria's western shores 2,500 years ago. Over the next 800 years they spread around the lake and by 1,500 years ago they completely surrounded Lake Victoria. During the following millennium they absorbed some Kalenjin communities, and about 400 years ago some Bantu speaking people pushed into the land north of the Winam Gulf. Shortly after this, some Maasai and Nandi communities in the area became Bantuised, and then some more Bantu speakers were pushed into the area by the Teso expansion in Uganda. And thus the Luhya came about, in the lands south and west of Kitale; they are Kenya's second most populous people.

It is believed that Bantu settlers in the area around the Winam Gulf built the remarkable stone structures known as Ohingas. There are over a hundred of these, and the largest, with stone walls over 4m high, is called Thimlich Ohinga. The name is Dholuo (the language of the Luo people), and means 'frightful dense forest', which is a superbly scary moniker but tells you nothing useful. Reminiscent of Great Zimbabwe, Thimlich Ohinga is a stone-walled complex 15,000 square metres in area, within which is a series of smaller stone enclosures. The walls are beautifully crafted, dry-stone walling at its most accomplished, with perfect rounded ends, neat little doors with huge lintels, and rocks selected by size to produce a stable enclosure without the use of any sort of mortar. Elevated platforms inside enable defenders to shoot or throw things down at an external enemy. Although it is not known exactly what the Ohingas were used for, the design suggests a defensive function – a fortified and easily defendable enclosure. But even the official National Museum of Kenya website is blandly uncommunicative about their purpose. These structures are the oldest stone buildings in Kenya's interior, being more than 500 years old (and hence built before the Luo got there) and on account of their age, significance and intrinsic elegance they are well worth seeing, although getting there is difficult, on unsignposted tracks. If you get there, you invariably have

Tom Kadondi shows the height of the entrance at Thimlich Ohinga. (Stephen Spawls)

Slag from ancient iron workings, Thimlich Ohinga. (Stephen Spawls)

Joel Onyango and Tom Kadondi demonstrate the defensive elevated platform at Thimlich Ohinga. (Stephen Spawls)

the place to yourself, and can stand and wonder about exactly what went on there. The inhabitants smelted iron, that we do know; the enthusiastic site guardians will show you the slag.

There is much speculation about the origins of the Kikuyu. They are believed to have originated in the Lake Chad area and entered Kenya from the south, possibly via the Tana River. Kenya's most populous people, at present their fertile heartland, stretches from Nairobi north into the land between the Aberdares and Mt Kenya. The Kikuyu have a creation legend, and it is described by Jomo Kenyatta, Kenya's first president, in his anthropological study of the Kikuyu people, entitled *Facing Mt Kenya* (Kenyatta 1938). The legend tells how Mogai, God, on Mt Kenya, commanded a man called Gikuyu to build a homestead and provided him with a wife, Moombi; they had nine daughters, and then nine sons appeared, and things proceeded smoothly. More prosaically, archaeological evidence at Gatung'ang'a, a village a few kilometres south of Nyeri, indicates that there were settlements there between 800 and 600 years ago, by the Thacigu people, the original Bantu inhabitants of the area. Discovered by the eminent historian Geoffrey Muriuki (who wrote a history of the Kikuyu), the site has yielded iron working slag and Kwale pottery, and at a higher (and thus younger) level coarse potsherds and obsidian blade tools were found.

Sometime during the last 600 years the Kikuyu moved from the vicinity of the Nyambeni Hills, north-east of Mt Kenya, into the land between the Aberdares and Mt Kenya, through to Sagana and south-east of there. Legend indicates that they assimilated the Athi people and pushed out the 'Gumba' peoples, who were small inhabitants of the montane forests. This may be a myth, but in fact a number of small hunter–gatherers lived in Kenya's highland forest, and some are still there today. They were not short for any genetic reason, unlike the pygmies, but because of their poor diet. These hunting people are often mentioned in books by the early European pioneers, where they are called the Dorobo or Wa Ndorobo. These names are all variants of a Maasai term, Il Torobo, meaning poor people; which they certainly were in Maasai terms, as they didn't have cattle. More correctly known as Ogiek, or Okiek, their origins are varied, although a number are related to the Maasai

Traditional village life, Tigania, 1960s. (Harry Spawls)

and Samburu, some are related to the Kikuyu. It is speculated that the Okiek as a whole are not members of any discrete ethnic community, but were simply unfortunates who had lost their land, or stock, and left their fellows to seek their living off the land. They speak various dialects, and some of these, sadly, seem to be dying out; the Okiek who inhabited the forests east of the central rift valley used to speak a dialect called Kinare, which is now extinct. The Okiek survive by collecting honey, hunting and using the resources of the forest. As with nomads, their peripatetic lifestyle and lack of wealth are troubling to governments. Okiek hunters remain in the forests of high central Kenya, and on some of the forested mountains of the north and east.

The Embu people, according to tradition, also migrated to Mt Kenya from the Nyambeni Hills. The Embu, the Kikuyu and the Tharaka people all have an oral tradition of migration from the coast, from a possibly mythical place called Shungwaya, also known as Port Durnford. This is situated somewhere near the modern Somali town of Bur Gao, just up the Somali coast beyond the Kenya border. Some historians believe Shungwaya was actually a region, a strip of land that extended from Tanga north to Mogadishu. The migration dates do not fit because there is evidence of settlements in the central highland before the proposed date of dispersal from Shungwaya. A migration up the Tana River is possible; Green Mambas managed it, migrating from the coast up to the Nyambeni Hills. But unequivocal data on many of these movements are not forthcoming. Most accounts are based on oral tradition, and some historians will therefore not comment on them. In *The Peopling of Africa*, for example, the Kikuyu are shown moving across from Lake Victoria to the Mt Kenya area (Newman 1995). The historian Roland Oliver has a cautious view of migrations in oral history, believing them to be highly memorable events and thus overblown. Oliver notes that sometime in the past a group of refugees fled from Manda, to get away from Moslem overlords, and made their way up the Tana to settle amongst the people of Meru and Tharaka. He suggests that their remarkable tale was adopted as an oral tradition by the peoples of Mt Kenya; this story is recounted in some detail in Andrew Fedders and Cynthia Salvadori's excellent book *Peoples and Cultures of Kenya* (1981). But the Bantu speakers of central Kenya might just as easily come from the south; the Kamba people did, moving 500 years ago from lands around

Mt Kilimanjaro to the Chyulu Hills, thence to the plains around Kibwezi and finally to their present homeland, Ukambani, in the dry country north-east, east and southeast from Machakos almost to the Tana River.

The coastal Bantu speakers, the Swahili, the Mijikenda peoples and the Pokomo, almost certainly came up the coast from Tanzania. It is suggested they can be traced back to the hinterland between Mt Kilimanjaro, the Taita Hills and the Pare Mountains. And tangible history can be found on the coast. During the last 1,200 years the Swahili peoples there had trade, towns, architecture, literacy, Islam and a social stratification, earlier than most of the people in the hinterland. The coast had contact with visitors from the outside; the name Swahili derives from the Arabic *sawahil*, which means 'coast' (and is the plural of *sahel*, although this is also said to mean 'edge'). But the language itself is a Bantu language and had its origins between the Tana Delta and Lamu, in a language called Pro-Sabaki. It does contain many Arabic loan words, but these mostly entered the language in the nineteenth century; pure Swahili is an African language.

Stone towns were built on the coast. Some have yet to be definitely located, for example Rhapta, an ivory mart that may have disappeared in the Rufiji Delta, or Kambalu, or Serapion. Over a thousand years ago building started in the Lamu archipelago and at Kilwa in Tanzania; Kilwa became a city and port that dominated trade on the Mozambique Channel between 600 and 800 years ago. Beautiful Chinese and Iraqi pottery from the ninth century has been recovered from Manda Island, from a city estimated to house 5,000 people, and Chinese documents from the ninth century mention Mualini, which might mean Malindi. Islam came to the coast; the remains of a ninth century wooden mosque have been excavated at Shanga, underneath a tenth century stone one. The historian Al Idrisi wrote in the eleventh century of a man who charmed snakes at Malindi. Mogadishu was a thriving port in the twelfth century, with two great mosques. Gede thrived in the fourteenth and fifteenth century and was then abandoned, for no known reason; excavations there have failed to yield any human remains. Something went wrong but the residents all escaped. The coast was a centre of trade; materials brought from the interior included gold, ivory, tortoise shell, timber, rhino horn, shells and skins ... and also slaves. In return, the East Africans received iron weapons.

This was an unusual time; Kenya began to be connected to the world, although it seems that it was only the coast that got connected. Adventurers and traders obviously went inland, but nobody thought of building in the interior; all the towns of the East African coast (and there were probably over a hundred, although only a few are known about) are right on the shore itself. It is also not known how much involvement travellers from outside had in the development of the coast. This has led to a great deal of high level, angry debate, often involving some of Africa's finest minds. The debate arose, in the usual way, when Western historians looked at the fine architecture and culture manifest in the Swahili harbour towns and decided that, since no such architecture was present in the hinterland, the architects and guiding minds of the towns must have been foreigners, from the Middle East.

EXPLOITATION, EXPLORERS, COLONIALS AND INDEPENDENCE

This is a fairly short section. As natural historians, we need to bring our account to a halt; competent historians have covered Kenya's recent history. What remains are in many ways simply the inevitable consequences when a group of social, gregarious, hierarchical and aggressive primates, in societies that are to some extent xenophobic, start to jockey for position. The concept that all members of humanity are equal, and all have equal rights, has been a long time becoming a reality. In our modern world, sadly, there are still societies where women are treated as second-class citizens, people do not have the vote and membership of a particular religion, ethnicity or skin colour makes you more important than your fellow citizens. It is a depressing thought that some members of our supposedly enlightened race still think that they have the right to discriminate against, punish, or even kill, people who don't subscribe to their beliefs. A few dates and facts will suffice: Switzerland did not give votes to women until 1971, South Africa did not allow its non-pure-white citizens to vote until 1994, in some legal systems a

Gede, 14th Centry Islamic city, mysteriously abandoned. (Stephen Spawls)

woman's testimony is worth half that of a man's, some countries still do not allow their citizens to own computers or mobile phones. Some of us still have miles to go before we sleep.

Most of us are aware of the hideous cruelty of the slave trade on the western side of Africa, a trade started in the fifteenth century when Portugal began to import slaves. By the time it ended, more than 17 million Africans had been carried across the Atlantic by westerners and sold into slavery; many millions more died before they reached the Americas. In Britain the trade was abolished in 1807and slavery itself in 1833, although it took longer in the Americas; the American Civil War, largely concerned with the right to keep and use slaves, took place between 1861 and 1865.

A more ancient slave trade existed in eastern Africa. It is recorded that in 868 AD slaves from Zanj (a term used by inhabitants of the Persian Gulf to describe Black Africans, sometimes spelt Zinj) revolted against their masters whilst engaged in draining the marshlands of southern Iraq. It has been argued, with some truth, that the East African slave trade was more humane than the West African one. Not in the actual capture and transport of the slaves, which was cruel – one needs only to read David Livingstone to be aware of that – but in the fact that those who reached the Middle East were often treated with comparative kindness. They were used as domestic slaves, whilet the victims of the West African trade were used in gangs on plantations. In 1873, Sultan Barghash of Zanzibar was forced to sign a treaty making the slave trade illegal between all his ports on the African coast and Middle East.

There is no record of penetration to the interior by Arab or Swahili traders before the eighteenth century (although, as we saw earlier, coastal material did penetrate inland several thousand years ago). But in the 1830s Seyyid Said, the Imam of Muscat, sent caravans to the interior to hunt elephants; the traders went in with beads, copper wire, hats and gunpowder, and brought out ivory, a trade we shall refer to again in Chapter 13 on conservation. The Portuguese controlled parts of the Kenya coast between 1498, when Vasco da Gama arrived in Mombasa, and 1698, when their garrison at Fort Jesus was overwhelmed by an army of the Sultan of Oman. Fort Jesus itself was built in 1593. From 1698 until 1856, the government at Muscat in Oman was in control of the Swahili harbour towns.

Kenya's future, youngsters near Homa Bay. (Stephen Spawls)

Western exploration of the interior of Kenya essentially began in the nineteenth century. On a map of Africa dated 1821, in Thomas Myer's *New and Comprehensive System of Modern Geography*, the interior of East Africa is totally blank, and the only recognisable Kenyan places are the island of Pate, Malindi ('Melinda') and Makupa ('Macupa'). Dr Ludwig Krapf made his way to within sight of Mt Kenya in 1849 and other Western explorers followed. Some were indeed imperial colonisers hell-bent on subjugating the local inhabitants, claiming land for their masters or shooting the wildlife. But not all. Many were brave men, with noble motives, and Africa killed many of them. Individuals of importance to Kenya include Krapf, Joseph Thomson, Count Teleki, Lieutenant Von Hohnel, Sir Gerald Portal, Captain John Hanning Speke, James Grant, William Astor Chanler and Arthur Neumann. Richard Burton and David Livingstone never made it to Kenya, and Henry Morton Stanley made only a brief journey there (although a famous Nairobi hotel was named after him).

Between 1876 and 1912, seven European countries – Britain, France, Belgium, Spain, Portugal, Germany and Italy – divided up Africa between them, in the 'scramble for Africa', an experiment in colonialism. Kenya became the British East Africa Protectorate on 15 June 1885. Only two African countries were not colonised: one was Liberia, which had become independent in 1847 as a home for freed slaves, and the other was Ethiopia, where Emperor Menelik and his fighting men had held back the Italians. A question that is often asked is, why did Europe conquer Africa, rather than the other way round? Why was it that eventually Europeans took over the African continent, rather than the rhino-mounted Black Africans defeated and subjugated earlier simple European empires? This question is explored by Jared Diamond (1997) in *Guns, Germs and Steel*. In essence, Diamond suggests that Africa's long north–south axis overlying the tropics, wildly variable climate and tropical diseases impeded the spread of crops, domesticated animals and technology whereas in the colder east–west orientated lands of Europe and Asia humanity had less of a struggle against the elements. In recent centuries this enabled them to use their economic and technological advantages to conquer other continents using their 'guns' and 'steel'.

In 1896, construction began at Mombasa of what was called 'the Uganda Railway', although some British cynics called it 'the Lunatic Line'. Man-eating Lions caused trouble as the railway pushed through Tsavo in 1898 and 1899 (mentioned again in Chapter 6). On 30 May 1899, the railhead reached the swamp that the Maasai called Nakusontelon, and is now called Nairobi, and on 20 December 1901, the last rails were laid, bringing the railway to Port Florence on Lake Victoria, as Kisumu was called then, named in honour of Florence Preston, wife of Ronald Preston the head engineer. Over the next 50 years a number of white settlers arrived in Kenya. Some fighting took place in the extreme south-east of the country in the First World War, but most East African action was in what was then German East Africa, later Tanganyika and now Tanzania, as the British and their allies chased the wily and gentlemanly general Paul Von Lettow Vorbeck, who was never defeated. In the Second World War the only fighting in Kenya took place in the north-east, although there was a major campaign in Ethiopia, and a few random bombs were dropped on Malindi by Italian aeroplanes. In 1920, the East African Protectorate officially became Kenya Colony. In 1952, the Mau-Mau uprising began as the Kikuyu asserted their demands for the return of their lands and freedom and a state of emergency was declared on 20 October 1952. On 21 October 1956 a notable Mau-Mau leader, Dedan Kimathi, was captured and later hanged. The last hangings in connection with the Mau-Mau took place in 1958. By then the age of empires, for so long the norm throughout much of the world (Niall Ferguson has controversially suggested that grumbling about empires is akin to grumbling about the rain) was drawing to a close. The idea that one country or people might be in charge of another, without causing resentment, might have lasted three thousand years, but in a gradually enlightening world it had come to the end of its time. Freedom beckoned. Most African countries had gained their independence by the 1970s, with a few late exceptions. Kenya became independent on 12 December 1963. The story is that as the Kenyan flag was raised, Britain's Prince Philip turned to Mzee Jomo Kenyatta, about to become first president, and asked him 'Are you sure you want to go through with this, old chap?' History does not record the great man's reply.

THE LITERATURE

For an overall survey of Africa's history, including the early times, you cannot do better than the eight volumes of the *Cambridge History of Africa*, published in the 1970s and 1980, by Cambridge University Press. They are not cheap; initial volumes were about £40 each and they have retained their value. In places they are a little dated but they are essentially sound and each volume has sections on East Africa. Also excellent is the eight-volume UNESCO *General History of Africa*, published at around the same time as the Cambridge History, with an abridged version published in 1990. Most of the authors are professional historians of African origin, who thus brought the correct perspective to African history; much of the work was done under the able stewardship of noted Kenyan historian Professor Bethwell Ogot. Ogot's autobiography is also excellent. Entitled *My Footprints on the Sands of Time* (Anyange Press 2003), it meticulously and revealingly documents the way politics and academia were interwoven in Kenya in the years before and following independence.

If you want an effective summary, Roland Oliver and John Fage's *A Short History of Africa* (Penguin 1990) is not too dated and fits the bill. For a broad general summary from an African perspective, Professor Ali Mazrui's *The Africans, A Triple Heritage* (Guild Publishing 1986) is good value; Mazrui is a distinguished Kenyan historian who gave the BBC Reith Lectures in 1979.

East Africa has no major historical work to compare with Ajayi and Crowder's two-volume *History of West Africa* (Longman 1987). However, sound material is available. A thorough summary is provided by *Zamani: A Survey of East African History*. Published by Longman in 1968 and revised in 1973, edited by Bethwell Ogot, this is a really good work, with eighteen chapters on East African history, each by an expert in that field. It covers the lot and is only slightly dated. If you were to get just one book on Kenyan prehistory and history, we would suggest *Zamani*. Out of print, it is usually available at reasonable prices through the Internet. For continental-wide coverage, from hominins up to the arrival of writing in East Africa, a superb book is David Phillipson's *African Archaeology*, the third edition published by Cambridge University Press in 2005. Phillipson is a conservative, hard-bitten professional of African archaeology, and this book contains no speculation, just the facts in diamond-hard text. James L. Newman's *The Peopling of Africa*, published in 1995 by Yale University Press, is an excellent summary of how Africa got its present population, with clear summative maps; Newman is a professor of geography who has brought his interpretive skills to the often-mysterious movement of Africa's peoples.

Peter Robertshaw's *A History of African Archaeology* (James Currey 1990) is an exhaustive summary of the archaeological field in the continent, documenting the people as well as the fieldwork. John Sutton's *A Thousand Years of East Africa*, published by the British Institute in Eastern Africa in 1990, is a succinct summary of some interesting periods of history, particularly the kingdoms of the Great Lakes, the history of the Rift Valley and the Swahili harbour towns. As mentioned above, Jared Diamond's *Guns, Germs and Steel* (Jonathan Cape 1997) addresses the question as to why literate societies with metal tools have conquered or exterminated other societies; his chapter entitled 'Why Africa became black' is very revealing. A really superb book is *Peoples and Cultures of Kenya* by Andrew Fedders and Cynthia Salvadori. Originally published by Rex Collings in 1981, it has been reprinted in Nairobi by TransAfrica. Every ethnic group is treated, in terms of their origins, languages, where they live and their culture, accompanied by nice (but unfortunately often poorly reproduced) pictures; all Kenya's peoples are here, from the Kisii soapstone carver labouring in the sun, to the Njemps woman with her catch of tilapia, and the Sanye, shown dismembering an elephant. Every Kenyaphile should have this book.

The scramble for Africa is thoroughly documented in Thomas Pakenham's 740-page book of that name, published by Weidenfeld and Nicolson (1991). The First World War in East Africa is detailed in Charles Miller's *Battle for the Bundu* (Macmillan 1974), and the construction of the Uganda Railway (and the exploration beforehand and beginning of the colonial era afterwards) are equally well covered in Miller's book *The Lunatic Express* (Ballantine 1976). Monty Brown's nicely illustrated 1989 book on Lake Turkana and environs, *Where Giants Trod* (Quiller Press), covers most of the nineteenth and early twentieth century Western explorers who adventured in northern Kenya, and opens with some excellent material on the oral legends and migrations of the peoples of northern Kenya. A dispassionate, scholarly book on the Mau-Mau is David Anderson's *Histories of the Hanged* (Weidenfeld and Nicolson 2005); all students of Kenya's history should read this. The political scene after 1963 is exhaustively covered in Charles Hornby's book *Kenya: A History since Independence* (Tauris 2012).

The bulk of important material is found in the scholarly journals. Much has appeared in *Science* and *Nature*, or in periodicals such as *Antiquaries Journal*, *Ethnohistory*, *Current Anthropology*, *Journal of World Prehistory*, *World Archaeology*, *Journal of Archaeological Research*, and so on; a journal that has consistently produced seminal papers on humanity's rise in Africa is the *Proceedings of the National Academy of Sciences of the United States of America* (PNAS). There are several historical journals devoted purely to Africa, or East Africa, including *African Affairs*, *Journal of African History*, *History in Africa*, *African Archaeological Review*, etc. Two of particular importance to East Africa are *Azania*, the journal of the British Institute in Eastern Africa, published since 1966, and a recent arrival, the *Journal of Eastern African Studies*, which is producing some excellent and thought-provoking papers; try 'But the coast, of course, is quite different: academic and local Ideas about the east African littoral', by Pat Caplan (Volume 1, number 2), or 'In the grip of the vampire state: Maasai land struggles in Kenyan politics' by Parselelo Kantai (Volume 1, number 1), both published in 2007. The British Institute in Eastern Africa has also produced a series of excellent (albeit rather expensive) memoirs on East Africa.

From all this literature, a few papers we would single out are 'Why did modern human populations disperse from Africa *ca.* 60,000 years ago, a new model' by Paul Mellars (2006) (PNAS;103 (25):, 9381–9386), 'The genetic structure and history of Africans and African Americans', by Sarah Tishkoff and 24 co-authors (2009) (*Science* 324 (5390): 1035–1044) and 'Lake Turkana archaeology: the Holocene', by the redoubtable archaeologist Lawrence H. Robbins, published in 2006 in *Ethnohistory* (53 (1): 71–92, a readable and yet scholarly description of humanity's arrival and development in north-western Kenya. All three are available as free downloads from the journals websites.

Chapter 4

The Landscape, Climate and Weather

Riu ni thatu, no riu ringi ni mbura ya mahiga ('today it is misty, but afterwards it will rain hailstones') – Kikuyu proverb, meaning that troubles lie ahead

Siku njema huonekana asubuhi ('a good day becomes evident in the morning') – Swahili proverb

This chapter will focus on Kenya's landscape and the processes that have created it. It looks at Kenya's soils and their importance, and examines the country's climate and weather. The literature section offers a good range of publications for those wanting more detail than can be given here. The sections are self-contained and can be read separately.

THE LANDSCAPE AND HOW IT HAS CHANGED

All the varied forms [of landscape] are dependent upon … three variable quantities … structure, process and time – William Morris Davis, 1889

Landscapes are not static, they change. They are built by geological forces, by the movement of plates or intrusive igneous rocks, and they are worn down by weathering and erosion. Both are very slow processes, although their effects can be dramatic, so we intend to start with a summary of Kenya's landscapes, and a brief discussion of weathering and erosion.

Kenya's topography can be quite easily summarised. Much of the east and north is low altitude (below 1,000m), old, largely flat and eroded, covered by recent sediment, with some hard, ancient crystalline hills sticking up; the coastal strip is a narrow ancient shallow seabed. The west and south-west is an elevated plateau (above 1,000m altitude), uplifted by relatively recent volcanic activity associated with the huge rift valley that bisects the country south to north. Some hard igneous and metamorphic rocks remain as hills and mountains, and the biggest ones are angular and very young, geologically speaking. The rift valley contains a chain of lakes in basins and their levels have fluctuated dramatically in the past. Only two large rivers, the Tana and the Athi–Galana–Sabaki, drain into

KENYA RELIEF MAP

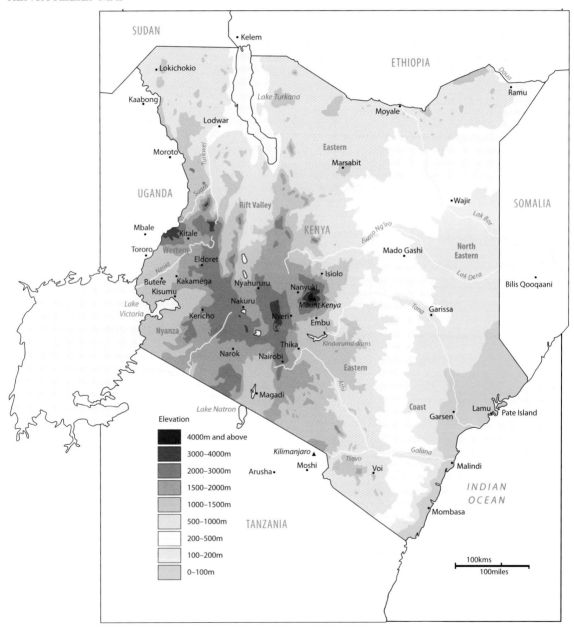

the sea, a few small rivers run down into the rift valley, others drain into a large shallow lake in the extreme west, Lake Victoria. The Kenyan geographer Francis Ojany called north and eastern Kenya 'lowland Kenya', and the south-west 'upland Kenya', an apt summary. However, this rather bald description conceals a dramatic landscape; no other African country has such a beautifully varied combination of snow-capped mountains, desert, coral coastline, rainforest and rift valley.

So the Kenyan landscape is constantly being altered by weathering and erosion; it is being denuded, to use the technical term. And the big deal about weathering is that it not only changes the appearance of the landscape, it produces, in time, sedimentary rocks and soil. Without soil, you don't get plants – or not many plants anyway – and thus you don't get food. So we need soil and we also need to conserve the soil we have.

Weathering and erosion gradually wear a landscape down. In lowland tropical Africa, where it is hot and wet, weathering can be rapid. Oddly, it can also be rapid at high altitude, where the freezing of water to ice in rock fissures can shatter them. But in general, these processes operate over huge time frames. Physical weathering gradually rounds off a landscape. Hence we conclude that the spiky peak of Mt Kenya and the angular hills and volcanoes of the rift valley have not been subject to a lot of weathering, so they are young, whereas the rounded hills of Tsavo and the Homa Bay area have; they are older. Chemical weathering decomposes rocks by chemical action. Carbon dioxide dissolved in rainwater forms carbonic acid, which dissolves limestone. It has had its effects along the coast, creating deep caves and sink holes, which then drain rainwater down into the rock. Hence small rivers are often absent, for example no stream crosses the Shimoni peninsula; the rain that falls there sinks into the ground. Even living things can aid weathering; plants can crack rocks. The odd catastrophic event may sometimes occur. Cut the trees on a hillside and you are asking for a landslip; remove too much surface vegetation and you will get sheet wash, flooding, formation of gulleys (also known as wadis, or dongas) and badlands. Occasional catastrophes like lava flows, volcanic explosions, earthquakes and tsunamis do occur. But in general, the landscape has been formed by the (relatively) gentle processes of slow uplift, weathering and erosion.

Badlands formed by soil erosion, Kerio Valley. (Stephen Spawls)

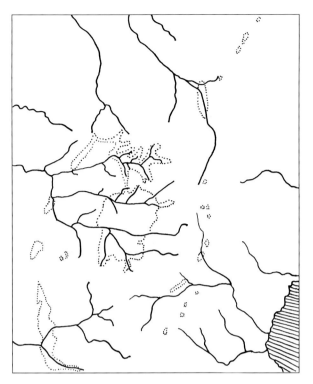

Figure 4(1) East African drainage before the final stage of rifting; the outline of the present lakes is shown. The big river across the northern half of Lake Victoria has been dubbed the Proto-Katonga, that across the southern half the Proto-Kagera

Countries need water. Kenya has more lakes than most African countries, but it has not got a lot of water. Only two big rivers reach Kenya's coastline, the Tana (710km) and the Athi–Galana–Sabaki (547km). These two are also the only permanent rivers in Kenya that are more than 300km in length. A fair number of little rivers (the Malewa, the Sondu, the Yala etc.) feed the lakes, but most are not very long. Africa is not without huge rivers, of course, with the Nile, at around 6,670km from source to delta, being the longest river in the world.

Kenya's longest river, the Tana, doesn't even get into the top ten of African rivers, but it wasn't always so. In the past, or so we believe, long rivers crossed Kenya, and some went all the way to the west coast. When the big collision took place under central Kenya, more than 500 million years ago, it created a watershed that ran up the centre of the country, with several large rivers running from the centre eastwards, out to the sea, and another set of rivers running west to the Zaire/Congo system. The central watershed persisted for a long time, scientists think, until disrupted by the rising volcanics of roughly 30 million years ago, creating new rivers, as shown in Figure 4(1). A huge river flowed from central Kenya, north through what was Lake Turkana and joined the Nile system. During the last 30 million years, as the Kenya dome rose up, the drainage pattern changed to some extent. The Gregory Rift Valley now dominates Kenya's drainage: rivers east of there drain eastwards, rivers west of there drain westwards. And across to the west, the rise of the Albertine Rift means that virtually every drop of rain that falls on Uganda and does not evaporate or get used makes its way to the Mediterranean, not the Atlantic.

Kenya actually has five riverine drainage or catchment basins (see Figure 4(2)), plus a number of blind river systems that end in swamps (we expand on these in Chapter 10 on fish). The basins are: the Lake Victoria basin (drains westwards from the Mau to the Cherangani Hills), the Rift Valley, which drains internally, the Athi River, the Tana River and the Northern Uaso Nyiro, all draining eastwards (some consider the Pangani drainage a separate entity, it is a cluster of small rivers in the extreme south-east). The southern section of the Lake Victoria catchment area includes Kenya's most high-profile river, the Mara, which runs off the Mau and flows across the flat plains of the Maasai-Mara national reserve, and thence into Tanzania. It is this river that features in one of the world's greatest wildlife spectacles, as it is crossed by millions of wildebeest *Connochaetes taurinus* in their shambling charge northwards on migration. Other rivers in this basin include the Nzoia, which the Boer trekkers crossed as they made their way up through Kenya at the end of the nineteenth century, seeking distant freedoms in wild places, and hence the land beyond was called the Trans-Nzoia, literally 'across the Nzoia', like the Transvaal, 'across the Vaal'.

The rift valley is a complex internal drainage system. No river flows out of it any more. Two major (but seasonal) rivers run north into Lake Turkana: the Turkwell and the Kerio. The Suguta River also flows northwards, but does not reach Lake Turkana; it runs into the swamps south of Lake Logipi. There are seven lakes south of there (Baringo, Bogoria, Nakuru, Elmenteita, Naivasha and Magadi, plus Natron in Tanzania). All have small feeder rivers associated with them, draining off the rift valley's internal slopes. The longest of these rivers is the

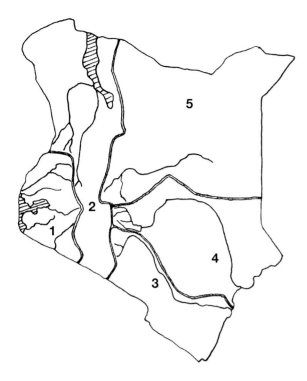

Figure 4(2) Kenya's catchment areas; 1 = Lake Victoria basin, 2 = Rift Valley, 3 = Athi River, 4 = Tana River, 5 = Northern Uaso Nyiro

Right: Flash flooding erodes the road, Kerio Valley. (Stephen Spawls)

southern Uaso Nyiro, or Uaso Ngiro (the name is Maasai, and means 'brown river', *uaso* being Maasai for 'river'), nearly 150km long, which starts on the south-western Mau escarpment. It flows south, picking up several large tributaries including the Seyabei, the river just east of Narok, passes to the west of Lake Magadi and then ends in a swamp just north of Lake Natron in Tanzania.

The northern Uaso Nyiro is the largest catchment area in Kenya but also the driest. The main river and its feeder, the Uaso Narok, the 'black river', are usually permanent in the upper reaches but the lower Uaso Nyiro sometimes dries up. This has catastrophic consequences for the fauna and flora along its banks, as visitors to the Samburu and Shaba Reserves have found; in the 1950s large Buffalo herds occurred along the river, but in 1951 the river dried up and the Buffalo died; they need to drink every day. It stopped flowing again in the mid-1980s and in 2009, with similar effects. A number of small seasonal rivers enter the Uaso Nyiro from the north. The Uaso Nyiro usually ends in the Lorian Swamp, south of Habaswein, but in really wet years may flow further to the south-east, to become the Lak Dera, a seasonal river that flows across southern Somalia. There is a large permanent river on Kenya's far north-eastern border, the Daua, (a tributary of the Juba) and a few small seasonal streams flow northwards into this.

The Tana is permanent, as befits the country's largest river, and is fed from streams on the eastern side of Mt Kenya and the Aberdares. The Tana is important in conservation: at least seven national parks and reserves are on its banks. It has a spectacular and yet rarely visited waterfall, Grand Falls, it feeds huge irrigation schemes and hydroelectric power stations, and its delta is an internationally important wetland. There are two blind rivers within the Tana catchment area, the Tiva and the Thua. The Tiva flows eastwards across Tsavo East National Park and ends in a series of swamps, much loved by elephants; in very wet years these swamps extend eastwards. Ojany and Ogendo believe that the Thua River, which also runs east, north of the Tiva, used to enter the Tana, and call the original river the proto-Thua. South of the Tana is the Athi River catchment area, fed by the rivers of the southern Aberdares and the high land south and west of Nairobi. The Athi is known by three names. It is called the Athi River in its upper reaches, after it is joined by the Tsavo River north of Voi it becomes the Galana, and it is called the Sabaki as it approaches the sea. Also very important, the Athi–Galana and its tributary, the Tsavo, supply water to Kenya's biggest national park complex, Tsavo East and West. The Pangani drainage is the smallest drainage in Kenya, essentially some small streams running off the southern and south-eastern flanks of Mt Kilimanjaro, and thence into Tanzania, but this drainage also includes Lake Jipe, and Lake Chala, on the border, Lake Jipe is suspected to have an underground connection with the Pangani system.

So, there are quite a lot of Kenyan rivers, but most are in the south-west and east. Large areas of the country are very dry, and the rivers carry very little water anyway. The Zaire/Congo discharges 1,400 billion cubic metres of water per year. The Nile, although far longer than the Congo, has a much smaller discharge, only 85 billion cubic metres, but this is because a lot of water is used before it gets to the mouth, and it flows a long way through dry lands where evaporation is rapid. The Tana discharge is nearly 300 times smaller than the Zaire/Congo, at 4.8 billion cubic metres per year, the Galana 1.5 billion. Interestingly, some of the rivers in western Kenya, although much shorter than the Galana, and with much smaller catchment areas, have relatively large discharge volumes. For example (using the same unit, billion cubic metres per year) the Sondu discharges 1.8, the Gucha 1.9 and the Mara 1.9; this is because western Kenya gets a lot of rain. The eastern river volumes also fluctuate a lot; the flow at the Tana mouth can vary from 60 to 1,100 cubic metres per second, meaning it can increase by a factor of eighteen, but this pales beside the Athi–Galana–Sabaki, which in the dry season can be as little as 0.5 cubic metres per second, and in times of high flood can surpass 750 cubic metres per second, increasing by an insane factor of 1,500. These violent fluctuations are extremely significant in terms of erosion and conservation, as well as the generation of Kenya's electricity. It is hard to control a river that gets 1,500 times bigger. Fish and crocodiles do not like it either, and change their behaviour – fish moving into backwaters and crocodiles taking food from the banks.

Kenya also has a lot of lakes, but most are small and not all are fresh. There are two significantly large ones: Lake Victoria and Lake Turkana. Lake Victoria is the world's second biggest lake, some 67,500km^2 in area, although flooding can increase this to 75,000km^2 or more (Lake Superior, in North America, the world's largest freshwater lake, is 82,466km^2). Only 5.5 per cent of Lake Victoria's area is within Kenya's borders. Most of Lake Turkana, formerly known as Lake Rudolf, lies within Kenya; its area is 6,400km^2, of which about 2 per cent lies within Ethiopia. Lake Turkana is the 23rd largest lake in the world. After that, Kenya's lakes are small. Baringo (130km^2) is the largest lake that lies totally within Kenya's borders, Naivasha is 120km^2 and Magadi, 105km^2. The rest are all less than 100km^2 and include Logipi, Bogoria and Elmenteita in the rift valley, Jipe and Chala on the border and shared with Tanzania; plus a few tiddlers such as Lake Ol Bolossat near Ol Kalou and a few glacial tarns in the mountains, including Lake Rutundu on Mt Kenya, near which Prince William proposed to Kate Middleton (continuing the association between Kenya and significant events in the lives of the British royal family). The list concludes with three big man-made reservoirs on the Tana: Masinga, Kindaruma and Kiambere. There are also a couple of seasonal lakes, flat areas that fill up in a good wet season – Lake Amboseli and the former Lake Stephanie, now Chew Bahir, shared between Kenya and Ethiopia.

As far as we know, none of these lakes existed more than 30 million years ago, and they all have stories, none more unusual than Lake Victoria. As we saw in the last chapter, the Luo settled beside it, bringing their fishing and boating skills. The story of Victoria's discovery by Western explorers is equally fascinating and is tied in with the search for the source of the Nile, which dominated nineteenth-century African exploration. Sir Richard Burton and John Hanning Speke had reached Ujiji, on the banks of Lake Tanganyika, in 1858. Burton was ill, and Speke went ahead and was shown the lake. His altitude calculations indicated that Victoria was higher than Lake Tanganyika, and he was certain this was where the Nile began. Burton did not agree and the men fell out.

Lake Victoria fishing boats, Rusinga Island. (Stephen Spawls)

Eight years later, in 1864, with the controversy still unsettled, Burton, Speke and Livingstone went to debate their theories publicly at a meeting in Bath, England. Speke unwisely suggested that if Burton turned up he 'would kick him'. Burton, on hearing of this, was determined to go. Speke became nervous; Burton was a formidable debater while Speke was a poor and hesitating speaker. On the fateful day, as Burton approached the podium, he learnt that Speke had died the previous afternoon, shot with his own shotgun while climbing a low stone wall. Accident or suicide, no one knows, although Burton is reported to have said, 'By God, he's killed himself'. There is a postscript. In the 1970s the East African countries changed many of their colonial names. Lake Rudolf became Lake Turkana. In the Albertine Rift, Lake Edward was changed to Lake Idi Amin Dada and Lake Albert to Lake Mobutu Sese Seko, but with the increasing unpopularity of those two men both lakes have now reverted to their original Western names. A group of East African academics debated changing Lake Victoria, but could not agree on a name: the Kenyans wanted to call it Lake Kenyatta, the Tanzanians Lake Mao Tse Tung, the Ugandans Lake Gadaffi. Perhaps it's fortunate they couldn't agree.

Controversy still surrounds Lake Victoria. It is very shallow, with a maximum depth of 69m; compare this with Lake Tanganyika, which is nearly 1.5km deep. A chart of Lake Victoria, surveyed by Commander Whitehouse and Mr C. S. Hunter, of the British Admiralty, around the turn of the nineteenth century, indicates nowhere deeper than 65m and most of the Winam Gulf at less than 25m deep. In those days, the Winam Gulf was called the Kavirondo Gulf (the term Kavirondo, told to Western explorers by their Swahili porters, is said to mean either 'people who sit on their heels' or 'people who stand on one leg', a mild abuse of the Luo, hence the subsequent change of name). The shallowness of the lake is connected with the fact that it is not very old. It is simply flooded savanna. Rivers flowed across the area where the lake is now, draining into the Zaire/Congo system, shown in Figure 4(1). But starting 10 million years ago the hills bordering the Albertine Rift were gradually uplifted, and by 800,000 years ago the rivers could no longer flow west and a lake began to form. Its irregular spiky shape is absolutely distinctive of flooded land; look at the map of Emin Pasha Gulf or Smith Sound. Victoria then overflowed, into Lake Kyoga, which is also clearly a series of interlocking drowned valleys, thence to Lake Albert and out to the north, to boost the Nile. Some weird effects resulted from this uplift. The angles that their tributaries join them on the eastern side of the watershed suggest that two Ugandan rivers, the Kafu and the Katonga, are flowing westwards to Lake Albert and Lake George. In reality, they are actually flowing east. Their direction was reversed by the upwarping of the land. As a result, along the course of both rivers there is a swampy region, on the west side the rivers flow west, on the east side of the swamp they flow east. Not many countries can boast a river that flows both ways.

It does not stop there. Studies by Tom Johnson and his colleagues (including the wonderfully named Ugandan geologist Immaculate Ssemmanda) have shown that some time between 18,000 and 14,000 years ago, during a time of aridity, Lake Victoria dried out completely, and was dry for more than 500 years. The cores that Johnson and his colleagues brought up from all over the lake bed showed a paleosol – that is an ancient soil that had been exposed to the sun – 15,000 years old. If the soil was exposed to the sun, the water must have gone, and the flow into the Nile stopped. There is no arguing with this, although a few scientists are unhappy with the implications – that the 700-odd endemic species of cichlid fish (Tilapia and their friends) known from Lake Victoria really started to speciate only in the last 15,000 years. This is a remarkable story of evolution, complicated by the fact that not only has the lake changed from a mesotropic lake (low productivity and dominated by diatoms) to a eutrophic lake (high productivity, dominated by cyanobacteria). The introduction of Nile Perch into the lake affected the ecology (see Chapter 10).

There is also reputed to be a monster within the lake, a creature known to the local people as the Lukwata. Tales abound of its ferocity. According to C. W. Hobley, an official from the foreign office, Sir Clement Hill, saw it just after it had attempted to grab one of the crew of the boat he was travelling on, in the Winam Gulf near Homa Mountain. Hill said it had a small head and a huge neck. Hobley also tells of a similar monster, the Dingoneck, seen on a tree trunk floating down the Mara River; he described it as 16ft long, covered with scales and with a head like an otter. The likely candidate in both cases might just be a large python, seen under frightening circumstances; there are some big pythons in western Kenya. A bevy of similar tales has been gathered together by Bernard Heuvelmans (1995) in his entertaining book *On the Track of Unknown Animals*, in a section entitled 'The Terrors of Africa'.

Victoria is a big lake; despite its shallowness, there is enough water in it to cover all of Kenya to a depth of more than 4m. The depth can fluctuate a lot; between 1960 and 1965 the lake rose 3m, and the subsequent flooding of the surrounding areas, especially the swamps of the northern and western shore, increased the effective size until it was temporarily larger than Lake Superior, and thus gained brief notoriety as the largest freshwater lake in the world. The water is fresh, as the lake has both inflows and outflows, the salinity is given as 0.093 parts per thousand, this means less than 0.1 part salt per thousand parts of water (sea water is 35 parts salt per thousand parts water, fresh water is defined as having less than 0.5 parts salt per thousand parts water. A lot of evaporation occurs in the lake; 85 per cent of its water escapes that way, the rest flows down the White Nile (which, incidentally, only contributes 10 per cent of the water that reaches Egypt, the other 90 per cent comes from Ethiopia via the Blue Nile). The evaporation and huge surface area means Lake Victoria is big enough to create its own weather systems. The surrounding hot land, warming up during the day, creates thermals that suck moist air off the lake onto the land and afternoon storms result. Compared with temperate lakes, which can get very cold in the winter, there is very little temperature fluctuation, as you might expect with a lake lying on the equator. The water temperature remains at a steady 24–26°C near the surface almost throughout the year; although naturally it cools as you go down, but not by very much, it rarely drops below 23.2°C.

The volume of dissolved oxygen also varies considerably: near the surface it is around 7 milligrams per litre, at the bottom it can drop to 1 or 2mg/l. This is very low compared to temperate lakes, which are colder and can thus dissolve more gas; concentrations of 18mg/l in cold temperate waters are not unknown. This is different from solids, where the hotter the liquid the more solid will dissolve; with gases, the hotter the liquid the less gas will dissolve; which is why fish tend not to be abundant in warm water. The actual concentration of oxygen changes greatly during the 24-hour cycle; by day it is boosted by aquatic plants, which produce oxygen as they photosynthesise. At night no photosynthesis occurs – it needs sunlight – and the oxygen concentration drops. However, Lake Victoria has a rich aquatic fauna, and this is connected with its shallowness. The oxygen concentration is boosted by water mixing due to wind. Although in tropical Africa winds speeds tend to be relatively low, compared with temperate areas, which have greater temperature variation and hence greater pressure variation, a wind speed of about 10m/s (36km/hr, 26 knots) over a fetch (open water distance) of more than 50km will cause satisfactory mixing to a depth greater than 30m. Lake Victoria is also subject to occasional powerful storms, as well, which assist with water mixing and oxygenation.

There are some strange dangers in tropical water bodies. Lack of wind can be fatal. In warm eutrophic (nutrient-rich) lakes, there is a risk of oxygen depletion if the air is still and the upper layer contains lots of algae and other organisms that stop light penetrating to any depth. So after a few calm hot days, a situation can arise where only a thin upper layer of water contains oxygen, and down below the water is deoxygenated. Then the fish have to swim near the surface. A situation like this arose in Uganda's Lake George in October 1957, when a calm windless period was followed by a violent storm. This caused thorough mixing of the deoxygenated water below with the very thin oxygenated layer at the top and the overall result was that there was not enough oxygen to support life anywhere in the water. The fish were seen gasping at the surface and then they died; over 1.3 million large edible fish perished and countless millions of smaller ones. However, this is unlikely to happen in Lake Victoria. Although there are some fears about the effect of the blue-green algae (correctly called cyanobacteria), which has increased a lot in the last few decades, the lake is big enough to create its own weather, with wind, rain and sharp storms; this should mean that the aquatic living things that breathe dissolved oxygen are in a safe environment, hopefully.

Lake Turkana is Kenya's desert lake, and with the exception of the outlier of Lake Chew Bahir or Stephanie, is the northernmost lake in the rift valley chain; all of the Kenyan rift's lakes were formed after rifting commenced,

Tide of blue-green algae (which are really bacteria), Lake Victoria. (Stephen Spawls)

creating basins that trapped water. The lakes of the rift valley are a consequence of the uplifted land that lies on their eastern and western sides. Few have any sort of outflow, and many have what Thomas Schluter describes as 'abnormal' concentrations of ions, in particular metals such as sodium and calcium, and halogens such as fluorine and chlorine. Lake Turkana was formed more than 20 million years ago. Known to the Samburu as Bassu or Basso Narok (black lake) and to the Turkana as Anam, in the dry north-west, at an altitude of 375m, this long narrow lake gets less than 250mm of rainfall per year; technically it is surrounded by desert. The mean maximum temperature on its shores is over 30°C (often over 35, and it can peak at 40), and its mean minimum is over 20°C. It is in a trough inside the rift; no rivers run out of it. One big permanent river (the Omo, in the north) and two big seasonal rivers, the Turkwell and Kerio, run into it, plus a few smaller seasonal ones; the Turkwell is fed by the Suam, which runs off Mt Elgon. When the Omo floods, carrying down water from the Ethiopian highlands after July, the lake level can rise by as much as a metre. Baking hot, shadeless and harsh, the haunt of gigantic crocodiles, scorpions, Red Spitting Cobras *Naja pallida*, abundant Carpet Vipers *Echis pyramidum*, hyenas, lions, tough nomads and bandits, Turkana is a lake of danger and mystery, and has attracted some remarkable visitors. Count Teleki and Ludwig Von Hohnel got there in 1888, and shot a number of elephants that had been forced down to the lake by drought. The elephant hunter Arthur Neumann walked up the eastern shore in 1885, and found what he thought was a new hartebeest, but it later proved to be just a variety of the Lelwel Hartebeest. He also lost his right-hand man, Shebane, to a crocodile on the Omo. The explorer Sir Vivian Fuchs led an expedition to Lake Turkana in 1934, to study the geology and natural history of the area. Two of his team, Dyson and Martin, landed on South Island, and signalled by fire to the mainland, that they had arrived, but neither man was ever seen again. George Adamson, lion expert and game warden, visited the lake, also in 1934, with Nevil Baxendale, searching for the legendary 'Queen of Sheba's mines'. With a home-made boat, Adamson and Baxendale crossed the lake; and managed to make landfall after dark just before one of the violent storms that regularly plague the lake got up (although the Kenyan game warden Ian Parker, in his book about sailing round Lake Turkana, is sceptical of Adamson's story). In 1955, Adamson returned to the lake with his wife, Joy, and managed to land on South Island, locally known as the 'island of no return' (the Samburu believed it was haunted by demons). It turned out to be haunted with Speckled Sand Snakes *Psammophis punctulatus* and Red Spitting Cobras; George found a cobra in Joy's bed. South Island also proved to be a major breeding spot for crocodiles and there were herds of goats on the island. The Adamsons found traces of Dyson and Martin's camp, and some of their equipment.

The journalist John Hillaby, author of a number of books about walking, made a safari from Wamba to the eastern shore Lake Turkana in 1963 and later published a most entertaining book, *Journey to the Jade Sea* (1964). The lake has this name because in the late morning, after the wind has dropped, algae rise to near the surface and turn the lake green; at other times when the wind is up the lake is sapphire blue. Hillaby was accompanied for part of his walk by the game warden Stan Bleazard, who later claimed that Hillaby liked walking naked, an idiosyncrasy on which Hillaby himself is curiously silent. The retired BBC journalist Tom Heaton, who settled in Kenya, at Ngong, walked in stages from Tanzania up through Kenya and went up the eastern lake shore, following Count Teleki's route, hence his book (Heaton 1989) is called *In Teleki's Footsteps*. Heaton met, worked with and travelled with ordinary and extraordinary Kenyans on his safari, and met the frightening bandits known as *ng'oroko*. His book gives wonderful insight into Kenya as a whole. John le Carré's thriller about skulduggery in the pharmaceutical industry, *The Constant Gardener* (2000), opens with a murder at Lake Turkana and finishes with one. Lake Turkana was also the scene of some of the most astonishing debacles in the history of aid: 30 million dollars of Norwegian aid money funded a fish-freezing factory, a road and a fishing vessel, which never functioned, 200 million dollars went on assisting a few thousand Turkana farmers to each grow about 100 dollars worth of vegetables for a couple of years, before the scheme failed. It was calculated that if the money had simply been given to the farmers, instead of setting up a massive non-functional scheme, each one could have had 25,000 dollars.

The saga of Kenya's desert lake has by no means ended; at present a dam is being built on the Omo, in Ethiopia, which will take a while to fill. It has been suggested that this could lead to a dramatic drop in the level of Lake Turkana, with unforeseeable consequences for the natural history of the area.

So much for humanity's attempts to make sense of this savagely beautiful region. Some facts. The lake has an area of 6410km², (according to Ojany, but A. C. Hamilton reckons it to be 7,500!) and is 120m deep at its deepest point. The water is surprisingly alkaline, about 9.5 on the pH scale. This gives the water a 'soapy' feel. The salinity varies from 1.7 to 2.5 parts per thousand. This fits the definition 'brackish' and the water tastes odd but is drinkable. Theoretically, seeing the lake has had no outlet for thousands of years, it should be much more saline, so water must be seeping out somewhere, possibly into the flats west of Ferguson's Gulf. In the past the lake has fluctuated considerably in height, up to 80m higher and 40m lower; the water level fell by 10m between 1975 and 1993. It has had outflows; it flowed into the Indian Ocean more than a million years ago, and has been connected to the Nile more recently. The limnologist Tom Johnson has calculated that in the last 200,000 years Lake Turkana has been in contact with the Nile at least six times; it may just have been part of a huge river that flowed from central Kenya northwards to the Nile. Johnson got his data from raised beaches and lake sediments, but other evidence is also available. In the lake live the Nile Soft-shelled Turtle *Trionyx triunguis*, the Nile Perch *Lates niloticus* and the Tiger Fish *Hydrocynus forskahlii*, all three species occur in the Nile and are not naturally known from anywhere else in Kenya (although the perch was introduced into Lake Victoria).

The lake is also subject to violent windstorms; it was probably a storm that drowned Dyson and Martin and nearly caught George Adamson. There is a reason for these storms. The bare land surrounding the lake gets violently hot in the day and very cold at night but the water is usually fairly steady at around 29°C. So during the day air rises rapidly over the land and a strong onshore breeze blows, at night the air rises rapidly over the water and a strong offshore wind blows. Deadly to sailors, these winds do thoroughly oxygenate the water from top to bottom, and this benefits the abundant fish fauna. The strong winds also prevent the growth of aquatic plants, which are confined to sheltered bays.

Just south of Lake Turkana is a curious small salty lake with a high pH, Lake Logipi, with no outlet. It was much bigger about 10,000 years ago, and was part of what the geologist Thomas Schluter calls Paleolake Suguta, a big lake 140km² in area and 200m depth, which extended from the barrier volcanoes south to the western flank of the Samburu Hills. It was a wetter and cooler time; rainfall was high and sediment cores obtained by drilling from the vicinity of the huge ancient lake contain pollen of junipers and *Podocarpus*, indicating there was cool montane forest in the vicinity. There is none there now. Lake Logipi was reached in 1897 by the dashing young English explorer and old Etonian Henry Cavendish, who named one of the barrier volcanoes there 'Andrew' after Lieutenant Andrew, who was his travelling companion. In 1929 George Adamson made a road in the Suguta Valley, and observed Lesser Flamingos *Phoeniconaias minor* breeding at Lake Logipi. He also recorded a shade temperature of 120°F, which is just under 49°C, if he had reported this to the meteorological authorities it would have trumped Kenya's highest official temperature, which is just over 46°C, at Garissa.

Lake Baringo lies a good way south of Lake Logipi, between the Tugen Hills and the eastern rift valley escarpment. Baringo is actually the largest lake that lies entirely within Kenya (most of the time, although Naivasha can flood to a larger area), it is 130km², and at the low altitude of 975m. It is not very deep; the average depth is about 6m, the maximum is 8m. Gregory, whom we met in Chapter 1, was one of the first Europeans to reach Baringo. The lake has observable outflow, although several rivers, including the Mukutan and the Molo, flow in. However, Baringo is fresh, although slightly alkaline at pH 8.8, so the water must leave somewhere, presumably through faults in the volcanic base, as below the sediments is trachyphonolite lava; it must have cracks. Geothermal heat is present under the lake, the hottest hot springs in Africa are on Ol Kokwe Island; they blast out water at temperatures of 94–97°C.

The Njemps, or Il-Tiamus people live on the lake shores; they are Nilotes who speak a language very similar to Maasai. Gregory found them a friendly people. Unlike most Maa-speakers, they eat fish which they catch from little boats with a very small draught. These boats are made from saplings of Ambatch or Balsa Wood, *Aeschynomene elaphroxylon*, a tropical African tree with a very light wood found around Lake Baringo; it actually grows standing in the water. The wood is less dense than cork. The fishermen compete with large numbers of relatively small Nile Crocodiles *Crocodylus niloticus* that also call Lake Baringo home. Baringo also

African Spoonbill and troublesome water hyacinth, Lake Naivasha. (Stephen Spawls)

has Hippos, and there are Hippos in Lakes Turkana and Naivasha, but they are rarely present in the smaller lakes, which are too salty.

Set in a natural depression not far south of Baringo is Lake Bogoria, formerly Lake Hannington, originally named in honour of Bishop James Hannington, killed in Uganda in 1885 on the orders of the then Kabaka (king), Mwanga. Bogoria is a small shallow lake, nowhere deeper than 12m, of area 34km². It lies in a harsh, hot landscape; the western shore slopes gently to the lake but the huge Siracho Escarpment dominates the eastern shore. Bogoria has no outlet; it is fed by the Waseges/Sandai River, which enters a swamp at its northern end and by groundwater from Lake Naivasha, via a number of hot springs. On the southern half of the lake there are more than 200 hot springs, plus geysers and steam jets, in fact eighteen geysers have been documented, the highest concentration anywhere in Africa. It is a highly alkaline and saline lake. The water is meromictic, meaning it is stratified, not by temperature but by salinity; as you get deeper it gets more saline and the lowest level has a salinity of 100 parts per thousand, that is ten parts salt to 90 parts water, which is far more salty than sea water and fits the definition of brine.

Between Baringo and Bogoria there are a number of swamps, and it appears that in the past the lake levels have varied considerably; it has been at least 9m higher and a lot fresher, and might have been united with Baringo. The high salinity means there is little vegetation and very few freshwater arthropods live there, although there are good growths of *Arthrospira* cyanobacteria, which the lake's flamingos eat. The daytime temperature on the shores can exceed 42°C and visitors have been known to describe a visit to Lake Bogoria during the heat of the day as a 'descent into hell'. It was proclaimed as a national park in the 1970s, primarily to protect the flamingos that breed there, although there are also some Greater Kudu *Tragelaphus strepsiceros* (a rare species in Kenya) living on the Siracho Escarpment. There are a few African Fish Eagles *Haliaeetus vocifer* there, which in the absence of fish have taken to eating flamingos.

The most recent chapter in Lake Bogoria's history concerns accusations of biopiracy, which conjures up a vision of a pirate with an eye patch, cutlass and lab coat. In 1992, ornithological researchers found ancient Archaean

Lake Nakuru, the high water level and saline crust clearly visible. (Stephen Spawls)

bacteria, capable of withstanding high temperatures, in the water. Such tough organisms are known in the trade as extremophiles, and are unusual in that their enzymes can function at very high temperatures. All living organisms have enzymes, they speed up reactions and keep you alive, but most are destroyed at temperatures above 45°C. This is why your body reacts violently if a fever takes you near that maximum. If you didn't convulse, sweat and cool, you would die. Thus an enzyme that functions at temperatures well above 70°C is pretty unusual. The lake bacteria turned out to be able to digest biological material, and thus remove stains, soften fabric and eat indigo dye, all at high temperatures; they were thus usable in a washing machine. A California-based company, Genencor, purchased the enzymes, and used them in a washing product that created the 'stone-washed' or' faded' look in jeans beloved of fashion icons. The company made a lot of money but Kenya did not benefit. The Kenya Wildlife Service subsequently launched a lawsuit, stating that the researchers had had no permit to take samples, but no money has been forthcoming yet.

South of Bogoria is Lake Nakuru. The meaning of 'Nakuru' depends on who you consult: some say it is Swahili for 'place of waterbuck' (*kuru* or *kuro* is Swahili for that antelope), others say it is a corruption of the Maasai expression *en-akuro*, place of dust devils. The lake is surrounded by the most-visited national park in Kenya, but it has had an even grander history; in the past it was part of a massive lake that stretched from the southern slopes of Menengai Crater to the northern slopes of Eburru, a lake that was more than 200m deep, and incorporated Lake Elmenteita. It absolutely dwarfed the puddles that remain today; the maximum depth of Lake Nakuru now is about 3m and it averages 2m, Elmenteita is no deeper than 3m anywhere. A fishing community lived on the shores of the giant lake. But both Elmenteita and Nakuru have dried up in the past, so in some ways we are lucky to have them, even if they are but a shadow of their former size. The guidebook to Kenya's important bird areas says the lake is 33km^2 in area, but earlier records indicate 45km^2; and it can shrink to zero. Nakuru is saline, averaging 45 parts per thousand, saltier than seawater. Menengai crater, overlooking the lake, was the site of a battle around 1850, between the Laikipiak and Il-Purko Maasai clans, ostensibly over the Laikipiak clan's disrespect for the *laibon* (seer) Mbatian, but also over grazing rights. The story is that many of the defeated Laikipiak Maasai were hurled over the edge of the crater, and consequently some Nakuru residents are superstitious about the crater rim and will not approach it.

There is a superb (although somewhat artificial) national park completely surrounding the present lake. It was set up to protect the remarkable bird life of the lake, in particular the flamingos, which do not breed there but feed on cyanobacteria. It is also a Ramsar site, recognised as a wetland of international importance (Ramsar is the Iranian city where the convention was signed). According to Beadle (1974), in *The Inland Waters of Tropical Africa*, the lake level has been low for the last 4,000 years, but was high between 10,000 and 7,000 years ago, as mentioned. In the early 1970s the lake was solid with algae, a bloom of *Spirulina platensis minor* algae accounted for 98 per cent of the biomass of the primary producers (photosynthesising organisms, i.e. plants and algae) of the lake and 94 per cent of the total biomass (weight of all living things) in the lake. There obviously were many fish in the lake when our ancestors lived there 8,000 years ago, but during the twentieth century there were none, until 1953, when the heat-tolerant, algae-eating Tilapia, *Alcolapia grahami*, was introduced from Lake Magadi (and re-introduced in 1959 and 1962), and these fish did well, providing food for a huge number of Great White Pelicans *Pelecanus onocrotalus*. In 1974, the level of the lake fell by about half a metre, causing a rapid increase in the salinity; this slaughtered the *Spirulina* algae, and dramatically changed the relative proportions of the arthropods living in the lake. This has continued to be the story of Lake Nakuru, because it is small, shallow and 60–70m below the level of the town. In the 1970s and 1980s pollutants from Nakuru town (which is Kenya's fourth largest and a hive of light industry) were cheerfully pumped into the rivers entering the lake, and the lake itself. This has been controlled, but this is a fragile ecosystem with a small volume lake, a lot of wildlife and a huge town in close proximity. The usual Kenyan problems continue to apply; the latest threat to this protected area is a proposal to drive a bypass across the north end of the park to relieve traffic congestion in the town.

Sir Charles Eliot liked the next lake south of Nakuru, and wrote 'Of the three lakes ... by far the most beautiful is Elmenteita'. Eliot was governor of the protectorate between 1900 and 1904 and his book, entitled, not unsurprisingly, *The East Africa Protectorate*, is a charming tome, packed with information. Eliot comes across as a thoughtful and sensitive man, which is at odds with Charles Chenevix Trench's description of him in *Men who Ruled Kenya* (1993) as 'cold and reserved'. Lake Elmenteita is just over the ridge south-east of Lake Nakuru and is similar to that lake in several ways. It is shallow, with lots of cyanobacteria, acacia woodland and feeding flamingos. Its name is derived from *muteita*, or *Ol-muteita*, a Maasai word meaning 'place of dust'. Joseph Thomson seems to have been the first Westerner to see this lake, in 1883. Ten thousand years ago, at the time when Lake Nakuru was at its highest, Elmenteita would have been 160m deep, part of the superlake. Now it is nowhere deeper than 2m and is alkaline and highly saline, 43 parts per thousand or 4 parts salt per hundred parts water, and with a pH of 10.5. It occasionally dries up, but is usually about two-thirds the size of Lake Nakuru. The lake is fed by two little rivers, the Kariandusi (which rarely flows) and the Mereroni. At the southern end, the Kekopey hot springs flow into the lake. Kekopey is a Maasai word meaning 'white through green', a reference to the thick white deposits of diatomite that are mined in the area and exposed in the gulleys. Diatomite is a soil or unconsolidated sediment made up of the silicon dioxide skeletons of diatoms, tiny single-celled algae; diatomite forms as a sediment on the beds of algae-rich lakes. It has a number of uses: as a filter, an abrasive and an unusual insecticide (it sucks water from insect exoskeletons, so kills them without contamination) but its most high-profile use was in dynamite, as discovered by Alfred Nobel. Before dynamite, the explosive of choice was nitro-glycerine but it was horribly unstable and liable to go bang if dropped. Nobel discovered that if you absorb nitro-glycerine into diatomaceous earth you get a stable mixture, dynamite, that will only go bang if you use a detonator. Fascinatingly, the profits he made from dynamite funded the Nobel Prizes. The white diatomite beds can be nicely observed at the Kariandusi Prehistoric Site, just east of the lake.

The western side of Lake Elmenteita is part of the Soysambu estate, acquired by Hugh Cholmondeley, 3rd Baron Delamere in 1905. His descendants still live there and the lakeshore is maintained as a wildlife conservancy; a beautiful place with a lot of game and acting as an important adjunct to the Lake Nakuru National Park. The

Diatomite, stabiliser for dynamite, from the bed of an ancient lake, Kariandusi. (Stephen Spawls)

farm Kekopey, surrounding the springs, was owned by Lord Delamere's brother-in-law, Galbraith Cole. The story of Cole's adventures and the remarkable people associated with the farms during that period are well told in Elspeth Huxley's beguiling book *Out in the Midday Sun* (1985). Like Lake Nakuru, Lake Elmenteita has also been declared a Ramsar site. When the lake levels are high, it creates small rocky islands on the north-west shore and up to 8,000 Great White Pelicans have bred there, as have both flamingo species, but in small numbers. There are no fish in the lake, and thus the pelicans fly to Lake Nakuru each day to feed. A waterbird count in 2008 found in excess of 39,000 Great White Pelicans and 220,000 flamingos there. Although the western shore is pretty well protected by the conservancy, the same cannot be said for the eastern shore; there is no nearby town but the lakeshore is threatened by development, deforestation, effluent from a timber-treatment plant and unregulated mining; all the usual factors of a land-hungry country. The Important Bird Areas book states hopefully that 'tourism … is a force for conservation' around Elmenteita. Let's hope so.

Lake Naivasha is the highest of the rift valley lakes, at 1,884m altitude; it is cold there at dawn. There are actually three lakes in the area: the main lake, a small lake on the south-western shore, Oloiden, and a tiny lake inside an explosion crater, the Crater Lake. The deep pool enclosed by Crescent Island is occasionally isolated and regarded by some as a separate entity. Dramatic mountains surround the lake: the Aberdares rise up to the east, Mt Longonot to the south, while the rugged slopes of the Mau and Eburru tower above the western shores. Naivasha is a beautiful freshwater lake of importance to Kenya; to be out there in the yellow and green acacia woodland on the shores, with a pair of binoculars under a clear blue sky, with the call of the African Fish Eagle ringing through the timber, is a magical experience. Naivasha supports a large agricultural industry, it is a major site for tourism (the birdwatching is among the best in the world, with over 350 species recorded), and it supplies, by seepage, groundwater to lakes Magadi, Elmenteita, Nakuru and Bogoria; a volume of up to 45 million cubic metres has been estimated to flow out underground per year. It is this that keeps the lake fresh as no river runs out of it. The groundwater also feeds the nearby geothermal power plant in Hell's Gate. And no other lake, or

Lake Naivasha, beyond it the great mass of Eburru volcano generating weather. (Stephen Spawls)

indeed site, in Kenya, has suffered as much human interference and daft introductions as the lake that the Maasai knew as Enaiposha (and said to mean swampy, rough or rippling water, take your pick). The Athi River Tilapia *Oreochromis spilurus* was introduced in 1926, in 1928 the American sport fish Black Bass *Micropterus salmoides* were introduced (following the earlier suggestion of Teddy Roosevelt, it is rumoured), both disappeared when the lake was low in the 1940s. In 1953 the Blue-Spotted Tilapia *Oreochromis leucostictus* was introduced and in 1955 the Redbelly Tilapia *Tilapia zilli*. In 1970, Louisiana Red Crayfish, *Procambarus clarkii*, were released in the lake. A kaleidoscope of vividly coloured strangers. In the 1960s, some Coypus, *Myocastor coypus*, big South American semi-aquatic rodents being commercially farmed on the Kinangop for their fur (known as nutria), escaped and made their way into the lake. These wreaked havoc with the cane and lucerne fields and in an attempt to control them, a number of Rock Pythons were released. In the 1980s, Bat-eared Foxes *Otocyon megalotis* got onto Crescent Island, on the lake's eastern shores, and wiped out a population of Kenyan Striped Skaapstekers *Psammophylax multisquamis*, attractive grassland snakes, whilst the extensive agriculture on the southern shores has decimated the formerly thriving population of Mole Snakes *Pseudaspis cana*, big harmless and useful snakes; humans and big snakes – even harmless ones – do not co-exist well.

It continues. In 1961, the floating aquatic fern *Salvinia molesta* appeared in the lake, and proceeded to cover large areas of the water surface. In an attempt to control this weed, a semi-aquatic South American grasshopper, *Paulinia acuminata* was introduced; the salvinia was also sprayed with herbicide. This met with limited success so a South American weevil was introduced, which was more successful. In 1988 the South American water hyacinth (genus *Eichhornia*) appeared, and wind-driven rafts of this invasive plant now choke large areas of the south-western lake. Two more foreign weevils were introduced to deal with the hyacinth. The shores of Naivasha were originally shrouded by papyrus plants and had many water lily lagoons. In the 1960s the Purple Swamphen *Porphyrio porphyrio* was common in the reedbeds. The swamphens have gone now, and the volume of papyrus and water lilies has fluctuated considerably; at present both have almost gone.

The development of huge flower and vegetable farms on the southern shore have, without question, brought employment to the area and economic benefits to Kenya, but they have also caused large amounts of vegetation to be removed, various alien chemicals introduced to the lakeshore and their demands for water, directly from the lake and via boreholes, has brought the lake levels down. To add insult to injury, the only indigenous fish in the lake, a species endemic to Lake Naivasha, the Naivasha Lampeye, *Aplocheilichthys* sp., was exterminated. It was presumably eaten by the introduced fish, but no one knows; sadly it was never given a formal scientific name as it was confused with another species, *Aplocheilichthys antinorii*, an Ethiopian species. John Gaudet and his colleagues from the botany department at the University of Nairobi made a major study of the lake in the 1970s and early 1980s and noted that the plant and aquatic invertebrate life was very diverse, unlike the fish fauna; indicators that the surface water may have dried up in the past but the ground remained essentially moist. At present, the lake's future hangs in the balance and if the ecosystem collapses then everybody loses. Many people who flocked to the area because of the horticulture, finding no work, resorted to fishing, and the lake's fish population is endangered. Some took to violent crime; many people have been murdered at Naivasha. In 2010 top Kenyan birdman Don Turner wrote movingly of the lake's plight in *Swara*, the magazine of the East African Wildlife Society. The heavy rains during early 2011 restored some of the water.

Naivasha is quite a large lake, varying between 120 and 195km^2; at maximum it becomes the biggest lake that lies entirely within Kenya. It has a distinguished history. It was part of a gigantic lake that, at its highest, stretched from the vicinity of Ilkek Station, 15km north of the present northern shore, south to Hell's Gate. This superlake was over 120m deep. There is some debate about whether it was connected with the superlake around Nakuru; informed opinion is that it wasn't. Naivasha is now rarely deeper than 6m anywhere, with the exception of the Crater Lake formed by the curve of Crescent Island, which is 18m deep in places. Some time between 9,000 and 5,600 years ago, when rainfall was 50 per cent greater than today, the lake overflowed, down Njorowa Gorge in Hell's Gate, creating the magnificent pass that you can drive through today; if you go there imagine a huge river swirling south from Naivasha past the basalt cliffs. Some 3,000 years ago the lake disappeared and remained dry for several decades, and Maasai legend tells that Naivasha also dried out for a few years around 1840. But it was high near the end of the eighteenth century; C. W. Hobley (1929) describes being forced to climb the cliffs, where the railway runs today, to avoid the water as his party made their way up the eastern shore in 1894. The cliff base is around 1,900m, indicating the lake was 10 or so metres higher than it is today.

Since the 1960s the level has fluctuated by up to 4m, but the general trend is downward. The horticultural industry actually occupies more than 100km^2 of the former lake bed. As mentioned, the lake has no surface outflow, the bulk of incoming water enters the lake via the Malewa River, which drains off Kipipiri Hill and the south-western flank of the Aberdares; this flow is threatened by a take-off intended to supply Nakuru with water. A lot of smallish streams run off the Mau escarpment and these all sink into the volcanic soil of the western shore, and seep into the lake. Experts have calculated that the horticultural industry and the geothermal plant at Olkaria in Hell's Gate use the equivalent of a year inflow from the rivers; couple that with evaporation and you have got net loss of water. And despite the presence of big water towers to the east and west, the lake and immediate surroundings actually lie in a pocket of erratic, low rainfall, less than 500mm per year. No wonder the lake level is declining.

Before westerners arrived, the Maasai fought over the area, with battles between the Uasin Gishu and the En-Aiposha clans. Joseph Thomson got there in 1883 after climbing Longonot, Gustav Fischer was there earlier, both men had problems with the Maasai; those warriors actually turned Fischer back. The lake lay slap on the border of the East Africa Protectorate before 1902. All the land between Naivasha and Lake Victoria was within what was called the eastern province of Uganda; it was Eliot who brought that land into the protectorate. The government then set up an experimental zebra farm on the shores, but the stock died of parasite infections. Between 1937 and 1950 the lake was used as a landing site for the huge flying boats of Imperial Airways, on their trans-Africa flights from Southampton to South Africa. Joy Adamson lived on the shore for a number of years and her home, Elsamere, named after the famous lioness, is preserved as a museum and is a delightful place to stay. A number of

Kenya's prominent conservationists live or have lived on Naivasha's shores, and there are some fine resort hotels, the oldest of which is the famous Lake Hotel, now the Lake Naivasha Country Club. These havens have protected some of the lovely acacia woodland and pastures of the shores.

South of Naivasha, the rift floor drops down, and dry country flora and fauna appear. Kenya's southernmost rift valley lake is Magadi. Like Bogoria, it is blisteringly hot and is at low altitude, 579m (other sources give its altitude as 590, and 604m), in the southern rift valley. It is quite a large lake, roughly 100km^2, lying in a graben, surrounded by low ridges and semi-desert, and it can be fairly said that Magadi is Kenya's most unusual lake, for a number of reasons. No permanent river enters or leaves it, only a few temporary streams, but there are over fifteen hot springs that pump their chemical outflow into the lake, some at temperatures of close to 90°C. These springs are fed by groundwater seeping down from Lake Naivasha. The southern Uaso Nyiro drains the Nguruman escarpment and runs close by the lake, but continues south into the swamps on the north side of Lake Natron. The depth of Lake Magadi is unknown but it is shallow in most places and during the dry season, a pinky-grey chemical crust lies above most of the water, a crust so thick you can drive across it. This crust is largely a mineral called soda or trona, a 'double salt' mixture of sodium carbonate and sodium hydrogen carbonate; older books call it sodium sesquicarbonate. The word *magadi* is Maa for trona, and Maasai men used to add it to their powdered tobacco to make a blinding snuff.

This mineral is important to Kenya. Gustav Fischer, that secretive German explorer, was the first Westerner to reach Magadi, in 1893, and by 1903 a group of influential British and South African businessmen, the East African Syndicate (men who clashed with the Governor, Eliot, and were instrumental in his resignation) had investigated Magadi's mineral potential and they liked what they saw. Between 1911 and 1913 a spur to the main Uganda Railway was constructed, from Konza down to the lake, to truck out the trona. It continues today: most days two trains each carrying 700 tonnes of trona go out. Tension still exists over the trona. The Maasai, the original inhabitants, have never received any royalties; the clause in the Magadi agreement that gave the government the right to mine trona was inserted after the Maasai had signed it!

At present Magadi's water is concentrated brine with a massive salinity. In the past, the lake has been both dry and far larger; and the lack of evaporate beds in the sediments below the high water mark indicate that it has been quite fresh. So why is the lake so briny? The water from the hot springs has filled Magadi up with chemicals, and the source of these chemicals is probably connected with Ol Donyo Lengai, the prominent carbonatite volcano that stands sentinel at the south end of Lake Natron. Around and below the lake is a mixture of volcanic and lacustrine (lake-generated) rocks; the volcanics are lowest, and above them is a layer of trona up to 48m thick, which accumulated during the last 9,000 years. That's a lot of trona and rapid deposition, one centimetre of salts every two years. And for an obvious reason; nothing flows out of Magadi and the water coming in contains a lot of dissolved material. What evaporates is pure water; the salt gets left behind. And so for nigh-on 100 years Magadi has been supplying trona to the world and money to Kenya's exchequer, and the deposit has not been diminished. Two new minerals were discovered in the Magadi beds in the 1970s, appropriately they are called magadiite and kenyaite.

What of Magadi itself, and the wild country around it? One of the world's most remarkable fish, the Lake Magadi Tilapia, *Alcolapia grahami* (originally called *Tilapia grahami*) lives in the streams between the hot springs and the lake, in spring water as warm as 43°C, and a similar species, *A. alcalicus*, lives in Lake Natron. Few fish can tolerate such temperatures; their enzymes can't take it. It is not the only strange inhabitant of the area. In 1913, an engineer called Hickes was travelling along the Magadi railway one morning when, about 16 miles from Magadi, he saw what he thought was a hyena running ahead. As he got close, he realised it was not a hyena. To quote his own words, it was 'as high as a lion … it was tawny … with very shaggy long hair. It was short and thick-set in the body, with high withers … a short neck and stumpy nose.' Hickes had spent 18 years in Africa, building railways, and had never seen an animal like it. Some have speculated that he might have seen a Brown Hyena *Hyaena brunnea*, a southern African animal that can look quite weird when running with its raised cape of long hairs – there is some evidence that this species might have existed in Kenya – but Hickes was convinced that he

Flat semi-desert of northern Kenya; at Hadado, west of Wajir. (Stephen Spawls)

had seen the fabled Nandi Bear. Lesser Flamingos breed on Lake Magadi; this was discovered by the prolific and irascible ornithologist Leslie Brown in 1962, and 8,000 young were ringed.

Many of Kenya's lakes span borders; we have mentioned Victoria, Chew Bahir and Turkana, three that are rarely mentioned are Lakes Amboseli, Chala and Jipe. The Tanzania border runs across all three. Lake Amboseli now lies along the north-west border of the game reserve of the same name; it used to be deep within the reserve, which was huge and included the big hill Ol Donyo Orok ('black mountain') but the reserve was shrunk to the present smaller conservation area in the 1970s. When flooded, and beloved of flamingos, Lake Amboseli can be up to 44km^2 in area and 0.6m deep, but it is usually dry. Lakes Chala and Jipe are both small but very different. Chala is a deep, circular explosion crater on the slopes of Mt Kilimanjaro, Jipe is a long shallow lake just east of the North Pare Mountains, enigmatic Tanzanian hills that are beautifully evoked in the British poet Francis Brett Young's book about the 1914 East African campaign, *Marching on Tanga* (1917). Chala is a parasite cone, formed by a volcanic explosion; there is a legend that when it exploded a Maasai manyatta (settlement) was blown up, and the ghosts of the warriors may be heard in the woodland around the lake. There is also a legendary monster in the lake. Although the surface area is only 4km^2, Chala is more than 90m deep; an endemic fish lives there, the Lake Chala Tilapia, *Oreochromis hunteri*. The lake is fed by underground seepage from Mt Kilimanjaro. Part of Lake Jipe lies just inside Tsavo West National Park. It has magnificent reed beds and is fed by the Lumi River, draining off Mt Kilimanjaro, but human activity on its shores have changed the lake's water quality and its small fishing industry is fading.

So much for the water. What about the land? We might start with Kenya's most prominent feature, Mt Kenya, the snow-capped massif that dominates the central highlands. The highest point, Batian, is 5,199m (Mt Kilimanjaro, in Tanzania, is Africa's highest, at 5,896m; Mt Everest, the world's highest, is 8,848m). The mountain gives the nation part of its identity; its rugged and yet homely image appears on all manner of advertising. There's a lot of debate on the origin of its name, and hence the name of the nation. Krapf saw the mountain in 1849 and

Mt Kenya in snow-capped glory, 1960s. (Robert Drewes)

called it Kima ja Kegnia, which he said was the KiKamba name for the 'mountain of whiteness'. some KiKamba speakers say it should be 'Kilima Nyaa', and it means 'mountain of the cock ostrich', the theory being that the black and white mountain resembled the black and white plumage of that bird. The British administrators supposedly shortened it to Ki-nyaa, and hence the nascent colony got its name, initially pronounced 'Keen-yah'; after Independence people began pronouncing it 'Ken-ya', to match Kenyatta. Another theory is that Kenya is a corruption of *erukenya*, a Maasai word for 'mist-shrouded mountains'. Sir Halford Mackinder, the first man to reach the peak, said that the Maasai name was Donyo Egeri, meaning striped mountain; the mountaineer Iain Allan notes that the Maasai call it Ol Donyo Eibor, 'the white mountain'. The Kikuyu name for the mountain is Kirinyaga (sometimes spelt Kerinyaga, or Kere-Nyaga); the meaning of this term is equally debated: 'mountain of whiteness', or 'God's resting place', but if we turn to its most famous son, Jomo Kenyatta, we find that in his book *Facing Mt Kenya* (1938) he says it means 'Mountain of Mystery'. And who should know better? The Kikuyu word for snow is *ira*, and in the old days, during initiation/circumcision ceremonies, a white powder said to be from the mountain, and called *ira*, was rubbed around the facial orifices and navel.

A number of Western explorers had been onto the massif before it was successfully climbed. In 1897 Teleki and his stalwart Somali bodyguard, Mohammed Seif, ascended to higher than 4,000m on the moorlands. Gregory got up to the glaciers in 1893, and the enigmatic German adventurer George Kolb climbed to the base of the peaks in 1896. The stage was then set for the arrival of the British geographer Halford John Mackinder, later Sir Halford, and according to some, the founder of modern geography. On 8 June 1899 Mackinder and three companions left London; he was back in London on 30 October of the same year, having ascended to the peak of Batian. Mackinder named many of the features of the mountain, including the Hinde Valley, the Gregory Glacier, and the peaks Batian and Nelion. Mackinder went on to become a most distinguished geographer, but the full diary of his journey did not appear in print until 1991. It is a simply superb read, by a man in tune with his environment, full of natural history notes, acute observations on the people and the landscape, illustrated by Mackinder's little geographical sketches.

The mountain was not climbed again until 1929, when the dynamic British climber Eric Shipton, along with Wyn Harris of the Kenya civil service, ascended Nelion. On the way they met the African adventurer Vivienne de Watteville, and found hundreds of locusts embedded in the Lewis Glacier, carried up by the wind. The following year, Shipton climbed again with the taciturn Major Bill Tilman and managed the first ever traverse between the peaks. The following day Tilman had a serious fall on the spike known as Midget Peak, but survived. Tilman became a distinguished soldier and international climber; he sadly disappeared at sea off the South American coast, at the age of 80. Another remarkable climb took place during the Second World War, when Felice Benuzzi, an Italian prisoner of war in POW camp 354 at Nanyuki, escaped with two compatriots and attempted to ascend Batian. To orientate themselves they used a diagram of the mountain off a corned beef tin. They were equipped with food and improvised climbing kit they had scrounged in the camp. Sadly, the three brave adventurers were defeated by the weather and only made Point Lenana. But Benuzzi's lovely, humorous and philosophical book – *No Picnic on Mt Kenya* (1953) – is a classic of both mountaineering and Kenyan literature. In 1959, Kisoi Munyao, from Machakos, was the first Kenyan to make it to the peak, as far as we know, unless some early adventurer got there before the *mzungus*. This might sound unlikely, but a Kikuyu contemplative from Chogoria named Ephraim Mk'iara, according to Iain Allan, regularly ascended Nelion in the 1970s, wearing only plastic shoes; the body of another solitary religious man was found on the Lewis Glacier. Munyao had been taught to climb by the Nairobi photographer Arthur Firmin. Kisoi Munyao then ascended the peak again, with the Kenyan flag, to celebrate Independence Day in 1963. Even with modern equipment, the two high peaks of Mt Kenya remain a formidably tough climb.

Mount Kenya is a stratovolcano, a tall conical volcano built up of many layers. As we mentioned earlier, it was formed between 3.5 and 2 million years ago and has been cooling and eroding ever since, although it continued to spew out lava until about 40,000 years ago. Celia Nyamweru reckoned the mountain was originally more

Mist is a common feature at Kiandongoro Gate, Aberdares. (Stephen Spawls)

Lewis Glacier, Mt Kenya. (Robert Drewes)

U-shaped glacial valley, Mt Kenya. (Robert Drewes)

than 6,000m high, while the geologist Nayan Bhatt, who described the geology in the Mountain Club's excellent guide, reckons it was over 6,500m. J. W. Gregory reckons it was over 7,000m but no one can be sure. Over the last 2 million years, physical weathering has been grinding the mountain down. Mount Kenya offers Kenya's people the opportunity to see glacial and montane landscapes in their own country.

Glaciers, rivers of ice, are rare in Africa. They form from compacted snow, and creep downwards, a few millimetres a day, but sometimes faster. The fastest achieve 5m in 24 hours; if you see one coming you must stroll for your life. On Mt Kenya 100 years ago the glaciers were moving at 40mm a day, or 15m per year, but by the 1990s they had slowed to 8mm per day. Some of the East African glaciers have gouged out U-shaped valleys. With the glaciers gone, the little rivers that remain in these valleys are cruelly called 'misfits' by geographers; in these politically correct days you can still be rude about rivers it seems. At the edges of the glacier, rocks that are being carried by the ice tend to be dropped, forming moraines.

There are only three glaciated mountains in all of Africa at present, and all are in East Africa: the Rwenzoris, Mt Kenya and Kilimanjaro; the area glaciated on their peaks is inversely proportional to their heights. This seems odd, but it is because of the relative amounts of snowfall. The Rwenzoris get a lot of snow, more than 1,000mm per year, mostly from the west. On Mt Kenya the snowfall is usually less than 500mm, on Mt Kilimanjaro the saddle plateau gets less than 250mm. All the East African glaciers are shrinking; we know this from studies started in the final decades of the nineteenth century. The glaciers on the three big Rwenzori peaks, Mounts Baker, Speke and Stanley, have shrunk from 6.5km^2 in 1906 to 1.7km^2 in 1990. The Rwenzoris still have valley glaciers; Mt Kenya had them in the past but has only cirque glaciers now. The meteorologist Stefan Hastenrath documents twelve extant and six extinct glaciers on Mt Kenya. The Lewis Glacier on Mt Kenya was 0.6km^2 in 1890 and Mackinder estimated it to be 1.6km long; by 1949 it was 0.36km^2 and by 1980 it had shrunk to 0.26km^2. Mount Kilimanjaro has only a few short radiating glaciers, although it has several large ice fields around Kibo.

There is vigorous debate on the topic of why the East African glaciers have shrunk so much since 1880. Studies in the Rwenzoris indicate that between 1880 and around 1910 there was a large decrease in rainfall, of the order of 150mm per year, and this was associated (due to smaller amounts of cloud) with a big increase in incoming radiation from the sun, particularly the short-wave type (ultra-violet, x-rays and gamma rays; infra-red waves are not so affected by cloud). This increase in solar radiation seems to have been part of a large-scale, global happening. The earth experienced a 'little ice age' between 1550 and 1860, when it was colder than it was today. Glaciers destroyed farms in Switzerland, the river Thames in Britain froze regularly between 1607 and 1814. In Africa, there was increased rainfall, or, in the case of mountains, snowfall. Potential culprits for this ice age include low solar activity, increased volcanic activity, the slowing down of big ocean currents, and the decrease in human population following major diseases such as the black death; allowing reforestation, increased carbon dioxide uptake and hence less global warming. Whatever the reason, when the little ice age ended, it started getting warmer and the East African glaciers began melting; they have continued to do so ever since.

The Landscape, Climate and Weather

Tarn in the freeze-thaw generated landscape, Mt Kenya. (FLPA)

Mount Kenya is thus a young and spectacular showcase of the effects of low-temperature weathering. The freezing and thawing of ice on steep slopes shatters rocks, which then falls, forming scree or talus slopes, a pile of broken rock, sorted by gravity so the biggest lumps are at the bottom; this explains the heaps of boulders seen below any steep slope in the mountain's alpine zone, especially around the edges of the tarns. The layers of volcanic material that formed the crater let the water in along the layers and have thus been heavily eroded, but the closely knit crystals of the igneous rocks of the central plug do not admit water so easily. So we are left with that magnificent jagged central massif.

Some of Kenya's other high peaks have features that indicate they were glaciated in the past including the Aberdares (on Ol Donyo Lesatima) and Mt Elgon. The name Elgon, incidentally, comes from the Maasai Ol Donyo Ilgoon, the breast-shaped mountain. Elgon's summit is a huge caldera, 8km across, and moraines extend from just below the summit of Wagagai (the highest point, 4,320m) to as low as 3,400m. Elgon has huge dark caves on the Kenya side, in the National Park, that elephants used to enter to obtain minerals. These caves became notorious in the 1980s, as detailed in our chapter on mammals. Sir Fredrick Jackson ascended Mt Elgon to the oddly named peak Sudex, in 1890, as part of his expedition to western Kenya; an entire chapter of his enjoyable book *Early Days in East Africa* (1930) describes the expedition's time on the mountain, with some enchanting nature notes.

Coming down from the mountains, it's time to turn our attention to the flat(ish) land known as savanna, or steppe, a landscape that makes up most of Kenya. Not everyone likes it. Much of northern Kenya is a harsh and dry landscape; there are some spiky hills but most of it is flat and with less than 500mm of rain per year. But is has proved, like the desert to the prophets of old, a magnet to many dreamers and adventures, and some fine writers, like Gerald Hanley, Charles Chenevix Trench, Wilfred Thesiger and Monty Brown – all have been beguiled by its austere beauty. The old name for the north, the 'Northern Frontier District', invariably shortened to NFD,

Sheldrick Falls, Shimba Hills (Stephen Spawls)

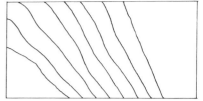

Davis's theory: the angle of slope gently declines, worn down over time, moving left

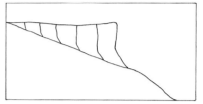

Penck's theory: the free face becomes smaller as the scree builds up, moving to the left

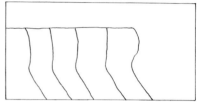

King's theory: the escarpment retreats to the left, leaving a plain

Figure 4(3) The three theories of slope development

conveys a vision of a remote, lawless, unmonitored land; inhabited by warriors, spiny plants and dangerous animals, policed from lonely outposts – and to a large extent, it is.

Africa is a land of flat savannas. In Kenya, in general, there are several ways that flat land is created and its age depends upon the circumstances. It may be young, the surface of a lava flow that spilled out like water and solidified, for example the Kapiti Plain. It may be a land that has been covered by waterborne or windblown sediment, as is much of north-eastern and eastern Kenya. It may be very ancient igneous rocks that have been gradually worn down by erosion and weathering, as in Ukambani, where the flat land is the old land. It may be a combination of any of these methods. Flat landscapes have fascinated geomorphologists and three theories as to how plains developed have held sway. The American William Morris Davis described a cycle of erosion, as elevated land was gradually worn down, the Austrian Walther Penck suggested that uplift and erosion can occur simultaneously, the South African Lester King suggested that uplift was followed by 'scarp retreat'. Any good geographical textbook will expand on these for you. Figure 4(3) shows a general diagram of these three theories of slope development. All had their merits but all are now considered invalid; they tried to use a single model for a many-faceted process. Much of the flat lands of eastern and northern Kenya are denuded ancient crystalline rock. Harder rocks remain, standing proud above the surface. Other plains form from sediments laid down in seas and now covered to a large extent by recent unconsolidated sedimentary materials (sands, soils, gravel) dumped there by water or air. Ex-marine landscapes tend to create a flattish sedimentary basin; the

Exfoliation, caused by heating and cooling, Mudanda Rock. (Stephen Spawls)

retreat of water (or uplift of the land) leaves behind a level or gently sloping landscape. Hills surrounded by flat country may have been emplaced relatively recently, formed by igneous material pushing its way through, or are tougher than the surrounding underlying rock and have resisted erosion. Some of the hills in eastern Kenya and also those around Kisumu resemble a dome or a jumble of boulders; terms used for these hills include bornhardts, inselbergs, tors or kopjes. In general, these hills are made of harder crystalline rock. When such hot rock cools, it contracts, and this can form both horizontal and vertical cracks, called joints. The rock then weathers along these cracks, forming big balanced boulders. The fluctuation of temperature in the surface layers can cause the layers to split off like an onion skin, as mentioned in the earlier section on weathering; this can be nicely seen at Mudanda Rock in Tsavo East.

So the development of a landscape can be summarised easily enough: tectonic activity may elevate the land, and igneous rock pops up here and there; weathering and erosion by wind and water wear it down but the harder rocks resist and form hills; the climate changes and the presence or absence of vegetation, the volume of precipitation and the temperature regime affect the speed at which this happens.

KENYA'S SOILS

Soil is important stuff; it is a resource. The rich loams of central and western Kenya are the foundations of the agricultural industry that feeds Kenyans and brings in outside money. Kenya needs its soil. And knowledge of its structure and properties are of great importance to the farming industry, and thus the country as a whole. It was lack of knowledge of the soil that contributed to the failure of a gigantic colonial project, the groundnuts scheme in 1947, in what was then Tanganyika. The administrators did not realise that many African soils have not been transported any distance, but developed in place from largely igneous rocks; the sharp quartz grains in the Tanzanian soil rapidly blunted the discs of the ploughs being used. Such discs had performed well in Europe, where the soil had formed from weathered sedimentary rocks, and the quartz grains had been rounded.

Soils can take a long time to form; no one is very certain as to exactly how long, because it is in the order of hundreds if not thousands of years and soil science is a relatively young discipline. Lester King, a man always willing to stick his neck out, was ready to hazard a guess; in a paper on the geomorphology of central Africa King stated that in tropical Africa a decent soil could form in a few hundred years, and would be at its best 'after about a thousand years'. He then went on to say 'between 1,000 and 5,000 years it may reach its optimum', which makes you wonder what difference King saw between 'best' and 'optimum'. Finally, King reckoned, the fertility 'slowly declines to about 30,000 years' and by 100,000 years a stiff and unworkable clay is left (King 1978). So there you are: it's no good waiting for soil to develop, it takes too long. Kenya's soil is a resource it needs to hang onto; that great brown stain that appears in Ungwana Bay when the Tana is in flood represents a loss to the country that needs to be stopped.

At its simplest level, soil can be classified by particle size; sand is the largest, then silts, then clays. Any soil with a reasonable mixture of all three is called a loam. But it gets more complicated than that, and many soil scientists feel that soil has been overclassified. The soil map in the *National Atlas of Kenya* (1962) indicates 38 different soil types, plus various mixtures. The western side of the map is a kaleidoscope of colours; with names like dark red friable clays, brown calcareous loams, dark red loamy sands, brown clay, light yellow-brown sandy loams, dark peaty loams … you get the picture. All of which led Pratt and Gwynne (1977), in their book on rangeland management in East Africa, to comment acidly that 'it is not surprising that soil classification is still relatively imprecise'. In many cases, one type gradually shades into another, a transition that soil scientists call a catena, Latin for 'a chain'. However, there are some clear general trends shown by Kenya's soils. The south-western quarter of Kenya, the land enclosed by a line connecting Mt Elgon east to Mt Kenya and south to Mt Kilimanjaro, has fairly good rainfall, and lots of recent volcanic activity, and its soils are generally well-drained loams, formed from basaltic lavas. Such soils are fertile, they hold water, but not too much, and have good air spaces for the development of roots. In many cases, if the rainfall is heavy, as the slightly acidic rainwater runs through these soils the highly soluble metal ions such as calcium, magnesium and potassium dissolve in the rain and are carried away, leaving iron oxide, which gives the soil a characteristic red colour. These soils belong in a group known as laterites, of which more shortly.

In some areas, particularly on the plains south of Nairobi, around Nanyuki, to the west of the Mara and around Kisumu, these soils have become waterlogged due to poor drainage, due to the impenetrable underlying lava. The soil pores fill with static water and there is no oxygen. A reaction occurs that changes the red, ferric iron compounds to blue/grey ferrous iron compounds, and the soils become darker. Such soils are also found on floodplains, in valley bottoms and riverbeds, anywhere where the soil is periodically flooded for long periods, and are known as gleysols. In Kenya they are often called 'black cotton soils' as they are good for growing cotton, but they are clay-rich, sticky and slippery at the same time, and trying to drive across them in the wet season can be a hellish experience, as many a driver on the plains south of Nairobi, caught in a rainstorm on a minor road, has found to their cost. The waterlogging also tends to kill large plants, as their roots cannot absorb oxygen when under water. So large areas of black cotton soil tend to have only small and spindly plants, that can pick up enough oxygen near the surface. They form extensive cracks when they dry.

Out in the northeast and the north, Kenya's soils are mostly sandy. The lack of rainfall has had two main effects: firstly there is little vegetation, and hence little leaf fall, fewer animals live in or on the soil, and so the humus content of the soils is low. Secondly, although the ambient temperature is high, there have been fewer water-stimulated reactions and the grain size is large. Such sandy soils can be quite fertile, if water is available, as many of the nutrients have not been washed out. However, they do not hold water well and tend to drain quickly. Along the rift valley, many of the soils are a mixture of salts and volcanic ash and are not very deep. They will grow crops but they need care; if they get too dry they will dehydrate readily and blow away.

Roots prevent soil erosion, Kibwezi. (Stephen Spawls)

Violent flash flood, Mokoyeti Gorge, Nairobi National Park. (Stephen Spawls)

Many areas of Kenya have soils known as laterites, the predominantly red soils mentioned earlier; some texts call them latosols or ferrallitic soils. They are usually associated with tropical areas, with clear-cut wet and dry seasons and a fairly high rainfall level. Since many of the soils of eastern Kenya, now a very dry area, are laterites, this indicates that the east was formerly wet. So now the east is red and the elephants of Tsavo are often red as well, from dust bathing and coating themselves with red mud. The name laterite derives from the Latin *later*, meaning 'a brick', because laterite has been used to make bricks in the past. During the wet season, rainwater percolating down through the soil leaches it, dissolving out salts of soluble metals like calcium, potassium, magnesium and sodium, but leaving behind oxides of iron and often aluminium, which are reddish and give the soil its predominantly red colour. During the dry season, provided the water has not all drained away, water creeps up to the surface by capillary action, an attraction between the water molecules and the solid soil molecules. At the surface, the water evaporates, leaving behind the dissolved salts, which form a hard crust, which may then be washed away in the next rainy season. In Kenya, laterite soils, under the technical name of murram, have been used to build durable roads, although the high proportion of clay in some laterites can lead to the road surface becoming very slippery when wet. Surface evaporation, especially in areas that have been over-irrigated and are poorly drained, can also cause salts to rise to the surface and be left there. This is called salinisation; a white crust appears, and such salty crusts can often be seen near the rift valley lakes. Saline soil is often no longer usable for agriculture, because the high proportion of dissolved salts in the soil means that water will not enter the plant by osmosis.

Soil erosion is a major problem in Kenya. Soil is unconsolidated stuff; it's not held together by any great forces and the best way of preventing its being blown away by the wind or washed away by rainwater is vegetation. The usual problems apply: over-use of land, gullying, periodic droughts and floods, felling of timber, overstocking and

Red laterite termite hill forms a road hazard, near Wajir, northern Kenya. (Stephen Spawls)

domestic animals that tear up roots and debark trees. Even on well-maintained farmland crops tend to have bare soil between the plants, and a heavy storm washes the soil away. Studies on agricultural land in Tanzania indicate that the rate of soil erosion by water on a hectare of bare land is in excess of 100 tonnes per year. Similar work in the Kenya lowlands gave a figure of over 30 tonnes per year. Every bit of that precious material that flows away into the rivers, into the Indian Ocean or the lakes, is a loss of potential growing land.

WEATHER AND CLIMATE

Any decent guidebook will tell you that most of Kenya has two rainy seasons, with rainfall peaking between March and May (the long rains) and another smaller peak between September and November (short rains), but western Kenya gets rain most of the year. The book will tell you that the highlands are often cloudy in July and August, with Nairobi getting around 4 hours of sunshine per day, whereas Lodwar gets on average over 10 hours of sunshine every day of the year. You will learn that the highlands are pleasantly cool, with an average daytime temperature of 15–20°C, the low country is often hot and dry, with an average temperature of 25–30°C, and the coast is humid and hot, but not too hot. And you will be told that the north-east trade winds blow from November to March, and the south-east trades from June to September, that January and February are the driest months in most places, and that 75 per cent of Kenya gets less than 500mm of rainfall per year and is arid. Your guidebook will also undoubtedly give you a sprinkling of record weather statistics, like these:

- In Lodwar, in 1928, only 2mm of rain fell in the whole year; in 1961 498mm of rain fell there. Lodwar is Kenya's driest town, averaging about 170mm of rain per year. It is also Kenya's sunniest town, with nearly 10 hours of sunshine per day throughout the year.

- Kakamega is Kenya's wettest town, averaging around 1,900mm rainfall per year. Lamu, however, averages around 350mm of rain during the month of April.
- In October 1953, at Vanga on the southern tip of Kenya's coast, 380mm (15 inches) of rain fell in one day.
- The highest shade temperature recorded in Kenya is 46.1°C, at Garissa. The lowest temperature is -12°C on Point Lenana, Mt Kenya. The mean maximum temperature in Nairobi is 25.2°C, and the mean minimum 13.6°C; the highest temperature on record for Nairobi is 31.8°C and the lowest, 6.1°C.

The picture thus presented is all largely true, although oversimplified in places. What the guidebooks hardly ever tell you, however, is *why* this is so. This is the point that we want to address here.

There is something very strange about Kenya, compared to other countries that lie along the midriff of Africa. If you fly along the equator, west to east, for most of the journey you are passing over thick forest, but as you cross the Albertine Rift the savanna appears, and as you cross Kenya, the landscape changes from forest and high grassland to arid semi-desert. As Francis Ojany, Kenyan professor of geography, points out, this arid near-desert belt is the largest tropical dry region of the world. The overall explanation for it is fairly straightforward. The elevation, within the last 30 million years, of the rift valley has created a rain shadow. Those heavily forested slopes along the eastern border of the DR Congo, the home of the Mountain Gorillas *Gorilla gorilla*, catch the last of the rainfall that comes off the Atlantic, as the wind and clouds coming from the south-west are pushed up, into the colder, higher air. The particles coalesce and much rain falls on the windward side. By the time the airstream has crossed Uganda and western Kenya, it has very little moisture left. And the wind that comes to Kenya from the north-east, which might compensate by bringing some rain from the Indian Ocean, has not crossed very much open water, so it is not carrying much moisture and is diverted by the Ethiopian Highlands. Hence eastern and northern Kenya lose out on rain. Africa also tapers, it narrows north to south, and so as you move north, rain-bearing airstreams, once they leave the coast, have further to go to get to the far side. That's the big picture.

Weather and climate are a consequence of the sun heating up the land, the water and the atmosphere, coupled with the rotation of the earth. The earth rotates on its axis every 24 hours, and every year it makes its way around the sun. And the earth is not perfectly upright; its axis of rotation is tilted by more than 20 degrees from the vertical. This has major climatic implications. At present, as observed from the 'side', in July, the earth is tilted towards the sun, which is directly overhead at noon in northern Europe and it is summer there; the days are long. In December, the earth is tilted away the sun, which is directly overhead in South Africa at noon. It's the southern summer, so in Europe it's winter and the days are cold. In between, in mid-March and mid-September, the time of the equinox in Europe and the United States, the 'tilt' is effectively parallel to the sun, and it is straight above your head at noon in Kenya.

So Kenya is hot for two reasons. Firstly, the sun is directly overhead twice a year and does not move very far away from that position, whereas if you live a long way away from the equator the sun is only overhead once a year and six months later it's a long way away. At really high latitudes the sun does not even rise in midwinter. So Kenya gets two hot and two fairly hot seasons, Canada gets one hot, two fairly hot and one very cold season. Secondly, as the sun moves away from directly overhead, its warming rays have to go through a greater thickness of air, which weakens them. So tropical regions get more heat energy. This heats the ground, the ground stays warm and it is ground heat that warms the atmosphere. On average, it is warmer in Mombasa than Cape Town. Nowhere in Kenya is much more than 5 degrees away from the equator. But the moorland of the Aberdare Mountains is also pretty near the equator, and yet it's a lot colder up there than it is at Mombasa. Why should this be so, especially since the moorlands are nearer the sun? The reasons are rather interesting. Firstly, ascending onto the Aberdare moorlands does not bring you that much closer to the sun. That warm star of ours is 150 million kilometres away; so ascending 3km nearer to it does not make much difference. And while the moorlands can get pretty warm in the daytime, and bring the Montane Vipers *Montatheris hindii* out to bask, there is not much air up there; the pressure is a lot lower than at sea level. And it is often cloudy and misty on those moorlands, so not only is there less heat trapped in the air up there, but the ground gets less sunlight, and much of the land is tilted. Thus it's colder.

The sun is the driver of atmospheric and oceanic circulation. When sunlight hits the earth, several things may happen. It may strike bare soil or rock and these warm up quickly, whereas vegetated land takes longer. Or the sunlight may fall on water. Water is more dense, it needs more heat, and hence warms up slowly. So a few hours after sunrise, the land surface can be very hot. It heats the air directly above it, which expands and becomes less dense, and rises. As it rises it cools, and so slows down and sinks. This is called a convection current. The same thing can happen in water, it just takes longer. If the land being heated is beside a lake or on the seashore, during the day the land is hotter than the water, the air rises over the land and cooler air blows onto the land from the water to replace it, the so-called sea breeze. At night, the land loses its heat more quickly, so as the night wears on, the water becomes warmer than the land, and then the warm air starts to rise over the water, and air to replace it blows offshore, the 'land breeze'. Now you know why you get a strong breeze blowing onto the land in the afternoon on the shores of Lake Victoria. The tops of hills and mountains also tend to cool down quickly at night; they cool the air in contact, which becomes denser and rolls down the hill – a katabatic wind, which you may experience if you are at the base of a hill or escarpment. If that falling air is moist and the temperature drops below dew point as it sinks, fog can form, and such valley fog is not uncommon anywhere in Kenya above 1,500m altitude in the rainy season. By mid-morning, the katabatic wind stops, and if a hill slope is heated by the sun, the rising air will cause an uphill wind.

The weather of the tropics is affected by where the sun strikes the land, and the position of the intertropical convergence zone or ITCZ, also known as the monsoon or near-equatorial trough. The ITCZ is where the circulation of the two hemispheres meet, the bit of land that is directly below the sun at midday, and thus gets the most heat. The land becomes hot; the air at ground level in the ITCZ is warmed and rises. Up it goes, and this causes a fall in pressure. The air rises, moves away, to the north and to the south, and comes down. As it cools and sinks, it is rather stable, and creates a zone of high pressure, the subtropical high-pressure belt, a little bit away from the line of the ITCZ. And then air at ground level blows back towards the ITCZ, towards the low-pressure zone, and this circulating system of air is called a Hadley Cell, named after the English lawyer and amateur meteorologist George Hadley.

The movement of the ITCZ and the airflow within the Hadley Cells has major significance for the weather of Kenya (and other tropical regions, of course). In the southern summer, in December and January, the ITCZ is roughly positioned across southern Africa, down towards the tropic of Capricorn. Winds blow southwards from the northern half of Africa towards it (and also from the southern tip of African northwards), so airstreams over Kenya in January are largely moving south. By March, the ICTZ has moved up to the equator, and by April and May it has moved north, so air is rising over the land near the tropic of Cancer, and the airstream over Kenya is moving north.

The situation is complicated by the existence of the Coriolis force, which makes winds in the northern hemisphere rotate clockwise and anticlockwise in the southern hemisphere; hence storm tracks rotate clockwise north of the equator. The Coriolis force is a consequence of the two components of air mass movement: perpendicularly away from the equator and eastwards with the rotating earth. As the air mass moves away from the equator and the lines of longitude get closer, the relative eastward movement gets greater and the air mass spirals. Entrepreneurs will willingly demonstrate this to you on either side of the equator line at Nanyuki. However, their demonstrations owe more to their manipulative skills than to the actual Coriolis force, which is effectively non-existent at the equator; thus hurricanes and typhoons do not affect the Kenyan coast. To get hurricanes, you need a combination of warm water, to power the storm, plus a reasonably strong Coriolis force; a combination that occurs only in a narrow latitudinal belt. The occasional hurricane hits Madagascar, often bringing heavy rains to the Kenya coast. Over Kenya, large-scale air movements and associated wind speeds are usually relatively slow and do not tend to get into the tightening spiral that can cause much damage, although in East Africa local wind speeds in front of an advancing front can be very fast. Little spiral windstorms, locally known as dust devils or willy-willies, also exist; on a hot day you may see them, particularly on flat plains at low altitude. These miniature tornados form when air rising over a hot bare patch is pushed sideways by a pressure difference and starts to spiral. The momentum of the moving air mass (the product of its mass and speed) is conserved, so as the spiral gets tighter, it speeds up considerably, and air molecules at the centre tend to be caught up in the circulating air mass and pulled out, so there is an 'eye' of low pressure at the centre. Powered by ground heat (and hence not seen early or late in the day), dust devils can pick up material and the rising air whirls it up to a great height. A car caught in one can suffer damage to its paintwork. They can also take off corrugated roofs and cause small structures to explode due to the low pressure vortex inside them. One that hit a chicken farm near Emali deposited the unfortunate birds over 3km away.

When the ITCZ has moved north of the equator, the winds blow northwards over Kenya, but the Coriolis force means they also tend to curve to the west south of the equator, so in April and May, the wind in southern Kenya arrives from the south-east, having crossed the Indian Ocean. This wind is called the south-east monsoon (from the Arabic *mausim*,

Dust devil, Amboseli. (Stephen Spawls)

Kusi rainstorm, Shimba Hills. (Stephen Spawls)

Marsabit Mountain, huge volcano in flat northern Kenya. (Robert Drewes)

meaning 'season'), the south-east trade wind or Kusi. Since the winds have come a long distance over the sea, the air masses have picked up a lot of evaporative water, making them humid. As they rise over the land, they are cooled. When the air temperature falls, the air and water molecules move more slowly, there is less energy available. They may stick and fall, causing rain. Atmospheric air usually contains some water. Warm air can hold more water than cold air, and the percentage of moisture in the air, compared to the total amount it could hold, is called the relative humidity. As air heats up during the day, the actual amount of moisture in the air may not change but it could hold more, so the relative humidity decreases. Likewise, as air cools in the evening and night, the relative humidity increases. The greatest relative humidity thus occurs when the air is coolest, which in Kenya is around 06:00 hr, and it is least when the air is hottest, between 14:00 and 15:00 hrs. If the air cools enough, the relative humidity may increase to 100 per cent; and the temperature at which this happens is called the dew point. Cool the air any more, and it cannot hold all the water, which condenses. If this happens at ground level you get dew, if it happens higher up you get cloud and possibly rain.

In November and December, the ITCZ is way south of Kenya and the wind is blowing southwards and from the north-east. However, these air masses have not crossed large areas of open water, because the Indian Ocean, unlike the Atlantic, is closed off on its north side by the huge landmass of Asia. So the winds arriving in north-east Africa carry less moisture in October and November. Hence, although they bring some rain and also used to bring trade (this wind is called the north-east trade wind, or Kaskazi), the rain they bring is less than in April.

Parts of Africa are very arid. As the air in the Hadley Cell moves away from the equator and sinks, it gets warmer, heated by the earth. This means that it could hold more water so the humidity decreases and the air becomes relatively dry. The collapsing air column is also being pushed down from above, creating high pressure, and this means that at a distance of about 20 degrees away from the equator, a sinking dry air mass not only pushes air away from its zone of descent, but is not likely to generate any rain itself, and the high pressure means no wet air is likely to blow in. Thus are deserts created – the Sahara to the north and the Kalahari/Namib to the south.

Near-desert, the Dida-Galgalu north of Marsabit. (Robert Drewes)

The ITCZ brings rainfall to Central Africa, where there is a great deal of rainforest. Moisture is present in the ground, and in and just above the forest itself. The trees act as pumps, water constantly flows upward through them, in at the roots and out through the leaves. Warm air rises over the forest and if a large mass rises rapidly into a cold layer, past the dew point, its water coalesces and creates rain, of a type known as convectional precipitation. It is common in the tropics, and in Kenya is likely to happen anywhere where the ground and vegetation are wet or very humid and the sun is shining; an initial big storm, even over a dry area, creates the conditions for more storms to follow. The big storms of central Africa are convection storms; forests really do generate their own weather, and this has conservation significance. Cut down the forest and you remove a self-sustaining source of rainfall. The water reserves of the Central African forests are augmented by moist airstreams moving north up the Atlantic, following the ITCZ. As they cross the equator the Coriolis force swings them across to the east, over the forest. In January they only penetrate a short way inland as the ITCZ is far to the south, but in July and August these airstreams cross the Atlantic, cross the equator and turn quite steeply round eastwards over Central and West Africa. They go right across to north-east African to drop rain on the forested escarpments of south-west Ethiopia.

But, as we mentioned, as the airstreams reach the wall of the Albertine Rift, and are pushed upwards into colder air, they dump their moisture there; this type of rainfall is called orographic precipitation (*oros* is Greek for mountain). So Kenya does not benefit very much from Atlantic moisture; Rwanda and Burundi do. However, occasionally these airstreams do reach western Kenya. They thus combine with evaporation from Lake Victoria, the associated convection thunderstorms and the forests and hills that ring the lake shore, to make the west Kenya's wettest area overall. Kakamega, Kisii, Kericho and Kisumu all have decent rainfall in almost every month of the year, the first three having more than 100mm of rain most months. Western Kenya is a moist and fertile place.

The overall significance of the ITCZ to Kenya is that it brings the winds onshore from the east, particularly after March as it moves north, pulling moist air and hence rain from the southeast Indian Ocean right the

West flanks of the Shimba Hills get slightly less Kusi rain than the east side. (Stephen Spawls)

way across Kenya. The western airstream does not cross the rift valley, but the airstream off the Indian Ocean does, and this is why the eastern slopes of Kenya's central mountains get most rain. The south-eastern slopes of Mt Kenya, the south-eastern flanks of the Nyambeni Hills and the south-eastern sides of the Aberdares often get in excess of 1,800mm of rain yearly. Much of this is orographic rain, augmented by the forests that are already there (although these are fast disappearing), and a great deal of it falls in April and May. Orographic rain is, of course, also important to the isolated forested mountains of Kenya: in April the Taita Hills get more than 200mm of rain, Mt Marsabit gets more than 150mm. Marsabit and the central highlands, from the Nyambeni Hills southwards, also get a fair amount of rain in November, from the air streams that have come from the north-west Indian Ocean, but this moisture is limited and little reaches the west. Rainfall, of course, combined with altitude dictates what sort of vegetation you are going to get. The low land, medium altitude plains and the high peaks and hills of Kenya have hence attracted rainfall in varying quantities, leading to the astonishingly varied and beautiful landscapes of the country, and its equally varying flora and fauna.

We mentioned the dew point, where the air is 100 per cent saturated. The base of a cloud is usually at dew point. If you stand on a hillside and look out across country on an afternoon when there are many cumulus clouds (the fluffy ones), you will observe that their bases all seems to be at around the same height; the height where the air is sufficiently cold to have condensed. On an average day in Nairobi, in the early afternoon, the temperature is likely to be around 25°C, the relative humidity about 50 per cent and the cloud base close to 2km above the land. At dawn, the temperature is much lower, around 13°C, the humidity around 90 per cent and the cloud base is likely to be lower, around 1.3km above ground level. In Nairobi, which at an altitude of 1,660m is fairly high for Kenya, the average daytime temperature ranges from around 12–14°C at dawn, up to 24–26°C between 14:00

Kusi storm drifts in from the south-east onto the coastal strip. (Stephen Spawls)

and 15:00 hrs, when the maximum temperature is reached. In general, in Kenya, for every 1,000m that you go up, the mean temperature drops by about 6–9°C (A. C. Hamilton gives it as a precise 6.3°C); as you go down it increases by a similar amount; this is called the lapse rate. At Lodwar, at about 500m altitude, at dawn it is around 24°C, by 15:00 hrs it has risen to 35°C. Up at Eldoret, at just over 2,000m altitude, it is about 9°C at dawn and warms up to 23 degrees in the day. Up on the Aberdares at 3,500m it drops to near freezing at dawn, and warms to 13 or 14°C during the day. This affects the fauna and flora; the colder it is, the fewer species there are and the shorter the trees get.

On high land, which in Kenya tends to be moist, a low temperature and fairly high humidity may mean the cloud base is actually at ground level, creating fog (or mist, although technically the two differ depending on how far you can see). When fog lifts, especially if there is a layer of warmer air up above (a temperature inversion, it normally gets colder as you go up), the rising air is unable to penetrate through the topmost cold layers and spreads out like smoke hitting a ceiling, at the same time the moisture remains, forming a flat layer of cloud known as stratus. Temperature inversions are caused by cold ground; the low air is cooled and a warm layer settles above. Above isolated mountains like Mt Kenya, temperature inversions are not uncommon. This is why there is often a layer of stratus, the so-called 'cloud deck', above 3,000m on Mt Kenya.

The coastal towns and cities are not as warm as those in the north and east, even though they may be at lower altitude, because they are exposed to the mitigating influence of the nearby sea, where the water temperature varies from 26 to 30°C. Hence at Mombasa the average daily temperature varies from about 22 degrees at dawn up to around 30°C at 15:00 hrs. Mombasa is lower than Lodwar but does not get so hot. It varies a bit, of course, depending on the season, but just about everywhere in Kenya is coldest and has the least rainfall in July, with an airstream mostly coming from the cold and dry south.

HOW THE KENYAN CLIMATE HAS CHANGED

The eminent British geographer L. Dudley Stamp once said that Africa's greatest curse is irregularity of rainfall. Kenya suffers badly in this respect: not only do the north and east not get enough rainfall to grow crops, but droughts (interspersed with floods) have occurred regularly. In the previous century alone Kenya had eighteen periods of drought, including one lasting from 1907 to 1911 and one from 1952 to 1955. They were mostly in the low rainfall areas, but a drought in 1921 killed 50 per cent of the livestock in the Baringo area, the 1928 drought in Kikuyu areas caused the 'great famine of Thika' and the 1952–5 drought killed 80 per cent of the cattle in the Maasai rangelands. In Kalenjin areas a drought in 1933 caused a famine known as *kimouito* ('skin time') because the people were forced to eat animal hides, the 1942 drought in Kipsigis areas was called *rubet ab sigirok* ('famine of the donkeys') because the starving people killed and ate their pack animals. The 1961 drought was followed by floods, with villages in the Tana River area and western Kenya submerged and the villagers evacuated; the flooding in 1997/1998 affected 1.5 million people in Kenya. Life is tough in Kenya; the country is all too often at the mercy of the weather. But what has caused this wildly irregular climate?

In the really long term, the uplift of the rift valley over the last 30 million years is significant, and has lead to major changes in the climate of the country. But that was a permanent change; it made north-eastern Africa largely arid and it won't change again until the mountains have been eroded away. But during the last 150,000 years, the climate has fluctuated, back and forward, from arid and cold to hot and wet. For example, between 150,000 and 130,000 years ago Africa was cold and dry, between 130,000 and 115,000 years ago it was hot and wet. Up to 13,000 years ago East Africa was very arid, then it gradually got warmer and moister, with brief arid interruptions, and from about 3,000 years ago it gradually got more dry. We know this from various strands of fossil evidence; in particular the rise and fall of ancient shorelines, pollen from cores of recent land sediments, marine sediments and the distribution patterns and accumulated changes of various plants and animals. Climate scientists have put forward several theories to explain this change. The big driver of climate fluctuations is the wobble of the earth and fluctuations in its orbit; their combined effect is called orbital forcing. The earth does not travel round the sun in a perfect circle, but a varying ellipse. The earth's tilt also varies, between 22 degrees and 24.5 degrees, over a period of 41,000 years. The tilt variation and ellipse have their own intrinsic effects and may act in tandem. For example, when the earth is nearest the sun and tilted towards it, that pole nearest the tilt will get a hotter summer, but a colder winter, than the pole at the other end. Overall, orbital forcing means that the earth's surface receives differing amounts of sunshine from year to year. This affects the climate, and has caused ice ages. A thorough study of these effects was made by the Serbian academic Milutin Milankovitch, and his results – the 'Milankovitch Curves' – indicate how the earth's climate has changed and how it will change. However, there is much debate about how safely his data can be extrapolated. Such predictions are also complicated by the gradual increase of carbon dioxide in the atmosphere, from 200 parts per million in 1900 to over 375 now. Carbon dioxide tends to cause the atmosphere to retain heat. The Milankovitch Curves seem to indicate that we are in a large-scale trend of gradual cooling, but within that downward trend we are actually on a short branch upward.

So orbital forcing affects Kenya's climate. But what does it actually do? One suggestion was that northern ice ages created large surges of icebergs that drift into the north Atlantic, and chilled the oceans, causing colder and dryer conditions in Africa. In the Indian Ocean, zones of warm and cold water in the north-east and south-west are believed to drive the fluctuations in climate. These hot and cold zones are called the Indian Ocean Dipole (IOD) and they are connected with the El Nino events. Much talked about in recent years, the El-Nino-Southern Oscillation (its full name) or ENSO is a curious phenomenon whereby warm sea-surface temperature zones migrate across the Pacific Ocean, from west to east, rather than east to west. It goes back a long way; ENSO events were first recorded by the Spanish in South America in 1576, and during the twentieth century they occurred every 5–7 years. However, the relationship between ENSO and IOD is poorly understood and disputed by some climatologists. During the periods between the blowing of the Kusi and the Kaskazi, the dipole affects the speed of both the equatorial westerly winds across the Indian Ocean, and the Indian Ocean eastward equatorial jet, a narrow fast-moving eastward current in the Indian Ocean. During 'positive' dipole events, the sea off the East African coast is unusually warm, the sea between Indonesia and eastern Australia is unusually cold, and this creates an easterly wind towards the Kenya coast. This slows down both the westerly winds blowing across Africa and the eastern jet current in the ocean, and causes Kenya to have good rains. It is believed to be a 'positive' event that caused the widespread flooding in 1997 in Kenya; during the same season many of the corals of the Seychelles were cooked. During 'negative' dipole events, the sea off the East African coast is cold, the sea between Indonesia and eastern Australia is warm, and this creates a westerly wind away from the Kenya coast. This speeds up the westerly winds blowing across Africa and the eastern jet current in the ocean, and causes Kenya to have poor rains, pushing away the gentle onshore winds that would normally bring rain. According to the meteorologist Stefan Hastenrath, the decades before 1880 were characterised by slow westerly winds and eastern jet currents, indicating that there was a long-term positive dipole event in place, and since 1880 the situation has been dramatically reversed, with fast westerlies and eastwards currents, causing a long-term drop in lake levels, gradual loss of forest and recession of glaciers in East Africa. And it continues so, although Kenya has the occasional good year. The next step in East African climatology is to find why these hot and cold spots appear – it may be connected with orbital forcing, as yet no one knows. And so we end this chapter on a note of mystery.

THE LITERATURE

In this chapter we have covered a great deal of geography; some of it has been necessarily rather briefly discussed. For those who want the background material on things like weathering, geomorphology and geology, glaciers, soils, weather, climate, clouds, rainfall and wind, any good higher-level school geography textbook should do; one we have used extensively is *Geography in Focus* by Cook, Hordern, McGahan and Ritson (Causeway Press 2000), which has lucid text and diagrams. The geography of Africa as a whole, with much that is relevant to Kenya, is covered in A. T. Grove's *Africa South of the Sahara* (Oxford University Press 1984). Colin Buckle's *Landforms in Africa*, published in 1978 by Longman, is a clear and lucid guide to the development of the African landscape. Essential also is the *National Atlas of Kenya*, published by the Survey of Kenya in 1962; its large maps and accompanying text showing soils, geology, rainfall and temperature provide an instant summary. No Kenyaphile should be without this book. Although somewhat dated, Francis Ojany and Reuben Ogendo's book *Kenya: A Study in Physical and Human Geography* (Longman Kenya 1974) is still the best geography book the country has seen, containing a mass of impeccable data and clear maps. The geomorphology of central and southern Africa (and thus covering Kenya) is summarised in a nicely opinionated essay of the same title by Lester King, in the comprehensive 1978 collection of essays entitled *The Biogeography and Ecology of Southern Africa* (edited by M. J. A. Werger and published by Dr W Junk in The Hague).

The waterways of Kenya, as well as the rest of sub-Saharan Africa, are covered in L. C. Beadle's major work *The Inland Waters of Tropical Africa*. Published in 1974, by Longman, this is a scholarly summary that has yet to be bettered. Equally excellent, although rather dense and not easy reading, is the multi-authored *East African Ecosystems and Their Conservation*. A product of Oxford University Press in 1996, it is packed with data, diagrams and lists of pertinent references; it was edited by Tim McClanahan and Truman Young, two hard-bitten professionals of the Kenyan ecological scene. Although somewhat dated, *Rangeland Management and Ecology in East Africa* (edited by D. J. Pratt and M. D. Gwynne and published by Hodder and Stoughton in 1977) is a valuable summary of East African land resources and their significance in terms of agriculture.

From among the literature of Mt Kenya, four books stand out: *The First Ascent of Mt Kenya*, by Halford John Mackinder (Ohio University Press 1991); *Upon That Mountain* by Eric Shipton (Hodder and Stoughton 1943), which covers many mountains including those of Kenya and the Rwenzoris. *No Picnic on Mt Kenya* by

Felice Benuzzi (E. P. Dutton 1953); and finally *Guide to Mt Kenya and Kilimanjaro*, published by the Mountain Club of Kenya and revised by Iain Allan in 1998, which is essential reading for climbers. A really excellent book on African meteorology, centred around Zimbabwe but very relevant to Kenya, is Jack Hattle's *Wayward Winds*, published by Longman Zimbabwe. The environmental history of East Africa is superbly summarised in A. C. Hamilton's scholarly and yet eminently readable book of the same name, published by Academic Press in 1982; anyone who wants to know more about how Kenya's highlands have changed during the last 2 million years should read this.

Many good papers have recently been published on Kenya and its climate, including some excellent material by Stefan Hastenrath, meteorologist at the University of Wisconsin, for example 'Variations of East African climate during the past two centuries', published in 2001 in the journal *Climatic Change* (volume 50, pages 209–217). Much useful material on the climate of north-east Africa, including Kenya, appears in papers by the climate scientist Peter B. deMenocal of Columbia University; two classics examples are 'African climate change and faunal evolution during the Pliocene–Pleistocene', published in 2004 in the journal *Earth and Planetary Science Letters* (Issue 220, pages 3–24), and 'Africa on the edge', published in 2008 in *Nature–Geoscience* (volume 1, pages 650–651). Good work on the vicissitudes of Lake Victoria has been done by Thomas Johnson and colleagues, at the University of Minnesota. Putting the names of these scientists into a search engine will reveal appropriate material. Jonathan Adams at the Oak Ridge National Laboratory in Tennessee has published some fine material on the African rainforest and the climate during the last 150,000 years, which is available as free downloads. An article in the *African Journal of Ecology* (volume 45, pages 4–16) in 2007, by Rob Marchant, Cassian Mumbi, Swadhin Behera and Toshio Yamagata, describes the Indian Ocean dipole; and the association between the eastward equatorial jet and the IOD is explained in Peter Chu's paper 'Observational studies on association between eastward equatorial jet and Indian Ocean dipole', published in 2010 in the *Journal of Oceanography* (volume 66, pages 429–434).

Chapter 5
The Vegetation and Habitats

Mwanzo kokochi, mwisho nazi ('the beginning is a bud; the end is a coconut') – Swahili proverb

What do you want here, a botanical garden? – Angry supporter of high elephant densities to Kenyan botanist Quentin Luke, Shimba Hills

This chapter deals with the plant life of Kenya. We look at the origins of plants and how Kenya's plant landscape developed, and describe the present landscape and habitats. We will move on to look at the history of botanical exploration in Kenya; and then focus on plant identification, ethnobotany, conservation, hotspots and endemics. We suggest some useful literature in the final section. As always, each section is self-contained and can be read independently.

Of necessity, our plant coverage will be superficial to some extent. Martin Ingrouille (1992), in his book *Diversity and Evolution of Land Plants*, commences by saying 'It seems a hopeless task to try to describe and explain [the plant kingdom] … there are … more than 300,000 species … any single text is impossibly limited'. And then he still has a decent shot at covering it, in 330 pages. Africa has spawned some noble botanical efforts. *The Flora of Tropical East Africa*, aiming to cover 13,000 species, with some 120 authors, has run so far to 130 volumes, published out of Kew Gardens, with each volume costing between £15 and £75. *Flora Zambesiaca* documenting the plants of the southern third of Africa, began in 1960 and over 30 volumes have now been published (also by the Royal Botanic Gardens at Kew), and the *Flora of Ethiopia and Eritrea* represents a similar labour of love by several collaborating institutions. Jean-Paul Lebrun and A. L. Stork's massive effort, centred on the Congo, *Tropical African Flowering Plants*, published by the Conservatoire Botanique de Genève, has now run to five volumes, each costing £70 or more.

So Kenya's botany is not all here, but we have tried to provide a decent overview of Kenya's plants in 30-odd pages. Here and there you will find some terminology and brief background science; skip that if you want to, but bits of it are (honestly!) fascinating. We hope that you will get some of the flavour, the big picture, on these fascinating organisms. If you want more, the literature section will point the way. You can certainly obtain good material, but have your credit card ready.

IN PRAISE OF PLANTS

Kenya's landscapes are defined, sometimes dominated, by plants. Drive on a track, away from the main Nairobi–Mombasa highway into the bush near Kibwezi, and you will find Baobabs, At dusk, as the sun slides below the horizon, darkness falls and nightjars squawk like little dogs, these weird, gigantic barrel-bodied trees seem to take on a life of their own. Standing huge and sinister against the dying light, like some obese frizzy-headed monster, they appear to be about to suddenly lurch out of the ground and lumber away across the savanna. Or pick a quiet spot in the depths of Kakamega's forest, and sit in the silence of the midday. You are effectively in Central Africa, surrounded by the giant trunks of huge Central African forest trees, and the call of Central African birds. The rainforest retreated westwards around 20 million years ago, but at Kakamega it left a pocket of that massive forest, for Kenyans to enjoy. Kenya's tallest tree is here, the Mukusu, *Entandrophragma angolense*, up to 50m high, In the dry north another plantscape awaits, tough, sparse and well adapted. Near Wajir, in the flat country, the trees are short, plants are widely scattered, there seems to be more sand than vegetation, but the vegetation blocks your view, the low, grim, iron-grey Commiphora trees stop you seeing more than a few yards in any direction. You can get lost here and in trouble quickly; the dry air sucking the moisture from your body. Or try making your way quickly across a field of *Festuca* and *Carex*, the huge tussock grasses and sedges that characterise Kenya's montane moorlands. You fall into gaps between the huge tussocks. It is the wrong place to meet an elephant.

But Kenya also has domesticated plantscapes. Up above Limuru, or north-east of Kericho, is a landscape dominated by tea and the hardworking people who pick it. Tea, originally from China (its scientific name is *Camellia sinensis*, the 'Chinese camellia') is a sterile crop; so few Kenyan animals can exploit it that it can be grown without pesticides. These virulently green hill slopes are providing not just work and money to many Kenyans, but a much-loved drink to the world. The same might be said about coffee, which originated in Ethiopia; some aficionados say that Kenyan Blue Mountain Coffee is the best in the world. Gaze across the great blue–green sisal fields west of Voi, or around Vipingo, the pineapple fields of Thika or the cane fields around Lake Victoria, they might not be aesthetically pleasing, but sisal, fruit and sugar bring much-needed foreign exchange to Kenya. The Mexican Cypress, widely planted in highland Kenya, provides timber to the country. The plants in glasshouses and poly-tunnels along the Lake Naivasha shores form a major part of Kenya's economy; whatever the shortcomings of this industry, it creates both food and employment. Most of the world's population are kept alive by grasses, that humble family of monocotyledons (the term means 'single embryonic leaf') that include such important plants as maize, rice, wheat, barley, millet and Finger Millet, all grown in Kenya. That simple alcohol, ethanol, C_2H_5OH, that enlivens life for many members of the human race, comes from plants as diverse as grapes, palms, maize, potatoes and bananas. Kenya derives a massive but unofficial income from the sales, to north-east Africa and the Middle East, of young shoots of the Miraa Tree, *Catha edulis*. Known in southern Africa as the Bushman's Tea Tree, its Kikuyu name is Muirungi. The shoots, known as khat or chat, contain a powerful stimulant, the only drug permitted to Muslims. In 1945 colonial legislators banned the growing of miraa, unaware that it was indigenous to Kenya.

Plants don't have brains, of course, or a nervous system, and it is no good talking to them, despite some theories to the contrary, but they do show some remarkable responses. The beautiful savanna inhabitant, the Pyjama Lily, *Crinum macoweni*, can anticipate rainfall, and produces flowers shortly before the rains begin. Plants can

The Vegetation and Habitats

Agnes Cheruiyot picks tea in the Nandi country; she earns 5 shillings per kilo (4p/6¢) and on a good day picks 60 kilos. (Stephen Spawls)

Wild Date Palm, Phoenix reclinata, *always beside rivers. Palms are an impoverished group in Africa. (Stephen Spawls)*

count. The Venus Flytrap *Dionaea muscipula*, which eats insects, closes when its trigger hair is touched not once, but twice. Plants can detect light, following the sun during the day and then turning east at sunset, to catch the dawn light. Plants can communicate; when antelope or Giraffe begin to browse on acacias the plants send out a warning to other acacias by releasing ethylene gas, and at the same time start to produce bitter chemicals called tannins, making their leaves unpalatable. Trees that are downwind detect the gas and also start producing tannins. The browsers respond to this by never feeding on a tree for very long and always feeding upwind. Plants can defend themselves in a number of ways, both passive (spines, poisons, itchy hairs) and active (stinging, collapsing when touched).

We need plants. Mark Twain, that famous laconic American humourist, was once asked what he thought men would become in a world without women. 'Mighty scarce, Sir', was his reply. In that same vein, it's worth saying a few words in praise of plants, for without them we, and virtually all other life on earth, would also become mighty scarce. Animals could not exist. Plants provide oxygen for us to breathe, they suck carbon dioxide out of the atmosphere and they provide us with food; every food chain starts with plants. They protect our soil; the roots stabilise it, the shoots and leaves cover it and prevent erosion. Plants give us a vast range of drugs and other beneficial chemicals; they provide us with renewable and sustainable energy in the form of firewood. Plants created the essential gases of our atmosphere. They even create rain; their huge pumping systems sucking water up from the soil through the roots, out through the leaves, making the atmosphere humid enough to generate a microclimate and rainfall. Plants give us shade, timber, resins, oils, charcoal, diverse habitats and beautiful scenery. When your eye rests on one of Kenya's beautiful landscapes, remember, you are looking not just on a peaceful scene; you are looking at the nation's general store.

In the next section we explain where plants came from, and why they are important. There is some science involved and we have tried not to make it too detailed, but if you find it heavy going, skip the section.

ORIGINS OF PLANTS

At some point after life began on earth, something momentous happened: a replicating organism began to use sunlight to make food. This is a process called photosynthesis. Light energy, plus water, plus carbon dioxide, create glucose and oxygen. The organism involved was probably one of the cyanobacteria; they are still in existence and tend to be (incorrectly) called 'blue–green algae'. The oxygen that our green cell created dissolved in water and initially reacted with dissolved iron, to form a rock called banded ironstone. Once all the free iron had reacted, the water became saturated with oxygen, and it began to gas out, into an atmosphere that contained mostly nitrogen and carbon dioxide. And thus the time of atmospheric oxygen began. Plants continue to suck up carbon dioxide and produce oxygen.

The photosynthesising organisms then evolved from single-celled to multicellular, which meant they could get large. Algae appeared. However, life on earth was initially small, through the time of the ice ages we call 'snowball earth', between 750 and 580 million years ago. But after the thaw, algae got out of the water and survived. Four hundred and eighty million years ago, plants were on land. Then plants with vessels to transport water appeared. Initially plants spread by spores; 360 million years ago the ferns and horsetails grew tall. When you admire the beautiful Tree Ferns of the Taita Hills, recall also that they have been on earth 1,800 times longer than *Homo sapiens*. Seed plants, gymnosperms and angiosperms were also on the rise. Gymnosperms have 'naked' seeds that are formed on modified leaves; they have no flowers. Typical gymnosperms include the cycads and the Pencil Cedar, *Juniperus procera*, Kenya's second tallest tree at 43m high and the only indigenous juniper in Africa. Other gymnosperms in Kenya include various exotic trees like pines and cypresses, which have been introduced; if you see a pine, fir or cypress anywhere in Kenya, you know it's an alien. The ancestors of gymnosperms were the first woody trees; they grew tall, in the battle to get the light, and many gymnosperms are very old. The *Podocarpus* of highland Kenya is an ancient species of a family that originated in Gondwana. The tough durable *Podocarpus* wood was used in both South Africa and Kenya for railway sleepers. The cones of the cycads found on the top of Mt Ololokwe, *Encephalartos tegulaneus*, are eaten by the Samburu people in famine years, although they are actually toxic. The name *Encephalartos* means 'bread in the head'.

Some time after 300 million years ago, flowering plants, angiosperms, appeared. They have both flowers and fruits, and they are pretty successful despite having arrived late on the plant stage. Most of Kenya's plant species, over 86 per cent, are angiosperms. Angiosperms are the plants we know and love; from the Nandi Flame tree *Spathodea campanulata* to the Baobab to the acacias, and most grassland flowers. Worldwide, there are more than 240,000 angiosperm species, but only just over 600 gymnosperms. In Kenya the contrast is even starker,

The Vegetation and Habitats

Tree fern, a plant of ancient lineage, Taita Hills. (Stephen Spawls)

Cycad cone. (Stephen Spawls)

with more than 5,900 angiosperm species and fewer than ten gymnosperms. Spare a thought the cycads of Mt Ololokwe, or the magnificent *Podocarpus* of the high forests; they were the first plants to go down the seed route, which proved so successful for others, but they did not diversify that much, and in Kenya, as a group they are at much greater risk than the angiosperms.

HOW KENYA'S PLANT LANDSCAPE DEVELOPED

Not a lot of fossil plants are known from Kenya, until we get near to the present, but the record is augmented by some valuable material from both fossil pollen and ancient DNA, as we shall see shortly. A few stromatolite fossils (sediment layers bound together by primitive cells called prokaryotes, probably filamentous algae) are known, some around a billion years old from the Bukoban rocks of Uganda, others from the late Permian (260 million years ago) from southern Tanzania. Fossil wood more than 300 million years old was found in the Mazeras formation, so there were woody trees in Kenya (or what was to become Kenya) then. We then leap forward in time, to a real hoard – a load of fossilised plant material of just under 20 million years old from Mfangano and Rusinga Islands. It is mostly tree material (leaves, fruits and seeds) of twenty or more families, including euphorbias and olives, with affinities to existing West and Central African trees. The original description by the paleobotanist Katherine Chesters suggested it was formed in a riverine forest at medium altitude; the palaeontologist John Van Couvering later reinterpreted this flora as coming from a wet lowland forest. So Rusinga Island had a rainforest on it (but it was not an island then). Important fossil plant finds in the Tugen Hills north of Kabartonjo, dated at around

12.5 million years ago, were documented by botanist extraordinaire Christine Kabuye, who did sterling work at the East African herbarium for 25 years. The Tugen Hills fossils included agaves similar to *Dracaena*, the Dragon Tree, plus grasses, euphorbias and olives, and indicated a mixture of forest, woodland and wooded grassland. A slightly older find at Fort Ternan (14 million years ago) was of fossil grasses typical of woodland with open grassed areas.

So Kenya's fossil botanical record is patchy. However, a number of fossil plants are known from other areas of Africa, and with these, plus evidence from fossil pollen and what we know about continental drift, researchers have been able to hazard some reasonable guesses about ancient climates. During the last 100 million years, Africa has drifted about 15 degrees northwards, gradually closing a massive seaway, the Tethys Ocean, that lay between Africa and Europe and covered parts of the Sahara. Between 75 and 55 million years ago, the northern half of Africa, including the Sahara, was covered by a lowland rainforest, with a patch of montane rainforest in the centre, and the area of Central Africa that is now under forest was savanna woodland then. As Africa shifted north, the rainforest moved south, so that between 30 and 25 million years ago, in the Oligocene epoch (have a look at the simplified geological timescale in Chapter 1), the rainforest covered most of Central Africa, the Sahara was savanna woodland. By the late Pliocene and early Miocene (15 million to 7 million years ago), the rainforest was more or less where it is now. The Sahara, north-eastern and eastern Africa were under savanna woodland and fingers of dry country vegetation were extending across the Sahara. During this time, Africa's northward drift plus the rising of the areas of highland along the rift valley had created a gradually widening arid zone.

Some very useful data on exactly what took place in Kenya between 9 million and 3 million years ago, a crucial time, comes from sections of sediment known as cores, obtained by drilling down into the sea floor in the Gulf of Aden, a research operation now on hold due to the activities of modern pirates. The scientific details of this research are quite dense, but the principles are fairly straightforward: lipids (fats and oils) from plant leaves are abraded and evaporated by winds and carried out to sea, where they sink and are deposited in sediment. The sediment is analysed using a mass spectrometer, and certain crucial chemicals, known as biomarkers, are identified. The variation in the quantities of certain biomarkers indicate how the land vegetation changed over time. Work by the geologist Peter deMenocal and colleagues indicates that between 9 million and 4 million years ago, the vegetation of north-east Africa consisted largely of so-called 'C3' plants, between 4 million and 3.4 million years ago, the proportion of C3 and 'C4' plants fluctuated a lot, and after 3.4 million years ago, as we approach the present, the proportion of C4 plants increases dramatically.

A note of explanation is needed. Most plants are C3 plants, that is, regular plants that live in wet places. During the Calvin cycle, the second part of photosynthesis, they start by making a compound with three carbon atoms, hence C3. C4 plants are adapted for near-desert landscapes; many Kenyan grasses are C4 plants. They start the Calvin cycle by making a compound with four carbon atoms; more importantly, they can function when it is hot and dry, when the holes in the underside of the leaf (stomata) are partially closed and carbon dioxide concentration is low, as occurs in arid places. So the change from C3 to C4 plants since 3.4 million years ago means that north-east Africa was getting dryer.

Some useful fill-in data on the last million years comes from fossil pollen, the male sex cell in sexually reproducing plants. (Plants can reproduce both sexually and asexually, asexual reproduction involves growing a new bit, a runner or a root, that becomes an individual in its own right; some plants reproduce both sexually or asexually, for example the strawberry.) Pollen is transported from male to female organs by a variety of ways, including by insects, bats, birds, gravity and by the wind. Plants that produce pollen no longer rely on water to transport the sex cells, a constraint that hampers sexually reproducing animals. It means that any animal that has moved onto land has to have internal fertilisation, as sperm need water to survive and move. Sperm cannot survive in air. Imagine if they could! Some distant male could release gametes into the air, that would drift downwind and stick on the exposed sex organ of a receptive female. You could get pregnant without knowing anything about the father, except that he was somewhere upwind. Fanciful though this sounds, it is what plants do. Since it is a rather hit-and-miss technique, wind-pollinated plants release rather a lot of pollen (as hay fever sufferers know) and some may land in

A tree that characterises Africa: Acacia tortilis, *the Umbrella Thorn. (Stephen Spawls)*

a suitable place for becoming fossilised, such as an acid swamp. The tough cellulose-based outer coat of the pollen, the exine, aids preservation. You can often identify a plant by the distinctive appearance of its pollen grain. Grass pollen is often spherical, like a tiny ball, *Podocarpus* pollen has wings. Fossil pollen provides valuable evidence of past climates; lots of grass or acacia pollen indicates a fairly dry climate, big tree pollen indicates a forest. The best places to find pollen from the last few million years are in the ancient muds of long-established swampy areas or lakes. You drill down deep into the mud using an augur, then you analyse the cores you bring up; gently breaking them apart, staining them and spotting the fossil pollen under a microscope. A. C. Hamilton (1982), in *The Environmental History of East Africa*, summarises the pollen data from East African mountains. Those little grains, preserved deep in the ancient mud, enclosing bundles of DNA, from a number of plants, including *Podocarpus*, olives, begonias, ferns, grasses and heather, provide some interesting data. It seems that much human disturbance, the cutting down of montane forests and their replacement by grasslands, took place in the last 2,000 years in Uganda, but not so much in Kenya; that forests were more widespread in East Africa up to 3,700 years ago; that it was wetter between 10,000 and 3,700 years ago, that between 12,500 and 10,000 years ago East Africa was relatively dry and cold, and between 28,000 and 20,000 years ago it was warm and wet.

And we end in the present day with the astonishing phenomenon of a zone of near-desert dryness that encompasses most of north-east Africa, even though it lies on the equator; a gigantic landscape of plants that are protected against heat, desiccation and browsing animals, by means of stiff leaves, tiny grey-green leaves, thorns and spikes, active only in the brief rainy season; iron-grey or brown in the dry season. This is the vegetation of most of northern and eastern Kenya. The big rainforests are no longer with us; they lie out west of Kenya, although we are left with a few pockets of rainforest flora and fauna; the Rhinoceros-horned Vipers *Bitis nasicornis*, Mukusu tree and Great Blue Turacos of Kakamega Forest were left behind when the forest retreated. What Kenya also has is some nice savanna, some very wet, some not so, with spectacular wildlife that can be still seen, and a relatively huge area of highland, home of a lovely, specialised flora and fauna. And all this the result of altitude, and rain or its absence.

Kakamega Forest, home of the central African flora and fauna in Kenya. (Stephen Spawls)

THE PRESENT LANDSCAPE, ITS PLANTS AND HABITATS

The flora, or plant communities, of the habitats we have mentioned – forest, woodland, savanna, semi-desert, and mountains – have come about as a result of their position, the changes that have occurred and (recently) the amount of disturbance they have been subjected to. Plants need air, water, warmth, light and nutrients, how much they get of these resources dictates what sort of plant community you will get. Vegetation is also affected by the pH (acidity or alkalinity) of the soil. The underlying topography is significant: a slope will support a different plant community to that of an adjacent flat valley bottom. Climate and weather are important (how much rain, and how it is distributed in time, the effect of wind), as are fire and the presence of herbivorous animals, especially destructive feeders like elephants. No other large animal – save humanity with its tools and machines – can damage a plantscape, like an elephant, with its grasping trunk and huge bulk. During the drought in 1984, six bull elephants spent 24 hours in a 2km^2 acacia woodland in the Mara. When they left they had destroyed or fatally damaged 34 per cent of the trees and left 22 per cent with multiple broken branches.

The changing nature of the African landscape is also significant. Plants can be long-lived: the Bristle-cone Pine *Pinus longaeva* of California lives for more than 4,000 years. Some Baobabs (*Adansonia digitata*) are over a thousand years old. A landscape can change and leave behind some plants from an earlier time, as the isolated and dying junipers of the Ndoto Mountains show. The occasional huge Nandi Flame tree, Croton, or Silver Oak, *Brachylaena*, preserved in gardens, indicates that the higher suburbs of Nairobi were once part of a big mid-altitude forest. Plants have strategies for coping with differing climates, as indicated by their height, the types of leaves and roots they have, their methods of reproduction, the space between them and the variety of species. Worldwide, plant communities, or biomes, show evolutionary similarities, or convergence. The rainforest

VEGETATION ZONES IN KENYA

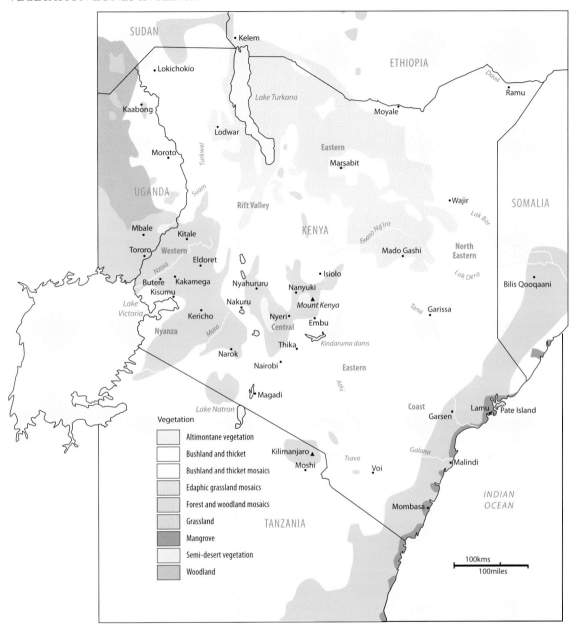

of Central Africa, with its huge trees and leaf litter, resembles that of South-east Asia and northern South America, the nutrient-poor heaths of South Africa look very similar to those of Australia; only an expert botanist could tell you that they have different species of plant. But the similarities are a result of adaptation and evolution. Plants are shaped by their environment.

Plants like it best where it is warm, well lit and wet. Under such ideal conditions, trees tend to dominate. There is a reason for this; plants need light, if you can get up there and seize it, you will prosper. In a forest, the plants have grown tall, all striving to be there on top. At ground level, there is little light; it has been said that every mote of

Job Milimu, forester, Kakamega forest. (Stephen Spawls)

sunlight that reaches ground level in a forest represents a failure by some plant to intercept it. Rainforest trees keep their leaves all the year, leading to them being called evergreen. A big tree needs a lot of water; a reasonable-sized Meru Oak *Vitex keniensis* will suck up through its roots around 800 litres of water per day – close on 300 tonnes annually – which is why true rainforests only occur in places that get in excess of 1,800mm of evenly distributed rainfall every year (and have a temperature of 26–31°C). The trees are pumping a lot of water out of the ground and into the air. They have to, that's their transport system: nutrients enter via the roots and are moved up the plant in water. Liquid flows one way only in a plant, it doesn't recirculate, unlike in animals where the transport fluid flows around and around. And this is how forests create their own climates and rainfall; a big tree in Kakamega forest will have put the best part of a cubic metre of water into the air at the end of a day. At night, as the temperature falls, that aerial moisture may cool to form mist, or drop at dawn as rain, or dew, or it may rise during the morning and generate an afternoon rainstorm. The forest depends on the trees. If you cut down the forest trees, less water is pumped into the atmosphere. The climate changes. Less will fall, as dew, or rain. This is what is happening on the Mau, one of Kenya's most important 'water towers', to use modern jargon. It was on the Mau Escarpment, as he scrambled through the emerald twilight, that Winston Churchill famously remarked how vital it was that Kenya's forests should not be laid waste by reckless hands, as described in his 1908 book *My African Journey*. The Mau sits squarely in the centre of Kenya's most-intensely populated heartland. If no rain falls there, the rivers feeding south-western Kenya dry up. No further argument is needed – except that a lot of relatively poor people would like to live there, and in the short term, they need the space …

Botanists have classified the various types of Kenya forest. Botanical classification is a widely debated and often contentious topic, but a broad classification indicates that Kenya has lowland forests, for example, those at the coast, mid-altitude (or sub-montane) forests, like those around Nairobi, along the shoulders of the rift valley and in the west, and high-altitude forests, as found on the Aberdares and eastern slopes of Mt Kenya. As you ascend, the average height of the forest trees decreases; few trees in Kenya's montane forest are over 25m high, whereas in the lower forests some trees reach over 30m. The cynical observer may detect here an attempt to compartmentalise what is a cline (a point we shall return to), rather like classifying people as short, medium-sized and tall, but it is an interesting fact that in Kenya, you rarely find a forest occupying a huge range of altitude. Most of Kenya's forests are at relatively high altitude, and as such are a conservation headache, as 75 per cent of the population and 80 per cent of the forests occur in the same place, the highlands, which are only 20 per cent of the total land area. To poor people, a forest is a resource, providing, at the most basic level, fuel with which to cook your food, wood for building your home and forage for your stock. One can see a reasonable range of Kenya's forests at high, fairly high and mid-altitude by driving eastwards down the Aberdares salient, from giant heath through *Podocarpus* and *Hagenia* down to mid-altitude deciduous forest. However, a complete transect, from high to low altitude forest, such as can be seen in the Eastern Arc Mountains in Tanzania, or by driving southwards off the Sanetti Plateau in the Bale Mountains in Ethiopia, cannot be seen in Kenya.

Typical mid-altitude forest, Aberdares salient, the red mud on the trees marks the passage of elephants. (Stephen Spawls)

Hidden stash of timber on the Mau, waiting to be removed after nightfall. (Stephen Spawls)

Lichen on highland rock. (Stephen Spawls)

In the savanna, the battle for light is just as intense but the rainfall regime will not support a lot of trees; there is not enough water, and it arrives irregularly. Thus trees in the savanna are not so tall. They get the light they need by spreading sideways, classic examples being acacias, savanna trees *par excellence*. Space is available; a savanna may be defined as a place where the gaps between the trees are larger than the diameter of the average tree canopy. Almost everywhere in Kenya that gets less than 1,500mm of rain per year is savanna, not forest, and the trees are widely spaced, rarely over 15m high, although within a savanna you may get pockets of forest along rivers or anywhere else where trees can suck up ground water; here there is competition with a diversity of plants and the trees get tall. If the trees get big enough and their roots go down deep enough, they gain the ability to survive drought. Along some of Kenya's rivers, especially those that run through the semi-desert, there are forests of acacias that are hundreds of years old, without any young trees, forests waiting to die; they are testament to a wetter past when the water table was nearer the surface and the young trees could get their start. Savanna areas get between 250 and 1,500mm annual rainfall, woodland between 1,500 and 1,800mm. In general, forests have a closed canopy, or 100 per cent canopy cover. This means that if you look down on a forest from above you cannot see the ground, and the trees – in Kenya, anyway – average between 7 and 40m height. In woodland, the trees average between 6 and 15m, and the canopy cover varies from 20 to 90 per cent. Another difference between forest and woodland is that the trees interlock at several levels in a forest; in woodland only the crowns interlock.

The term tree, incidentally, describes in general an angiosperm or a gymnosperm with a single woody stem or trunk. Trees do not form their own families, but have evolved many times in different families as a response to the demands of the environment, and are thus a nice example of parallel evolution. Trees get tall in order to catch more light, have a bigger volume and live longer than a season. Consider the Leguminosae, the legumes, a wonderful family that mostly enrich the soil (their roots have nodules containing bacteria that 'fix' or extract atmospheric nitrogen and turn it into useful compounds). This huge family contains such well-known trees as the Fever-tree or Yellow-barked Acacia, *Acacia xanthophloea*, and the erythrinas, Lucky-Bean Trees, but the family

Acacia erioloba *flowers and thorns, Voi. (Stephen Spawls)*

The Sausage Tree, Kigelia africana. *Local belief is that a fruit in your house keeps away whirlwinds. (Stephen Spawls)*

also contains the peanut, and the clover, which are not trees. The genus *Euphorbia* is largely represented by big cactus-like plants (there are no true indigenous cactuses in Africa, apart from the epiphyte *Rhipsalis*, although several exotic species are present), the Candelabra Euphorbia (*Euphorbia candelabrum*) grows to 17m high, with a woody trunk, but you can find little yellow-flowered herbs that are also euphorbias in your garden. Trees are just the giant members of the various families. A plant becomes woody if its water-transporting tubes, made of xylem, become strengthened by the incorporation of lignin, a tough carbohydrate polymer. This gives strength to the plant and allows it to grow both tall and tough, and as it gets older the central xylem tubes die but retain their toughness, becoming the 'heartwood' of the tree. It may form obvious rings and these can sometimes be counted; in temperate latitude these rings tell you the number of years, in parts of Africa the number of wet and dry seasons, or growing seasons. The outer wood is called the sapwood, and a really big tree may have an entirely dead interior – it lives by means of its outer layers. This is why ungulates that eat bark, especially goats, are such a threat to the landscape; any animal that ring-barks a tree kills its transport system and the tree will usually die. Some trees with thick bark can survive having a lot of their bark removed, but others cannot. A group of introduced Rothschild's Giraffes *Giraffa camelopardalis* are gradually killing the Yellow-barked Acacias of Lake Nakuru by eating the bark.

Soil warmth is crucial for trees to grow, in Central Africa trees will not grow if the average soil temperature is below 8°C or so, a value that delineates the timberline. And the timberline occurs at around 3,200m on Mt Kenya and the Aberdares, but is lower, 2,800m, on the drier and thus colder Mt Kilimanjaro. In really arid areas, with less than 500mm annual rainfall there are fewer trees; in areas with less than 250mm of rain there may be hardly any. The canopy cover can vary from zero to around 20 per cent –not much shade. In such areas, the structure of the flora is extremely interesting. There is usually a preponderance of annual plants, which avoid drought by dying in the dry season; the species survives by passing the dry times as dormant seeds, classic examples being grasses of the genus *Aristida*, known as Needle or Spike Grass, for reasons that walkers in Kenya's dry country know about. There are also usually geophytes, plants that avoid being wiped out in the drought by dying above the ground and surviving using underground water-storage organs, and there are desert plants that tolerate drought, such as aloes and stem succulents, that have few stomata and store water. If you walk in Kenya's low rainfall areas in the dry season, the land looks barren. Leafless bushes and succulents appear to be the dominant life form, but they are not. Return in the rainy season and annuals dominate; the landscape is transformed and looks lush. A wet or 'emerald' season safari has its disadvantages – damp and lots of cover – but it offers the visitor a view of the flora in all its glory.

In some areas there may be grass but no trees. If the rainfall is high enough to support tree growth then there will be a reason why there are no – or few – trees. Grasses are common plants, and annual species of grass are found everywhere except for true rainforest. In Africa, grasslands are not usually regarded as climax, or final stage, vegetation. Few African grasslands are as extensive or permanent as those of, for example, the Russian steppes or the central United States. Kenyan grasslands are usually a consequence of one of the following: (a) far too little rainfall to support trees, as for example the dry grass plains west of Wajir; (b) periodic flooding, which kills non-annual plants, such as that found on any flood plain; (c) intense cold, which inhibits tree growth, as on the Aberdare moorlands; or (d) regular fires, which kill off woody shrubs, herbs and young trees; the grazing of ungulates may also be significant. The lovely grassy valleys of the Maasai Mara are there as a result of periodic fires, coupled with the cycle of wet and dry seasons, human activity and the eating habits of herbivores. Grasses may have remarkably deep, convoluted root systems. An illustration in Pratt and Gwynne's *Rangeland Management and Ecology in East Africa* (1977) shows a scale diagram of the root system of the common grass *Panicum maximum*; the grass heads are 80cm above ground level and the roots go down 2.4m! A supply system is under your feet.

As you move across Kenya, from east to west, along a line for example from Wajir to Kakamega, from semi-desert to dry to moist savanna and thence to woodland and forest, and the biomass of vegetation increases with almost every mile, you are actually following a rainfall cline. A similar effect can be seen by driving from Lake Naivasha up into the Aberdares via the Mutubio Gate. There are some specialised plant communities in Kenya; for example swamps, although these are not so extensive, unlike Uganda where around 10 per cent of the country is covered by papyrus *Cyperus papyrus*. Kenya has mangrove forests and isolated montane communities. These

Snake Bush, Euphorbia tirucalli, *the white latex sap is extremely toxic. (Stephen Spawls)*

montane communities are intensely interesting; they are often hotbeds of speciation, for example the Taita Hills, the Mathews Range and Mount Marsabit. Before we expand on plant communities, however, it may be useful at this point to have a look at how botanists over the years have classified Africa's, and Kenya's plant communities, and also the use of the term 'communities'.

Plants give the landscape its unique character. In 1845, Charles Darwin wrote 'a traveller should be a botanist, for in all views plants form the chief embellishment' (in the lengthily titled 'Journal of researches into the natural history and geology of the countries visited during the voyage of H.M.S. Beagle round the world'). A landscape is defined by its plants. And yet no two places are exactly alike, and this has created problems with the definition of plant communities. The American ecologist Frederic Clements argued that plant communities reached a final stage of equilibrium, called a 'climax community'. The British botanist Arthur Tansley disagreed, arguing that differences in soil, topography and water supplies created many different plant communities that are stable within a region; a drive from Shimoni through Gede to Garissa will demonstrate the truth of this. The ecologist Henry Gleason saw plant communities, however, as merely a coincidence, not superorganisms but just a random assemblage of species found in the same area, simply because they happen to have the same abiotic (non-biological) requirements, such as temperature, soil type and rainfall regime. Gleason argued that disturbance keeps plant communities from reaching a state of equilibrium, and that they change constantly, depending on the circumstances and the versatility of individual plants, as any long-term resident of the Lake Naivasha shore will appreciate. All three models have been replaced by one in which plant communities are constantly changing, the 'non-equilibrium' model – a good model, as anyone who has lived for the last 40 years in Isiolo will agree. Disturbances tend to lead to greater species numbers as changes open up niches, a phenomenon called the 'edge effect'. Whether or not that is a good thing depends on the circumstances; driving a road through a forest means all sorts of plants that could not exist within a forest get a toehold along the open spaces of the road, but this is not necessarily beneficial in the big scheme of things.

Analysis of plant communities provides plenty of data in a short time. Unlike, say, a herpetologist studying the snakes of an area, who may take days before they find a single specimen, and many months to put together a representative collection of snakes from a discrete locality, an expert botanist can find dozens of species per day in the field; provided the season is right, the plants are there – they don't hide or run away. Kenya National Museum botanist Quentin Luke's record is 300 documented species in a single day. There are difficulties, of course; to identify many species you need both flowers and fruit and these may not be available for long periods of time. Plant communities change, as mentioned, with both the time and the season, and deciding exactly where to sample, and what, over what area, can present problems, as any student of botany knows. Put down a sampling quadrat at one spot and you will get a certain number of species and individuals, another quadrat 10m away will produce a different result. However, baseline data on plants can be gathered in large quantities.

This abundance of data is reflected in the wealth of analyses of Africa's floral kingdoms, or phytochoria (which means plant geography) by botanists such as J. F. Schouw, Jean Paul Lebrun and Theodore Monod. In 1959, R. W. J. Keay authored a vegetation map of Africa south of the Sahara, and this was popular and widely used until the appearance of Oxford professor and cricket enthusiast Frank White's fine series of papers, between 1965 and 1993. White summarised his work in his classic 1983 *Vegetation Map of Africa*, a map accompanied by an excellent book, published by UNESCO. White's analysis is still widely used, particularly by zoologists looking for affinities between the fauna and the flora, as it is straightforward, fairly simple, avoids over-analysis and is easy to use.

White classified Africa's vegetation into eighteen major zones and about 80 different subcategories. The major categories are: eight centres of endemism, nine regional transition zones or mosaics and one centre of impoverishment. Kenya's plantscapes fall into seven of these categories. Five of these are 'centres of endemism', meaning they have a lot of plants (and animals, incidentally) found nowhere else. The five are the Guinean–Congolian Forest, which includes a few forests in extreme western Kenya, the Sudanian Region (savannas of north-west Kenya), the Zambezian Region (savannas of central and southern Kenya), the Somali–Maasai Region (dry north-east and eastern Kenya, and arid areas in the south, for example around Lake Magadi) and the Afromontane Region (high country, for example the Mau, Aberdares, Cherangani Hills, Mt Elgon, Mt Kenya). The Lake Victoria Region, a mixture of woodland, forest and savanna grassland around the lake, is a regional mosaic, as is the Zanzibar–Inhambane Region, a mixture of forest, woodland, thickets and grasslands along the coast; included among these are Kenya's 500km^2 of mangrove forests, most of which are centred on the Lamu archipelago, and standing in tidal or brackish water (we discuss mangroves in Chapter 12). Some authorities think the Zanzibar–Inhambane category is too broad and split it; they regard Kenya's coast as part of the 'East African Coastal Mosaic'. White's model remains fully functional.

Kenya has had its own vegetational analyses. D. C. Edwards, in a 1940 paper in the *Journal of Ecology*, produced the first comprehensive vegetation map of Kenya, in 1936. Other vegetation maps include Pratt, Greenway and Gwynne's 1966 analysis of rangelands; a modified and somewhat confusing version of which was used for the vegetation map in the *National Atlas of Kenya*. In 1983, Kiilu, Dean and Trump, under the auspices of the Wildlife Planning Unit, published a map of the biotic communities of Kenya, splitting the county into 24 vegetational zones. Recent works, including Henk Beentje's *Kenya Trees, Shrubs and Lianas* (1994), use the K-botanical numbering system, in which northern and eastern Kenya are the K-1 area, the highlands are K-3, the coast and lower Tana River are K-7, and so on, although in reality these are just based on old colonial boundaries.

So, as we have seen, botanists have had fun analysing, or dare we say overanalysing, Kenya's, and Africa's vegetation, and hence their analyses are not all the same. Such work reinforces a point we made earlier: different scientists view the flora and fauna in different ways. Some of the flavour of this is summed up in Jon Lovett and Samuel Wasser's fine 1993 book, *Biogeography and Ecology of the Rain Forests of Eastern Africa*, in Lovett's chapter on the Eastern Arc moist forest flora. As Lovett cheerfully points out, what he calls upper montane forest, i.e. forest above 1,800m altitude, with more than 1,200mm of rainfall per year, other authors, eminent botanists all, have variously called montane forest belt, bamboo zone, moist montane forest, higher altitude types of submontane moist forest, wetter low canopy types of montane forest, higher altitude types of upland rainforest, montane mossy forest, upper montane forest, wetter types of undifferentiated Afromontane forest, Afromontane

The Vegetation and Habitats

Pyrenacantha malvifolia *or monkey chair, a fat succulent from the dry country. (Stephen Spawls)*

bamboo and finally, the beautifully named subalpine elfin forest. It comes down to a problem that pervades natural science: the attempts to compartmentalise a continuum. It is probably easier to stick with montane, mid-altitude and low-altitude forest.

Alternatively, many biologists and naturalists prefer the simple term 'habitat', used by ecologists precisely as the range of environments in which a species occurs. Plants define a habitat well, and the term gives us an opportunity to describe areas that are otherwise hard to define, such as habitats created by humanity. Kenya's habitats include desert, semi-desert, grassland–savanna, woodland, forest, montane, wetlands, coasts and islands, and man-made habitats. Deserts and semi-deserts are defined as areas with little rainfall and sparse vegetation, the term grassland–savanna essentially describes open country with trees but where the tree canopy is not continuous, woodland areas have a single-level continuous canopy. Forest has a closed canopy and a multi-layered plant community. The term 'montane habitats' nicely describes what a botanist would split into high-altitude forest and moorland (we have expanded on this in Chapter 7 on birds). Wetlands include lakes, rivers and swamps. Coasts and islands include a range of habitats, as detailed in Chapter 12, and the useful category of man-made-habitats includes gardens, farms and other agriculture, reservoirs and urban landscapes. These are extremely useful terms, especially for describing where certain species occur – particularly birds – and are going to remain so for a long time to come.

Being tropical, Kenya has quite a lot of plants – as already mentioned, it has around 6,800, more than four and half times as many as Great Britain – and it is not known how many plant species may have become extinct in Kenya, although one tree, *Karomia gigas*, was known from a single specimen that has now been felled. But the Kenyan total is set to rise, for three reasons. Firstly, there are still many areas in Kenya that have yet to receive the attention of a botanist. When they get there, they are going to find new forms. When Quentin Luke studied the Shimba Hills between 1987 and 2003 he found 1,396 plant species, including twelve species that had never been recorded in Kenya before, despite the area being fairly well known. He has also recently discovered a

Commiphora at Kajiado, the peeling papery bark is distinctive. (Stephen Spawls)

completely undescribed 15m forest tree in the Ngaia Forest near the Nyambeni Hills. Secondly, the taxonomy, or classification, of plants, is undergoing active changes, and this nearly always leads to an increase in the number of known species by a process known as splitting (of which more later). And thirdly, plants themselves are quite adept at hybridisation, which is when two related species breed to produce a third.

Due to the remarkable variation in altitude and climate, Kenya has plants that are associated with near desert, semi-desert, savanna, mangrove swamps and tropical and montane forest. Any particular area can also have a lot of plants. A square metre of soil in a fertile area can have 20 or 30 species of plant growing there; a 20m x 20m plot in a grazed area on the outskirts of Nairobi had 115 species of plant on it. A fair number of checklists for various areas have been published by assiduous botanists: 448 plant species are known from Mt Nyiro, more than 520 species of plant are known from Meru National Park, more than 575 species have been recorded from Lake Nakuru National Park, more than 650 from the Tana Delta, 708 plant species are known from Kuki Gallmann's ranch, Ol Ari Nyiro, near Rumuruti, and 986 from the Kakamega Forest. As mentioned above, the Shimba Hills at present tops the list, with 1,396 species known. And so on (although no one seems to have done a repeat study, to see if numbers are changing). The numbers are large, and hard to get to grips with; one has to admire botanists.

Figures for Kenya as a whole are also large. One hundred and sixty-eight mosses and liverworts are known from the Taita Hills. Oscar Wilde didn't think much of grass, but around 1,500 species of grass are recorded from tropical Africa and 590 species of that useful monocot are known from Kenya. One hundred and forty-two African grasses belong to a single genus, *Eragrostis*, and these include *Eragrostis tef*, called Tef also in Amharic, a grass that keeps the majority of highland Ethiopians alive. It is plucked, ground, fermented and used to make injera, the staple flat pancake-like bread eaten throughout highland Ethiopia. Malcolm Coe said there were 44 species of acacia in Kenya, the latest figure is 50 and the African total is 142. In Africa there are at least 110 species of *Commiphora*, the Paper Bark, most of them trees, and more than 30 commiphoras are known from Kenya.

The Sickle Bush, Dichrostachys cinerea, *widespread in Africa, Asia and Australia. (Stephen Spawls)*

There are several hundred species worldwide of the genus *Impatiens*, including 110 or so in Africa; many of which have showy purple or pink flowers. Yes, numbers will increase: Jan Gillett and P. G. McDonald's checklist of Kenyan trees, shrubs and lianas listed eight species of aloe from Kenya in 1970, but a major trade review of that useful plant listed 55 Kenyan species in 2003, 24 of them endemic. In 2006 Emily Wabuyele and Len Newton said there were 'about 60' aloe species in Kenya. One of them, *Aloe ballyi* (named after Swiss botanist Peter Bally, who was Joy Adamson's second husband) smells like a rat! Aloes are plants that are sold abroad and bring in some money to Kenya. We all know of the benefits of *Aloe vera*, but between 1999 and 2001 Kenya exported 200 tonnes of derivatives from *Aloe ellenbeckii* to China. We mentioned hybridisation earlier; Len Newton has documented seven aloe hybrids in East Africa alone. At the other end of the species number scale, Kenya has only three species of African Violet and one species of baobab (Madagascar has six). Incidentally, large baobabs are over 1,200 years old, possibly older. Popular legend says that God planted baobabs upside down, and their flowers are inhabited by spirits; anyone plucking one will be eaten by a Lion. Baobabs can be 25m in height but up to 28m in circumference.

There has been some violent debate about species numbers in the world of plant taxonomy. In 2010, a three-year project involving a group of botanists worldwide created a provisional list of all the world's plants. The initial list had more than a million species but some forms had been described more than twenty times in different parts of the world! This is something unusual about plants; they often have huge distributions, unlike most vertebrates. For example, only one snake found in Kenya, the Flowerpot Snake *Ramphotyphlops braminus*, is found outside Africa, and that is only because it gets transported in soil and is parthenogenic (that is to say, only females exist and give birth to other females without fertilisation). African snakes are only found in Africa, and this goes for most other animals. But many plants occur worldwide: the Sickle Bush *Dichrostachys cinerea*, that dryland plant with the lovely feathery yellow and purple flowers, is found not only in Africa but in Asia and

Australia; the Sodom Apple *Solanum incanum* with its distinctive yellow fruit (a reputed cure for warts) occurs in Africa and Asia. Trees of the genus *Prunus*, which includes the Peach, the Flowering Cherry and Kenya's distinctive forest giant, *Prunus africana*, the Red Stinkwood, are found in Asia, Europe, the Americas and Africa.

So there had been a lot of duplication, and the team of botanists eventually announced that they had pruned the plant list down from a million to fewer than 400,000. Many conservative botanists put the true number nearer 250,000. Of these, 220,000 were seed plants, some 54,000 in continental Africa, and 30,000 found in tropical Africa. Kenya's total, as we mentioned earlier, is just under 7,000, of which nearly 740 are endemic. This exceeds Somalia, which has some 500 or more endemic species, but Ethiopia has over 1,000 endemics and all of them pale beside the Cape Floral Kingdom (CFK), that remarkable array of plants that inhabit the small mountains and deeply dissected valleys of southern South Africa, more than 8,700 species of plant have been recorded there, and over 6,000 of them are endemic.

The status of some plant species has caused controversy. Species is an essential term because it defines the organism that we are talking about. A species is a group of interbreeding organisms that are each other's closest relatives. Some species are easy to define. The African Baobab is a species. Species are described by their Latin names, *Adansonia digitata*, for the Baobab. But other organisms are not so easily defined. Is the olive tree found in Kenya's dry highland forests a full species, *Olea africana*, or is it just the European Olive *Olea europaea*? Some botanists have described literally dozens of different species of closely related plants, on the basis of tiny little differences, classic examples being oaks and dandelions. Splitting has happened quite a lot with plants. Normally, with sexually reproducing organisms, the DNA gets shifted around a lot in the gene pool, during meiosis and crossing over, and this leads to a fair amount of homogeneity. But some seed plants choose to reproduce asexually, and this can lead, oddly, to a slight variation from the norm being perpetuated, since in asexual reproduction all the offspring are clones, identical to the parent. A genetic mutation might give the parent a slightly different leaf edge, for example. If it reproduces asexually, all its offspring will have this variation. There is no exchange of genetic material to bring it back to the norm. And then a botanist spots the group of plants with slightly different leaf edges and describes them as a new species.

This has led some radical botanists to suggest that the old style taxonomy should be scrapped, replaced by a new system without species, in which only lines of evolutionary descent can be traced. A thoughtful article in *Nature* (Rieseberg *et al.* 2006) examines this proposition. Such iconoclastic botanists tend to be computer experts. A new system, the phylocode, has been proposed by a small but growing band of techno-taxonomists; traditional taxonomists hate the idea. Developments will surely occur, but in the end species have to have Latin or scientific names, otherwise you can't be sure what you are dealing with. The name 'Flame Tree' means one species in highland Kenya, another on the coast. Or consider that beautiful (and deadly) tree of rocky land, small and chunky, with grey bark and vivid pink and white flowers, source of a virulent arrow poison. In Kenya its Swahili name is *utupa*, *mdiga*, *mwandiga* or *madiga*; in English it may be called the Desert Rose, the Elephant's Foot, the Mock Azalea, the Coral Tree or the Impala Lily; in southern Africa it is called the Sabi Star. And thus botanists avoid confusion by giving the Desert Rose a scientific name, *Adenium obesum*. It may not be a very elegant name, a mixture of Latin and Greek that translates as 'fat gland' (or maybe 'fat acorn'), but it means that botanists worldwide can discuss the plant with no possibility of confusion. The world of plant taxonomy, though, is in a state of flux, due to the advances made in molecular taxonomy. Molecular taxonomy involves finding relationships between organisms by looking at chemical sequences, particularly in the nucleic acids such as DNA or RNA. Some of the relationships that have been exposed, or elucidated, have upset traditional taxonomists, and even the ordinary naturalist. One such botanical hot potato concerns the acacias. Recent decisions at two botanical congresses to restrict the generic name *Acacia* to the acacias of Australia have enraged African botanists and naturalists alike. We should keep calling them acacias. If there is any plant that symbolises the African savanna, it is the acacia. In South Africa, in the late 1980s, shortly before that land experienced true freedom, the Standard Bank chose as its logo a backlit acacia. It proved a remarkable Rorschach ink-blot test: various interested parties claimed that in the gaps between the branches, so typical

Aloe volkensii, *the unusually tall tree Aloe, Narok.* (Stephen Spawls)

Desert Rose, Elephant's Foot or Sabi Star. (Stephen Spawls)

of many acacias, they could see the outline of a clenched fist, a Kalashnikov rifle and Nelson Mandela's head, amongst other things. The logo was hastily redrawn.

The name savanna comes from a Central American word, meaning 'land without trees'. The gentle drift of Africa northwards, coupled with the rising of the land along the rift valleys and their margins, has turned what – theoretically – was forest into a relatively arid region. The capture of the rainfall by the higher land in the west and the lack of rain on winds from the north-east means that there is little forest in Kenya nowadays, save on the high land of the south and west, or close enough to the coast to pick up evaporation from the sea. The interior is largely dry, and as it dried out it created savanna habitats; there was no longer enough rain to sustain the forest. In the new habitat, fire has become significant, as have the habits of ungulates, and the habits of people. The savannas of eastern Africa are lands of grass, shaped by fire, grazers and humanity. Grass is a relatively recent arrival on the scene, having got going only in the last 90 million years, but grass is a survivor. There are many species of grass, and it's tough: its leaves contain phytoliths, little chunks of silicon dioxide, SiO_2, that make it both sharp and indigestible (phytoliths have been found in fossil dinosaur dung 67 million years old, indicating that some dinosaurs ate grass). Grass grows from the base, not from the tip, so it can be grazed down by herbivores and grows straight back, unlike most plants. When a grazing animal grips it and pulls, the plant is not ripped up by the roots; there are nodes of weakness along the stem so the blade pulls out, leaving the growing base and important roots behind. Try this yourself: grip a grass by the upper blade and pull. You get the top bit, but the roots and base remain enabling the blade to grow back. This is why your lawn needs mowing so frequently. And if grass dries out and dies, or is burnt off, it has two more survival strategies: fire-resistant seeds and a deep root system. In areas with less than 300mm rainfall, most grasses are annuals; they die off when it gets dry, surviving as seeds. In areas with more than 300mm of rain, the grasses are usually perennial and their roots are long. And grass is near the earth, where the warmth is, and it spreads rapidly. Its extensive roots knit the soil together, and if the going gets tough it accelerates its growing speed; it can set seed very rapidly if conditions are harsh. It is something of a super plant, which is why the only two places you don't find it in Kenya are high on Mt Kenya, above 4,900m, where it is too cold, or in deep forest where no direct light reaches the ground. There are even grasses in the sea, which are monocotyledonous angiosperms although they are not true grasses and belong in different families.

So during the last 30 million years or so, ridges of high land associated with the rift valley gradually created a rain shadow over eastern and north-eastern Africa. The vegetation changed, there were fewer trees and they were less tall, there was more open space, more grass and low plants. At the same time, grazing animals were gradually spreading southwards and affecting the landscape, having entered Africa from the north. Our ancestors were also spreading out across this dryer land, and at least 250,000 years ago, possibly earlier, they began to use fire for cooking. Before that, natural fire, from lightning or decomposition, began to have its effect on the landscape. Fire is uncommon in deep forest, it is usually too wet for that, but woodland and savanna are another matter; dry grass and low bush can burn fiercely and if there is enough of it the fire can spread and keep spreading, often at fairly high speed, especially if driven by wind.

A paper by the paleobotanist Bonnie Jacobs (2004) indicates that grass-dominated savanna began to expand in eastern Africa about 16 million years ago, and by 8 million years ago was widespread. The fossil pollen from Fort Ternan, roughly 14 million years old, contains more than 50 per cent grass pollen; the Tugen Hills fossil plant material indicates a mixture of dry woodland and savanna, and fossil horses entered the area around 9 million years ago; horses are grazers of open country and eat grass.

There seems to be general agreement that these four factors (humanity, grazing and browsing animals, increasing dryness and fire) created the east African savanna landscapes. There is some debate about how significant each was, however. Many think fire was the prime mover; Africa has been called 'the fire continent'. Some good papers have been published on the effects of fire. A 2005 paper on anthropogenic, human-started, fires in Namibia by Asser Sheuyange and co-workers in the *Journal of Environmental Management* suggest that fire is more significant than any other factor in creating savanna landscapes, particularly in better-watered savannas, although in dryer savannas fires may not be so important and termites may have more effect. Fire acts like a generalised herbivore; it reduces the amount of wood by killing small unestablished trees and shrubs, as ground temperatures in a bush fire are often between 300 and 500°C, but are less than 100°C at heights over 1.5m, so the bigger trees survive the fire. Seeds also survive. Studies at Mpala Research Centre on Laikipia found that acacia seeds that were buried a few centimetres largely survived the fire. Even a few, 2–3 per cent, actually lying on the ground were still viable although the surface is the hottest place in a fire. Fire does not appear to change the number of species present, but may change the relative numbers, with the more fire-sensitive species decreasing. The time since the last burn is also significant: if a long time has elapsed then the fire is fiercer and hotter. Fire also benefits ungulates by killing ticks. Experiments show that if savanna is protected from fire, trees become more abundant and thicket or woodland results. At one time it was believed that without fire savannas would have supported thick (or even forest) vegetation and thus savanna was a result of human activity, as humans in general started fires. This now seems unlikely: humans may be significant in the later development of the savanna but they are not the only factors. A fire causes the rapid release onto the surface of elements such as potassium, calcium and phosphorous. It also reduces moisture content, and kills certain animals that live on the surface. However, as any savanna dweller knows, a burn has the effect of causing perennial grasses to regenerate and fresh grass seedlings to sprout. These seedlings are highly nutritious and attract grazing animals; they may cause changes in the normal movements of such species.

There is much discussion about the effect of grazing and browsing game animals and stock on savanna, particularly since the situation is dynamic, with wild game gradually disappearing in many places in Kenya.

Mara grasses; from left Chloris, Dactyloctenium, Harpachne, Sporobolus. *Kenya has nearly 600 grass species. (Stephen Spawls)*

Not everything is in decline, however; one and a half million wildebeest still roam across the Serengeti–Mara ecosystem. They eat the grass, but they also urinate and defecate, providing fertiliser. But often sheep and goats have replaced the game, and the owners of such stock care deeply about them, and make strenuous efforts to keep them alive and increase their numbers. This tends to lead to overstocking and in much of Kenya the rainfall is variable. In dry times, goats are destructive feeders; they pull up plants by their roots and strip the bark off woody plants. This leads to loss of vegetation cover, the scorching of the soil and erosion. Such degradation, particularly around water points, can appear devastating, although studies in South Turkana indicate that such degradation is not necessarily permanent. Dry landscapes can bounce back, once rainfall returns.

In recent years, there has also been much debate about the effect of elephants in the savanna. Elephants are also present in forests, of course, but their effect is less because the primary production (the number of tonnes of vegetation produced per year) is far greater in a forest – between 20 and 60 tonnes per hectare – while in an African savanna it is between 5 and 10 tonnes. Elephant numbers have fluctuated dramatically; they have been poached but elephant populations are also capable of rapidly springing back. Some ecologists believe that in Africa elephants are the primary agents of habitat change, although it has to be said that since elephants do not exist in most areas of Kenya, their effect is no longer significant there. A study in the Arabuko–Sokoke forest found that in areas where there were elephants, there was less leaf litter, fewer beetles, and more flies. Birds like the East Coast Akalat *Sheppardia gunningi* were less common, as the elephants tended to open up the understorey, which the bird didn't like. In the Serengeti–Mara ecosystem in the 1980s, some very interesting developments took place, and the scientists were there to record it. In those years, the Serengeti lost a lot of its elephants, with numbers dropping from about 2,500 to a few hundred; 1,500 were poached and 500 or so escaped into the Mara. Elephant numbers in the Mara rose. The result was obvious quite quickly: the Mara lost parts of its thickets and woodland, and is now largely grassland, and the Serengeti became much more heavily wooded. Some ecologists have noticed the same effect at the bottom of the Aberdares Salient, where trapped

Grassland, Maasai Mara, controlled by elephants. (Stephen Spawls)

grazers have created grassland. The fluctuation between woodland and savanna may also be significant in the story of the Cheetah *Acinonyx jubatus*, a savanna animal *par excellence*. Cheetahs need to chase game at high speed and thus must have open country. Genetic evidence shows that some time in the last 20,000 years the world's Cheetah population dwindled to just a few hundred animals. It is tempting to speculate that it may have been confined to a savanna that shrank (and it is equally tempting to speculate that its extinction in North America resulted from the arrival of early humans who killed all the small game that it lived on).

The effect of humanity on the botanical landscape is hard to quantify. As mentioned, savannas were widespread in eastern Africa 8 million years ago. Presumably fires also occurred. Humanity was not present then, that was about the time that the hominin–chimpanzee split occurred and, according to palaeontologist Pat Shipman, there is virtually no evidence of humanity's controlled use of fire before 250,000 years ago. And savannas existed in the Americas before humanity got there, so it seems certain that humans didn't create the savanna. However, they may have had some late effects. The pastoralists reached Kenya 4,000 years ago, and pollen evidence indicates that, in Uganda at least, anthropogenic forest destruction has been taking place for at least 2,000 years, clearing the trees to produce fuel wood, grow crops and graze livestock.

One unusual botanical habitat, almost unique to Kenya, is that of forested mountains sticking up out of dry country. A rather nice paper, entitled 'Islands in the Desert', was published in 2002 by the German botanist Rainer Bussman. He thoroughly examined a number of such isolated mountain forests, including Mt Nyiro, the Ndoto Mountains, Mt Kulal near Lake Turkana, Marsabit and the Mathews Range. These are beautiful mountains, isolated and ancient igneous and metamorphic monoliths, often with huge cliffs and rounded domes, looming up starkly out of the plains and attracting rainfall by virtue of their height. They are Kenya's equivalent of Yosemite, with equally testing climbing. Bussman and his colleagues found some interesting things. It seems that the underlying rocks (gneisses, granites, volcanics) did not influence the vegetation; what was much more important was the altitude and consequent amount and duration of rainfall (although a study in the Arabuko–Sokoke forest found that the vegetation *was* affected by the underlying geology). All the mountains had juniper forests, but the African Mountain Bamboo *Arundinaria alpina*, grew only on Mt Nyiro. It provided a link between the bamboo forest of Mt Kenya and those of southern Ethiopia, otherwise separated by more than 1,000km of dry and bamboo-less vegetation. This might be connected with the fact that bamboo has a peculiar – one might say lethal – habit in that every 40 years or so, all the bamboo plants flower and then die. Where do they recruit from? Many of the hilltop forests contained huge specimens of the cycad *Encephalartos tegulaneus*, up to 10m tall, giving them a primeval appearance. These forests also contain unique animals. In 2006 a new chameleon was described from Mt Nyiro, and the curator of herpetology at the National Museum in Nairobi, Dr Patrick Malonza, recently found a totally unknown, large-eyed tree snake in the high forests of the Mathews Range; it is not even certain what genus it belongs to. These mountain forests are visually stunning, are reservoirs of unique organisms and have great potential for tourism. They deserve protection but few of them have any, as yet. It is time for action.

African Mountain Bamboo, Arundinaria alpina, *with typical inhabitant, Guereza Colobus. (Stephen Spawls)*

THE HISTORY OF BOTANICAL EXPLORATION IN KENYA

Kenya has not been graced, as yet, by giants of plant collection, such as Sir Joseph Banks, Sir Hans Sloane or the Hookers, father and son. However, some unusual, even notorious, botanists have collected and worked in Kenya. Joy Adamson, the famous author and Lion lady, eventually murdered under mysterious circumstances, painted many of Kenya's plants for Ivan Dale and P. J. Greenway's classic work, *Kenya Trees and Shrubs* (1961). Her second husband, before she married George Adamson, was the National Museum botanist Peter Bally (it was Bally who suggested she change her name to Joy – she was christened Friederike Victoria). The hobby of Sir Michael Blundell, remarkable Kenyan politician, was wild flowers and in 1987 he published a field guide to the wild flowers of East Africa. The cattle baron Lord Delamere collected a number of plants in northern Kenya for the British Museum, before he settled at Lake Elmenteita. Count Eugenio Ruspoli collected a large number of Kenyan plants, before being killed by an elephant in Ethiopia in 1893. After his death his assistant Domenico Riva bravely returned with the specimens to the Florence Museum. The National Museum herbarium in Nairobi was first curated by the redoubtable Evelyn Napier, second daughter of Lord Napier of Magdala. Napier was the man who led the campaign to free a group of missionaries held hostage by the Emperor Tewodros in Ethiopia in 1868; he used Indian elephants for transport.

The first botanical collections from what is now Kenya seem to have been made in the 1820s by the Bohemian collector and naturalist Wenceslas Bojer, who visited Mombasa; a glorious golden weaver bird is named after him. Bojer was followed in 1848 by the appropriately named French botanist Louis Hyacinthe Boivin. The explorers John Hanning Speke and James Grant brought back some 750 plants from their 1860 expedition from Zanzibar to Lake Victoria and thence down the Nile. The two captains did not actually enter modern-day Kenya but their plants (it was actually Grant who did all the collecting) represent the first major East African botanical collection. It is nicely documented in an appendix to Speke's 600-page book and gives something of the flavour of East Africa's botanical riches, although rather spoilt by dismissive comments by a Dr T. Thomson, acting curator of what was then the Hooker Herbarium (the plants are now at Kew). 'The small number of plants indicates a poor flora, and … a comparatively dry climate' sniffed Thomson. The mountainous interior of east Africa was obviously still *terra incognita*.

Three German adventurers made their mark in Kenyan botany in the latter half of the nineteenth century. Baron Karl Klaus von der Decken explored the country around Lake Malawi in 1860; the following year he went inland from Mombasa to explore the Kilimanjaro massif, and returned the following year. In both years he attempted to climb the mountain but was frustrated by poor weather. The Baron collected a number of plants, including a lobelia named after him, *Lobelia deckeni*; he also is honoured in the names of three birds he collected, the attractive eponymous hornbill, a subspecies of the Olive Thrush *Turdus olivaceus* and a subspecies of the Northern White-crowned Shrike *Eurocephalus rueppelli*, although unfortunately this latter is no longer regarded as valid. Sadly, Von der Decken was killed in 1865 in Somalia after his boat foundered on the Juba River. He got ashore but some bystanders murdered him, proof that there are such things as guilty bystanders.

Johan Hildebrandt, originally from Dusseldorf, collected close on 5,000 Kenyan plant specimens that are now in Kew and the British Museum. A medical doctor, Hildebrandt saw himself not as an adventurer or an explorer, but a meticulous scientist, and he once said 'whilst … explorers push relentlessly forward … and do not let the grass grow under their feet, it was my task to collect this very grass'. His first expedition was to Aden, Eritrea and Somalia. He visited Mombasa in 1875, and in 1877 he set off inland, heading for Mt Kenya, which he did not reach, although he spent several months in Ukambani. His notes are fascinating; he describes a meeting with the 'Ariangulo', that is, the WaLiangulu or Sanye. He was eventually turned back beyond Kitui by hostile warriors. His final expedition was to Madagascar, but he died there of internal bleeding. Hildebrandt is commemorated by a number of plants, including the lovely cycad *Encephalartos hildebrandtii*, as well as a little bark snake, an entire genus of ornate frogs, *Hildebrandtia*, a fish endemic to the Athi River and three species of birds – a francolin, a firefinch and a beautiful red-breasted and vividly red-eyed starling, denizen of dry country. And also by an irritating but stingless little bee, the Mopane bee, *Plebeina hildebrandti*. What a collector!

Our third German, Gustav Fischer, we have already met in Chapter 1 in the context of geology. Fischer collected a number of plants in northern Tanzania and southern Kenya. Like both Von der Decken and Hildebrandt, he died young; Fischer was 38 when he died, Von der Decken was 32, Hildebrandt was 34. Joseph Thomson, whom we encountered earlier and who also collected plants, died at the age of 37. All gave their lives for African exploration and the furtherance of science, worth remembering when you gaze on one of the beautiful organisms named after them.

In 1884, Sir Harry Johnston, consul-general in British Central Africa and author of an excellent book on the region, made a scientific survey of Mt Kilimanjaro, and collected 450 species of plant for Kew, including the Giant Groundsel, *Senecio johnstonii*. Johnston also found three new species of sunbird, including the gorgeous Scarlet-tufted Malachite Sunbird *Nectarinia johnstoni* and the Bronze Sunbird *Nectarinia kilimensis*. He also has a vivid three-horned Ugandan chameleon named for him. Johnston had a most distinguished career in Africa; he was appointed vice-consul of the Niger Delta and Cameroon, and Imperial Commissioner for the British Central Africa Protectorate. His books are a delight, containing a wealth of accurate biological information, and illustrated not only by his photographs but also by his superb pen-and-ink sketches. Johnston's great moment occurred in 1899, while he was on a visit to the west of the Semliki River, and heard stories of a strange striped horse that lived in the forest. He managed to get some strips of skin and sent them to the British Museum. Johnston had discovered the Okapi, and it was named after him, *Okapia johnstoni*, by Sir Edwin Ray Lankester, director of the Natural History Museum in London.

The hard, alluring and visually stunning country of northern Kenya attracted a number of botanical collectors. We have mentioned Count Ruspoli and the 3rd Lord Delamere. Another remarkable collector was the American doctor Arthur Donaldson-Smith. A short but highly enthusiastic young man, he started his African adventures with a hunting trip to Somalia, and in 1894, at the age of 28, arrived back there for his next adventure. Frustrated by Emperor Menelik's refusal to let him enter highland Abyssinia, as it was then, he made his way southward across the Ogaden and thence to Lake Turkana, finding a new species of nightjar and collecting several hundred plants, which were sent to the British Museum. From Lake Turkana he went south, almost dying of thirst as he approached Mt Kulal. The party then made their way to Marsabit and eventually Lamu. Donaldson-Smith returned for a second expedition in 1899, and went beyond Lake Turkana to the Nile; thereupon he lost interest in Africa. Although Africa did not wreck his health, alcohol later did, and he died penniless and unknown in Pennsylvania.

We have already met John Walter Gregory. Gregory collected some 300 plants for the British Museum during his Kenya travels in 1892 and 1893, and these are documented in a seventeen-page appendix to his book. In the late nineteenth and early twentieth century, a number of adventurers collected plants in Kenya, including G. F. Scott-Elliot, Theodore Kassner and the missionary Georg Scheffler, who was based in Kibwezi. The historian Sir Claud Hollis, who wrote classic books on the Maasai and the Nandi, collected a number of plants in Maasailand in 1904. In 1909, Dr Edgar Mearns, a magnificently bearded all-round naturalist, accompanied ex-president Teddy Roosevelt on his great African collecting expedition. Mearns, who was an assiduous collector of everything living, obtained a number of plants. In the 1920s it was estimated that at least 10 per cent of the specimens in the Smithsonian had been personally collected by him.

In 1902, the German administrators in what was then German East Africa, now Tanzania, set up a research station, herbarium and scientific institute at Amani, in the Usambara Mountains. The institute was taken over by the British in 1927. The entire Amani collection (apart from those that had been taken to the Berlin Museum, where they were destroyed by a bomb in the Second World War) was relocated to Nairobi in 1950, while the research centre moved to Muguga, near Nairobi, to become the East African Agricultural Research Institute. In the 1930s, the Coryndon Memorial Museum, now the National Museum, had appointed a botanist, Evelyn Napier, as mentioned earlier; in 1935 she married and left the herbarium, although she continued to collect plants. Her successor was Peter Bally. Under the directorship of Louis Leakey, many improvements were made to the museum. The original plant collection was started by Edward Battiscombe, chief conservator of forests, and in fact he wrote what is probably Kenya's first botanical work, *Trees and Shrubs of Kenya Colony*, describing some 80 species. Just after the war, a purpose-built large herbarium was constructed just behind the main museum block, and all the botanical material was moved there, under the charge of Peter James Greenway, who had originally been based at Amani, and Peter Bally. Bally is an interesting man. He was never formally trained in botany, he taught himself plant taxonomy and did it so well that he was in charge of the herbarium for 20 years, from 1938 to 1958. Bally encouraged Kenyans to collect plant material, and he was a highly productive taxonomist; he described more than 120 new species of plant, including ten aloes. One species that he named, *Euphorbia tanaensis*, from Witu, was known from ten trees alone, of which only five remain now. Bally also named a species for his wife, *Euphorbia joyae*. He died in Nairobi in 1980.

In 1958 the decorated British soldier and botanist Jan Gillett was nominated as botanist in charge of the herbarium. Gillett had already been to Kenya; he worked on plants on the Kenya/Ethiopia boundary commission. But in 1958 the colonial government vetoed his appointment; Gillett had been a committed communist and at the tail end of the Mau-Mau emergency it was considered politically risky to appoint him. Bernard Verdcourt was appointed instead, and remained in charge until 1964; after his retirement he published a guide to the common poisonous plants of East Africa. After independence, in late 1963, Gillett did become director, with the Ugandan botanist Christine Kabuye as his assistant; she succeeded Gillett in 1971, but Gillett remained as adviser to the herbarium, paid and then unpaid, until 1984. Christine Kabuye was later unfairly shown the door by someone who wanted her job; she now lectures at Makerere University in Kampala. After a hiatus, the herbarium was headed by the assiduous and flamboyant collector Dr Geoffrey Mwachala; he then became deputy director and Dr

Euphorbia Robecchii *in Tsavo East, a distinctive Somali-Maasai plant species. (Stephen Spawls)*

Itambo Malombe took over. A modest foursquare building behind the museum, the herbarium is nevertheless the second largest in Africa after the National Herbarium in Pretoria, it has close on a million specimens, a library and a plant conservation and propagation unit. The specimens are in the process of being computerised. We would encourage those who are enthusiastic about plants to use the herbarium. They pride themselves on being accessible to the public; take in your photographs, or some leaves, flowers and fruit.

Other prominent botanists associated with the Kenya flora include Professor Len Newton, at Kenyatta University, the Swedish Professor Olov Hedberg, expert on the mountain flora and author of a major paper on Afro-Alpine plant ecology, Quentin Luke, self-taught expert on Kenya's plants, and Professor John Kokwaro, who taught botany for many years at the University of Nairobi and published a number of extremely useful books.

ETHNOBOTANY, ENDEMICS, CONSERVATION AND HOTSPOTS

The study of local knowledge of plants and their uses is known as ethnobotany, and some major ethnobotanical studies have been carried out in Kenya. There are a wide range of potential uses for plants: for example, use as food for humans and for stock, for firewood, construction, as tools, for dyeing, in mystical practices and rituals, in social life, as currency, for carving, as cosmetics and as medicines. Or as poisons! The bark and roots of the tree *Acokanthera schimperi* are used to make a deadly poison used on arrows by generations of Kamba hunters, hence its English name of Poison Arrow Tree. In KiKamba it is called Muvai. A standard test for the potency of arrow poison was to make a cut and let the blood run down, then apply some poison to the streak, bubbles should run up it. When Rainer Bussman carried out his study of the ethnobotany of the Samburu people on Mt Nyiro, he found that the local experts could identify 448 species of plant. Of these, the Samburu reported that 199 species

The glorious sacred Blue Water Lily, Nymphaea caerulea, *on Lake Victoria. (Stephen Spawls)*

were of no use, 180 could be used for fodder, 80 were of medicinal use, 59 species could be used as firewood, 42 for construction, 31 for tools and weapons, 29 for food and 19 for ceremonies. This is a clear indication of a people in tune with their environment. Oddly, though, only five species were used as medicine for livestock. Similar expertise can be found among other peoples; Maasai ethnobotanists are known to have more than 400 herbal remedies.

Plants are one of the nation's resources. A tree like *Markhamia lutea*, Markhamia, produces a tough timber and various remedies are obtained from its bark and leaves. Crushed seeds of the Desert Date *Balanites aegyptiaca*, can be used to purify water; they kill the snail that carries bilharzia. The Neem Tree *Azadirachta indica*, admittedly an exotic species but widely planted in Kenya, is practically a general store: its leaves contain a powerful insecticide which, if placed in clothes, will protect them for months and will keep insects away from stored grass, a solution from the leaves sprayed on crops will protect them from pests and is not toxic, the fruit contains an aromatic oil that can be used to treat skin problems and can be used as a paraffin substitute, the wood makes a hardy timber. There are also some interesting superstitions attached to plants; most involve magical beliefs but they may be significant. A lovely belief in Kenya is that trees along a watercourse are sacred and must not be cut; the same philosophy applies to the Mijikenda kayas at the coast and to trees in churchyards in highland Ethiopia. So such places are havens of wildlife and indigenous plants. Some people believe that burning the wood of *Erythrina abyssinica* (Lucky Bean Tree) will attract lightning and in eastern Kenya there is a belief that anyone who drinks water in which Baobab seeds have been stirred will never be eaten by a crocodile. Many older people in Kenya are aware of the benefits and uses of various plants, but their knowledge is usually stored in their heads, not written down. With increasing urbanisation, the loss of habitat and the desire of young people to move to towns, such knowledge is being lost. This is where the botanical garden and museum becomes of prime importance. Such gardens should be present in all main Kenya towns, and should consist of living stands of local plants, with local

It may not do what it says: ambitious medical claims in Entebbe's botanical garden. (Stephen Spawls)

Left: Botanist extraordinaire, Christine Kabuye, in Entebbe Botanical Gardens. (Stephen Spawls)

names and lists of potential uses. There is an excellent one in Entebbe. Visitors to such gardens, especially older knowledgeable people, should be invited to contribute.

There is a vexed question attached to ethnobotany; this relates to the use of plants in medicine, so-called 'phytomedicines'. The scientific use of plants as a source of drugs is well known and documented; one thinks of plants like *Aloe vera*, the benefits from the Opium Poppy *Papaver somniferum*, source of morphine, the Madagascar Periwinkle *Catharanthus roseus*, compounds of which are used in the treatment of leukaemia, or the derivation of quinine, for malaria treatment, from *Cinchona* bark. The bark from *Prunus africana* (Red Stinkwood, Muiri in Kikuyu, also called *Pygeum africanum*), a huge forest tree found in Kenya, produces a drug used in the treatment of prostate cancer. There has been skulduggery in the past over the exploitation of this tree in Kenya. It is now officially 'vulnerable'. There is a massive industry based around plants and their potential use in medicine. At the same time, a lot of the 'traditional' medicinal uses attached to various plants are ineffective, even highly dangerous, although there is no doubt that some are effective, and others may have the usual placebo effect. Often the active chemicals involved are unknown, the material is unsterile and may be contaminated and few have been subject to well-controlled double-blind clinical testing. And yet traditional medicine is hugely popular. Very often people with problems will visit a herbalist first and a hospital second, although some of the popularity of local remedies no doubt lies in the fact that most villages will have a traditional healer, whereas a clinic or hospital may be many hours or even days travel away. There is actually a scientific publication, the *African Journal of Traditional, Complementary and Alternative Medicine*, whose business includes the rigorous analysis and testing of such medicines. Its principles, set out in an editorial in 2004, are entirely scholarly and professional. However, the situation is complicated. Oddly, many thoroughly scientific books concerned with African botany uncritically describe these traditional remedies, even though they have never been tested and some of them are pretty strange. For example, the use of a root decoction (meaning it has been boiled for a long time) of plants of *Senna didymobotria*, the Candle Bush or Koka Lupin, is believed by the Luhya to 'clean the stomach', as well as curing gonorrhoea and syphilis. A lot of traditional medicine is connected to traditional beliefs, one prevalent such belief being that most illnesses are due to 'pollutants' that block or inhibit digestion. Many remedies are hence effectively purgatives, things that induce vomiting or diarrhoea. Some traditional remedies may well work, but many are untried and unlikely to actually do anything useful, other than – possibly – reassure the victim. A classic example is the slew of local remedies for snakebite.

Snakebite is a major medical problem in rural Africa, and everyone yearns for a simple cure. No such thing exists. When a venomous snake bites, it injects, into the bloodstream a complex cocktail of proteins, designed to paralyse nerves and destroy tissues; its purpose is to immobilise the prey and start digesting it. If the amount injected when a snake bites a human is greater than the body's immune system can cope with, death may follow. The only thing that will save the victim – apart from drastic remedies such as slashing open the bite site, or amputation of a bitten digit – is the injection into the bloodstream of a sterile chemical, anti-snake-venom serum, that will circulate, come into contact with and neutralise the various toxins. No other cure exists. And this means that any suggested remedy

Sodom Apple, purported cure for warts. (Stephen Spawls)

for snakebite that involves pastes being smeared on the wound or any sort of preparation (tisanes, decoctions, macerates etc.) to be eaten or drunk, is not going to work except as a palliative. And a snakebite victim certainly doesn't want some non-sterile concoction introduced into their bloodstream. So no traditional medicines work for snakebite, save as a placebo. which means, sadly, that any cure promoted from the same source for any other malady is equally suspect.

There is something interesting here in Bussman's findings among the Samburu. They used 80 species of plant for human medicine but only five for stock. Humans are able to tell you what effect medicines are having upon them, and a placebo effect exists; most of the preparations used are probably harmless or at least non-lethal. But stock are not susceptible to superstitious or placebo effects and the mistreatment of an animal with a dangerous plant that causes the death of stock is something any pastoralist is desperate to avoid. This indicates that the plants used for treating stock may well be genuinely tried and trusted. Or maybe it's just that goats can't tell their owners that they have a headache. Plants are, without doubt, the biggest potential source of useful drugs known to us, but it is worth remembering that medicine, health and nutrition is still fraught with misinformation and bedevilled by charlatans. This is so even in the most highly industrialised countries, with vigorous legal controls over who can administer what, as anyone who has read Ben Goldacre's *Bad Science* (2008) will know. In Kenya, the 1925 Witchcraft Act outlawed traditional medicine; parts of the law were repealed following independence. What is needed is scientists to record, collect and test it these preparations before the plants and the knowledge disappear. But remember, a great deal of traditional botanical medicine, or any other traditional medicine for that matter, is untested. The moral here if you want to try traditional botanical medicine is *caveat emptor*.

So where are Kenya's plants found and where are the rare ones? What do we need to conserve? As we discussed earlier, even a relatively small area may have several hundred species of plant. Kenya has quite a large network of protected areas. Within these areas, plants theoretically receive as much protection as any other organism, although some of the forests – especially those with few visitors – are under greater threat than other protected areas. Fortunately, introduced plants do not seem to have become the nuisance in Kenya that they have in other countries, although *Salvinia molesta* and Water Hyacinth *Eichhornia* spp. are still a problem, as are the Mesquite Bush *Prosopis* spp., Witchweed *Striga* spp., Lantana and the invasive Central American weed *Parthenium hysterophorus*.

Endemic is the important word where the protection of biodiversity (the variety of life) is concerned. An endemic species is confined to a particular area or region, often a political area, a country. Around 740 species of plant are endemic to Kenya, found nowhere else, that's just over 10 per cent of the total. This may be a somewhat meaningless statistic, in that the distribution of many plants extends beyond Kenya's borders, but only within a certain botanical region; for example Kenya has around 300 Zanzibar–Inhambane endemic plants, coastal specialities. However, the documentation of strict Kenyan endemics is a powerful tool in conservation, enabling the conservation authorities to prioritise crucial areas. It has to be said that Kenya is low on endemics compared to Tanzania, where the ancient forest of the Eastern Arc Mountains supports a rich endemic flora. Central Kenya has only eight endemic moist forest tree species, whereas the Eastern Arc has 67. In Kenya crucial botanical areas for the protection of endemic plants are the coastal forests, in particular the sacred forests known as kayas,

Forester John Mwadeghu in Ngangao Forest, Taita Hills. (Stephen Spawls)

the Taita Hills, the dry, moist and riverine forests of central Kenya, coastal thickets and non-forested areas, dry forests of north-east Kenya and the isolated hills of northern Kenya. But few of those areas are formally protected. Although a fair amount of material exists, relatively few areas have been intensely surveyed. Some plants may have wider distributions than are known at present, but we willl never know unless we check; an atlas of Kenya's plants would be useful.

A study of plants in the Taita Hills area provided pointers. These hills are home to ten endemic plant species. An endemic cycad *Encephalartos kisambo* and an African violet *Saintpaulia teitensis* growing there are known from only a handful of sites, and the violet is pollinated by a single genus of bees, *Amegilla*, according to research by entomologist Dino Martins. These plants might be more widespread, but they need to be looked for. The Taita Hills do not have any proper protection. The level of protection of forest reserves varies. A 1995 IUCN report on Kenya's indigenous forests says rather blandly 'management interventions are generally low key … [the] forestry department does not have the … capacity to look after many of the small forests that are gazetted, let alone those which are not' (Wass 1995). A few years ago, an aerial survey of the Mt Kenya Forest, probably Kenya's most prominent and important forest outside of a national park, revealed that huge areas within the forest had been felled, Camphor Trees *Ocotea usambarensis* had been cut and local entrepreneurs were growing *Cannabis* there. Anyone who drives the smaller tracks inside the Mau Forest is liable to come across big stashes of cut timber, not well concealed, waiting for a truck to come in the night and pick it up. Forests may be poorly guarded, and there is often tension between the guards and local people; one individual is paid to protect the forest, the whole village wants to use it. Trees have to be high on the list of protected plants, for the obvious reason, they are extremely

Botanist Quentin Luke with his new tree, Ngaia Forest. (Stephen Spawls)

Strangler Fig. (Stephen Spawls)

saleable and useful. The average citizen is not likely to exploit the African Violet (although some specialised botanical collectors might; there is a lucrative trade in plants like African Violets, orchids, aloes and cycads) but to an ordinary Kenyan, a nearby tree may represent timber for construction, for cooking, for sale to others or for making charcoal, which can also be sold. But is it fair to say to the rural poor, you cannot cut that tree down, even if you have little money and want to provide for your family? At present, deforestation in Kenya averages 1–2 per cent of the total forested area per year.

So what needs protecting, and how are the ethics of conservation to be balanced with the wants, and rights, of the ordinary citizen? Although many areas are under threat, there are certain reasons to be cheerful about plant conservation. Some really important areas are already protected: forests such as Kakamega, Mt Kenya, Aberdares, Arabuko–Sokoke, Shimba Hills and Chyulu. And some effective plant conservation can be practised in quite small areas, unlike wildlife conservation. (If you want to protect Cheetahs, for example, you need a big area for them to run around and chase gazelles, and there needs to be a decent-sized gene pool, thus a big population.) A few dedicated individuals can protect plants effectively in some quite small plots. However, there can be a problem with plants that need high humidity, for this decreases dramatically within 200–300m of the edge of a forest. In a large forest, there may be plants, or other organisms, that need 95 per cent humidity. If this sort of humidity only occurs once you get 3km into the forest, it follows that any forest that is less than 6km wide will not have anywhere that is more than 95 per cent humid. Thus small forests cannot act as a reservoir for such delicate species.

Plants can also be protected as seeds, carefully stored and grown in nurseries and botanical gardens. Any attempt to protect the bulk of Kenya's Bongo *Tragelaphus euryceros* by rounding them up and storing their frozen sperm, or putting them all in a couple of big enclosures, or sending them to another country, would cause an outcry, for aesthetic reasons; we want to see them wild in their natural habitat. But no one minds if plant seeds are collected and stored, or sent to botanic gardens and nurseries for propagation. Some of Kakamega's lovely trees will look just as nice in a botanical garden in Nairobi. Plants are not so vulnerable to quick destruction; a couple of hundred bullets could decimate Kenya's wild rhino, but it isn't so easy to wipe out Kenya's Red Stinkwood trees. But at the same time, it is important to protect original habitat; one thinks immediately of those lovely hilltop forests, moist islands towering above the dry savanna below, or the sacred coastal forests, the kayas. They are remarkable places, havens of humidity-loving plants. Studies in Tanzania showed that these dense little sacred forests were often 10°C cooler than the surrounding thicket and savanna, and remained 80–90 per cent humid for days following a rainstorm, where the surrounding savanna became dry quite quickly.

In the 1980s, the famous ecologist Norman Myers (who worked in Kenya, and regularly visited Nairobi Snake Park) coined the term 'biodiversity hotspot' to define certain critical areas that had high numbers of plant and animal species and were under a high level of threat. The numbers are interesting: the hotspots encompass only 2.3 per cent of the world's land area but are home to more than half the world's plant species and a third of its land vertebrates. There are just under 40 such hotspots, seven in Africa. Bits of Kenya lie within three of the hotspots: the horn of Africa, eastern Afromontane and the coastal forests of eastern Africa. Most of north-eastern Kenya lies within the horn of Africa hotspot. Typical animals of this region include the Gerenuk *Litocranius walleri*, Grevy's Zebra *Equus grevyi*, the Masked Lark *Spizocorys personata* and the Speckled Sand Snake. But there also are around 2,800 species of vascular plant endemic to the area – thirteen genera alone are endemic to Socotra Island. Kenya has virtually no national parks in the north-east, save Malka Mari on the Ethiopian border, and no protected forests there, although some threatened forest trees occur there. It is a hostile area, with little rainfall, and most species found there tend to be hardy with wide distributions. It should not be too difficult to protect some of it for botanical purposes. Some would say it doesn't need protection; vast areas of near-desert are hard to overexploit. The eastern Afromontane hotspot includes the Aberdares, Mt Kenya, Kakamega, as well as the Eastern Arc Mountains, the land along the Albertine Rift down to Malawi and the Ethiopian and south-western Arabian highlands. This really is an area of high endemicity, with many organisms with minute ranges. The whole hotspot is estimated to contain some 7,600 plant species, of which about 2,400 are endemic. Unlike the horn, there are lots of protected areas here, but there are also lots of people. Within this hotspot, the Ethiopian highlands is also a Vavilov centre, named after the Russian botanist Nikolai Vavilov. These are regions that were centres of plant domestication and thus contain valuable original genetic plant material with potential for domestication. The coastal forests hotspot extends from southern Mozambique north to coastal Somalia, and this contains some 4,050 plant species, of which 1,750 are endemics. Kenya has a number of protected zones inside this region. The hotspots concept has come in for criticism, with some scientists arguing that they overemphasise the importance of plants and certain localities and ignore smaller significant areas. It's as if one might say that it doesn't matter if the Kakamega Forest is felled, as the keynote species within it occur in huge numbers elsewhere and we should concentrate on them. But to Kenyans, the Kakamega Forest is important. And the crucial point about the hotspots concept is that it increased awareness of the variety of life in the earth's wild places, and the importance of protecting it. This can only be a good thing. The original hotspots book was recently revised. It is a superb publication, with stunning photographs, but weighing 4 kilograms it is not a field book!

One problem with plants in Kenya is that there are almost certainly species out there that have not been found. More survey work is needed before they disappear. A huge undescribed forest tree was recently discovered in the Ngaia Forest, near Meru National Park; a new tree (*Vangueriopis shimbaensis*) recently described from the Shimba Hills is so far only known from four individuals. Who knows what other riches may be out there, and how they may be useful? Land hunger is a major factor in Kenya, as the population increases. We need surveys, to document the vegetation; we need small, carefully sited botanical reserves in crucial areas, botanical gardens and research on plant products. Crucial stands of vegetation need protecting. In Egypt, a country with little vegetation, the government has tried to prevent vegetation being cut for fuel by supplying subsidised bottled gas. Kenya at present has no gas reserves but supplies could be piped in from neighbouring Tanzania. Some sort of scheme whereby commercial plantations produce wood for charcoal-making has to be better than the present system; the devastation of the woodland between Makindu and Mtito Andei for charcoal is heartbreaking. And why not allow local people to harvest forest products sustainably? The public need more information and enthusiastic botanists need to be talking to Kenya's young people, either through the museum/botanical gardens system or on outreach programmes. But time is against us; a burgeoning population and their demands means that action has to be urgent. A gloomy survey in 2010 by scientists at Kew Gardens estimated that one in five of the world's plant species face extinction. It would be unfortunate if a cure for certain cancers, or a product that could keep our skin from cracking, or treat malaria was in the leaves of one of those threatened plants.

THE LITERATURE

For the professional botanist, the 'Floras' are essential. If you want to do botanical work in Kenya then you have to have the *Flora of Tropical East Africa*. The illustrations, keys and facts are all there. And for peripheral work you will need *The Flora of Ethiopia and Eritrea* and *The Flora of Somalia*. But you also need deep pockets and determination; these are expensive and dense scientific works, without colour illustrations, to be used for reference, never light reading. They are produced in many volumes by various botanical institutions and details can be found through the Internet. Other solidly reliable tomes include the 404-page *Sedges and Rushes of East Africa*, by Richard Haines and Kare Lye (published by the East African Natural History Society in 1983), Andrew Agnew's massive *Upland Kenya Wild Flowers*, published in 1974 by Oxford University Press, or Min Chuah-Petiot's 274-page illustrated guide, *Mosses, Liverworts and Hornworts of Kenya*, privately published in 2003. The enthusiastic naturalist would be best to start with a small well-illustrated guide, with colour pictures, and there are several books that fit the bill. The *Collins Guide to Tropical Plants* by Wilhelm Lötschert and Gerhard Beese (1999) covers some 250 species. A very useful book is David J. Allen's *A Traveller's Guide to the Wildflowers and Common Trees of East Africa*, published by Camerapix in 2008. This cover about 110 species, nearly all indigenous, slightly biased towards Tanzanian species, arranged in groups according to rainfall and altitude. The photographs vary wildly in quality, but this is a good book, easily slipped in a pocket, with a decent selection of prominent species and some thoughtful opening essays concerning plants in the landscape.

Another useful book, with close on 900 colour photographs and covering some 330 species, is Najma Dharani's *Field Guide to Common Trees and Shrubs of East Africa*, published by Struik in South Africa in 2002. It is quite a handy size for the field and is tough and well bound. Allen's book and this one are the ones to have in your car in the field. However, there is a curious feel to this latter book. Many exotic species are described and yet some prominent species, such as Miraa, are not mentioned. Twenty exotic palms appear here, but several of Kenya's eight indigenous palm species are missing.

A lovely book of diamond-hard material is Tim Noad and Ann Birnie's privately published *Trees of Kenya*. It covers 300 of Kenya's trees, largely indigenous species, with black and white drawings for every species, showing the trees, the leaves and the fruit. Sadly, it has only sixteen pages of colour plates. This is a book that every aspiring Kenyan botanist should have, and it cries out to be reprinted with more colour plates. A book with lots of colour pictures is *Orchids of Kenya*, by Joyce Stewart and colleagues, published in 1996 by Timber Press. The high-quality pictures of Kenya's 280 orchids in this book were largely taken by Bob Campbell, the *National Geographic* photographer who took the famous 'first contact' picture of Dian Fossey touching hands with a wild gorilla.

In 1969 Collins published *The Common Poisonous Plants of East Africa*, by Bernard Verdcourt, ex-director of the herbarium, and E. C. Trump. It is a rather dense book, but botanists will find it useful. It contains a wealth of botanical detail, including some most interesting material on symptomology and local use of poisons. However, it has no colour plates and only a handful of black and white drawings. The standard text on poisonous plants is John Watt and Maria Breyer-Brandwik's *The Medical and Poisonous Plants of Southern Africa*, published in Edinburgh by E. & S. Livingstone in 1932 and last reprinted in 1962.

Sir Michael Blundell's *Collins Photo Guide to the Wild Flowers of East Africa* (1987) has 864 rather small but nice colour pictures, mostly of flowers but with a few trees. It also lists the numbers of species of the 10,000 or so seed plants known from East Africa. The classic work on Kenya's plants is, of course, *Kenya Trees and Shrubs* by Ivan R. Dale and P. J. Greenway, covering about a thousand species. This was first published in 1961, by Buchanan's Estates in Kenya, and illustrated with 31 beautiful colour plates by Joy Adamson. In 1994 a revised edition by the Dutch botanist Henk Beentje appeared as *Kenya Trees, Shrubs and Lianas*, retaining the colour plates but covering 1,850 species; it was published by the National Museum. This book has something that virtually no other East African plant book has, as far as we know – distribution maps. These are very useful to the amateur. The book also has lists of local names, and contains not just dry botanical stuff but lots of fascinating snippets of information, like the fact that the fruits of the Borassus Palm are eaten with enjoyment by both elephants and Lions. This is a book to move on to when you know most of the plants in the picture guides. But it is hard to get hold of. It needs to be reprinted as a softback with more colour plates.

John Kokwaro's book *Flowering Plant Families of East Africa*, published in Nairobi by East African Educational Publishers in 1994, is an extremely good book, although it has no colour pictures. It has a very useful opening essay on botanical exploration in East Africa, a section on plant taxonomy, a really good glossary and twenty-plus pages of diagrams showing the essential features of plant identification. It also has keys to all the gymnosperm and angiosperm families, plus keys to the groups of dicotyledons, and keys to the genera. After that, you are on your own, but there are keys to a few species, for example the figs. It's a great book for a budding botanist.

There are a handful of books covering neighbouring countries that are useful in Kenya. Reichard Fichtl and Admasu Adi's book *The Honeybee Flora of Ethiopia* illustrates some 400 flowering plants and 100 trees with large high-quality colour photographs. Eberhard Fischer and Harald Hinkel's book *La Nature du Rwanda/Natur Ruandas* covers some 200 flowering plants, with a colour photograph of each. Fischer has also recently published on the flora of the Nyugwe Forest and Rwandan orchids. The pick of the crop to date, where popular African botanical books are concerned, is *Trees of Southern Africa*, by Keith Coates Palgrave, published by Struik in 2002. This is nearly a thousand pages, with more than 300 colour illustrations; some 700-plus species are described, and each has a distribution map showing its range in southern Africa.

Gardening in East Africa, a Practical Handbook is a book of historical interest. Edited by Arthur Jex-Blake and first published in 1934, the later editions contain a number of Joy Adamson's colour plates. The 1957 edition had 20 plates by Joy, plus three by Peter Bally. Bally was no mean artist himself; his pictures of succulents are gems. But Joy's paintings jump off the page; the quality of the reflected light is glorious. This isn't really a book for botanists, it's about gardening, but it has a lot of material of relevance to the botanist, for example an essay on succulents by Bally and one on flowering trees and shrubs by Lady Muriel Jex-Blake. These books have become collectors' items and the originals have kept their value; an updated version was reprinted by the Kenya Horticultural Society in 1995. Arthur Jex-Blake was a well-known Nairobi heart surgeon, and he and his wife endowed Nairobi with a lovely little botanical garden on the lower slopes of Nairobi Hill.

Four books that have good botanical material, giving background information on the vegetation and the environment, although none is any good for identification, are Pratt and Gwynne's *Rangeland Management and Ecology in East Africa* (Hodder and Stoughton 1977), *East African Ecosystems and Their Conservation*, by McClanahan and Young (Oxford University Press 1996), *East African Vegetation* by Lind and Morrison (Longman 1974) and Lovett and Wasser's 1993 work, *Biogeography and Ecology of the Rainforests of Eastern Africa*, published by Cambridge University Press. A somewhat dated but still fascinating work on the vegetation of Mt Kenya is Malcolm Coe's book *The Ecology of the Alpine Zone of Mt Kenya* (Dr W. Junk Publishers 1967). Coe worked on the mountain for four years; this book is a summation of his PhD and much of it is given over to studies of the mountain flora. Coe also co-authored a guide to Kenya acacias, with Henk Beentje, and a very readable book (*Islands in the Bush* published by George Philip in 1985) on the activities of a Royal Geographical Society research team at Kora National Park. The chapter entitled 'God-forsaken Wilderness' gives an overview of both the vegetation of the reserve and the way a botanical research team operates in the field. Also of historic importance to Kenya botany is *A Numbered Checklist of Trees, Shrubs and Noteworthy Lianes Indigenous to Kenya*, by Jan Gillett and P. G. McDonald, published in 1970 by the Kenyan Government Printer. Good material on Kenya's forests and their status can be found in a 150-page report by Peter Wass entitled *Kenya's Indigenous Forests: Status, Management and Conservation*, published in 1995 by IUCN. Somewhat dated but extremely sound on East African montane vegetation and how it has changed during the last 2 million years is A. C. Hamilton's *Environmental History of East Africa* (Academic Press 1982).

The *Study of African Plant Taxonomy* conference in 1994 at Wageningen, in the Netherlands, on the biodiversity of African plants, resulted in a book of the proceedings that probably represents the best overview of African botany to date. It was published by Kluwer Academic Publishers in Dordrecht, and contains literally dozens of useful papers, including one by Henk Beentje on centres of plant diversity in Africa, one by Jon Lovett and Iib Friis on the patterns of endemics of the plants of north-east and East Africa and a paper on the conservation of plant biodiversity in Kenya, with some very useful statistics, by Stella Simiyu. Good material is also available for the subsequent conferences, held in Harare and Addis Ababa, among other localities. Some other equally significant papers on East African botany are contained in the 1978 book *Biogeography and Ecology of Southern Africa*, published by Dr W. Junk Publishers and edited by M. J. A. Werger. This book contains a lot of material relevant to Kenya. Among its many thoughtful essays are one by Frank White on the Afromontane region, a nice description of the Quaternary vegetation changes in Africa by that expert palynologist E. M. van Zinderen Bakker, and a really excellent summation of Africa's recent and distant botanical history, entitled 'Late Cretaceous and Tertiary Vegetation History of Africa', by D. I. Axelrod and P. H. Raven. Serious African botanists need both these works.

It has to be said that many of the books we have mentioned are out of print and hard to obtain. They also tend to hold their value and you are not likely to find a copy of Beentje's book, or the original Dale and Greenway, in a second-hand bookshop. The occasional bargain does appear on the Internet.

The East African Natural History Society (now Nature Kenya) has published, over the years a lot of thoroughly sound botanical papers on Kenya. The analyses of the flora of the Shimba Hills, Kakamega Forest, Meru National Park, Lake Nakuru National Park and Ol Ari Nyiro all appeared in the EANHS journal and they published Haines and Lye's monograph on the sedges and rushes of East Africa. Another good paper is Henk Beentje's 1998 work on an atlas of the rare trees of Kenya, published in *Utafiti*, the National Museums of Kenya (NMK) journal. Kenya grasses are covered in K. M. Ibrahim and Christine Kabuye's 1988 paper 'An Illustrated Manual of Kenya Grasses', published by the Food and Agriculture Organisation of the United Nations (FAO) from their Rome headquarters.

Finally, for those interested in Africa-wide material, there is a powerful searchable database, the 'Plant Resources of Tropical Africa'. Find it at http://database.prota.org/. It is fairly easy to use, in French and English, and there are pages of information and a simplistic map (showing distribution by country) for most species.

Chapter 6

The Mammals

Meyek olenkaina ilala lenyena ('the elephant does not get tired of its tusks') – Maasai proverb

Murunguru utuhaga na ime ('the wild cat skips in the dew') – Kikuyu proverb

Mammals are the reason most visitors come to Kenya. A recent tourism survey found that over 80 per cent of visitors came to see large game animals; the two most hoped-for encounters were with elephants and Lions. An average one-week safari to Kenya's national parks will yield sightings of between 30 and 40 large mammal species; a total that cannot be bettered outside sub-Saharan Africa. About 400 mammal species are known from Kenya, but only 130 or so of these species weigh over half a kilogram. The provisional Kenyan list includes 38 species of shrew, about 100 species of bat and 108 rodent species. Kenya also has 37 species of antelope; more than any other African country. Four species of Kenyan mammal are critically endangered: the Tana River Red Colobus Monkey *Pileocolobus rufomitratus*, the Black Rhinoceros *Diceros bicornis*, the Hirola or Hunter's Hartebeest *Beatragus hunteri* and Aders's Duiker *Cephalophus adersi*.

Kenya's mammals can provide us with many superlative statistics:

- The world's largest land mammal, the African Elephant *Loxodonta africana*, is widespread in Kenya. Adult males have been known to weigh over 6 tonnes (6000kg). Elephants also appear to be the longest-lived mammal (after humans), reaching at least 70 years. The world's largest living mammal, the Blue Whale *Balaenoptera musculus*, which can weigh 100 tonnes, has been recorded off Kenya's coast.
- The world's tallest mammal was a Maasai Giraffe *Giraffa camelopardalis* from Kenya that was at least 5.48m tall.
- Kenya's most numerous large mammal, apart from humans, is probably the wildebeest *Connochaetes taurinus*. In 1977 one and half million wildebeest entered the Maasai Mara on migration from Tanzania.

The Mammals

One of the world's greatest wildlife spectacles: the great migration. (Stephen Spawls)

- The Cheetah *Acinonyx jubatus*, still fairly common in parts of Kenya, is theoretically the world's fastest land mammal although there is much debate about its top speed, with estimates varying from 70km/hr (44mph) to 114km/hr (70mph).
- One of Kenya's mammals, the Naked Mole-rat, *Heterocephalus glaber* differs from every other mammal in that it is an ectotherm (like a reptile), receiving its heat from outside. Most Naked Mole-rats spend all their lives in one burrow, and have been recorded giving birth to 28 offspring – the largest number known for any mammal.

This chapter will discuss the origins of Kenya's mammals, and fossil mammals in Kenya, and the inventory, ecology and development of Kenya's mammal fauna. There are also sections on zoogeography, and on interactions between mammals and humans in Kenya. The chapter finishes with an overview of literature and research on mammals. As usual, each section is self-contained.

MAMMALS, THEIR ORIGINS AND FOSSIL MAMMALS IN KENYA

We all know what mammals are. We ourselves are mammals; cats, whales, mice and elephants are mammals. Mammals have backbones and four legs, or the remains of them. They also have certain unique features: a control centre called the neocortex in the brain, hair for insulation, sweat glands, and they nurse their infants with milk. Their blood is constantly warm (Naked Mole-rats excepted), meaning they do not need to warm up and are always ready for action; they can live in ultra-cold places. Mammals have a single bone in the lower jaw, three sound-transmitting bones in the ear, and some have movable external ears. They have teeth of differing types, some for cutting, some for piercing and some for crushing (and some shrews have a venomous bite!).

Naked Mole-rat, a bizarre ectothermic semi-desert mammal with a record number of offspring. (Glenn and Karen Mathews)

Although some biologists dislike rating evolutionary development, mammals have some advanced evolutionary strategies. They keep their young inside them until they are fairly well developed, and when they are born the mother produces milk to feed them. These are clever anti-predator strategies. If danger threatens the mother runs away with the babies inside her; the bigger they are at birth the fewer things can eat them and the mother doesn't have to leave the newborn baby to find food for it. There is a downside: parental care is necessary and the loss of an offspring is distressing. (Contrast this with the mother Spitting Cobra, who leaves her eggs in a warm, deep termite hole and goes off without a care in the world.)

There are other disadvantages. There is a limit to how many babies you can have developing inside you before you get too large to move. And being warm-blooded means you need a lot of fuel; 80 per cent of energy generated from food goes on keeping mammals warm. If that food is low in energy, then you need to eat frequently. The average African Elephant spends most of its day eating 150–200kg of food, of which 40 per cent is usable. And if you are a small mammal, with a big surface area to volume ratio, you lose heat relatively quickly and have to eat a lot of food; a small shrew eats five times its own weight in food daily.

In terms of species numbers, mammals are not very diverse; there are about 5,500 species (of which more shortly), less than any other vertebrate group. But in terms of use of habitat, the ability to modify one's environment and adapt to changing circumstances, mammals have to be regarded as a highly successful group. And finally, one mammal has evolved to become the most successful species of living animal known; a species that has, despite its recent evolution (the present species is less than 200,000 years old) spread from pole to pole and even left the earth. But that species is also the most destructive animal this world has ever seen; it has destroyed huge areas of this world and has invented the potential to destroy itself. In the wise words of Edward Rodwell, one of Kenya's most charismatic journalists, take a look at that mammal in the mirror tomorrow morning.

Mammals evolved from reptiles and became hairy and warm-blooded. A group of mammal-like reptiles, the cynodonts, with complex cusp-patterned teeth, lived 260 million years ago. The major radiation of mammals is believed by some to have occurred when the dinosaurs disappeared, 65 million years ago. However, the primitive egg-laying monotremes split off nearly 200 million years ago (some monotremes are venomous. But the Crested Rat *Lophiomys imhausi*, found in Kenya, manufactures its own poison; it chews the bark of the poison-arrow tree and slathers the poison onto hollow hairs on its back, to protect it from the bite of predators). The Australasian marsupials became isolated some 130 million years ago. Recent work indicates that, when Africa split from South America, about 103 million years ago, the ancestral placental mammals split into three groups. These were the Xenarthra (anteaters, armadillos and sloths), the Afrotheria (elephants, hyraxes, dugongs and manatees, aardvarks, golden moles, tenrecs and elephant shrews) and the Boreoeutheria (everything else). The Afrotheria developed in isolation in Africa, the Xenarthra never left America. The 'supergroup' Boreoeutheria dispersed into North America, Europe and Asia, and became the Laurasiatheria and Euarchontoglires. These two groups contain many animals that are in Africa today, but were not there originally, such as primates, equids, carnivores, hippos and rodents, but they radiated first throughout the Americas and Eurasia and nipped into Africa when it bumped into Eurasia 30 million and then 17 million years ago. Today the African mammal fauna consists of both original inhabitants and upstarts that came from the north.

Strange and unfamiliar creatures roamed over Africa 20 million years ago, such as *Brachyodus*, an extinct ungulate, which looked like a huge, heavily built horse with a giraffe's head, *Pterodon*, a wolf-like carnivore with massive jaw, a variety of hyraxes and some strange elephants, including the pig-snouted *Archaeobelodon* and the mammoth *Eozygodon*. Various species of elephant and hyrax, some relatively large, roamed the northern half of Africa until the arrival of the intruders from the north. The hyraxes then disappear from the fossil record; and it is presumed that immigrant carnivores ate them and immigrant ungulates displaced them from their niches. The only hyraxes that remain today are small, living amongst rocks or trees, where the grazing animals do not go. Possibly the demise of the hyraxes was also connected with their inability to regulate their body temperature (they need to bask, and huddle together for warmth) or their rather primitive brain.

Records from the Lake Turkana sites show a huge variety of mammals. A paper by Kay Behrensmeyer and co-workers (1997) looked at more than 130 species of mammal that lived in the area, between 4 million years ago and now, including pigs, antelopes, shrews, rhinos, elephants, primates and rodents. Most of them appeared between 4.5 million and 3 million years ago, and by 800,000 years ago nearly all of them had gone; only a handful of species, like Grevy's Zebra *Equus grevyi*, Greater Kudu *Tragelaphus strepsiceros* and Black Rhino, are still alive. There seems to have been three cycles of extinction, at 2.8, 1.7 and 0.8 million years ago. This is probably connected with climate change – north-east Africa gradually getting dryer.

The hypothetical big picture here is a number of large mammals finding a suitable area of woodland or savanna, speciating rapidly and then dying off as the land dried up. There were at least eleven species of elephant present in northern Africa (including the Asian (Indian) Elephant *Elephas maximus*), only one is left. There were a number of fossil pigs (at one time 77 species had been described, but John Harris and Tim White (1979) reduced the total to around fifteen species). Even so, only four species of pig survive today in East Africa – the two warthogs *Phacochoerus sp.*, Bush Pig *Potamochoerus larvatus* and Giant Forest Hog *Hylochoerus meinertzhageni*. Pig diversity has plummeted. A giant giraffe-like mammal, *Sivatherium*, some thirteen species of horse or zebra and five species

Bush Hyrax, representative of a once-huge group decimated in Africa when the carnivores arrived. (Laura Spawls)

of hippopotamus were present. Relatively few carnivore fossils have been found, but they were exciting. One was a huge 500kg shaggy bear, *Agrotherium*; no bears are found in Africa nowadays. Three big 'sabre-toothed' or scimitar cats also lived in Kenya and have become extinct; it is reasonable to suppose that they too lost out as a result of the climate change, which would have meant their prey became much harder to find. Fossil Cheetahs are known from this time.

A study of fossils from the lower Omo River, in Ethiopia, indicates similar extinctions, with a decrease in forest mammal species and a rise in grassland species between 3.6 million and 2.4 million years ago, with a steep increase in arid country forms between 2.6 million and 2.4 million years ago. A major paper by climate scientist Peter deMenocal (2004) documents the appearance and disappearance of fossil and living bovids (antelopes) across Africa during the last 7 million years; a graph shows details of 133 species of antelope, of which only 31 are extant; some 45 species have disappeared in the last 3 million years. Some believe our hunting ancestors were to blame, but this cannot be quantified. Many mammal species that were of no importance to hunters (rodents, shrews and so on) also disappeared. It does seem more likely that climate change was to blame.

In contrast, close to 90 species of fossil mammals of Miocene age, 18 million to 15 million years ago, have been found on Maboko and Rusinga Islands in Lake Victoria, recovered during nearly 80 years of diligent field work. The plant assemblage indicates that the dominant vegetation there was wet lowland forest. The fossil mammals include some primitive carnivores, a giant hyrax, extinct hares, rhinos and antelope. The islands also yielded a superb collection of bones of the extinct primate genus *Proconsul*, four species of which are known. It is presumed that, unlike the Lake Turkana fossils, the Rusinga and Maboko mammal fossils represent a forest or woodland assemblage, but the problem is that many of the species described are extinct, and we cannot be sure of their affinities, and the length of succession is not as long as that seen around Lake Turkana. In general, moist and forest areas are not good for fossilisation; the warmth and moisture contribute to decomposition, and suitable sites with exposed strata are hard to find; they are concealed by vegetation. This probably explains why no chimpanzee or gorilla fossils are known. In the end, however, one has to remember that fossils are rare. Not everything becomes fossilised, and the more things are found, the more our perception of what went on in the past will change.

KENYA'S MAMMAL FAUNA: THE INVENTORY AND ECOLOGY

So during the last 4 million years, the Kenyan mammal fauna has changed dramatically and this change is probably due to fluctuation in the climate. If more than 70 per cent of the mammal species that were alive in Kenya over 2 million years ago are now extinct, what are we left with? David MacDonald's massive and scholarly 2001 work *The New Encyclopaedia of Mammals* lists some 4,680 species of mammal worldwide, while Campbell and Reece's superb undergraduate text *Biology* (2008) states there are 5,300 known mammal species. The mammal entry on Wikipedia reckons there are 5,400, Wilson and Mittermeier's massive handbook of world mammals (2009) says there are 5,490 species. Why the difference? Can the mammal numbers have increased by 810 in a decade? Is biodiversity really increasing? Probably not. A few small changes may be due to recent discoveries. New living mammals are being found all the time, even big ones; the Saola or Asian Unicorn *Pseudoryx nghetinhensis*, a 90kg ungulate from the mountain forests of Vietnam and Laos, was only discovered in 1992, and the first live one was captured in 2010. However, many changes – in particular increases – in species numbers are due to scientists (and sometime non-scientists) splitting populations, changing subspecies into species. Some authorities believe there are two species of African Elephant, Forest and Savanna, others say there is only one. And this is a huge, prominent, well-studied species; there have probably been more major papers published on the African Elephant than any other wild mammal, more than a thousand, in fact. There is similar debate about the status of the Northern and the Southern White Rhino *Ceratotherium simum*, the Bush Pig and the Red River Hog *Potamochoerus porcus* and the Lowland and the Mountain Gorilla. And these are large animals.

White Rhino never occurred naturally in Kenya in human memory. (Stephen Spawls)

At the very small level, there is even more taxonomic confusion. A classic example is a little short-nosed African rodent, known from Kenya and widespread in Africa, *Dasymys incomtus*, the Shaggy Swamp Rat. This innocuous little grey-brown swamp dweller has in the past been described as seven different species (*Mus incomtus* and *Dasymys gueinzii*, both from Durban, *Mus bentlyae*, from Zaire, *Dasymys medius* and *Dasymys montanus*, both from the Rwenzoris *Dasymys helukus*, from Mt Sergoit near Eldoret, and *Dasymys orthos*, from Bunyoro in Uganda). That mix-up has now been resolved, and with the ease of access to taxonomic material on the Internet and free exchange of museum specimens, it is less likely that the same species will get described twice, but it still happens. For this reason we have often been cautious with numbers and avoided being precise. There is often acrimonious debate about numbers. One expert will say that there are 35 species in an area; another will angrily say there are 38. Often both are right; they are just counting in different ways. Or one expert will include some species that are likely to occur in an area, another scientist will refuse to accept records without actual voucher specimens, i.e. animals deposited in a museum. And, it has to be said, there are a few taxonomists who are not purely motivated by the ethics of good scientific practise; some have split variable species into subspecies, or full species, or described new species, for personal, political or financial purposes.

In raw terms, the number of mammal species recorded from Kenya is not particularly startling; just fewer than 400 species are known, in fifteen orders (27 orders are known worldwide). If we stick conservatively with MacDonald's figures above, then about 9 per cent of the world's mammal species are found in Kenya (the figure for birds is 11 per cent). There are about 1,150 species of mammal in Africa, according to Jonathan Kingdon (and who should know better), so nearly 35 per cent of Africa's mammals are found in Kenya. There are certainly a few mammals yet to be discovered in Kenya, particularly among the smaller rodents, insectivores and bats; there might even be a little carnivore or two, or an antelope. In 2010 an unknown colour form of elephant shrew was spotted in the Boni-Dodori Forest; it is probably a new species.

A rare sight in the wild (this is captive), the Spot-necked Otter. (Stephen Spawls)

Compared with other African countries, Kenya has a fairly rich mammal fauna; Uganda has about 330 species, Tanzania about 350. It is the tropics, not just the variety of habitats, that create this species richness. Botswana is slightly larger than Kenya but only has 160 species of mammal, South Africa is twice the size of Kenya, and has a huge diversity of habitats, but has about 270 species. Great Britain has only 56 mammal species, of which several are introduced, including two of the commonest, the Grey Squirrel *Sciurus carolinensis* and the European Rabbit *Oryctolagus cuniculus*. Kenya has been largely spared, so far, the spread of introduced and problematic species, although a few introductions have taken place; White Rhino into several national parks and other conservation areas, Red Deer *Cervus elaphus* on Kipipiri (they died out; the day length never shortened and it is this that stimulates ovulation in females), chimpanzees into the conservancies around Mt Kenya, European rabbits at Meru, Coypu *Myocastor coypus* around Lake Naivasha, also now extinct, and of course, those commensals of humanity, the Brown Rat *Rattus norvegicus* and the Black Rat *Rattus rattus*, widespread in the highlands. There have also been a few translocations: Hirola from the Tana River into Tsavo East, Roan Antelope from the Ithanga Hills into Shimba Hills (they didn't survive) and Giraffe into the Shimba Hills, not counting translocations of mammals from one area to another within their natural range, like elephants from Mwaluganje to Tsavo.

The Kenya list is as follows, with numbers to be taken with a pinch of salt for the reasons mentioned: 38 shrews, the Giant Otter Shrew *Potamogale velox*, a hedgehog and a golden mole (order Insectivora), 10 fruit bats and 90 insect-eating bats (order Chiroptera), 19 whales and dolphins (order Cetacea; not land mammals), 108 rodents (including 10 squirrels, one flying squirrel, two cane rats and three porcupines, plus 70 mice and rats, eight mole rats, 10 gerbils, order Rodentia), 19 primates (monkeys, bushbabies and the Potto *Perodicticus potto*, order Primates), 34 carnivores (order Carnivora), three odd-toed ungulates (zebras and rhinos, order Perissodactyla), 44 even-toed ungulates (37 antelope, the Giraffe, one hippopotamus, one buffalo and four 4 pigs, order Artiodactyla), plus a ragbag of other animals that include five elephant shrews (order Macroscelidae), three pangolins (order Pholidota), four hares (order Lagomorpha), the Aardvark *Orycteropus afer* (order Tubulidentata), one elephant, (order Proboscidae), five hyraxes (order Hyracoidea) and the Dugong *Dugong dugon* (order Sirenia). That's fifteen out of nineteen orders of placental mammals (no marsupials, or pouched mammals, occur anywhere in Africa). The only placental mammal orders that are not represented in Kenya are the seals and sea lions, tree shrews (a group of weird little mammals found in south-east Asia), colugos (flying lemurs, also from south-east Asia) and the Xenarthrans (anteaters, sloths and armadillos of South America). One might be inclined to say that the whales and dolphins shouldn't really be counted as part of Kenya's mammal fauna as they live in the sea, and this would reduce the count to around 380 total. You could apply the same logic to that strange fat Afrotherian mammal the Dugong; it lives in the sea, but Alex Duff-MacKay, of the National Museum, once observed a Dugong in the lower Galana River, near Kisiki Cha Mzungu village.

Kenya's 400 mammal species found are not all found in one area, of course. The Black-and-white Colobus or Guereza *Colobus guereza* is found in the wet forests of the northern slopes of Mt Kenya, the Yellow Baboon *Papio*

Elephants in the mist in the Shimba Hills. (Stephen Spawls)

Hidden hare, wise in African mythology. (Stephen Spawls)

Grant's Gazelle, an East African speciality. (Stephen Spawls)

cynocephalus occurs just down the hill in dry savanna near Isiolo, the two species are found within a few kilometres of each other but never in the same place, at least in Kenya. Each habitat tends to have a distinct mammal community. If the area is disturbed, with humans living in it, then the community will lack certain species, usually larger animals, or those that humans consider a nuisance. The high grassland on the eastern side of the Ngong Hills still has Grey or Bush Duikers *Sylvicapra grimmia*, hares, porcupines, baboons and a few nocturnal species, but there are no Lions *Panthera leo* or rhino left there; they can't co-exist with cattle herding and farming humans in large numbers.

However, Kenya still has a fair number of places where undisturbed mammal communities exist, usually protected places. These communities are changing, of course, and in general wildlife is being lost; with increases in population and urbanisation this is inevitable. Having said that, the structure of the mammal community in Nairobi National Park today is not very different to that recorded by Richard Meinertzhagen between 1903 and 1906. In his book *Kenya Diary* Meinertzhagen (1957) records six counts of game on the open land south of Nairobi and north of the Mbagathi River. The species he saw were: rhinoceros, Lion, Cheetah, Spotted Hyena *Crocuta crocuta*, Wild Dog *Lyacon pictus*, zebra, Giraffe, warthog, baboon, jackal, Eland *Taurotragus oryx*, wildebeest, Kongoni *Akelaphus buselarphus*, Thomson's Gazelle *Gazella rufifrons*, Grant's Gazelle *Gazella granti*, Impala *Aepyceros melampus*, Steenbok *Rhaphiceros campestris* and Grey Duiker. One hundred years have passed, but you can visit Nairobi National Park and, with a little luck, see all those species, with the exception of the Wild Dog. You might also see waterbuck, (two species) Bushbuck *Tragelaphus scriptus* and African Buffalo *Syncerus caffer* (generally referred to simply as 'Buffalo'). So the game has been preserved in places, although in others it has not.

Actual checklists of mammals for discrete areas of Kenya are not easy to find, because comprehensive surveys that include small fauna such as insectivores, bats and the smaller rodents have rarely been done. Adequate sampling means lots of work. To get a representative list may take years and few scientists seem to have had the time. You can hardly find a single comprehensive mammal checklist for any Kenyan conservation area, not even the Maasai Mara, Kenya's most well known conservation unit. The Internet and popular literature offer

A young Warthog meets its fate, Maasai Mara. (Stephen Spawls)

A huddle of Banded Mongooses; a social and diurnal species, Tsavo East. (Stephen Spawls)

Symbiotic relationship: the rhino stirs up insects, the birds scream if danger approaches. (Stephen Spawls)

plenty of information as to what big game is there, but thorough lists of all mammals from any protected area, which perforce must contain ten+ bats, a similar number of rodents and a few shrews, as well as the so-called 'megafauna', largely do not exist. The Kenya Wildlife Service 'official' list for Lake Nakuru National Park states that there are 56 species of mammal in the park; the actual total is twice that. A list of the mammals of Lake Bogoria, as part of the management plan, says there are 24 species of mammal there, none of which is a rodent, and yet a quick Internet check reveals the existence of a paper on the rodents of the area which mentions nine species. Nairobi National Park actually does have a decent list, with more than 100 species, although only 54 of them are 'large', i.e. over 500g in weight. If we count large mammals only, then 38 species of large mammal are known from the Aberdares, but the total for Samburu National Reserve is 53. The mountains are not a good place for mammals. In general, a decent-sized chunk of wild ecosystem in Kenya, say 100 square kilometres or more, with adequate protection, should have between 30 and 60 large and a total of between 75 and 120 species of mammal. In such a habitat, with a water supply, the large mammal community should usually contain at least two large cats, a hyena species, one or more types of jackal, two or more types of diurnal and one or two nocturnal mongooses, two or more small to medium cats (Serval *Felis serval*, wildcats etc.), one or two genets, at least two large species of primate (monkey, baboon) and one or two small species of primate (bushbaby), one or two odd-toed ungulates (zebra, rhino), one or two species of wild pig, a hare, two or more squirrels, a hyrax and at least eight or ten species of antelope. Other possibilities, depending upon the available resources, may include elephant and the Hippo, Giraffe and Buffalo, plus a selection of the smaller animals such as the small carnivores, rodents, bats and insectivores. The actual species numbers will vary according to the type of habitat. In any well-forested habitat the number of primates is liable to increase, there will be no jackals and only a single large cat, the Leopard *Panthera pardus*. In very dry areas there will not be any Hippos (for obvious reasons), Buffalo or Vervet Monkeys *Cercopithecus pygerythrus*, as they need to drink regularly. Numbers of

Deadly hidden danger, can you see the Lion? (Stephen Spawls)

smaller mammals are not so easy to quantify, but a reasonable-sized area should contain at least eight rodent species, ten or more bat species, and three or four shrews or more. We reiterate 'should' because, as we have said, very few exact numbers are known.

Kenya actually has the largest selection of savanna mammals in Africa; it only really loses out on certain forest mammals. For example the Democratic Republic of the Congo has between 27 and 33 species of monkey to Kenya's 11–13 (the variability depends on whether you count some forms as species or subspecies). But the dazzling array of more than 130 species of large mammal has benefited Kenya, because it attracts visitors who want to see the game and are willing to pay for the privilege. National parks exist worldwide, but there is nowhere outside of sub-Saharan Africa where a visitor can expect to see between 30 and 40 species of big mammal in a week's safari (and in fact precious few countries in Africa). It is also one of a handful of countries where one can see huge herds of game, and feel one is getting something of the flavour of wild Africa. Other countries in Africa that receive a reasonably large number of visitors coming to see big game include Uganda, Tanzania, Botswana, Zimbabwe, Zambia, Namibia and South Africa. Countries such as Rwanda, Ethiopia, Burundi, Malawi, Lesotho, Swaziland, Senegal and Ghana get a few tourists coming to see wildlife. The national parks in most of those countries are set in savanna and semi-desert. The DR Congo has seven or eight national parks, which are the only place to see the Congo's endemic monkeys and the fabulous Okapi *Okapia johnstoni*, but very few visitors go there owing to the political instability and lack of infrastructure. In addition, the parks there are hard to get to, and visitors to forest parks rarely see much if they drive in a vehicle; they have to walk. But in Kenya, one can drive comfortably or fly to see the wildlife, ride in a comfortable vehicle, stay in a nice lodge, and eat well and safely. And the diversity of Kenyan mammals is so unusual; a trip that takes in the Maasai Mara, the Aberdares and Samburu reserves exposes the visitor to typical plains game, to some montane specialities and to the unusual animals of the Somali–Maasai fauna, and all within some really spectacular scenery. The visibility is significant; Kenya's most popular conservation areas are in savanna or semi-desert. The country has the unique combination of a wide range of visible mammals living in beautiful, open country.

Before moving on, we need to mention a few things about the structure and numbers of a mammal population. This will be obvious to readers with ecological training and you may wish to skip this bit and go to the end of the section. In any faunal community, there are going to be far fewer predators than prey. In Robert Ardrey's words, there is one tiger to a hill (Ardrey 1961). When you go out for a drive in the Maasai Mara, you will see a vast amount of grass, a lot of Topis *Damaliscus lunatus*, or zebras, or wildebeest, but not more than one or two Leopards. A walk in a highland forest will produce plenty of trees and a couple of dozen forest birds, but you are unlikely to see more than one Crowned Eagle *Stephanoaetus coronatus*. Why is this?

In the 1920s, the Oxford ecologist Charles Elton recognised the connections between the living things in a community. The basic structure can be shown in what is called a food chain, for example: grass → zebra → Lion. This is read as: grass is eaten by zebras, zebras are eaten by Lions. The arrows show the direction of energy flow; the zebra gets energy from the grass, the Lion gets energy from the zebra. Food chains can be longer: for example grass stem → cricket → Striped Grass Mouse → Puff Adder → Secretary Bird → Serval. The various food chains in any community interconnect to form a web, a typical such web is shown in Figure 6(1). As you can see, all food chains start with plants – the arrows always go away from them – and most end with a predator, sometimes called a 'top' or 'apex' predator, an animal that eats other animals. The arrows mostly point towards the predator. Organisms in a habitat are connected. The numbers involved are also very interesting, particularly as they also apply to a crucial human activity, farming. Consider the zebra. A browsing stallion on the plains may eat 5kg of grass in a day. If it lives for 17 years, it will consume 17 x 365 x 5 = 31,025kg of grass, that's over 31 tonnes. Now the Lion. If the average Lion lives 12 years, and kills a zebra once a week, that means that it needs a total of 12 x 52 = 624 zebras. There is, of course, an element of uncertainty in these calculations; not all Lions live to 12 years, several Lions may share a zebra, they may kill a Buffalo or a dik-dik, but the big picture indicates that for every Lion, there must be over 600 zebras or their equivalent, and 600 x 31 = 18,600 tonnes of grass. Ecologists show these figures in what is called an Eltonian Pyramid, named after Charles Elton.

These figures often enter into the debate about arable and pastoral farming where the return is much greater from crops than from stock, and hence pertinent in the global debate about crops or cattle. And they are significant in the preservation of protected areas. If you want to protect a wild place and retain large predators in it, then you need a correspondingly large number of herbivores for the predator to eat. If there are not enough wild herbivores, then the predators may either die out, move away or start taking other animals, usually stock, with the usual consequences. Hence there are no Lions in parks like Hell's Gate or Ruma, and this is why Nairobi National Park maintains a relatively large Lion population only with the dispersal zone to the south, a buffer rapidly closing,. A classic paper published by William Newmark in 1987 regarding extinctions of mammals in American national parks found that if parks are too small, they don't preserve large predators. This lead to the SLOSS debate in conservation circles: the acronym stood for 'Single Large or Several Small'. The answer really is that you need both: a small park will protect plants and smaller animals (provided things like humidity and temperature are not too different to their original values) but if you want to keep the big animals, then you need to protect ecosystems. Big game needs big parks.

The inter-connectedness of organisms in a food web leads to the question, what happens when one or more organisms are removed from the system? The wilderness is shrinking, but the tourism industry needs big game. Two of the toughest mammals to maintain in an ecosystem are Lions and elephants, because their activities easily affect humans, and yet these are the two mammals that most visitors hope to see. A first encounter with such animals can be a life-changing experience: one is reminded of Robert Ruark's comment (1967) that there a few more impressive sights than a city man's first glimpse of a maned Lion in wild country. If Lions and elephants disappear, the system will not collapse. Ecosystems that have had elephants and Lions and no longer do so exist all over Africa. But the ecosystems will change (and fewer visitors will come). Consider Great Britain; during the last thousand years the country has lost its bears, lynxes, beavers, boars, wolves and reindeer.

The Mammals

Striped Grass Mouse, Lemniscomys, *a genus of ten hard-to-identify species. (Stephen Spawls)*

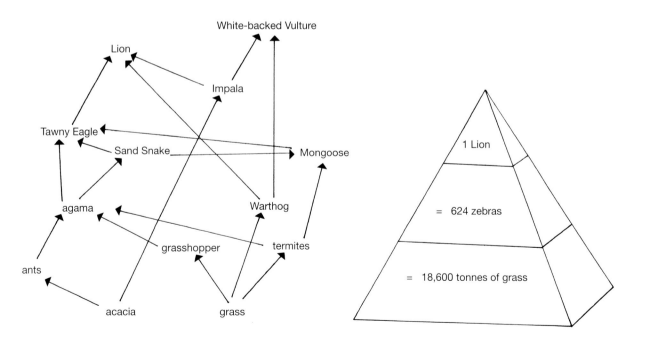

Figure 6(1) Food web in the savanna

Figure 6 (2) Eltonian pyramid (not to scale)

Its ecosystems still exist, although how they have changed is a mystery. But you cannot walk across an English moor, no matter how remote, and see wolves.

There have been many studies of mammal introductions; we are all aware of what happened in Australia when rabbits were introduced. Alien introductions are often bad news. But what happens if you take an organism out of an ecosystem? This isn't so well documented. The Black Rhino has disappeared from the Kapiti Plains and Ukambani, where it was once common. It is hard to say what effect this has had, largely because those ecosystems were not well studied at the time that the rhino was disappearing. Black Rhino chew up a lot of young acacias, but the Kapiti Plains look as open as ever. The effect of removing a top predator has been documented elsewhere; when wolves were exterminated from Yellowstone National Park in the USA the numbers of elk boomed, and they became incautious; restoration of the wolves had the desired effect. But the disappearance of the lynx from parts of Canada did not lead to a major surge in the numbers of its main prey, the Snowshoe Hare, because other predators compensated

Kenya has not yet lost its top mammalian predators, but it has lost almost all its Black Rhino. And many of Kenya's ecosystems are under threat. Often rather nightmarish scenarios are put forward, suggesting that something drastic will happened when certain important, 'keystone' species are removed. We cannot be sure – yet – what the effect will be if we lose top predators, although a long-term study of a park like Hell's Gate, where there are no Lions, might be useful. The problem, as mentioned, is that the conservation areas of Kenya lack comprehensive lists of their mammal species, let alone all their vertebrates (and certainly not their invertebrates). Too little work has been done. If we don't know what's there in the first place, we cannot say anything about interconnections, whether things are increasing or decreasing, and we certainly can't work out what will be the effect of removing an organism, particularly a small one.

Some data do exist. We can draw conclusions on the effect of removing elephants, because we have information resulting from the ivory trade. An interesting paper on the topic was published by Thomas Hakansson in the journal *Human Ecology* in 2004. The trade in African elephant ivory has been going on for more than 4,000 years. Those who lived in the interior didn't document what was happening to their elephants, but Hakansson points out that much ivory hunting was taking place in Kenya's interior by the sixteenth century. In the 1840s, there were still elephants in the high country between Kilimanjaro and the Cherangani Hills, and they were still common in Tsavo. But by 1890 elephants had been exterminated as far inland as Voi, and were rare between there and Makindu. There were good stocks deep in the Kenyan interior, but the extermination of elephants near the coast forced the price of ivory up threefold. When Colonel Patterson passed through Tsavo with the construction team of the Uganda Railway and struggled with the man-eating Lions, he saw no elephants; he doesn't mention them at all (Patterson 1907). But by the 1950s, the elephants had returned, and although their numbers have fluctuated, particularly in the 1980s a result of poaching, they are still there. In connection with this study, an important paper was published by Lindsey Gillson in 2004; she looked at the changes in vegetation in the Kanderi Swamp, near Voi, by drilling down into the sediments

Elephant path and elephant-ravaged landscape, Tsavo East. (Stephen Spawls)

Elephant spoor, front foot rounded, back elongate, the scuff at the front shows it was walking towards the top left. (Stephen Spawls)

and taking cores, which were analysed for pollen content. Gillson found that there was a fluctuation of trees and grasses, over a cycle of 250–500 years, and this was almost certainly due to local effects (fire, rainfall, effects of grazing animals and elephants). When grass pollen levels were high, tree pollen levels were low, indicating the area was grassed, and vice versa. About 430 years ago the amount of grass pollen rose dramatically and the tree pollen went down, and this might be due to the effects of increased numbers of elephants, possibly caused by them moving inland, away from hunters. This effect was also mentioned in Chapter 5, in connection with the Mara and the Serengeti. But 150 years ago, i.e. around 1850, the tree pollen levels rise sharply. This might have been due to the numbers of elephants decreasing, as a result of increased hunting in the Voi area. Elephants tend to pull down trees, and there is a big drop in the tree pollen roughly 60 years ago; this could possibly be elephants coming back. Elephants are a keystone species; their arrival means more trees are destroyed, their disappearance means more trees survive. And elephant numbers can be monitored without too much of a problem. But what happens when a mouse, or a bat, disappears? Nobody knows … yet.

Kenya has a few fabulous mammals (in the true meaning of that over-used word) that haven't made it into the records, including the Spotted Lion, or Marozi, the Brindled Cat, the Nandi Bear or Chemosit. Most of these legendary beasts are said to live in forests or on mountains – difficult areas to explore; those who study them are called cryptozoologists. Surprises do occasionally turn up. The Marozi, a small spotted cat, is said to inhabit montane forest and grassland. Its name might be a corruption of the Kikuyu *muruthi*, which means 'lion'. Many of the people who lived in the higher villages around the Aberdares and Mt Kenya knew of this animal, and in 1931 Captain R. E. Dent, the fish warden between 1926 and 1937, saw four small 'Spotted Lions' up on the Mt Kenya moorlands south of Meru. Later that year, Michael Trent, who farmed near the Aberdares, shot two small Lions on the moorlands of the north-western Aberdares, in the Pesi Valley at 10,000ft altitude. One skin, now in the Natural History Museum in London, is large and clearly spotted. It was believed to be a new species. The adventurer Kenneth Gandar Dower mounted an expedition to the Aberdares, accompanied by the charismatic Nanyuki rancher and polymath Raymond Hook, to try to find the Spotted Lion. They had no success, although Gandar Dower wrote an exciting book about their adventures. General opinion is that the Marozi is just a juvenile Lion that has retained its spots. Stories of the 'Brindled Cat', a fierce looking feline with stripes, roughly in size between a Leopard and a Lion, come from the Mau Forest and Mt Kenya. The origins of this legendary creature are probably based on hurried sightings of Striped Hyenas *Hyaena hyaena* or wildcats; it is hard to judge the size of an animal glimpsed through mist. And the WaKamba tell of a great winged beast that lives on the mountain and flies down to Ukambani at night to eat people; you can see the skid marks where it landed.

Kenya's most famous legendary animal is the 'Nandi Bear', as notorious as the Loch Ness Monster, Bigfoot and the Yeti. Many early adventurers and farmers recorded encounters with this fabulous beast, and many of the people of high western Kenya took its existence as a fact. The inhabitants of the west know it as Chemosit, or a Geteit. The sophisticated believe that it is a legendary beast, like a dragon, used to frighten children, but some older people insist that it is real. Encounters with the Nandi Bear are described from Sergoit Hill, Eldama Ravine, Londiani, Sotik, Kericho, the Mau forest and Elgeyo-Marakwet. Many of the descriptions of the Nandi Bear are rather fanciful – it is said to have one leg and nine buttocks, to be half man and half bird, with long black hair, a glowing mouth and a tail. It is believed to eat brains. However, several observers describe it as looking like a bear, and seen on a hillside at dusk or in the mist. Their stories occupy an entire chapter in Bernard Heuvelmans' book *On the Trail of Unknown Animals* (1995). In 1916, in the country near Londiani, a supposed Nandi Bear terrified the local people. It savaged ten sheep in a night, and tore out their brains. Over 10 days, the monster killed 57 sheep and goats, and many terrifying tales were told. But the beast was finally tracked down and killed by a posse of local spearmen; it turned out to be a large Spotted Hyena *Crocuta crocuta*. Some zoologists think that the Brown Hyena might have occurred in Kenya (or still occurs), and when this animal has raised its mane, it can look quite sinister. It might be the source of the legend. There is probably a prosaic explanation behind the stories of unknown creatures in Kenya's wild country; and not much tangible evidence. At a recent conference on the possible existence of the Yeti, delegates were told that there was 'irrefutable evidence of the existence' of this legendary creature, which prompted an Internet blogger to point out that the only irrefutable evidence of a Yeti is a Yeti, everything else is refutable. But when you are in Kenya's high hills, as dusk approaches, the mists creep down and the forest looms dark above and around you, it's not difficult to believe that monsters lurk.

MAMMAL ZOOGEOGRAPHY

Africa's mammals can be grouped according to where they are found; this type of grouping is called zoogeography. Zoogeographic groupings are often associated with certain altitudes, climate and vegetation and are useful biological and conservation tools, enabling workers to define a fauna. For example, the Gerenuk is a typical animal of the zoogeographical group known as the Somali–Maasai fauna, the animals that live in the low altitude, dry country of north-east Africa where those two peoples live, and it lives nowhere else. Whereas the Giant Forest Hog is a member of the zoogeographical group known as the Guinean–Congolian rain forest fauna, it lives in forest and, as with the Gerenuk, is also found nowhere else. And because they have different habitat requirements, you never find Gerenuk and Giant Forest Hog in the same place. This may seem obvious.

The world is divided into six major zoogeographical regions: South America is the Neotropics, North America the Nearctic, southern Asia the IndoMalayan, Australia and surrounding islands is the Australian region, the Sahara and the land north of it, including Europe and Asia, lies within the Palaearctic zone. Sub-Saharan Africa was originally called the Ethiopian region, but this term has fallen into disuse because of confusion with the organisms of Ethiopia, the country itself, and so it is more commonly called the Afrotropical region. There are a number of African mammals that have huge ranges, resourceful and tolerant Pan-African, or Pan-Afrotropical

The Mammals

Giant Forest Hog, big black swine, perhaps appropriately discovered by Richard Meinertzhagen. (Stephen Spawls)

Male Bushbuck, Nairobi National Park, found throughout African savannas and forests. (Stephen Spawls)

Egyptian Fruit Bats, Kakamega gold mine. There are many bat species, few are as easy to identify. (Stephen Spawls)

Grey or Bush Duiker, shy but tolerant of farmland. (Stephen Spawls)

species. Some of these animals live in both the forest and the savanna, for example the African Elephant, the Common Hippopotamus ('the Hippo'), the African Buffalo, the Leopard, the African Civet *Civettictis civetta*, the Honey Badger *Mellivora capensis* and the Bushbuck *Tragelaphus scriptus*. They are obviously adaptable to different habitats, and it is tempting to say that they have spread so widely on account of their size. Bigger animals disperse more easily; a Bushbuck can wander several kilometres in a day, a shrew can't. But a few smaller mammals have spread throughout the African forest and savanna, for example a couple of bats, the Straw-coloured Fruit Bat *Eidolon helvum* and Schlieffen's Twilight Bat *Nyctecius schlieffeni*, and some widespread carnivores like the Slender Mongoose *Herpestes sanguinea*, the Marsh Mongoose *Atilax paludinosus* and the Blotched Genet *Genetta tigrina*. However, there are only about fifteen such species. Among mammals, the ability to live happily in both forest and savanna is unusual. The same logic applies to reptiles; with perhaps the exception of the Rock Python *Python sebae* and Nile Monitor Lizard *Varanus niloticus*, virtually no African reptile is a true inhabitant of both savanna and forest. But this is not the case with birds; many African birds occur in both habitats.

The Bongo, a pan-African forest species with a very limited range in Kenya. (Paolo Torchio)

A slightly larger group of mammals, eighteen species or so, are Pan-African forest mammals, occurring throughout the big forests of Africa, both the huge central Congolian block and the Guinea forest west of the 'Dahomey Gap' (a point on the West African coast, along the coast of Ghana and Togo, where grassland extends to the sea and splits the forest of Central and West Africa). Large mammals in this category include Chimpanzees, the Giant Forest Hog, the Golden Cat *Felis aurata*, the Bongo and the African Palm Civet *Nandinia binotata*; smaller examples include Demidoff's Galago *Galagoides demidoff*, Lord Derby's Anomalure *Anomalurus derbianus* (a flying squirrel), the Mill Rat *Mylomys dybowski* and Emin's Giant Pouched Rat *Cricetomys emini*. A decent group, but a much larger group of mammals inhabit the entire African savanna (or did until humans occupied their land). The list of Pan-Afrotropical savanna dwellers includes such species as the Lion, Black Rhino, White Rhino, Grey Duiker, Oribi *Ourebia ourebi*, Roan Antelope, Zorilla *Ictonyx striatus*, Side-striped Jackal *Canis adustus*, Wild Dog *Lyacon pictus*, Spotted Hyena, Banded Mongoose *Mungos mungo*, Serval *Felis serval*, Giraffe and the warthogs. Some of these mammals occur – or did occur – outside Africa and diffused across that continent from the north, for example the Lion, Cheetah and Caracal; others occurred right across the Sahara into North Africa, for example the White Rhino and the Common Genet *Genetta genetta* (this little carnivore even gets into Spain and France). There are many more Pan-African savanna mammals than Pan-African forest mammals, and there are a number of species than can handle both habitats. This has led some zoologists to conclude that in the past the African forests shrunk to very small areas and the survivors were those who could tolerate savanna or both habitat types. Support for this idea comes from studies of reptiles; no forest habitat in Africa has more than 45 or so snake species, whereas parts of the South American forest can have more than 70 species of snake, and those forests are older and less disturbed. However, this is a contentious idea. But it may be worth mentioning that in terms of mammalian biomass (total weight of all mammals in a discrete area), forests lag far behind savanna; a square kilometre of forest contains less than 10 per cent of the mammalian biomass of a savanna.

Uganda Kob, an antelope typical of the Sudan savanna but virtually extinct in Kenya. (Stephen Spawls)

In Chapter 5 we mentioned Frank White's (1983) seven main vegetation zones (Guinea–Congolian rain forest, Somali–Maasai savanna, Sudanian savanna, Zambezian savanna, Lake Victoria forest-grassland mosaic, Zanzibar–Inhambane coastal mosaic and Afromontane rainforest and grassland). To some extent Kenya's mammals can be shoehorned into these zones. We have mentioned above the Guinean–Congolian mammals, plus two regional groupings that White, as a botanist, did not consider, the Pan-Afrotropical mammals, and the Pan-Afrotropical savannah mammals. Let's look now at some of the other zones. In Samburu, Shaba and Buffalo Springs, and points north, you can find creatures of the Somali–Maasai fauna. The mammals of this group include Grevy's Zebra, the Gerenuk, the Somali Galago *Galago gallarum*, the Naked Mole-rat, the Beisa Oryx *Oryx beisa* and the Desert Warthog *Phacochoerus aethiopicus*. Mammals of the Sudanian savanna that reach Kenya include the Patas Monkey, the Kob *Kobus kob* (which has just about disappeared in Kenya) and the Thomson's Gazelle *Gazella rufifrons*. Zambezian savannah mammals in Kenya include Burchell's Zebra *Equus quagga*, the Eland, Impala, Wildebeest and Southern Tree Hyrax *Dendrohyrax arboreus*. Coastal mosaic mammals are rather few, but include the Zanj Elephant Shrew *Rhynchocyon petersi*, the Red-bellied Coast Squirrel *Paraxerus palliatus*, the Zanzibar Galago *Galagoides zanzibaricus* and Ader's Duiker. Mammals of the Afromontane region include the Black and White Colobus Monkey, the Aberdare Mole-rat *Tachyoryctes audax* and the Mount Warges Pipistrelle *Pipistrellus aero* (a bat). Some of these are endemics, confined to Kenya, a group we will return to. But you will have noticed that a lot of Kenya's mammals don't seem to fit into any clear-cut zoogeographic zone, or into White's plant zones … and one of White's zones, the Lake Victoria forest-grassland mosaic, doesn't have any mammals that fit into it. Many mammal species resist classification; a similar phenomenon is seen amongst birds. And this is for a significant reason: mammals are adaptable and relatively mobile. Leave aside, for the minute, the problematical little things like bats, shrews and rodents, where short lifespans and speciation has created a lot of headaches, and consider an animal like Kirk's Dik-dik *Madoqua kirkii*, which has a disjunct distribution – it is found in Namibia, Angola and Kenya. You cannot shoehorn it into one of Frank White's zones. There are a handful of mammals (and other vertebrates) that have a similar range; they occur in south-western Africa and East Africa, with no connection in between, for example, such mammals as the Bat-eared Fox *Otocyon megalotis*, the Black-backed Jackal *Canis mesomelas*, the Spring Hare *Pedetes capensis*, the Oryx and the Aard-wolf *Proteles cristata*. There is a theory that at one time they did occur in the intervening country, in Zambia and southern Tanzania, that an 'arid corridor' connected the two regions and it has now disappeared. But let's look at a few other mammals. The Sitatunga *Tragelaphus spekei*, that mysterious swamp animal, lives mostly in the forest, but it goes up the Nile and down into the Okavango. The Klipspringer *Oreotragus oreotragus* occurs in southern, eastern and north-eastern Africa. What it needs is rocks, be they in woodland, savanna or semi-desert. The Yellow Baboon inhabits huge areas of the Zambezian savanna, the coast and the dry north-east; it's a Zambezian, Zanzibar–Inhambane and Somali–

The Mammals

The Oribi with its curious black gland neck patches. (Stephen Spawls)

White-toothed Shrew; a member of a genus of more than 100 species. (Stephen Spawls)

Kirk's Dik-dik, an 'arid-corridor' animal of northeast Africa and Namibia. (Stephen Spawls)

Maasai species. The Striped Hyena occurs across large areas of southern Asia, most of the desert region of north Africa, and most of East Africa. The Dwarf Mongoose *Helogale parvula* has a huge range along the eastern side of Africa. If we look for the big picture, it appears that many of Africa's larger mammals have big ranges and cannot be pigeonholed into zoogeographic groupings. This is not only hopeful in conservation terms, it is a clear indicator that the African mammal fauna consists mainly of immigrants.

So the mammals wandered into Africa, speciating, looking for niches, and as the land changed over the last 30 million years, some prospered, particularly the adaptable ones. Some found small handy habitats and remained there, while others couldn't cope with the climate changes, or other factors, like disease, evolving and new predators, or competitors who were better at surviving than they. Those became extinct. It is tempting to speculate on the future here. Leaving aside the effects of humanity (which in reality are likely to have the greatest influence on Africa's mammal fauna), consider a change in which the climate gradually gets more and more dry, and the forest disappears. We probably lose the Bongo, the Okapi and gorillas, for a start. What if it goes the other way? Things get wetter and wetter and vegetation gets thicker and thicker. The Dik-dik, the Oryx and the Cheetah need open country; they will become extinct … unless they can adapt. There is interesting evidence that Cheetahs have adapted in the past. In the Beaufort West area in South Africa, some specimens of Cheetah with long hair and shorter stouter limbs were collected in the late nineteenth century. It is tempting to speculate these were following the path of the Leopard in becoming a short, stocky carnivore that doesn't hunt by long high-speed chases. And in eastern Botswana, Zimbabwe and northern South Africa the 'King Cheetah' is found; it is slightly larger and heavier than the typical Cheetah and its spots have coalesced into bars. King Cheetahs have been successfully bred in captivity, and some wildlife husbandrists believe their extra weight and dappled markings are an adaptation to hunting in the woodland typical of their habitat.

The unusual Hirola or four-eyed antelope, critically endangered, possibly endemic, occasionally translocated. (Paolo Torchio)

As we've seen, mammals are adaptable. But the difference between big and small mammals is mobility and rapidity of breeding. With the exception of bats, which can disperse by flying and cover many kilometres in a night, smaller mammals in Africa tend to have small ranges. They also have shorter lives, and this means that genetic changes can accumulate more rapidly. It takes 50 years to have five generations of elephants but you can get eight generations within a year in some mice. So a genetic change can be rapidly perpetuated. Imagine a family of mice on Mt Kenya. Variation exists, some are rufous, some brown, some silvery-yellow. Their relatives in the dry red country of Tharaka come out at night, so the colour doesn't matter so much. But those on the mountain find it cold at night, so a few begin to operate in the twilight, or even by day. The Augur Buzzards *Buteo augur* start to get them, and the rufous ones are more obvious. They get eaten. The silvery-yellow ones, better camouflaged, survive. The structure of the gene pool changes until eventually only the silvery-yellow ones exist. In 500 years 4,000 generations pass. Then comes a mammalologist, who collects a few, notices they are different in colour to the ones from Tharaka, and perhaps a bit heavier. The mammalogist describes them as a new species, an endemic species. And it is for this reason that most of Kenya's endemic mammals (and most other endemic vertebrates) are small animals, with small ranges. They have speciated rapidly, adapting to a tiny habitat.

The number of endemic mammals in Kenya depends on whose version you accept: some say there are 21, others 17. It depends on the classifier, and what weight they give to differences between populations; and it's this classifying game that has prompted some angry zoologist to throw up their hands and say, let's not bother with species, let's just look at lines of evolutionary descent, – give them all numbers and have done. But we talked about this approach in Chapter 5, looking at problems with taxonomy of plants. There's logic to it but in the end you need names; you need to be able to say what you're dealing with. Let's look first at the list and then the taxonomic debate.

Named for Sir Frederick Jackson: Jackson's Hartebeest in Ruma National Park. (Stephen Spawls)

Kenya's most recent endemic species, and its largest, has been acquired by default; it is the rather strange-looking 'four-eyed' antelope, the Hunter's Hartebeest or Hirola. The experts cannot even agree on its scientific name; some say *Beatragus hunteri*, others *Damaliscus hunteri*. A pale-coloured smallish hartebeest, originally described in 1889 from specimens shot by the zoological collector H. C. V. Hunter on the north bank of the Tana River, its range extended from the Tana to the Juba River in Somalia. Since Somalia has descended into anarchy, all the Hirola there seem to have been killed, so by elimination Kenya has acquired an endemic species. Some would say that it should not be classed as such because there might be a few left in Somalia. The list of Kenyan mammalian endemics contains two monkeys, both also found in the forests of the lower Tana: the Tana River Red Colobus and the Tana Mangabey *Cercocebus galeritus*. All Kenya's other endemics weigh less than 500g (although the Golden-rumped Elephant Shrew *Rhynchocyon chrysopygus*, of the coastal forests, is over 400g in mass). The remaining list has four shrews, two bats, between four and seven mole rats (depending on whether or not you regard some forms as species or subspecies) and four rats. Three, a shrew, a mouse and a bat, only live on Mt Warges near Wamba in the Mathews Range. The two monkeys on the lower Tana are joined by an endemic bat known from Garsen, with the enchanting name of the Kenyacola Butterfly Bat *Glauconycteris kenyacola*. Most of the other endemics live on hills, apart from the Golden-rumped Elephant Shrew, which lives around Gede. Mt Kenya has an endemic shrew, a mole rat and a mouse, the Aberdares a mole rat and a shrew, the Nyambeni Hills shrew.

Many of these little endemics were described by Edmund Heller, a splendid-looking man who was one of the zoologists with Teddy Roosevelt's Smithsonian Expedition. The stout ex-president (his Kenyan workers called Roosevelt 'Bwana Tumbo', i.e. Mr Stomach, or Mr Fat) led a major collecting expedition from Kenya through Uganda to the Sudan, in 1909 and 1910, and collected more than 23,000 specimens, including 5,000 mammals, for the Smithsonian Museum. Even in those days, when hunting wasn't considered politically incorrect, many people were disturbed by the slaughter. A newspaper cartoon of the time shows a Lion pleading with Roosevelt; saying 'Please, Teddy, spare a few of my subjects'. Edmund Heller was responsible for the preservation of the mammals. Heller returned to Kenya for a further year, visiting the coast, Loita, all around Mt Kenya, Mt Warges and Ol Olokwe. He was an indefatigable taxonomic splitter and named literally dozens of new species and subspecies, some of which later turned out to be invalid on account of not being very different to other species or already described from elsewhere. He described a new subspecies of Banded Mongoose (valid), a new subspecies of Zorilla (might be valid), a new subspecies of African Clawless Otter *Aonyx capensis* (probably not valid), a new subspecies of Black-backed Jackal (invalid), a new subspecies of Grant's Gazelle (also invalid), two subspecies of waterbuck (neither valid) … the list goes on and on. Heller described two new subspecies of Lion, *Panthera leo roosevelti* and *P. l. nyanzae*; at one time there were 23 subspecies of Lion described from Africa. None is now considered valid.

The debate about whether Kenya has 17 or 21 endemics is to some extent based on the status of mole rats; at present the International Union for the Conservation of Nature (IUCN) regard four Kenyan mole rats of the genus *Tachyoryctes* (*T. annectens*, *T. naivashae*, *T. rex*, *T. spalacinus*) as being simply varied specimens of the species *T. splendens*, the Splendid Mole-rat. And splendid they are: fat chunky beasts with blunt noses, tawny pelts and huge incisors; inhabitants of Nairobi know them as they churn up their lawns, like moles do in Europe and America (there are no moles in Africa). Jonathan Kingdon says these four mole rats are full species, and he calls them root-rats. It's your choice who you go with. But even seventeen endemics is a respectable total; Botswana has no endemic mammals and Uganda has only three (all rats). Tanzania has 31, although eleven of these are shrews, five are bats and one is a sengi (elephant shrew).

MAMMALS AND HUMANS IN KENYA

Mammals, big and small, have had a lot of impact upon humanity in Kenya; the only animal group with more influence are probably the arthropods. The problems with big mammals is diminishing as they disappear, but little animals like rodents remain significant, as they eat stored food and carry diseases. And, of course, domesticated mammals such as the cow, goat, sheep, camel, donkey, dog and cat are of major importance to the average Kenyan. So throughout human history in Kenya, the mammals are involved as domestic animals, as predators, as prey, as crop raiders and as vectors of disease.

In very few countries nowadays are there still big wild animals that constitute a threat to quite a lot of the population. The adventurer Ray Mears once said that he really enjoyed walking in Africa's wild country, because it was a place where humanity wasn't the top of the food chain. This gave a buzz to his walk. There are wild places in Britain, but no carnivores that might conceivable consider you as a meal, or ungulates that could quite easily flatten you if you make them cross. Step out of your car in Amboseli, or Tsavo and you experience a remarkable frisson, a primeval one, that comes from knowing there are things out there that could kill you. This must be how our ancestors felt.

At the other end of the spectrum there are animals with which humans live comfortably side by side – Kenya's domesticated mammals. Cows, goats, sheep and camels provide meat, milk and income for the people. They may also be used for transport. However, the benefits they provide are diminished by several factors. Domestic stock, particularly in low rainfall areas, cause a lot of damage to the vegetation when feeding. People tend to overstock; numbers become important, carrying capacity is ignored. Stock owners are often reluctant to actually use the stock for meat and won't kill a beast, even when it is their interest to do so. Nomadic people with large herds cause instability, because it is very easy, with the use of violence, to transfer huge amounts of wealth. Rival groups of pastoralists tend to arm themselves to the teeth and get into murderous fights. This is why the anti-stock theft unit of the Kenya police is such a large department. Lastly, stock competes with game for fodder, space and water, and this causes conflict in protected and buffer areas. Many conservationists scratch their heads when it comes to dealing with the problems of stock, because you cannot deny people the right to keep cattle and goats.

Dealing with the problems of stock requires both a cultural and an economic perspective. For many Kenyans their livestock is their wealth and is not measured in monetary terms. In Kenya in 1965, there were 75 cows for every 100 people (7.58 million people and 5.7 million cattle), in Nigeria there were only were eighteen cows to every 100 Nigerians. The proportion has since dropped as the carrying capacity of the land has peaked, and in 1989 there were 63 cows for every 100 people (21.4 million Kenyans and 13.4 million cattle,). By 2010, the cows numbered 13.2 million and the people about 40 million, so the proportion had dropped to 33 cows for every hundred Kenyans. We are both losing cows and gaining people. With increasing urbanisation and the development of a middle class, it will continue dropping. Although one usually associates cattle with the great rift valley rangelands, the bulk of Kenya's cattle are actually within 150km of Lake Victoria or within 100km of the southern side of Mt Kenya. Similar effects are seen with goats and camels; there were 1.2 million camels in Kenya in 1965, nowadays there are only 830,000 and this is due to the instability and droughts that have cursed the main camel-herding areas.

The western Rothschild's Giraffe in Ruma National Park. (Stephen Spawls)

Let's look next at the wild predators and some of the dangerous large ungulates. An animal doesn't have to be big or aggressive to kill you directly, (many people are killed indirectly by mosquitoes). A visitor died after an encounter with a giraffe at the Aberdare Country Club. Robert Ruark was almost killed after a zebra he had shot got up and nearly kicked him in the head. However, the truly dangerous mammals in Kenya, ones that have killed a number of people, are members of what Kenya's most famous professional hunter, John Hunter, called the 'big five', the animals that visitors pay to see – Lion, Leopard, elephant, Buffalo and rhino. Also dangerous are the Hippo and the Spotted Hyena. Most other mammals are not directly dangerous, although stories of occasional incidents exist. General von Lettow-Vorbeck tells of a European in Tanganyika being killed by Wild Dogs, but the general had heard this story at second hand. Arthur Blayney Percival, Kenya's first 'ranger for game preservation', and later its first official game warden, describes how two Maasai youngsters who had captured some Wild Dog pups were severely bitten by the mother, but, as Percival dryly says, 'we must allow for the maternal defensive instinct, a fox terrier ... might do the same ...'

The early pastoralists must have suffered from predators, although it looks as though some of the really nasty ones, the sabre-toothed cats, machairodonts, became extinct before humans got going. A giant hyena *Pachycrocuta*, a metre at the shoulder, also existed in East Africa, but probably disappeared before modern humans arrived. Another big and potentially dangerous mammal that is no longer with us is the Giant Baboon *Theropithecus oswaldi*, which weighed up to 70kg. But the boot was on the other foot at times; the anthropologist Pat Shipman found evidence at Olorgesailie of the killing of 90 Giant Baboons by hominins, some time between 400,000 and 700,000 years ago; all had had their skulls smashed.

The Lion and the Leopard are Kenya's top predators; they can consider humans as potential prey. Rhino and Buffalo sometimes kill humans, but only in defence, and rarely consistently. Rogue elephants have been known to kill a lot of people. C. J. P. Ionides, who worked for the then Tanganyika Game Department before becoming an internationally famous herpetologist, hunted a rogue elephant that had been shot in the backside by a European

A successful cud-chewing ruminant, African Buffalo occurs in most of Kenya's parks. (Stephen Spawls)

hunter. The elephant killed 28 people before being shot by Ionides (generally known as 'Iodine'). But nothing can match the carnage that can be wrought by a big cat that has decided to start eating people. Some figures may be interesting here. The conservationist Charlie McDougal calculated that in India during a five-year period in the 1920s, over 7,000 deaths due to tigers were reported. In India and Nepal, the notorious Champawat Tigress killed 438 people, and the Panar Leopard is credited with 402 victims in India. Both were shot by the much-respected Colonel Jim Corbett, or 'Carpet Sahib', as he was known in India. Corbett has a connection to Kenya; he retired from India to Nyeri and lived in a cottage at the Outspan Hotel. He is buried in Nyeri. Corbett was a gentleman hunter of the old school, more interested in preserving game than shooting it; his respect for the natural world shines through in his books and a national park in India is named after him. Few African cats have killed as many as those notorious Indian man-eaters – a term used for convenience, women are victims too. In Africa Leopards do not seem to take adult men very much, but they will kill women and children. A female Leopard at Masaguru in Tanzania killed 26 women and children and didn't eat any of them. She was eventually shot by Brian Nicholson, a man who rose to fame in the Tanganyika game department and then settled in Kenya. 'Iodine' himself hunted a Leopard in the Ruponda area of Tanzania that had killed eighteen children. Iodine's gunbearer and right-hand man, Hemedi Ngoe, was convinced that a big male Leopard with a missing right claw was the killer, and a Leopard fitting this description was caught in a trap. The gunbearer was proved right, because killings ceased after this Leopard was shot.

Kenya's most notorious man-eaters were the two Lions that terrorised the workers building the Uganda Railway in 1898, around the Tsavo River crossing, between Mtito Andei and Voi. Known as 'The Man-eaters of Tsavo', they killed well over 100 people, mostly railway workers, Indian and African. For ten months the railhead camps were paralysed with fear. The two killers, both huge, tawny, maneless males, were eventually shot by Lieutenant-Colonel J. H. (John Henry) Patterson, the ex-Indian army engineer who had taken over the construction of the Tsavo River Bridge. The Tsavo saga was mentioned in the British House of Lords by the then Prime Minister, Lord Salisbury, and later spawned several films. Patterson's 1907 book, *The Man-eaters of Tsavo*, often reprinted, is a classic of Kenyan wildlife literature. One night Patterson sat up over a bait, a dead donkey, on an elevated platform, waiting for one of the man-eaters. He had no torch. The Lion arrived and began to stalk him. As Patterson sat in silence, in the black night, with this deadly cat creeping up on him, something hit him on the back of the head and he nearly fell out of the tree with shock. It was just an owl, which had mistaken the motionless hunter for a branch. But Patterson got his shot, and the Lion was found dead the next morning. The story of the Tsavo Lions occupies a major section in Charles Miller's excellent book on the history of Kenya's railway, *The Lunatic Express* (1976) and is also mentioned in Peter Hill Beard's controversial book *The End of the Game* (1965). The zoologists Julian Kerbis Peterhans and Tom Gnoske published an excellent paper on the science of man-eating, as exemplified by the Tsavo Lions, in 2001, in the Journal of the East African Natural History Society.

For a while, it was not known how many people were killed by the Tsavo man-eaters. Patterson only mentions the deaths of some 28 Indian labourers, although some years later, in a leaflet for the Field Museum of Natural History in Chicago (which bought the man-eaters' skins and skeletons, they remain mounted there to this day), Patterson (1925) states that 107 of the local inhabitants were also killed. Peterhans and Gnoske (2001) agree, with a figure of over 130 people. Both ordinary citizens and prominent personalities fell victim to Lions along this railway. In 1899 a Lion snatched a road engineer called O'Hara from his tent as he slept beside his wife and small children at Voi, and in 1900 the superintendent of Railway Police, Charles Ryall, was taken by a Lion from a railway carriage at Kima, just north-west of Sultan Hamud. Ryall was actually hunting the Lion, which had been terrorising the WaKamba people at Kima, but he dozed off and the Lion took the opportunity to leap into the carriage, break Ryall's neck and jump out of the window with the unfortunate superintendent in his jaws – proof, if it were needed, that it's not a good idea to snooze while hunting a man-eater!

Less notorious, but more dangerous Lions are known from East Africa. In their paper Peterhans and Gnoske (2001) mention the depredations of a group of Lions in southern Tanzania, hunted by the famous elephant hunter, and later senior Tanganyika Game Warden, George Rushby. It is estimated that these Lions killed more than 1,500 people. It is probably safe to assume that there has been conflict between humans and Lions wherever both occurred. In the times of our hominin ancestors, the activities of Lions would have had a major effect on where and when humans were active. At night the big predator is king, and the weak take shelter. The development of weapons changed the situation; if you have a gun and a good light, then moving at night is safer. Lions in Africa have now largely learned to avoid humanity.

The usual explanation for why big predators become man-eaters is that they have become sickly, slow or infirm and unable to kill their normal prey, and switch to the slower, weaker naked ape that is us; Corbett largely found this amongst the man-eaters that he hunted in India. Killing powerful, fast-moving prey requires strength and agility, and it is easy to get injured. There is an interesting story here. Initial studies of Lions were made on zoo animals, and they often lived longer than 20 years; it was assumed therefore that in the wild they would have a similar lifespan. But when the brilliant American zoologist George Schaller studied Lions in the Serengeti, he found that few lived beyond 10 years. When they got to eight or 10 years old, they became a little slower and sooner or later made a mistake. They got injured and then could no longer catch food and died. Lions are the only social cat species, as well as the only cat that is strongly sexually dimorphic, the males usually have manes. One of the benefits of social activity is that hunting can be co-operative and food can be shared, but genuine food sharing doesn't seem to happen with Lions. Other theories as to why Lions live in groups include the benefits of group babysitting and the ability of large groups to hold and maintain big territories.

Peterhans and Gnoske found that not all man-eating was done by infirm animals. One of the Tsavo killers did, indeed, have a traumatic injury, a shattered canine tooth that had led to a remodelling of the jaw, but the other did not (it had minor tooth damage but not enough to prevent it killing normally). Other factors that lead to man-eating include the disappearance of the usual prey, leading the predator to take livestock and, after coming into conflict with the defenders of livestock, humans. Changes in the habitat, as is taking place over much of Africa, are significant. Conservation experts reckoned that in the 1950s there were close on 200,000 Lions in sub-Saharan Africa, now there are reckoned to be only between 15,000 and 30,000. Peterhans and Gnoske note that in Tsavo,

Nairobi National Park Lion, clinging on despite the risk of being killed if they migrate into cattle country. (Stephen Spawls)

following the activities of ivory hunters and the effect of the ungulate disease rinderpest, the elephant and the African Buffalo disappeared and Tsavo became thick and woody. Few medium-sized ungulates such as hartebeest and zebra were around, and in addition, slave caravans that had passed through the area often left dead bodies along the route. Peterhans and Gnoske (2001) believe that these conditions created man-eaters. It has also been suggested that some predators start eating humans after an accidental encounter ends with the predator killing a human and finding that it is easy to kill people and their flesh tastes nice. And this habit, once acquired, may be passed from one generation to the next. Rushby thought that this was the case with the Tanzanian Lions; the cubs learnt from their parents that humans were tasty and easy to catch. In the final analysis, however, humans without weapons have always been defenceless against big predators. It remains so; between 2002 and 2004 some 35 people were killed by Lions in the vicinity of the Rufiji Delta in Tanzania. Not everyone appreciates how dangerous lions can be. In the early 1980s, a pride of Lions chased a wildebeest into the grounds of the Mara River tented camp and pulled it down. The manager, Alistair Dawson, emerged to find a newly arrived group of visitors coming out of their tents and walking towards the kill with their cameras, under the impression it had been set up for them. Dawson had to shout at them to get back under cover.

Some of Africa's killer Leopards seem to have regarded children as just another potential prey item. Leopards vary greatly in weight: a big savanna Leopard can have a mass of 90kg, although 50–60kg is more usual, and a study of Leopards in the Cape Mountains in South Africa found some adults weighed as little as 20kg. Like other cats, Leopards are solitary and eat a huge range of animals. They will subsist on small prey such as rodents, frogs and even fish; a Leopard trapped by rising waters on the lake behind the Kariba Dam fed itself entirely on tilapia, even though antelope were available on the island. A Leopard lived for some time in the House of Manji biscuit factory in Nairobi's industrial area and lived entirely on rats (and possibly biscuits). At the other end of the scale,

Leopards will take big antelope like kudu and hartebeest, weighing over 200kg, although they generally prefer medium-sized animals of their own weight or less. Thus a child would be the right sort of size while an adult man probably looks too big, although this doesn't explain why Indian killer Leopards regularly kill and eat adult men but African Leopards don't. This ability to subsist on small prey, the ability to hide behind a tiny scrap of cover and their solitary habits mean that Leopards can also persist in relatively small wild areas, where Lions cannot. Thus they cling on in pockets of bush, or even within totally urbanised places, like the biscuit factory Leopard. And yet they don't seem to kill a great many people.

Under certain circumstances, hyenas can be dangerous. Africa has three species of hyena: Spotted, Striped and Brown. The Spotted is a true Pan-African savanna animal, the Brown lives only in the south and the Striped Hyena comes in from Asia and the north, just reaching Tanzania. Brown Hyenas hurt nobody, and Striped Hyenas are not very dangerous, although every decade a few people are killed in India. But the Spotted Hyena is a different matter. Often described as scavengers, they will kill if prey is available, although around some African cities they live largely off garbage. Spotted Hyenas have conical bone-crushing teeth and can bite with a force 40 per cent greater than that of a Leopard. Their droppings are often almost pure white, with calcium from the bones they have eaten. In Ethiopia, where there are a lot of Spotted Hyenas and they may live close to habitation, a number of people are killed each year, usually old people or children. Spotted Hyenas are not dangerous in the way a Lion or Leopard is; they will not usually launch themselves out of a bush onto an alert human. What they do is prowl around areas where humans are, looking for the infirm or the unsuspecting. In the 1970s, doctors at Wajir general hospital treated several hyena victims every month; nearly all had had bites taken out of them while they were asleep or lying ill. Arthur Blayney Percival (1928) describes a child at Muranga (then Fort Hall) who had been bitten by a Spotted Hyena and survived but with the loss of much of his face. People who get drunk and collapse outside are particularly at risk, and in the Entoto Hills above Addis Ababa parents ensure that small children are safely indoors by the time the sun descends to touch the horizon. Occasionally, a big Spotted Hyena will get a taste for human flesh. A pair killed 27 people in Malawi in the 1960s, and both were relatively heavy when shot, over 70kg (in Kenya they average 40–60kg). It is said that if you run from a hyena, it will chase you. George Adamson in his book *Bwana Game* (1968) describes meeting a Samburu man who was chased up a tree by three hyenas, and had some severe bites on his legs as a result. In many remote areas of Africa hyena kills probably go unreported, and a fair number of people may be killed. However, in northern Ghana, the tables are turned; hyena skins are so widely used in local sorcery that they command high prices and consequently hyenas have largely been wiped out. Hyenas used to be common around Nairobi and were frequently seen in the National Park, but in the 1960s one of the wardens decided that they were too numerous and had more than 300 of them shot. Coupled with the spreading of poison bait in and around Nairobi, this devastated the population, and since then there have been few hyenas around the capital city. Perhaps this is for the best, although many people feel that their eerie cry of 'whooo-oop', heard after darkness, symbolises Africa.

No other African carnivore is really dangerous. There is no risk to humans from a Cheetah. Raymond Hook, who cross-bred zebras and donkeys, took Cheetahs to the UK to race at the greyhound stadiums. Hook caught his Cheetahs by chasing them with dogs until they climbed a tree, whereupon he donned a huge pair of gauntlets, climbed up, tied the Cheetah's legs, grabbed it by the scruff and put it in a box. You couldn't do that with a Leopard. But some of the big herbivores are dangerous. However, unlike carnivores, big herbivorous mammals do not usually make a habit of attacking people because they can't eat them. Most incidents involving a big mammal killing or injuring a human are essentially accidents, where the person inadvertently gets too close to the animal, forcing a 'flight or fight' response. This is why, if you are in the bush and don't want an encounter with a dangerous animal, you should make sure you advertise your presence. Move in the open, sing or whistle, and don't get too close to thick cover. Following this logic, it is also inadvisable to go out at night. Every so often in Kenya people are killed by a Hippo, elephant, Buffalo or rhino that they have accidentally stumbled upon and it simply panics and crashes into them. Even expert bush hands are occasionally killed. In 2003, Paul Kabochi, a highly experienced safari guide who had spent most of his life in Kenya's wild places, was killed in Tsavo by an elephant. Inexperience

Bull Sable Antelope eats on the run, in his last stronghold in Kenya, in the Shimba Hills. (Stephen Spawls)

and carelessness can be a factor, of course. A visitor from abroad on his first day in Kenya, at Aruba Lodge in Tsavo, heard a noise outside and saw an elephant scratching itself on the edge of the verandah. He went out to offer it a banana. The elephant panicked as the tourist suddenly appeared in its line of vision and ran over him as it fled, breaking his leg. At the Mara River tented camp, there was a habituated elephant called Windy who was fairly tolerant of visitors, until one posed for a photograph too close to him and was thrown into the Mara River.

Rogue elephants have occasionally notched up a number of victims; we earlier mentioned the Tanzanian rogue that killed 28. No elephant in Kenya has killed so many people, as far as we know. Headlines about rogue elephants are not uncommon in the press and a Google search for 'African rogue elephants' yields millions of hits, but this is because the media tend to call any elephant that becomes defensive or aggressive a 'rogue'. The original meaning meant an elephant that deliberately starts trying to kill humans, and in most cases it is because the elephant has been wounded by humans. Often the injury involves a tusk, and no wonder; imagine the pain with a tooth that big which has a huge nerve running down the centre. In parts of Africa it is believed that if you see that nerve, you will go blind. John Hunter's life story, entitled, appropriately and simply, *Hunter*, (1952) opens with a description of how he and his right-hand man, Saseeta, hunted a rogue elephant on the edge of the Aberdares. It is a simply riveting story, a classic of hunting: man versus the monster, the lengthy pursuit by the two professionals through the great mountain forest, ending as the bull comes out of the bamboo and Hunter just manages to shoot it at point blank range. And the rogue proves to have a musket ball embedded in the right hand tusk. The Samburu game scout Lembirdan, one of George Adamson's men, shot an elephant that had killed a number of Samburu people in the Mathews Range. One of its tusks was snapped off near the base and the nerve was decaying; the massive toothache had driven the poor creature mad.

Elephants are unusual, sentient animals. They have a complex social structure involving small herds based around a dominant female. They seem to have an understanding of death, and also appear to mourn friends and relatives who have died. They may pick up and carry bones of dead relatives, and will help another elephant that is in trouble. In his book *The Woodpecker Calls on the Right* (2010), the Kenyan professional hunter and sculptor Terry Mathews describes how a bull elephant that he had shot was helped up and escorted away by two companions. One can imagine an elephant that has been hurt by a human remembering the incident and bearing a grudge. But wild elephants are also capable of relating to humans. Iain and Oria Douglas-Hamilton, elephant experts, became familiar with a number of elephants in the Lake Manyara National Park in Tanzania; eventually they were able to take their baby daughter Saba (now a wildlife expert in her own right) to introduce her to the elephants. A bewitching photograph in their book *Among the Elephants* (1975) shows the elephant stretching its trunk out to the baby girl. Another picture shows Douglas-Hamilton handing some food to a wild elephant called Virgo. As they say in the disclaimers – don't try this yourself! In the same park, an angry group of elephants that Douglas-Hamilton dubbed the Torone sisters never lost their dislike of humans, and frequently charged the scientist's car.

One of the most astonishing stories of elephant behaviour is told by Dr Joyce Poole, who has carried out elephant research in Amboseli for over 30 years. Poole was observing an elephant she had called Tonie, which had given birth to a stillborn calf. She stood over it for a day; Poole was convinced that Tonie was grieving. After Tonie had been there for more than 24 hours, Poole knew she must be desperately thirsty. She placed a bowl of water near the grieving mother. Tonie drank it, then came over to Poole's car and placed a trunk on her chest and arm. That cannot be anything other than gratitude. It seems that, given the chance, many mammals that we think of as being relatively dangerous can become perfectly at ease with humans. Problems arise if they lose their temper, be it ever so briefly.

In the past, many people in Kenya were killed by Buffalo and rhino. Rhino used to be common, particularly in the country between Nairobi and the coast. Travellers like Meinertzhagen and Gregory encountered rhino frequently. Maasai youngsters used to play an astonishingly dangerous game with rhino: when a sleeping one was found, the boys would stalk up on it and one would place a stone on its flank; the next boy had to remove it, and so on. John Hunter was asked by Captain Archie Ritchie, the Kenya game warden, to shoot the rhino in the Makueni area, south-east of Machakos, so that the area might be developed for agriculture; and in the mid-1940s Hunter and his team shot around a thousand rhino there. There are fewer than that number today in the whole of Kenya; those rhino that remain are major attractions. Black Rhino can be truculent animals. If you get close to one, it may charge. A good number of rural dwellers have been killed as a result of an accidental encounter with Black Rhino in the past. However, genuinely rogue rhino are not common, although George Adamson shot a rhino near Wamba that had killed at least one person and attacked a number of others. It proved to have an old festering wound in its shoulder, probably caused by a spear. Likewise, rogue Buffalo are also rare, despite their reputation as a dangerous beast to hunt. Myles Turner, who was warden of the Serengeti between 1956 and 1972, after a stint as a professional hunter, shot over 900 Buffalo, and said he was only once in danger, when a wounded animal attacked him. However, in the early 1980s, an old Buffalo on the Mara River seriously injured the well-known guide Dennis Zaphiro and a client; the bull was then shot and turned out to have an injury. It isn't easy to surprise a Buffalo, as they have reasonable eyesight, good hearing and a good sense of smell, an unusual combination of senses. They are tough animals: a Buffalo that had been shot in the heart once ran more than 3 miles before dropping dead. But in places where they are not persecuted, they can become almost tame. Unlike the rhino, they don't have any body parts of great value (yet), so they have become very common in some places in Kenya. There is hardly a national park that doesn't have its complement of Buffalo. In places, they still have a reputation of being bad-tempered; it is said that the Buffalo of the rift valley and Mt Elgon are truculent, whereas those of the Aberdares and Mt Kenya are not. In 2004, the well-known wildlife artist Simon Combes was killed by a Buffalo at Soysambu, Lord Delamere's estate near Lake Elmenteita, and in 2010 Parmois Ole Kerito and Rick Hopcraft, both experienced hunters, were killed by the same Buffalo on Loldia Farm near Naivasha. Buffalo are certainly not just wild cows.

There is much debate about how dangerous Hippos are. A lot of authors routinely claim that the Common Hippopotamus is the most dangerous mammal in Africa and 'kills more people than any other large mammal'. It might not be true, but exact statistics are hard to find. There do not seem to have been any 'rogue Hippos'. Hippos can be dangerous, however, for several reasons. They are big and have huge teeth. They can be territorial,

This is a thin Hippo and the far one is dead; victims of the 2009 drought in Tsavo. (Stephen Spawls)

a dominant male may attack boats or swimmers in his territory, and a female can be aggressive if she thinks her young are threatened. Because they live in water and emerge at night, Hippos can move up a river into an inhabited area without being seen, and they also wander distances at night. In Kenya Hippos quite often turn up unexpectedly in dams or small rivers. And if you get between a Hippo and water, or bump into one in the bush beside a lake, you may be faced with an angry or startled 2 tonne monster with fearsome teeth. So take care in waterside vegetation –try not to walk there – and don't approach Hippo in a small boat. Hippos are not worried by size; Terry Mathews had a Hippo attack his LandRover, biting a huge hole in the door. The husband and wife photographic team of Alan and Joan Root were attacked by a Hippo while filming at Mzima Springs, in Tsavo West. They were actually in the water, which was slightly murky. The Hippo charged and Joan was knocked out of the water. Chillingly the bull's huge canines went right through the front of her diving mask; 3 inches further forward and they would have gone through her head. Alan was hit from behind, and the Hippo then got his leg, and tore open the calf. He survived but needed 17 pints of saline and 8 pints of blood. In the 1990s several visitors to Crescent Island, on Lake Naivasha, were attacked by Hippos and two were killed. It's not known if the same Hippo was involved. So although no Hippo appears to have become a habitual killer of humans, people who live beside the continent's great lakes and rivers are at risk from this big, territorial animal. Sir Samuel Baker, explorer of the Nile, gets the last word. 'There is no animal', he wrote, 'that I dislike more than the Hippo.'

Dangerous mammals are not confined to those who might eat us or crush us, however. Some carry diseases that we can catch; such maladies are known as zoonoses (singular, zoonosis). A number of zoonoses can be caught from animals other than mammals; for example Chagas' disease and dengue fever, which are transmitted by insects, or psittacosis and Newcastle disease, from birds. By definition, zoonoses can complete their cycles in animals other than humans, so diseases like malaria and bilharzia are not zoonoses because humans are essential for their life-cycle to complete. And diseases that we get from other humans, like influenza, are not zoonoses. However, since we are mammals and thus most closely related to other mammals, there are a fair number of diseases that can

Red-bellied Coast Squirrel in the Shimba Hills; a southeast African coastal species. (Stephen Spawls)

jump from non-human mammals to humans, and some of these are quite nasty. Zoonoses include tuberculosis, brucellosis and salmonella infections. You can get leishmaniasis from rock hyraxes, via the bite of a sandfly. The deadly bacterial disease anthrax is often carried by domestic animals, such as goats and cattle, but in Kenya has been recorded in zebra, rhino and elephant, among other animals. Occasional human cases occur, although almost always when poor people eat the meat of an infected cow that has died of the disease. Rift Valley Fever was discovered in Kenya in 1915, and is spread by infected mosquitoes from livestock to humans. An outbreak in Kenya in late 2006 and early 2007 killed nearly 150 people. In 2004 more than 140 cases of leptospirosis, including six deaths, were reported from Bungoma; this invidious bacterial disease is carried by rats, including the Giant Pouched Rat *Cricetomys emini*.

Some zoonoses are particularly frightening. One is rabies, the ghastly viral disease spread to humans by the bite of an infected mammal. Once symptoms of rabies appear, it indicates major inflammation of the brain and is virtually always fatal, although a small number of intensively treated victims in the United States have recovered after being put into deep sedation. Those who live in remote areas of Kenya are particularly at risk from rabies, as the vaccine is not always available at clinics. It is estimated that more than 24,000 people are killed by rabies every year in Africa; in Kenya the average is 30–50 fatal human cases detected per year. There will be some that are undetected, in the remote areas of the north and east. Most victims are bitten by domestic dogs, but other mammalian vectors in Kenya include Honey Badgers, Wildcats, squirrels, Leopards, hyenas and, most commonly, jackals; one of us (G. M.) was bitten by a rabid Bat-eared Fox at Wamba. Carnivores with rabies are a hazard because, in the final stages, they become furious and tend to bite anything with which they come into contact. In Kenya, quite a lot of cattle get rabies, and in Botswana the number of rabies cases is three times higher in cows than in dogs. A major study of rabies in the Machakos area found that approximately one dog in a thousand had rabies, that dogs accounted for 92 per cent of suspected rabies cases, and around 240 out of every 100,000 people in the Machakos area were bitten by rabid dogs in a year. So it is a significant disease in Kenya. In the 1950s and early 1960s the colonial government tried to curb the disease by putting out poisoned bait around towns, but this was not very successful; many innocent carnivores (and domestic dogs and cats) were killed, without having a significant effect on case numbers.

In the Americas, rabies can be carried by bats, including the vampire bats. For a long time, there seemed to be no evidence that bats in Africa could carry the disease, but in 2007 a Dutch visitor was struck by a bat in Tsavo West. She noticed a couple of small bleeding wounds on her nose, but was assured that bats didn't carry the disease in Kenya. However, she died of rabies after returning to the Netherlands. The idea of rabies scares many potential visitors to Africa; in the UK those intending to visit Kenya are urged 'to consider rabies vaccination'. In reality, the likelihood of being bitten by a mammal without knowing it is small, and if you are bitten you can have post-exposure vaccine, even if you have already been immunised.

In 1989 most (21) out of a pack of 23 Wild Dogs died of rabies in the Maasai Mara National Reserve, and in fact the Wild Dog has now disappeared from the Mara. Controversy attaches to this story. The pack was being intensively studied; several of the dogs had been anaesthetised and fitted with radio collars, others had been captured and vaccinated. Some scientists

Kitum Cave, Mt Elgon, famed for cave-dwelling elephants and the frightful Marburg virus. (Glenn and Karen Mathews)

suggested that this had traumatised the dogs and made them susceptible to diseases such as rabies; a point of view angrily rejected by those who had been responsible for the monitoring and actual handling of the dogs. One can sympathise with the expensive medical monitoring of a pack of wild animals in this case; Wild Dogs are charismatic high-profile animals and their disappearance from wilderness areas can have a significant impact on tourism revenues. However, a case in Tanzania, where a large number of baboons suffering from venereal disease were captured and treated with antibiotics provoked a furious reaction by the general public. Baboons are common, the costs involved in treating a pack of them in a country where many people do not have access to clean water and anti-malarial prophylactics outraged many.

Another little cluster of frightening zoonoses are those caused by three dangerous viruses: Lassa, Marburg and Ebola. Lassa fever and Ebola are not known from Kenya, but Marburg haemorrhagic disease is. It is also called Green Monkey disease, as a European outbreak was initially traced to a consignment of Green or Vervet Monkeys from Uganda. The actual virus is called the Lake Victoria Marburg virus. In two high-profile cases in the 1980s, two visitors to Kitum Cave on Mt Elgon caught the disease and both died, events dramatised by the American writer Richard Preston in his book *The Hot Zone* (1995). Initially it was suggested that they had died of rabies. In an attempt to identify the host of the virus, a United States army medical team visited Mt Elgon and captured a number of mammals, but did not find the virus in any of them. But in 2007 the virus was isolated from Egyptian Fruit Bats *Rousettus aegyptiacus* in Uganda, Kenya and the DRC (and it was also found in a few insectivorous bats). It has been suggested that the visitors who were killed breathed in urine or faeces from infected bats. That's a risk you run if you stand underneath a big bat colony in a cave or mine. But let's keep things in perspective; many thousands of visitors have been safely into Kitum Cave. Statistically, you run a greater risk driving to Mt Elgon than you do from going inside its caves.

We should make mention of HIV, the human immunodeficiency virus that causes AIDS. It might have evolved from the closely related SIV, simian immunodeficiency virus, found naturally in apes on the west side of Africa. Possibly a hunter was bitten by an ape, or cut himself whilst butchering one; this led to a disease that mutated into HIV. This is a hotly debated and contentious subject but it is of vital importance. Many people are HIV-positive or have AIDS and it causes enormous suffering. One controversial theory suggests an oral polio vaccine prepared using Congo chimpanzees infected humans with SIV, which mutated into HIV. Championed by the brilliant Oxford evolutionist Bill Hamilton (who died just after returning from a Congo expedition), this theory has been rejected by the scientific community. But there remains nervousness about primate zoonoses. Diseases tolerated by one species may be lethal to others; primates are our close relatives. Bits of dead primate or 'bushmeat' are often illegally transported out of West and Central Africa. In 2002, 5 tonnes of illegal bushmeat was seized in London, on flights from Lagos. One visitor was stopped when customs saw blood oozing out of his suitcase. It only needs one chunk to carry some dangerous pathogen.

Some mammalian diseases do not pass to us directly, but have had a major effect upon the ecosystem and on us via our stock. Pleuro-pneumonia, a bacterial affliction known by the Maasai as Ol Kipiri, affected stock and Buffalo in Kenya in 1911. Arthur Blayney Percival saw a huge herd of Buffalo, all dead, on a bend of the Athi River. The disease did reappear in 1917; many cows died but the Buffalo had become immune. But it was curbed by antibiotics in the 1950s. Another deadly disease was rinderpest, or cattle plague, spread by a virus. It affected Buffalo, Giraffe and warthogs, some antelopes (Kudu, Eland, Roan, Bongo and sometimes wildebeest and Impala) and also, importantly, domestic cattle. It has been suggested that the rinderpest virus passed to humans and mutated to become the measles virus, which is closely related; several other diseases followed the same route, from our animal friends to us, including tuberculosis (from cattle), smallpox (from cowpox), influenza and falciparum malaria (from birds). Distemper, a viral disease of carnivores, is closely related to measles and rinderpest, and in 1994 distemper caused a 20 per cent mortality of Lions in the Serengeti.

In the mid-eighteenth century, rinderpest wiped out huge numbers of cattle in Europe. Attempts by Dutch veterinarians to make a vaccine met with so little success that it was believed it was a punishment from God, and the vaccination trials were abandoned. In the 1890s, rinderpest spread through Africa from the north, killing over 90 per cent of the cattle, beggaring many pastoral people and causing starvation, moving from one herd to another, usually by direct contact between infected and uninfected animals, although there is some evidence that it can travel a short distance through the air. The first outbreak appears to have been around Loitokitok in 1890. C. W. Hobley saw its effects on the Tana in 1891, he saw thousands of dead and dying Buffalo and described the stench as nauseating. Maasai cattle raiders took infected stock from eastern Kenya to the rift valley and Laikipia and virtually all the cattle there died, leading to the weakening of the Laikipiak Maasai and their enforced removal from Laikipia in the early days of the twentieth century. Subsequent attacks occurred in Kenya in 1960, killing livestock and game in Isiolo district, there were more outbreaks in West Pokot, Marsabit, Kiambu and Kajiado at the end of the 1980s, and amongst the Buffalo in Tsavo in 1994. But the battle against rinderpest ends in success, because the British veterinary research scientist Walter Plowright, who was based at the East African Veterinary Research Laboratory at Muguga, developed a highly effective vaccine in the 1970s. Its use curbed the disease, the last known cases occurring in the Kenya/Ethiopia border country in 2001, and in 2011 the United Nations officially declared that rinderpest had been eradicated. It is only the second such disease, after smallpox, to be totally wiped out; a triumph for scientific technology and a boon to the Kenyan cattle farmer.

Much of the relationship of humans with wild mammals in Kenya, however, has centred around their slaughter. Kenya's people, before the foreigners arrived, would have had a fairly fraught time with predators and big ungulates, but they didn't kill them in any numbers;, they avoided them where possible. There was harmony of a sort. The arrival of outsiders with guns changed all that. Initially ivory was the target, and as we saw, it was the elephants between the coast and Kilimanjaro that suffered most. Following the arrival of colonial administrators, the game was either regarded largely as a nuisance, standing in the way of agriculture or development, or something for sportsmen to hunt. Many colonial or military personnel stationed in the highlands did a bit of hunting; some did a lot. With the

Lesser Kudu – a Somali-Maasai species. (Stephen Spawls)

arrival of the First World War, game animals were seen in some quarters as even more of a problem. Philip Percival was told to poison Lions along the railway, as they were a hazard to the military operation. Soldiers were instructed to shoot all the rhino, for similar reasons. Giraffes kept wandering into the telegraph lines, which were set at 14ft, and again instructions to kill were issued. Arthur Blayney Percival was particularly upset by this. His book *A Game Ranger on Safari* (1928) has a picture of Giraffes entitled 'victims of the war'. Percival estimated that 40,000 head of game were shot to feed the soldiers. In the 1920s and 1930s, game was freely shot in areas intended for agriculture. The Makueni rhino hunt saw more than 1,000 rhino killed. George Adamson was hired in 1939 to shoot game on the farms north of Mt Kenya and the Aberdares; he was told to simply eliminate the game as they were hindering development. The meat wasn't used; the zebra and oryx he shot were left where they fell. Similar manic schemes took place elsewhere; between 1932 and 1940 over half a million wild animals were shot in what was Southern Rhodesia (now Zimbabwe) in a pointless attempt to eradicate the tsetse fly. It is futile to blame the authorities; at the time they knew no better and wildlife was seen only as a problem. No one had realised that a time would come when people would travel to Africa simply to admire its wildlife and pay good money for that privilege.

THE LITERATURE AND RESEARCH

Alex Duff-MacKay, who at various times held the posts of both mammalogist and herpetologist during 31 years at the National Museum in Nairobi, once commented that research on Kenya's natural resources was unfairly biased towards the 'large, hot and hairy fauna', meaning the mammals. He was right, and there is a reason. Mammals are appealing. It is animals like gorillas, chimpanzees, Lions, Leopards, elephants and rhinos that have attracted the big money and the high profile researchers from prestigious institutes; large and charismatic birds like eagles and Lammergeiers *Gypaetus barbatus* might get a look in, but the little stuff just doesn't sell. The pages of eminent journals like the *African Journal of Ecology* (formerly the *East African Wildlife Journal*), the *Journal of the East African Natural History Society* and the *Journal of Wildlife Management* frequently have articles dealing with big carnivores, primates or charismatic ungulates. *Tropical Zoology, American Zoologist* (now *Integrative and Comparative Biology*), conservation journals like *Oryx*, and of course special journals dealing with mammals or particular groups of mammals, like *Mammalia*, or *Primates*, are full of material dealing with Africa's large furred mammals. By 1998 there were well over a thousand scholarly articles published on the African Elephant. It is Africa's most charismatic animal, and charismatic people have worked on them. The same is true of gorillas, Lions and rhinos. But could you name anyone who has worked with rodents? Frogs or catfish? How many papers are published on cisticolas, or pseudoscorpions? Not many. The same goes for television documentaries. Many of us saw the 'Big Cat' diaries, and Saba Douglas-Hamilton's 'Elephants of Samburu' on television; the 'Big Dung Beetle Diaries' would be a non-starter, as would have been Saba Douglas-Hamilton's 'Shrews of Samburu'. And returning to Duff-MacKay's prescient comment, he might have also added that most research has been done by Westerners. Many years ago, this imbalance was addressed in an exasperated editorial in *Africana*, now *Swara*, the magazine of the East African Wildlife Society. The editor expressed his view that it might be more appropriate to use funds to take Kenyan children to see their wildlife heritage and thus generate enthusiasm for the natural world, rather than funding research by some foreigner to prove that ostriches hide their heads in the sand because they can't bear to look at the erosion. In the end, though, we should welcome anything that raises the profile of Kenya's natural history. If somebody sees a documentary on cuddly Lion cubs and subsequently decides to visit Kenya, we have to hope that is a good thing.

So, mammals are popular and lots of literature is available. For the global view, David MacDonald's aforementioned *The New Encyclopaedia of Mammals* (2001) is excellent value, as sound as can be, with superb photographs, some good stuff on mammalian origins and lines of descent. The hardback weighs over 4kg, however. A new eight-volume handbook of the mammals of the world is in production, but expensive. Reay Smithers' *Mammals of the Southern African Subregion*, published by the University of Pretoria (1983) only covers those mammals that occur in southern Africa but is a lovely (although heavy) book. It has African-wide distribution maps for all the widespread species, and is extremely readable, with anecdotal information that is missing from other mammal guides.

The main man where East African mammals and their literature are concerned is Jonathan Kingdon. A big, dynamic bearded white African, born in Tanzania and trained in fine art at Makerere University in Uganda, Kingdon has spent most of his life researching Africa's mammals. His seven-volume work *East African Mammals; An Atlas of Evolution in Africa* (published by the University of Chicago Press throughout the 1980's) amounts to several thousand pages. Illustrated by Kingdon's inimitable pencil sketches, showing not just the animals in a whole range of poses, but often their behaviour, musculature and skeletal details as well; these are superb books and not too expensive. For those who want a summary, Kingdon has also published a handy 460-page field guide (1997 Academic Press) to all of Africa's mammals, showing all the big animals and a good selection of the smaller ones, with lovely colour pictures, with distribution maps. The only shortcomings of this excellent and portable book is that many of the smaller mammals have been lumped together (for example all 103 species of white-tooth shrew) and some of the paintings, although technically perfect, show the animals in slightly odd poses. But this is *the* mammal book to have with you in the field. Other good books on African mammals include Jean Dorst and Pierre Dandelot's *Field Guide to the Larger Mammals of Africa* (Collins 1970), for a long time the only comprehensive guide, its only shortcomings being simple maps and absence of small mammals. It's now out of print but can be obtained via the Internet. Collins also published a field guide to the mammals of Africa that included Madagascar, by Theodore Haltenorth and Helmut Diller (1995), which has rather stylised illustrations. More recently, the South African husband and wife team of Chris and Tilde Stuart have published a photographic guide and a pocket guide to the mammals of East Africa (Struik 2009).

In 1960, Charles Astley Maberly, a British wildlife artist based in South Africa, published and illustrated a very nice little book, on Kenya's mammals, rather oddly entitled *Animals of East Africa* (published in Cape Town by Howard Timmins). The book is based on an extended safari he made to Kenya's main reserves, which in the

1950s were the Aberdares, Tsavo, Amboseli, Mara, Marsabit and Mt Kenya. Most of his distribution notes refer to the occurrence of the mammals in those parks; so it gives a picture of what was there at that time. There are some strange omissions and mistakes – for example, Astley Maberly states that wildebeest are not found in the Mara – but the man was a simply superb illustrator; few artists before or since have captured the 'jizz' of African animals better than Astley Maberly. There are a number of books that deal with East Africa and Kenya's wildlife in a general fashion. Two worthy of mention Clive Spinage's *Animals of East Africa*, (Collins 1962) which despite its title is almost totally about mammals, and is illustrated by superb photographs, and Jonathan Scott's *Safari Guide to East African Animals*, (1997 Kensta) enlivened by Scott's fine pictures. An essential field book for those with time to observe is Richard D. Estes' *The Safari Companion*, subtitled *A Guide to Watching African Mammals*. (1993 Tutorial Press). Estes, an American zoologist, has worked for nearly 50 years on East African mammals, and this book is very unusual in that it tells you how to watch East African mammals, with a lot of cleverly encoded information on how they behave. Keep it on your lap as you watch.

A lot of researchers – mostly foreigners, it has to be said – have researched on mammals in Kenya and published books and papers on their findings. One thinks of Shirley Strum's books on the baboons of Gilgil and Laikipia, Anna Rasa's *Mongoose Watch*, (John Murray 1984) about the Banded Mongooses of the Taru desert, Joyce Poole's *Coming of Age with Elephants* (Hyperion Press 1996) and Cynthia Moss's *Elephant Memories*, (University of Chicago press 2000) about the elephants of Amboseli, Joy Adamson's books on her Lions, Leopards and Cheetahs, and Daphne Sheldrick's *The Orphans of Tsavo* (Collins and the Harvill Press 1966) . The scientific content of these books varies greatly. George Schaller's *Golden Shadows, Flying Hooves* (Alfred A Knopf 1973) is a lovely book by one of our finest zoologists, about the predators of the Serengeti; it is packed with hard zoological information and yet it reads as smoothly as a thriller. The lion research Schaller undertook was published as a separate book, *The Serengeti Lion* (University of Chicago Press 1973). Schaller's work on the Mountain Gorillas in the Virunga volcanoes, described in his book *Year of the Gorilla* (University of Chicago Press 1988) is equally valuable. Another good book is Cynthia Moss's *Portraits in the Wild*. Published in 1976 by Hamish Hamilton and possibly a little dated now, it is still a thoroughly decent summary of the research conducted on (and the researchers of) most of Kenya's large mammals, elephant, Giraffe, Lion and rhino included. Moss's bibliography gets you up to speed with all the professional material published before 1976.

Three compilations of papers about the Serengeti–Mara ecosystem, with much relevant research on mammals, are *Serengeti: Dynamics of an Ecosystem*, published in 1979, *Serengeti II*, published in 1995, and *Serengeti III*, published in 2008, all by the University of Chicago Press. Excellent though these compilations are, they haven't managed to stop the Tanzanians planning a major road across northern Serengeti, and one is tempted to wonder, no doubt cynically, if all the money spent on sponsoring and publishing the scientists' work might not have been better spent elsewhere, amongst the young people of Tanzania.

Some nice early material on the distribution of mammals in Kenya in the early twentieth century, can be extracted from Richard Meinertzhagen's 1957 *Kenya Diary* (Oliver and Boyd) and Arthur Blayney Percival's two books, *A Game Ranger's Note Book* (Nisbet 1924) and *A Game Ranger on Safari* (Nisbet 1928). Meinertzhagen was an army officer stationed in Kenya between 1902 and 1906 and his work is discussed further in the bird chapter. His nature notes from southern Kenya give interesting insight into what mammals were present then, and in what numbers. Unfortunately, he then shot a great many of them. Percival travelled widely in Kenya, as the ranger for game preservation, and later game warden, and his charming books give a careful record of what mammals he saw where. Although Percival was not averse to shooting things, he wasn't as trigger-happy as Meinertzhagen, and was often content to stay his hand and simply observe the animals. His books are gems and every Kenyaphile should read them.

The literature mentioned here is just a personal selection. There are literally hundreds of books on the various East African and Kenyan mammals. There might be only one book on the Naked Mole-rat, two books on African Buffalo and only ten or so on antelopes, but there are more than 50 books on African Elephants and the scientific journals are full of papers on African mammals. The literature reflects their popularity. Let us hope they continue to be both popular and visible.

Chapter 7

The Birds

The lake … [Naivasha] … was a marvel of bird life. I had never before seen anything like it…. a narrow strip of sandy foreshore was packed with Egyptian geese … yellow-billed ducks, pochards and pink-billed teal … the whole so totally unconscious of danger and strangers to a shot gun that it … was downright barbarous to molest them. Sir Frederick Jackson, *Early Days in East Africa*, 1930

Kenya has a stunning bird fauna. Just over 1,100 species of bird are presently known from the country, compared with 1,050 from Tanzania and just over 1,000 from Uganda. Around 2,100 species are known from sub-Saharan Africa, and about 10,000 species are known worldwide. Between eight and twelve species of bird are endemic to Kenya, depending upon which authority you follow.

About 860 of Kenya's birds are resident, just over 130 are migrants from Europe and Asia and about 30 are migrants from other parts of Africa. The remainder are sea birds and vagrants, two categories that may overlap. Many birds occur across a range of habitats, but nevertheless there are several distinctive bird faunas in Kenya; these include forest, mountain, savanna and semi-desert faunas. Most Kenyan birds are either seedeaters or insect eaters, and this is connected with their savanna habitat. African savannas are among the world's richest bird habitats.

In the 1994 World Birdwatch, Kenya recorded the highest total of any country in the world, 797 species seen by the teams in one day. One team saw 342 species alone at Lake Baringo. With its network of protected areas, accessibility, relatively open country and vast variety, Kenya has to be regarded as one of the best – possibly the best – places to watch birds in the world. And Kenya's rich birdlife is of economic importance to the country. Birdwatching is a major part of ecotourism; it brings in over 25 million dollars annually to South Africa. Kenya is currently spending 300,000 dollars on an initiative to boost bird tourism.

This chapter begins with a look at the features of birds in general, their origins and spread before going on to focus on the Kenyan bird fauna: species numbers, migration and zoogeography, ecology, communities and populations. Then there is a section on the history of ornithology in Kenya, with a look at some of the personalities involved, followed by an examination of the interactions between birds and humans. The chapter concludes with the relevant Kenyan bird literature and an outline of some current research. Each section is self-contained and can be read independently.

BIRDS IN GENERAL

Birds are very visible animals and everyone knows what one looks like – they have feathers, two wings and a beak. They have internal fertilisation, they lay and sit on eggs, keeping them warm. This reduces incubation time, which is less than two weeks in the Common Bulbul *Pycnonotus barbatus* and paradise flycatchers *Terpsiphone* spp., although some of the bigger eagles can take up to two months. Birds look after their young until they are fledged; a clever reproduction strategy, only mammals have improved on this. Many birds are monogamous, choosing one partner for a season (or sometimes longer) and in most species the male helps brood the eggs and feeds and defends the young; a good system, unlike mammals where it is the female that bonds with her offspring. Having said that, there are a few dastardly partners who don't do the right thing: in most duck, parrot and bustard families the female is left to care for the young, but among the coucals, jacanas, painted snipe and buttonquails it is the wicked female that deserts her mate. Birds sing, often to ward off other males from their territories and to tell females they are there. Most birds can fly; they are relatively large and easily seen. Some of them make gigantic journeys. Birds are warm-blooded, they don't need to bask and hence occur world-wide. There are birds in the tropical forest and birds in the deep Antarctic. For many people birds are a constant delight. Their colours are pleasing to the eye, consider the Lilac-breasted Roller *Coracius caudatus*, the Northern Carmine Bee-eater *Merops nubicus* and the Malachite Kingfisher *Alcedo cristata*. More people are interested in birds than in any other group of vertebrates; the lecturer who gives a public talk on birds attracts a far bigger audience than one who talks about reptiles (we speak from experience!). And our lives overlap with birds in a number of ways. For many people, birds are the only species of wild vertebrate that they see regularly. Glance skywards if you live in the heart of Nairobi, or Nakuru, and the Marabou Stork *Leptoptilos crumenifer* or Yellow-billed Kite *Milvus aegyptius* that you see is your contact with the free creatures of the natural world. There is no pleasure to compare with being out early on a sunlit morning, in wild country, with a pair of good binoculars and in contact with a mixed bird party. Kenya is a simply superb place for both 'birders' and the more enthusiastic 'twitchers' alike.

Birds are highly specialised and successful organisms, suited for their environment. To fly, to defy gravity, requires a number of important modifications, most of which have evolved to make the body light. Other vertebrates can be very large. Blue Whales weigh

Streaky Seedeater, one of Nairobi's most common birds. (Stephen Spawls)

Lilac-breasted Roller, unmistakeable. (Stephen Spawls)

100 tonnes, the heaviest modern reptile (the Salt-water Crocodile *Crocodylus porosus*) is over a tonne, the biggest fish (the Whale Shark *Rhincodon typus*) can weigh 36 tonnes, and even the biggest amphibian, the Chinese Giant Salamander *Andrias davidianus*, weights 45kg. But the biggest flying bird, a Kenya resident, is the Kori Bustard *Ardeotis kori*, at a modest 17kg (the word *kori*, or more precisely *kgori*, is Setswana, the language of Botswana; it means 'bustard'!) There are heavier birds, a cock ostrich, for example, can weigh up to 150kg, but they cannot fly. Birds are light. Sixty-seven percent of Kenya's mammals weigh less than 500g, but 74 per cent of Kenya's birds weigh less than 120g. Kenya's smallest bird is the Mouse-coloured Penduline Tit *Anthoscopus musculus*, which is about 7cm long and weighs less than 5g (the world's smallest bird is a Cuban hummingbird weighing under 3g). More than 50 per cent of Kenya's birds are between 10 and 20cm in length and weigh between 10 and 50g. Kenya's longest flying bird, the Great White Pelican *Pelecanus onocrotalus*, which can be over 1.7m long (although it rarely stands that tall) can weigh up to 8kg, and if you watch one of these birds trying to take off, you can see the problem associated with flying for a relatively heavy bird. The Goliath Heron *Ardea goliath* stands 1.5m tall, but only weighs 4kg and it can take off relatively easily. A flying bird gets its lift from muscle-driven moving wings; the food energy needed for this, and the oxygen to burn it, is supplied to the muscle cells from the bloodstream. In theory, there is a limit to how fast this energy can be supplied and how rapidly it can be processed, and hence no flying bird is very large. But there might be something wrong with this theory. Seventy million years ago a gigantic flying pterosaur existed, *Quetzalcoatlus northropi*, with a wingspan of 12m and weighing between 75 and 200kg. If such a huge reptile could fly, why is it that birds are now so small? Some despairing palaeontologists have calculated that *Quetzalcoatlus* theoretically could not fly, but its hollow bones and articulated wings suggest otherwise; a creature doesn't evolve big wings unless it's going to fly. So why don't modern birds get that big? It's a mystery, although some scientists have suggested that at the time of the giant flying reptiles there was a greater proportion of oxygen in the atmosphere. But this theory also has its shortcomings: 6-million-year-old fossils of a recently discovered and genuine bird, the Giant Teratorn, *Argentavis magnificens*, from Argentina, indicate that it

Elegant African Jacana, the male alone raises the young. (Stephen Spawls)

African Openbill Stork; the permanent gap enables the bird to grip the snails it eats. (Stephen Spawls)

had a wingspan of 7m and weighed over 70kg. Oxygen was present then in roughly the same proportion it is now. So we really don't know why modern flying birds are so small.

Modifications for flight in modern birds include hollow bones, absence of teeth and bladder, and lightweight feathers that insulate the body and increase the surface area of the wing. Birds have a highly efficient lung that is linked with air sacs. Air only flows one way through the bird's lung, unlike our lungs, where air is pumped in and out and is therefore less efficient. When a bird breathes in, air enters a tube with more than one exit; some air goes to the lungs directly, some bypasses the lung and flows into an air sac. As the bird breathes out, the air from the air sac also enters the lung by a second tube. As you will know if you've carved a chicken, birds have a large sternum or breastbone, and the purpose of this big flat bone surface is to provide an anchor for the powerful pectoral muscles that flap the wings. As the wings are lifted, they are twisted inwards close to the body to keep drag or air resistance low, they then spread out on the downstroke. The downward push and air flow over the wing creates a region of low pressure above the wing and high pressure below, and this causes uplift. Take off is assisted with a jump, courtesy of strong leg muscles, coupled with the wings going upwards and then a powerful downstroke. Bigger birds may run a short distance to assist with takeoff, and water birds may run along the surface.

Birds also have remarkable eyesight, as one might expect with a flying animal; sharp vision is essential if moving at high speed through the air. Bird eyes are relatively large and have four types of colour receptor (we have three); birds can thus see ultraviolet, which we can't. It is believed that ultraviolet has a role in signalling, possibly in sexual displays, and some birds – especially parrots – are also fluorescent; when ultraviolet falls on their plumage it is converted to visible light of a longer wavelength. Birds have very many receptors per square millimetres on the retina at the back of the eye; a buzzard has a receptor density five times greater than a human. Some birds have two foveas, the 'yellow spot' of acute vision. A Black-shouldered Kite *Elanus caeruleus* can spot an insect 5mm long from over 25m away.

So birds are well adapted for their environment and appear to be remarkably successful. But why? Being briefly unscientific, we shall split vertebrates into five handy classes: mammals, birds, reptiles, amphibians and fishes. There are about 57,000 species of vertebrate, a good number (although low compared with some invertebrates – there are over a million insect species already). The least diverse are the mammals(about 5,500 species), then amphibians (roughly 6,200 species), about 8,000 species of reptile, and some 10,000 species of bird. Fishes are the most diverse, with about 32,000 species known. So, despite being relatively young (amphibians appeared 340 million years ago) birds have diversified a lot. And, in exploiting the air as a habitat, they are highly visible. They are the most obvious and most diverse vertebrate group that the layperson comes across, but in some respects their apparent abundance is subjective. No bird lives permanently under the ground, or in water. None is very small. They move around rapidly, across the open sky, and perch in trees. Nearly all are diurnal. Birds live in the same sort of places that humans live, so we see a lot of them; it is our *perception* that birds are common. There are more species of fish, but we don't see them because we don't live in water. There are a lot of beetles, but we don't see them because many live concealed in ground cover and they are small. An enthusiastic amateur bird watcher in Kenya with a little knowledge can quite easily observe over twenty species of bird in a small park, or big garden. A competent bird watcher could spot 150 species of bird a day in Kenya; a real expert can see 300. There is no other group of vertebrates for which you could do that (although underwater on a coral reef you might get into triple figures in one day with reef fish if you know your species). But in the narrow habitat within which we humans live, birds are relatively abundant. Their evolutionary success has to be connected with their warm-bloodedness, in that they can handle a range of temperature regimes, and with their ability to fly, and thus exploit a range of habitats.

ORIGINS AND THE SPREAD OF BIRDS

Birds prove to be highly adapted reptiles – essentially they are surviving bipedal dinosaurs, the visual evidence for this is the scales on their legs. And a most interesting question is, why did birds survive while other dinosaurs became extinct some 65 million years ago, after a meteorite struck the earth and sent up huge quantities of dust

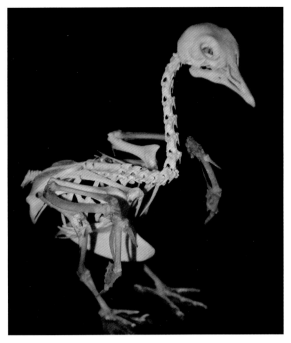

Pigeon skeleton, the huge breastbone extends well down and serves as anchor for the flight muscles. (Stephen Spawls)

Ostrich; the largest eye of any terrestrial vertebrate. (Stephen Spawls)

Serrated Hinged Terrapin Pelusios sinuatus *and Black-crowned Night Heron, Hippo Pools, Nairobi National Park. (Stephen Spawls)*

Water Dikkop in flight, Rusinga Island. (Stephen Spawls)

into the atmosphere. Layers of this dust occur worldwide and contain high levels of iridium, a element present in large quantities in meteorites. But the birds made it through. Maybe their warm blood helped them survive the subsequent fall in global temperatures, which did for the exothermic dinosaurs.

Birds evolved from feathered dinosaurs, a group that includes the famous *Archaeopteryx*. We usually think of feathers and flight as the defining essence of birds, but a crop of recent fossils from China indicate that feathers evolved before flight, so their initial function might have been insulation, camouflage or courtship. The small variation in bird DNA also indicates they all had a single common flying ancestor, which means that modern non-flying birds have lost their flying abilities. A major radiation of birds occurred between 65 million and 30 million years ago. One is tempted to speculate that with the big flying reptiles gone, birds got their chance to speciate into a freshly vacant habitat.

There are three main strands of evidence for this radiation: fossil bird bones, molecular taxonomy and diversity. Bird fossils recorded from East Africa include an ancient Turaco from Songhor, 10 million years old, a 20-million-year-old flamingo (named after Louis Leakey, as *Leakeyornis aethiopicus* and possibly closely related to the living Lesser Flamingo) from Rusinga Island, and a 10-million-year-old Marabou Stork from Baringo. Some bird fossils from outside Africa are interesting, fossil Secretary Birds from Tertiary strata and fossil mousebirds 40 million years old are known from France; a 15-million-year-old ground hornbill is known from Morocco. But the fossil record is poor. There are no fossils for many families. Evidence of the diversity and timing of evolutionary splits largely comes from analysis of their biomolecules. Classic early work on bird DNA by two controversial American ornithologists, Charles Sibley and Jon Ahlquist, elucidated the relationships and lowered the number of bird families from 27 to 23. Many ornithologists, and ornithological organisations, both professional and amateur, strongly disliked the new taxonomy, particularly its detachment from bird anatomy and refused to use it. A joke among traditionalists was that DNA stood for 'don't need anatomy'. More recent work, by a number of top molecular ornithologists, such as Per Ericson, John Fjeldsa, Joel Cracraft and Shannon Hackett, has uncovered

Flamingoes on Lake Amboseli. (Stephen Spawls)

more disturbing changes in bird relationships. For example, the falcons, birds of prey *par excellence*, are shown not to share a common ancestor with other birds of prey, but to be closely related to the passerines and the parrots! They evolved their killing beaks and claws independently of eagles. The cuckoos are more closely related to the bustards than any other birds, and flamingos are closely related to grebes. It makes birdwatchers grind their teeth.

The big birds are international: between 70 million and 60 million years ago they spread out from Gondwana, as that continent and Laurasia separated into what are now the big zoogeographical regions of the world – the Nearctic, Palaearctic, Afrotropical, Neotropical, Indomalayan and Australasian. Using figures for breeding (thus resident, not vagrant) species, those six regions each has between 15 and 25 herons and egrets, between 10 and 45 ducks and geese, between 23 and 60 raptors (excluding falcons), between 20 and 40 gulls and terns and between 11 and 40 species of owl. These birds are relatively large, and these figures indicate that these are old groups, that spread out from the great southern continent. Mobility can also lead to huge ranges of individual species; several bird species occur virtually worldwide, for example the Osprey *Pandion haliaetus*, Barn Owl *Tyto alba*, Black-crowned Night Heron *Nycticorax nycticorax*, Greater Flamingo *Phoenicopterus ruber*, Peregrine Falcon *Falco peregrinus*, Striated Heron *Butorides striata*, Great White Egret *Egretta alba* and Gull-billed Tern *Sterna nilotica*. Such huge ranges are not seen with other vertebrates, where there are no international species save those spread by humanity, like the Black Rat.

When you get to the smaller birds, the figures for the zoogeographical regions look very different; they fluctuate wildly and indicate that ancestors – possibly just one or two birds – got to certain areas and then speciated rapidly. This is supported by molecular data. Evidence indicates that passerines did not originate in South America, which has the greatest number of passerine species, more than 1,800. They originated in Australia, appearing less than 60 million years ago. Birds disperse quickly because they can fly. Present distribution patterns indicate that pioneer individuals reached distant places and speciated relatively rapidly there. An ancestral weaver bird got to Africa, liked it there and evolved. The offspring didn't bother going anywhere else.

THE KENYAN BIRD FAUNA: SPECIES NUMBERS, MIGRATION AND ZOOGEOGRAPHY

Around 2,100 species of bird are known from Africa. We cannot give a definite figure as species are constantly being added to (and sometimes deleted from) the continent's list. And the status of some species is debated; some ornithologists regard the ostrich as a single species, others as two. Thus all the figures we give in this section are approximate. Just fewer than 1,100 species of bird have been recorded from Kenya but it should be noted that several species of bird have become extinct in Kenya since the 1950s (although they survive elsewhere in Africa), including the Velvet-mantled Drongo *Dicrurus modestus*, Yellow-mantled Weaver *Ploceus tricolor* and the Forest Wood-hoopoe *Phoeniculus castaneiceps*. Although around 70 of Kenya's birds are vagrants, a list of more than 1,030 residents and migrants is impressive. Botswana, in southern Africa, is larger than Kenya, but has fewer than 600 recorded species; Ethiopia is more than twice the size of Kenya and has only 870 species, the Sudan is more than four times larger than Kenya but has only 950 species. In Africa, Kenya's total is exceeded only by the Democratic Republic of the Congo, which has just under 1,200 species, but the DR Congo is larger than all the countries of East Africa put together. Djibouti has only 330 or so. Some ornithologists believe that Tanzania's list exceeds Kenya's, but the African Bird Club's Tanzania checklist, at the time of writing, only had 1,050 species on it.

This huge species diversity in Kenya is real, not an artefact resulting from excessive monitoring or observation, as is the case in some countries. Vagrants make up less than 7 per cent of the total Kenya list. Great Britain is a case in point, where vagrant species are concerned. Just over 260 species of bird occur in Britain naturally (and only about 220 of these breed) but the 'British List' actually has 590 species on it. These extra 330 species, nearly 60 per cent of the total, are all freaks that have been blown or drifted to Britain and been spotted by birdwatchers.

Kenya's birds are divided into various 'status' categories. No such classification exists with mammals, reptiles or amphibians, which are largely land-based organisms; in general, they either live there all the time or they don't occur. The Black Mamba or Burchell's Zebra is either here, or it's not. You are not going to spot a vagrant Tiger *Panthera tigris*, or rattlesnake in Kenya. But birds are a very different matter. American birds have been seen in Kenya. The African Bird Club's excellent Kenya checklist, available as a free download from their website, offers us the following status categories: migrant (M), vagrant (V), resident (R), rejected (X). We can also choose various localities – African (A), breeds in Palaearctic (P), oceanic (O), Nearctic (N) – plus a few extra data points: winters (W), meaning it comes to Kenya in the northern winter, introduced species (I), breeds locally (B), and the special symbol §, meaning uncertain record. And the experts have combined the symbols to classify any Kenyan bird. Well-known shoreline waders like the Ringed Plover *Charadrius hiaticula*, the Sanderling *Calidris alba* and the Ruff *Philomachus pugnax* are classified PW, they come for the winter from the Palaearctic, they are migrants. The Ringed Plover spends the northern summer in Siberia, up there on the tundra, but come December–February time you can see it on Lake Victoria's shores. Happily, the bulk of Kenya's birds are classified RB, resident breeder, which means they live and breed in Kenya

It may be appropriate here to talk about bird migration, as migrants swell Kenya's species numbers. Migration and homing are still mysterious to some extent. Many creatures, if taken away from their homes, even species that do not normally wander, will find their way back. In the United States, studies of Black Bears *Ursus americanus*, White-tailed Deer *Odocoileus virginianus*, Moose (Eurasian Elk) *Alces alces* and Deer Mice *Peromyscus* spp. found that they could often make their way home after being moved into unknown territory. A house cat in Britain walked a distance of over 200km back to its old home. Leopards trapped in highland Kenya in the 1970s and released in Tsavo made their way back to where they were trapped. But exactly how they do it is not yet perfectly understood

By definition, migration in animals means a seasonal movement. Migratory animals include eels *Anguilla anguilla*, Humpback Whales *Megaptera novaeangliae*, wildebeest, sea turtles and Monarch Butterflies *Danaus plexippus*. Migration can occur for a mix of reasons, like finding a seasonal food source, finding somewhere warm, avoiding cold or bad weather, breeding, avoiding competition, predators or parasites. Most Palaearctic migrants to Kenya have come here to get away from the northern winter and obtain food; they don't breed in Kenya. Their

The tiny Sedge Warbler migrates six thousand kilometres. (Stephen Spawls)

numbers include some ducks, birds of prey, a lot of waders and a rag-bag of birds that include the White Stork *Ciconia ciconia*, swallows, martins and pipits. The list is concluded by a handful of tough little warblers, like the Common Chiffchaff *Pylloscopus collybita* and Willow Warbler *Phylloscopus trochilus*. To them, Kenya and Africa represents somewhere warm with food, but it isn't the place to raise a family. In Kenya there is no guarantee that if they bred they would find adequate food supplies for their young; the cycle of rainy seasons in the tropics is too erratic. Instead, they return to Eurasia for spring, March–May, when the emergence from winter guarantees warmth and small food items for their nestlings; possibly there are fewer predators as well. A little Willow Warbler in a Kilifi thicket faces danger from Black Mambas, sand snakes, various mongooses, small carnivores like genets, many different birds of prey and disease-carrying insects, and if it attempted to raise a brood it would also run the risk of safari ants and other arthropods. In a nice thickety bush in England the predators are far fewer, maybe just a Eurasian Sparrowhawk *Accipiter nisus*, Stoat *Mustela erminia* and domestic cat.

Migration is a remarkable thing. Consider a European Sedge Warbler *Acrocephalus schoenobaenus*, that skilful mimic of other birds' calls. It is only a 10g ball of fluff and feathers, with a bold eyestripe, and yet this little creature only just heavier than a Kenyan twenty shilling coin, sets out, cheerfully, twice a year on a dangerous five or six thousand kilometre journey, flying for hundreds of hours. For a long time, in Europe, the fact that such tiny organisms actually travelled away from their summer breeding grounds in Europe and Asia, down to Africa, wasn't known. It was thought that they hibernated in holes in trees, or in the mud of ponds.

Birds migrate over several routes. The big migrations are between the northern and southern hemisphere, following the seasons; birds go between North and South America, between the Arctic and the Antarctic and between Eurasia and Africa. Much of what we know about African bird migration is based on the work of a remarkable Englishman called Reginald Moreau (known to friends and fellow ornithologists as Reg). A sickly young man, in 1920, at the age of 23, he contracted rheumatoid arthritis and was told by his doctor he should go abroad. He obtained a posting to Cairo, and began to watch birds, and in 1928 he managed to get a job at the Agricultural and Forestry Research Station at Amani, in the Usambaras (appropriately enough, *amani* means 'peace'

Grey Crowned Crane, the bird on Uganda's flag. (Stephen Spawls)

or 'tranquillity' in Swahili). In his 18 years in those lush, forested Tanzanian hills Moreau became 'the' expert on Africa's birds, and after he returned to England he was offered both the editorship of *Ibis*, the journal of the British Ornithologists' Union, and a research post at the Edward Grey Institute of Field Ecology in Oxford. Ill-health dogged his later years, but he produced not just a string of scientific papers but three classic books, *The Bird Faunas of Africa and Its Islands* (Academic Press 1966), *The Palaearctic–African Bird Migration Systems* (Academic Press 1972) and (with B. P. Hall) *An Atlas of Speciation in African Passerine Birds* (British Museum 1970). Moreau's migration book was published in 1972, after his death in 1970. It remains a classic and African ornithology is forever in the debt of this cheerful bald-headed enthusiast. There is a lovely book, recently published, on the birds that migrate from Africa to England, seen from the British end. Written by the environmental correspondent of the *Independent* newspaper, Michael McCarthy, and called 'Say Goodbye to the Cuckoo', (John Murray 2009) it is a beautifully written, heart-wrenching account of how Britain's migratory birds are disappearing; Reg Moreau gets several mentions in it.

It appears there are several migratory routes into and across Africa. A number of species spend the summer in the high latitudes, right up in northern Scandinavia, Siberia and other far northern regions, and fly back down to Africa in September/October. Other birds come from Europe, and many come into Africa via the straits of Gibraltar or north-east Egypt; a few come down the length of Italy and thence across the Mediterranean. Although the migrants seem keen to avoid crossing large areas of open sea, there are species that come from eastern Asia and some may cross huge areas of the Indian Ocean; birds in this group include the Willow Warbler, the Northern Wheatear *Oenanthe oenanthe* and the Barn Swallow *Hirundo rustica*. But some birds just fly straight across the Mediterranean Sea and the Sahara, in non-stop flights that take between 40 and 60 hours. Some migrations are relatively short. Birds heading into West Africa from western Europe, for example the Melodious Warbler *Hippolais polyglotta*, travel only 2000–3000km, whereas the Ringed Plover may travel close to 10,000km from Siberia down to southern and eastern Africa. They flap as they fly, they don't soar, although larger birds like the White Stork, and the cranes that reach Ethiopia (but rarely Kenya) do, using the thermals (columns of hot rising air) to gain height.

Before a bird sets off for warmer climes, it eats lots of food and this is largely converted to fat; the body mass of a bird about to migrate may be 50 per cent greater than usual. The mathematics are interesting. A Marsh Warbler *Acrocephalus palustris* that normally weighs about 10g will increase her weight to 15g and then set off on a 6000km flight, climbing up to 2000–3000m or even higher. A bird of that size, when flying on migration, will burn fuel at a rate of around 1,500 joules per hour, at a power of just over 0.4 watts. The 5g of fuel she took on will provide energy; fat will supply 38,000 joules of raw energy per gram, so her total energy reserves are 5 x 38,000 = 190,000 joules. If she's burning it at 1,500 joules per hour, that means that the energy available will last 127 hours. A little warbler on a long distance flight is probably travelling at about 50km per hour, so she has enough fuel to carry her 6,350km. If the wind is at her back, she can go further. If the wind is in her face, it can create problems. Migrating birds sometimes come down, and it can be fatal. One of us (S. S.) once found nearly

The Black Kite, common around habitation, scavenger extraordinaire and scourge of the chicken farmer. (Stephen Spawls)

30 warblers under a rock in the Egyptian desert in April, all dead. They had obviously descended for some reason and frozen beneath the boulder.

In November, big flocks of migrants pass southwards over eastern Kenya and under cloudy conditions may descend onto hilltops; such group landings are called 'falls'. Regular falls take place on the Ngulia Hills in Tsavo West, particularly in November and December, if there is cloud hanging around the mountain; the birds see the lights of the buildings glowing in the haze and descend. Kenya's ornithologists take advantage of this, and on suitable nights set fine mist nets in the bushes around the lodge at Ngulia. They catch the birds and ring their legs (without hurting them at all) and then release them. Since 1969, more than 220,000 birds have been ringed here, and very occasionally a ringed bird is recovered (roughly one bird per thousand ringed), sometimes alive but more often dead. Each ring carries a unique legend that tells ornithologists where the bird has come from; it adds a little bit of data to the jigsaw of animal navigation.

There are a group of birds that actually migrate within Africa, the intra-African migrants. They move for the same reasons as the Palaearctic migrants, although obviously not so far and they are not avoiding extreme cold. About 600 species of African bird migrate within the continent, and since most of Africa lies north of the equator most intra-African migrants make their seasonal movements on the north side, but some species move across the equator, some examples being the Grasshopper Buzzard *Butastur rufipennis*, which goes north to breed in April to September, as does Abdim's Stork *Ciconia abdimii*. Some Black Kites *Milvus migrans* breed in Egypt in April but then move south to the Kenya coast. The intra-African migrants may have cultural significance in parts of Africa; for example in West Africa the appearance of the Black Kite indicates the start of the harvesting season. (There's a belief that if you see one and don't sing to it, the harvest will be poor.) Other prominent intra-African migrants in Kenya include the Violet-backed Starling *Cinnyricinclus leucogaster* and a number of cuckoos, rollers and nightjars; this is the reason why some indisputably African birds may be highly visible in some months of the year and seemingly absent in others. There is

Greater Blue-eared Starling; the subtleties of colour are hard to see if the light isn't right. (Laura Spawls)

also a rather curious type of migration that occurs from Madagascar to Africa. Some six species of Madagascan bird migrate to Africa in their non-breeding season and all of them come to Kenya.

Two skills are involved in migration: orientation and navigation. Orientation is getting the direction right; navigation is finding the right route. Birds use a mixture of cues: the stars, the sun (especially at sunset), the day length, the wind, sound, smell, topography and the earth's magnetic field. Birds watch the landscape; they recognise features like the Suez Canal, the Central American Isthmus and the Rift Valley. They follow flyways. Experience helps too; older birds guide younger ones. Of much interest is the use by migrating birds of the earth's magnetic field and its three components: direction, declination and strength. In an elegant experiment, very tiny magnets attached to the necks of homing pigeons stopped them finding their way home. Another study on Australian white-eyes *Zosterops* sp. used a strong magnetic pulse to alter the orientation of magnetite grains found in tissues above and behind the beak. The White-eyes were unable to navigate properly for some days. And it seems that birds can 'see' the earth's magnetic fields, using light of a particular wavelength; that's how they find their way through cloud.

Before looking at the zoogeography of Kenya's birds, it is worth expanding a little on the global situation. South and Central America (the Neotropical region) has the richest bird fauna in the world, with about 3,300 resident species (i.e. excluding migrants). Consider some species numbers. Outside of South America, the most diverse group of birds in any discrete zoogeographical region is the old world warblers, those little brown members of family Sylviidae, with 170 species in Africa (some ornithologists split this group). There are 160 species of honeyeater in Australasia, and 125 species of weavers and sparrows (family Ploceidae) in Africa. But in South America there are 367 species of New World flycatchers (Tyrannidae), 324 species of hummingbird (Trochilidae), 250 species of tanager (Thraupidae), 231 species of antbirds (Formicariidae), 218 species of ovenbirds (Furnariidae) and 193 species of bunting (Emberizidae). So there are six families in that one continent alone that are bigger than the largest family elsewhere. This appears to indicate that modern birds must have originated there but they didn't; they may have reached South America late, but they radiated explosively there, and this is probably connected with the size and old age of the South American forest. Forests are good places for birds. Birds are at their most vulnerable when they nest and they cannot carry away the eggs. It is a huge risk, nesting on the ground; the bird is relying on never being seen by a predator throughout incubation time and when chicks are in the nest, which is why most ground-nesting birds tend to have precocious young who can run and hide rapidly from early days. Birds are happiest when they can nest in thick vegetation (many of Kenya's birds take advantage of the vicious thorns of acacias and place their nests deep in them). The higher a bird can go, the safer it is, and if it can nest in a cavity, like a tree hole, it is safer still. So South America's breeding birds benefit from an ancient forest. A classic graph showing this direct relationship between the number of bird species on any continent and its area of forest can be found in Gordon Maclean's book *Ornithology for Africa* (1990).

Not all weavers are yellow; Vieillot's Black Weaver, typical bird of the Guinean-Congolian forest. (Stephen Spawls)

Africa (including extreme south-west Arabia) is the second-richest continent of the world in terms of bird species: it has just over 1,800 resident species. Africa also has the largest number of resident raptor species, 61 (excluding falcons). However, raptor numbers are pretty consistent throughout the world's six geographical regions. Africa's bird specialities are as follows: 22 hornbills (out of a world total of 47, all the others save one species in Asia), 17 bustards (eight elsewhere), 16 bee-eaters (out of a world total of 28), all the African barbets (perhaps not unsurprisingly), 57 larks (only 35 other lark species occur worldwide), 62 bulbuls (66 elsewhere, 63 in Asia and three in the Palaearctic), 78 sunbirds (43 elsewhere), 73 waxbills (60 elsewhere), 125 weavers and sparrows (26 elsewhere, in Asia and Europe). The fact that most of these families are shared with Eurasia indicates the close connection; for flying birds, the hop across the Red Sea is an easy one. Africa's tally is completed with the world's only Ostrich(es), the Secretarybird *Sagittarius serpentarius*, guineafowls, mousebirds, louries and turacos, helmetshrikes, whydahs, rockfowl and sugarbirds; representatives of all but the last two are found in Kenya.

There has been a fair amount of work done by scientists to analyse the zoogeography/biogeography of Kenya's fauna and flora; we discussed Frank White's work (1983) in Chapter 5. Reg Moreau did the early work; he concluded that most of Africa's resident birds didn't live in forest. In *The Bird Faunas of Africa and Its Islands* (Academic Press 1966), Moreau regarded the sub-Saharan, resident African birds as inhabiting four simple, broad ecosystems: montane and lowland forest, and montane and lowland non-forest. His species numbers for forest were as follows: lowland forest 250, montane forest 120, in both high and low forest 39; so total forest birds 409. Outside of the forest, his figures were; lowland non-forest 887, montane non-forest 74, in both high and low 69, so total non-forest (i.e. savanna and semi-desert) birds 1,030. There were 42 species that were uncertain. In other words, out of a total of 1,481 resident African species, most (60 per cent) inhabit lowland savannas and semi-deserts. It was a highly effective analysis, and it indicates that most African birds are of savanna origins. This looks initially surprising, in light of our earlier remarks about the advantages of nesting in a forest, species richness being connected with area of forest, and the fact that species richness usually increases with actual evapotranspiration (i.e. rainfall). But it is explicable in terms of their history. Birds arrived in Africa over the last 70 million years, and the passerines even later, and during recent years Africa's forests have shrunk. This probably caused the extinction of many forest species that could not survive and breed in the savanna. The habitat was open for savanna species and a reservoir of savanna and arid country species existed to the north and north-east of Africa, ready to fly in and fill the vacuum as the forests receded. More recent works regard Moreau's ecosystems as somewhat simplistic, but his work paved the way.

Recent zoogeographical research has tended to assign the Kenyan bird fauna into groups that fit into White's (1983) botanical regions. There are seven major biomes, as mentioned earlier: Somali–Maasai, Afrotropical highlands, East African coast, Lake Victoria basin, Guinea–Congo forests, Sudan and Guinea savanna and Zambezian savanna. This analysis is similar to that used for reptiles (but reptile zoogeography is slightly more detailed, there are no migrant, and virtually no vagrant reptiles). Bennun and Njoroge's *Important Bird Areas of Kenya* (1999) give the number of bird species that are distinctive of six of these ecosystems as follows (total number known from the entire biome, which obviously includes areas outside of Kenya, in brackets):

1. Somali–Maasai 94 (129): in Kenya dry savanna from Tsavo West north and east to the Somali borders; typical birds of this biome include Von der Decken's Hornbill *Tockus deckeni*, the brilliant yellow Golden Pipit *Tmetothylacus tenellus* and the White-headed Buffalo Weaver *Dinemellia dinemelli*.
2. Afrotropical highlands 70 (226): high country on either side of the central rift, Aberdares, Mt Kenya, the Mau, Cherangani Hills and Mt Elgon, plus the Taita and Chyulu Hills. The birds that typify the varied habitats of the high country include the Abyssinian Ground Thrush *Zoothera piaggiae*, the Tacazze *Nectarinia tacazze*, Golden-winged *Drepanorhynchus reichenowi* and Malachite Sunbirds *Nectarinia famosa* and the endemic Aberdare Cisticola *Cisticola aberdare*.
3. East African coast 32 (38): this is the coastal strip and the associated moist savanna and forest, extending 20–40km inland; it includes several important areas, including the Tana Delta, Arabuko-Sokoke Forest and the Shimba Hills. Important coastal birds include the Sokoke Scops Owl *Otus irenae*, two species of pipit and the endemic Clarke's Weaver *Ploceus golandi*.
4. Lake Victoria basin 9 (12): this is the warm and wet low savanna and grassland around the lake. Although an excellent birdwatching area, it doesn't have many typical species (and as we saw earlier, no typical mammals). Birds of the lake basin include the vivid Papyrus Gonolek *Laniarius mufumbiri* and the Red-chested Sunbird *Cinnyris erythrocercus*.
5. Guinea–Congo forest 43 (277): dense forests of western Kenya, including Kakamega Forest, North and South Nandi Forest, Mt Elgon and the forest remnants above the Soit Ololol escarpment, just west of the Mara. Many of the birds of this system typify what one might call the 'hard-core' forest avifauna, birds like the Grey Parrot *Psittacus erithacus*, the Great Blue Turaco *Corythaeola cristata* and the Black-and-white-casqued Hornbill *Bycanistes subcylindricus*.
6. Sudan and Guinea savanna 13 (55): savanna of north-western Kenya. Typical birds of this West and Central African biome include the Piapiac *Ptilostomus afer*, a little crow that rides elephants and other game like an oxpecker, the reddish Fox Kestrel *Falco alopex* and the Black-bellied Firefinch *Lagonosticta rara*.

This is a total of 259 species, and says something interesting. Of Kenya's 850 or so resident birds, 259 are confined to the habitats above. The remaining 600, if they cannot be classified, must therefore occur across a range of habitats. And they do, they are the so-called generalists, like the African Pied Wagtail *Motacilla aguimp*, the Hamerkop *Scopus umbretta*, the Brown-crowned Tchagra *Tchagra australis* or the Black-chested Snake Eagle *Circaetus pectoralis*. You might see a Hamerkop any place where they can find the frogs they eat; they are on the list for the Aberdares, Homa Bay, Nairobi, Tsavo and Malindi. However, many species occur in only two biomes. There are also no migrants in the ecosystem listings above, too, for obvious reasons: migrants range across continents.

However, Bennun and Njoroge's (1999) analysis is slightly odd in that there is no mention of the Zambezian savanna. This is a valid biome of southern and eastern Africa, and contains some 67 distinctive indicator species like Southern Ground Hornbill *Bucorvus leadbeateri*, the Red-necked Spurfowl *Francolinus afer* and the Golden Weaver

White-fronted Bee-eater with wasp; bee-eaters are born knowing how to safely kill bees. (Stephen Spawls)

Ringed Olive Thrush; the chances of finding the ring after the bird's death is about one in a thousand. (Stephen Spawls)

It is hard to distinguish between different lark species; this is a Rufous-naped Lark. (Laura Spawls)

Pipits are also hard to differentiate; this is the Plain-backed. (Stephen Spawls)

Ploceus subaureus. This region also contains distinctive animals like the Impala, Mole Snake *Pseudaspis cana* and the Leopard Tortoise *Geochelone pardalis*. Likewise, the Afrotropical highland ecosystem, as defined by Bennun and Njoroge, contains habitats that are very different, but are lumped together purely by virtue of being at high altitude. It is not a grouping that could be followed at lower altitude. Birds that Bennun and Njoroge list as being typical of Afrotropical highlands include birds that are only ever found in forest (Bar-tailed Trogon *Apaloderma vittatum*, White-starred Robin *Pogonocichla stellata*) and birds that are only ever found on open moorland (Moorland Chat *Cercomela sordida*, Sharpe's Longclaw *Macronyx sharpei*). Nevertheless, it is a functional system.

Endemic species are important in conservation as they are unique to a particular country and are useful in heightening awareness and pinpointing important areas. Kenya has between eight and eleven endemic species, depending upon which ornithological authority you follow (Tanzania has more endemics, nineteen, but Uganda has only one). Some of Kenya's endemics are accepted by almost all

Black-headed Gonolek at Homa Bay: a bird of the western savanna. (Stephen Spawls)

ornithologists; for example Williams's Lark *Mirafra williamsi*, named for John Williams and found in the dry north, Sharpe's Longclaw, a bird of the high central and western grasslands, Hinde's Pied Babbler *Turdoides hindei*, of the cultivated valleys around Mt Kenya, and originally more widespread, the Aberdare Cisticola of the high grassland of central Kenya, the Tana River Cisticola *Cisticola restrictus* (which may be extinct but some ornithologist doubt it was a valid species) from the lower stretches of that river, and Clarke's Weaver, from the coastal forests on both sides of the lower Galana/Sabaki. These six are genuine. More controversial are some of the other putative endemics. The Taita Thrush is one, found on the Taita Hills and Mt Kasigau. Its scientific name is *Turdus helleri*, named by Edgar Mearns for Edmund Heller, field companion of Teddy Roosevelt. Is it a full species, subspecies or just a montane colour form of the well-known Olive Thrush *Turdus olivaceus*? In the extremely useful *Birds of East Africa*, Peter Britton and his team (1980) regard it as a 'race', although giving it a subspecific name. Terry Stevenson and John Fanshawe, in their Helm field guide to the birds

Helmeted Guineafowl. (Stephen Spawls)

of East Africa (2002) regard it as a full species, as do Sinclair and Ryan (2003) in *Birds of Africa South of the Sahara*. In *Birds of Kenya and Northern Tanzania* (1996), Dale Zimmerman and his team cautiously hedge their bets by treating it as a separate entity but giving it the scientific moniker *Turdus (olivaceus) helleri* and stating it is considered a race of Olive Thrush by many authors. It is worth repeating Steve Jones's apposite comment here, that what is a mere variety to some is granted its own identity by others. Similar problems attach to other possible endemics, including the Kulal, Kikuyu and Taita Hills White-eyes, Taita Apalis and the Sokoke Batis.

Endemics often give some interesting pointers as to what areas may need some protection. Four endemic birds are found in the central highlands, two in the Taita Hills, two in the coastal forests, one in the dry north, one on the Tana and one on Mt Kulal – places thus worth protecting. Some remarkably similar patterns of endemism are seen with reptiles and mammals. This has led to the usual evolutionary scenario: that eastern Africa was populated with widespread forms 60 million to 50 million years ago, and the uplift of the central plateau provided new habitats. Species either moved into the new montane habitats or moved up with them. The high country attracted rain, plants got bigger, forests appeared. The creatures of the new forest adapted (or died), they were separated from the lowland populations and, reproductively isolated, began to accumulate changes. At the same time, climatic fluctuations caused the movement up and down of the treeline (both lower and upper); in some cases the forest shrank to tiny refugia. What tended to survive were the creatures that could cope with both forest and savanna conditions – and many such species remain today – or those that could cling on in little forests. Kenya's endemic mammals, reptiles and birds tend to be small, this may be connected with the fact that small animals have short lifespans and thus genetic changes accumulate quickly, or with the fact that they inhabit small areas, or both. Or maybe just because there are more of them.

East Africa's montane regions are hotspots of vertebrate endemism. When the refugium theory was originally proposed, in a classic paper by A.C. Hamilton, in 1981, it was enthusiastically adopted by many zoologists. There has been some debate about which is more important: was it the uplift, or the creation and existence of

Red-billed Oxpecker, protected by Sir Arthur Hardinge in 1900. (Stephen Spawls)

Pangani Longclaw, uncommon endemic of the eastern coast. (Stephen Spawls)

A frog-eating Hamerkop at Homa Bay. (Stephen Spawls)

the forests, or their fluctuations in size and the existence of refugia? This is interesting and topical debate. A recent paper on African forest robins by Gary Voelker, Robert Outlaw and Rauri Bowie (2010) comes down on the side of small populations getting isolated in montane forests. What is indisputable, however, is that our montane endemics need protection.

THE KENYAN BIRD FAUNA: ECOLOGY, COMMUNITIES AND POPULATIONS

This section starts with some ecology, the study of organisms and how they interact with their surroundings. A few technical terms will be useful here. A *population* is a group of individuals of the same species. A *community* is all the organisms that inhabit a particular area. The *environment* is the conditions in which an organism lives. An *ecosystem* describes all the organisms in a given place or area, and the abiotic factors (abiotic means non-biological, physical and chemical factors like heat, humidity and light) with which they interact. A *biome* is a major ecosystem, usually classified according to the dominant vegetation of that area.

The birdwatcher in Kenya quickly notices that different habitats have different bird faunas. In Samburu, the Vulturine Guineafowl *Acryllium vulturinum* is common, in the Maasai Mara you see the Helmeted Guineafowl *Numida meleagris*, you find the mop-headed and untidy looking Crested Guineafowl *Guttera pucherani* in the Aberdares. These three guineafowl species are similar in size and appearance and undoubtedly had a recent common ancestor, but each occupies what biologists call an ecological niche, a particular space and lifestyle, unique to that species. Something separates them. If you consider the biome or environment where the animal lives as its 'address', then its niche is its 'profession'. The main factors that define an animal's niche are its size, what it eats, where it lives and when it is active.

Martial Eagle nest in Yellow-barked Acacia. (Stephen Spawls)

White-backed and Rüppell's Vultures on giraffe kill, Maasai Mara. (Stephen Spawls)

To understand why the niche idea is important, consider eagles. An encounter with a big eagle is one to be savoured; they are at the top of the Eltonian pyramid. Visit a dense Kenyan forest and, with luck, you may encounter a huge, powerful mammal-eating eagle, dark above and light below, weighing over 5kg. Out in open, dry savanna, you may also come across a huge, powerful, mammal-eating eagle, dark above and light below, weighing more than 5kg. But they are not the same eagle. The forest bird is a Crowned Eagle *Stephanoaetus coronatus*, the savanna bird, a Martial Eagle *Polemaetus bellicosus*. They are subtly adapted for different environments; one is an absolute indicator of a forest fauna, the other an indicator of a savanna fauna. Like the guineafowl, each has its niche although these two eagles do sometimes overlap in space. Leslie Brown, (1970) in his classic long-term study of Kenyan eagles, found the Martial and Crowned Eagle breeding on the same hill, a complication arising from the fact that birds are mobile and adaptable. But despite the fact that these two eagles may come into contact, they hunt different prey in different places. As the day warms, the Martial Eagle heads out into the open country, to try its luck with a guineafowl or a dik-dik; the Crowned Eagle goes into the forest, hoping for a suni or a monkey. The two giants do not compete for the same resources. They don't occupy the same niche. They encounter each other only briefly.

This is a very important point. Each species has its own niche. By theoretical definition, similar but successful species living in the same habitat must occupy different niches. The niche difference may be very subtle. For example, White-backed Vultures *Gyps africanus* and Hooded Vultures *Necrosyrtes monachus* both occur in savanna and feed on carrion, but the White-backed is larger, relatively dominant and feeds at the centre of the carcass, the Hooded is smaller and tends to feed on scraps at the edge. Other niche differences are broader. The Malachite *Alcedo cristata* and African Pygmy Kingfisher *Ceyx pictus* are almost identical in size and appearance, and might be expected to compete. But they avoid competition by living in different habitats and eating different prey; the Malachite feeds mostly on fish and lives near water, the Pygmy lives in savanna and feeds mostly on insects.

Malachite Kingfisher, Lake Victoria. (Stephen Spawls)

The theory as to why different species need different niches is simple. If two species are very similar, and occupy the same niche, they will compete and, eventually, one species will either drive the other away, or cause it to become extinct. This is called competitive exclusion. So, if two similar species exist in the same area (if they don't overlap, there's no competition), and both are successful, then they must occupy different niches; for example, feed in slightly different places, or on slightly different prey, or be active at different times. However, a classic study in West Africa by Marie-Yvonne Morel (1980) on ring-necked doves of the genus *Streptopelia* (often three or more *Streptopelia* species of very similar appearance may occur in one area), found that they co-existed quite well. And this suggests that niche overlap and competition might not be quite as important as is sometimes believed.

Birds (and other organisms) that live in a particular habitat (for example montane forest) form a bird community, a number of populations of different species in a particular area. Strictly speaking, a community should include all species, but it is acceptable to talk of a bird community. The inhabitants of the community are connected, they interact. How many birds there are in a fauna, or more accurately a bird community, is a vexed question. It is hard to be certain because of the mobility of birds. You can accurately say, for example, if you have done your research, how many snake species there are in a particular biome, because snakes are not prone to vagrancy. They don't fly or get blown anywhere, although the occasional one can get carried in firewood, or released by someone who caught it elsewhere. But birds fly into areas where they do not normally live. It has sometimes been caustically said that the number of bird species known from an area reflects the number of birdwatchers and the time they have spent there; and there is some truth in this. (We saw this with the British list, which is – in ornithological terms – a nonsense, a country list of which 60 per cent are freaks that do not normally live there.)

The numbers of birds known from a territory steadily increases upwards. In 1980 John G. Williams and Norman Arlott said Kenya had 1,033, the 1986 East African Natural History Society checklist had 1,064 species (Backhurst 1986) , Bennun and Njoroge (1999) state that 1,089 species are known from Kenya, the African Bird Club 2011 Checklist (http://www.africanbirdclub.org/fpdf/testv10.php) has 1,109. Up and up it goes. This increase is due to several factors. Firstly, new species may be observed, both secretive and resident, or vagrants. For example, the spectacular Shoebill *Balaeniceps rex* was recently added to the Kenya list (or, strictly speaking, returned, since the species was deleted from the 1986 checklist, and then a vagrant specimen stopped by in Amboseli and Nairobi National Parks). In 2006 a specimen of the River Prinia *Prinia fluviatilis*, a species thought to be endemic to West Africa, was spotted at Lokichokio, boosting the Kenya list by one. Secondly, taxonomic work can cause some species to be split; others may be lumped. In Zimmerman, Turner and Pearson's *Birds of Kenya and Northern Tanzania* (1996), the Grey-headed Sparrow *Passer griseus* is treated as a single species, in Stevenson and Fanshawe's *Birds of East Africa* (2002) it is treated as five separate species. Thirdly, new species may be described. It is therefore difficult to be precise about how many species occupy a particular area.

However, bird species numbers in Kenya are high; so any reasonably large area will have a good range of birds. The increase in species numbers doesn't go up directly against area; there is what is called a logarithmic relationship.

A vagrant in Kenya, the spectacular Shoebill catches a lungfish in the Mbamba Swamp in Uganda. (Stephen Spawls)

For example, careful watching over a few weeks of a Nairobi garden of 4,000m² by a good birdwatcher will probably yield about 40 species of bird. An observer in a 4 hectare woodland plot in Langata, ten times the size, is not going to see 400 species, but they might see about 70. And in a 4km² area of upland woodland, one hundred times bigger, careful watching might reveal over 100 species. This logarithmic increase yields what is called a species–area curve. However, accurate censuses of actual bird numbers hardly exist for any area of Kenya. This isn't unexpected; birds are hard to count, their numbers fluctuate from day to day, they fly away, some are secretive. Waterbird counts exist, but birds beside an open stretch of river or lake are easier to count than skulking species. What you *can* find is lists for discrete areas, sometimes annotated with comments on the 'status' of the bird (i.e. how common it is), but even the finest scientific ornithologists are sometimes hazy about this. Professor Gordon Maclean (in *Roberts' Birds of South Africa, 1985*) says, laconically, 'terms relating to abundance are necessarily somewhat arbitrary', and Dale Zimmerman and co-authors (1996), in *Birds of Kenya and Northern Tanzania*, say something very similar; they use what they call 'relative terms', such as abundant, common, fairly common, uncommon and so on. Ken Newman, one of South Africa's top birdwatchers and bird artists (although originally from Basingstoke, in England) made a valiant attempt to quantify these terms in his excellent book *Newman's Birds of Southern Africa (1983)*. He defined an 'abundant' species as one seen 100 or more times a day in the right habitat, 'very common' as 50–100 times, 'common' as 10–50 times, fairly common as 1–10 times, uncommon as less than 30 times a month, and so on. Anyone with mathematical skills can see the complications arising: you get a nice flock of quelea and the number of sightings gives you 'abundant'; and is that the same flycatcher that keeps popping out or is it another one? One approach is to trap and ring them, or put on a coloured tag, and then calculate the population based on what percentage of ringed/tagged ones are seen. The most common surveying method is to choose a suitable plot and then use what are known as 'transects' – the ornithologist walks along a fixed path, recoding every species they see, their numbers and how far they are from the path. They may count all the way (line transect) or stop and count at specific points (point transect). They then calculate the size of the

The male Red-billed Quelea. (Paolo Torchio)

transect area (essentially, the length of the transect multiplied by double the distance the furthest specimens was from the path, this is the width of your monitored area), divide this into the total survey area and multiply this figure by the number of individuals seen. If regularly repeated, and producing consistent results, transects are a good way of surveying the avifauna. Surveys involving transects in the North Nandi Forest, and in some of Kenya's coastal forests, have yielded species lists of between 100 and 150 species. However, transect species totals are much smaller than actual lists. This is because there are some birds that transects will never reveal, and most birdwatchers, compiling a list, tend to revisit the area frequently, get data from more than one observer, move unsystematically around prime spots, and also record all species seen. As any enterprising birder knows, watching by water will boost your species list, and other techniques, like playing the call of an owl, will also increase numbers.

The ornithologist John G. Williams, who was curator of birds at the National Museum (when it was originally the Coryndon Museum) from 1946 to 1966, published a number of lists of birds from various conservation areas in his Collins *Field Guide to the National Parks of East Africa* (1978). There is no doubt that Williams knew his birds, but according to John Miskell, the noted expert on Somali birds, Williams's lists need to be treated with caution. Most are underestimated, and there are birds on some of Williams's lists that no subsequent birdwatcher has been able to find, like the Lammergeier he said he saw in the Maasai Mara or the Long-tailed Nightjar *Caprimulgus climacurus* from Lake Nakuru. It is possible that some honest mistakes with identification have been made, of course, and not just by Williams. A story is told of the fifth Pan-African ornithological conference at Lilongwe in Malawi. Some of Africa's top ornithologists were there, and were asked to make lists of the bird species they observed during the congress. The lists, later examined by Malawi bird experts, apparently contained a number of species never recorded near the city, before or since!

A number of lists for Kenyan localities have been published, in addition to Williams's ones. The grandfather of Kenya lists is the Nairobi one, with 605 species recoded from the city and environs, according to a man called Bill Harvey, of the British Council (1997). Nairobi National Park is the next highest, with 529 species. After that come

Silverbird in a Balanites Tree, Ruma National Park. (Stephen Spawls)

Lake Baringo and the Maasai Mara, both with 'over 500' species, all according to Bennun and Njoroge (1999). Peter Lack, Walter Leuthold and Chris Smeenk recorded 466 species from Tsavo East (1980) Williams lists 451 from the Mara, Bennun and Njoroge say 'some 450' from Lake Nakuru and '400+' from Elmenteita. Those are the 'big lists'. Other lists close to 400 include Brian Finch, David Pearson and Adrian Lewis's list of 380 from Samburu/Buffalo Springs (1989) and Williams's list of 373 from Mt Marsabit.

Not everyone agrees on numbers. John Williams listed only 59 species from Kakamega; Bennun and Njoroge state that the Kakamega Forest contains 194 'forest-dependent species'. This term – forest-dependent – comes from an interesting paper by Leon Bennun, Christine Dranzoa and the venerable Kampala-based ornithologist Derek Pomeroy, entitled *The Forest Birds of Kenya and Uganda* (1996). This is a pioneering work, particularly since bird-watching in deep forest is very difficult, and was produced for an important reason: forests are the most threatened habitat in Kenya, as a result of land hunger and poverty. The logic and conservation significance is obvious; forests

Brian Finch, top Kenyan ornithologist and safari guide. (Stephen Spawls)

The Bateleur, bird of game country, a bad omen if seen on the left. (Paolo Torchio)

have easily reachable resources and all members of the community can exploit them. So it's important to know, more than any other habitat, what is dependent upon forests for survival. This is a point of view echoed by Reg Moreau, who believed it was essential that the birds of the African evergreen forest, highly specific to their habitat, should be seen as a unique group. Bennun and his colleagues classified forest birds thus: forest specialists, found in the interior of deep forests, forest generalists, found in undisturbed and disturbed forests, and forest visitors, birds often found in forest but not dependent on it. Kenya had 110 forest specialists, 120 forest generalists and 105 forest visitors, a total of 335 of which 230 are dependent upon forest. The visitors include such well-known species as the Grey-backed Camaroptera *Camaroptera brachyura*, Hooded Vulture, Long crested Eagle *Lophaetus occipitalis* and Red-eyed Dove *Streptopelia semitorquata*. The forest generalists include the African Green Pigeon *Treron calvus*, the Bat Hawk *Macheiramphus alcinus*, Pel's Fishing Owl *Scotopelia peli*, the African Wood Owl *Strix woodfordii* and the Trumpeter Hornbill *Bycanistes bucinator*. But the forest specialists are unusual birds, the ones that are likely to disappear if the forest is disturbed: five star specials like the Buff-spotted Flufftail *Sarothrura elegans*, the African Crowned Eagle, Black-billed Turaco *Tauraco schuettii*, Blue-headed Bee-eater *Merops muelleri*, African Pitta *Pitta angolensis* and the Eastern Mountain Greenbul *Andropadus nigriceps*. These are the sort of bird that only the top birdwatchers see, and they live in places where humanity has as yet not chopped down more than a tiny proportion of the trees. They are vulnerable to the humidity gradients that we mentioned in the Chapter 5. Since Kenya has over a thousand bird species, of which 230 are forest-dependent, it follows that 750+ are savanna creatures. But bear in mind that what Bennun and his team called forest generalists may be found in quite open country. Species on the generalists list include the Bat Hawk and Ayre's Hawk Eagle *Hieraaetus ayresii*, both found in Tsavo and Samburu, where the only forests are the clusters of trees lining the rivers.

Many of the 'big' lists are simply uncritical lists of everything seen, and numbers vary wildly. As we just saw, the Kakamega list depends on who you quote: Williams lists 59 species, Fred Munyekenye and co-workers (2008) found 129 species, Bennun and Njoroge give 194, Professor Udo Savalli's online list (http://www.uky.edu/~cfox/Students/Savalli/KakaBirdList.html) has 358 species, but this included a number of vagrants that wouldn't normally live there, like the Bateleur *Terathopius ecaudatus*, Pink-backed Pelican *Pelecanus rufescens* and White-winged Widowbird *Euplectes albonotatus*. Williams listed 267 from Naivasha, Bennun and Njoroge say there are more than 350. Williams listed 21 species from Arabuko-Sokoke, John Fanshawe's (1994) list was over 230.

Not all workers have consulted the literature and what went before. In 1998, Herbert Schifter and Chum Van Someren published a list of the birds of the North Nandi Forest, and recorded 117 species. In 2010, Birdlife International funded a report on the avifauna of the North Nandi forest by a four-member ornithological team lead by Simon Musila. Their 22-page paper lists 108 bird species, but makes no reference whatsoever to Schifter and Van Someren's paper. The two lists are quite different. Some bird lists are very conservative and some very odd; Williams lists 16 species from the Olorgesailie area, including birds like the Kurrichane Button Quail *Turnix sylvaticus*, not at all easy to see, but doesn't include the Grey-capped Social Weaver *Pseudonigrita arnaudi* and White-browed Scrub Robin *Cercotrichas leucophrys*, which no keen birdwatching visitor could miss. Some 75 bird species are recorded from Mpala research centre near Nanyuki, 86 from Ol Donyo Orok and 94 species are known from the Kianyaga Valleys just south of Mt Kenya (although one was the five-star endemic Hinde's Babbler), causing Bennun and Njoroge (1999) to comment sadly that the bird diversity there was low.

If vagrants are omitted, it would appear that any discrete area of Kenya, of over 100km^2, with only one type of habitat, will have at least 80 species that actually live and breed there. The bigger and more diverse it gets, the more species will occur. As any birdwatcher knows, visit an area where there is open water and the number of birds on your list goes up. The greater the variety of habitats, the more species. Add a big hill with some woodland and up the totals go. The 300+ lists from Kenya's big national parks are largely genuine residents, not vagrants. There has been some discussion of the effect of altitude, which definitely affects reptile, amphibian and mammal species numbers, although it should be noted that this doesn't apply to actual individual numbers; which can be high. There are only about fifteen species of snake found around Lake Naivasha, compared with 45 in the Arabuko-Sokoke forest, but at one time the actual individual numbers of snakes per hectare on Naivasha's Crescent Island

White-eyed Slaty Flycatcher at Njoro. (Stephen Spawls)

was far higher than anywhere else in Kenya. Some 73 species of bird are listed for the Cherangani Hills, Williams lists 126 on Mt Kenya and just over 200 for the Aberdares (although this includes the relatively low salient). Consider these figures (which include vagrants):

PLACE	ALTITUDE	NO. OF SPECIES
Tsavo East	1,000m	466
Kakamega Forest	1,570m	358
Lake Naivasha	1,900m	about 350
Aberdares	2,500m	200

There is a definite decline in species numbers with increasing altitude, but the fly descends into the ointment with the figures for Arabuko-Sokoke (altitude just above sea-level, species numbers 230) and Nairobi (1,660m, 605 species). Does Nairobi really have that many or is it just that there are a lot of birdwatchers there? You would expect the hard-core resident bird species in Nairobi to be around 380, the rest are accidentals. As Simon Barnes, the award-winning British sportswriter and occasional ornithological columnist remarks in his enjoyable book *How to Be a Bad Birdwatcher*, the spotting of lost birds is not a scientific pursuit. To get a true picture of the bird fauna of a particular area, most zoologists would agree that vagrants should not be included, but the picture is complicated by the fact that what appears to be a vagrant might turn out to be a pioneer; the first example of a species that later moves into the area, or proves to be present but hitherto unrecorded. But listers are not troubled by vagrants, in fact they welcome them. Even thoroughly scientific ornithologists will add unusual species to their lists.

So, a list of birds in an area does not necessarily represent a coherent interconnected bird community. It is just a list. In parts of the world where distributions are pretty well known, for example Great Britain, one can say something about a bird community by considering the species that regularly breed there. Data like this do not really exist for Kenya, yet. If you are compiling a list and something unusual is seen perched in a tree or flying overhead, the inclination is to add it, rather than ask, 'is that bird part of the community?'. Numbers can be helpful, and some data here are available. The lists of birds produced by Fleur Ng'weno (the patron saint of Kenya birdwatchers) recorded over 40 years of bird walks in Nairobi, establish what constitutes Nairobi's 'hard-core' bird community. The 'top ten' are the Variable Sunbird *Cinnyris venustus*, Common Bulbul, Baglafecht Weaver *Ploceus baglafecht*, Speckled Mousebird *Colius striatus*, Streaky Seedeater *Serinus striolatus*, Red-eyed Dove, Singing Cisticola *Cisticola cantans*, Black Kite, Common Fiscal *Lanius collaris* and Bronze Sunbird. Simon Musila, Alex Syingi and Nickson Sajita found that the ten most common birds in the North Nandi forest were the Yellow-whiskered Greenbul *Andropadus latirostris*, Cabanis' Greenbul *Phyllastrephus cabanisi*, Yellow-rumped Tinkerbird *Pogoniulus bilineatus*, Black-collared Apalis *Apalis pulchra*, Lüdher's Bush-shrike *Laniarius luehderi*, Buff-throated Apalis *Apalis rufogularis*, Stuhlmann's Starling *Poeoptera stuhlmanni*, Grey-throated Barbet *Gymnobucco bonapartei*, Chubb's Cisticola *Cisticola chubbi* and the Cinnamon-chested Bee-eater *Merops oreobates*. There is no overlap with Nairobi at all and you have to be a serious birdwatcher to have seen most of the forest-dwellers on this list. Some nice data were obtained from the vicinity of Kichwa Tembo camp in the Mara by Alasdair Gordon and Nancy Harrison (2010), who recorded 'fifteen-species' lists. They recorded 111 species of birds in mixed-bird parties, and their top 10 were the Common Bulbul, African Paradise Flycatcher *Terpsiphone viridis*, White-browed Robin Chat *Cossypha heuglini*, Collared Sunbird *Hedydipna collaris*, Black-backed Puffback *Dryoscopus cubla*, Speckled Mousebird, Grey-backed Camaroptera, Black Saw-wing, Cinnamon-chested Bee-eater and African Citril *Serinus citrinelloides*. One overlap with the North Nandi list, the bee-eater, and two with the Nairobi list, the mousebird and the bulbul. So bird communities vary in numbers and content. The actual interactions of the members of Kenya's bird communities and the food webs that include and connect them have yet to be worked out, and there has been virtually no long-term monitoring of bird population trends in Kenya, as yet.

THE HISTORY OF ORNITHOLOGY IN KENYA

'Ornithology' is the scientific study of birds, from the Greek *ornis*, a 'bird', and for those who practise it Africa has been a dangerous place. Johan August Wahlberg, Eduard Schnitzer (Emin Pasha), Karl von der Decken, Hugh Clapperton and James Sligo Jameson (of Jameson's Wattle-eye *Platysteira jamesoni*), all died young in Africa, and the continent had a hand in the deaths of collectors like Gustav Fischer, Theodor von Heuglin and Johan Hildebrandt. But some reached old age: Sir Harry Johnston survived the rigours of Central Africa and lived to 69, Anders Sparrman reached 72, Sir Geoffrey Archer made it to the age of 80.

Sub-Saharan African ornithology started in South Africa, probably with Hendrik Claudius, the artist who accompanied Simon van der Stel's expedition to Namaqualand in 1685 and painted a number of birds, including the Sacred Ibis *Threskiornis aethiopicus* and the European Bee-eater *Merops apiaster* (although the earliest African bird art is undoubtedly the paintings done in Egypt at the time of the pharaohs. Horus was the god of the sky and he had the head of a falcon. His picture is found all over ancient Egypt. Some of the tombs at Thebes, more than 3,000 years old, show ducks, doves and an unmistakable heron). One of the first Western ornithologists to visit Africa was the Guyanan-born Frenchman Francois Le Vaillant, who collected in southern Africa between 1781 and 1784. Le Vaillant did not like Carl Linnaeus and his naming system, and gave French names only to a number of birds, for example the Bateleur, which means 'acrobat'; he called the Pale (Eastern) Chanting Goshawk *Melierax poliopterus* 'chanteur', or singer. Le Vaillant greatly admired the Khoikhoi people (known in those days by the pejorative term Hottentots) of southern Africa and Narina's Trogon *Apaloderma narina* is named after a beautiful Khoikhoi girl he knew. Klaas's Cuckoo *Chrysococcyx klaas* is named after one of his workers. Le Vaillant sent more than 2,000 bird skins to the natural history museum at Leiden, in Holland, where they were studied and described by the Dutch ornithologist Coenraad Temminck, after whom a courser and a stint are named (and a giant pangolin).

The golden age of bird collection in Africa was the nineteenth century, and much work was done by the Germans. Collectors sent specimens back and prolific museum taxonomists described them. Nineteenth century German bird taxonomists include Karel Hartlaub (1814–1900) of the Bremen Museum and the magnificently bearded Jean Louis Cabanis (1816–1906) of the Berlin Museum, where he took over from another ornithologist, Martin Lichtenstein. Cabanis was succeeded by his son-in-law Martin Reichenow, a long-lived German (1847–1941). A German zoologist who got into the field was Martin von Heuglin, who collected in Egypt and then went into the Sudan in search of Edouard Vogel, the German explorer who was killed in Chad. Other prominent European ornithologists and collectors of the eighteenth and nineteenth century include Hendrik Severinus Pel, who was governor of the Dutch Gold Coast between 1840 and 1850 and collected the fishing owl that bears his name, the Frenchman Louis Vieillot (1748–1831), who published privately on birds, and William Swainson (1789–1855), who described a number of species before emigrating to New Zealand in the 1840s.

The top man in nineteenth-century African ornithology is undoubtedly Andrew Smith, or to give him his full honours, Sir Andrew Smith, MD, KCB. Born a shepherd's son near Hawick, Scotland, in 1797, he chose medicine as his career and graduated from Edinburgh University. Achieving his initial licence to practise at the age of 16, he then joined the medical department of the British Army, and served from 1815 to 1858. A trim man with a superb beard, in 1821 Smith was posted with his regiment to the Cape of Good Hope, South Africa, where he remained until recalled to England in 1837. He met the young Charles Darwin there when he landed at the Cape in 1836 and they remained friends for life. While in South Africa he was appointed director of the newly created South African museum. The second-in-command there was Dr Jules Verreaux, after whom the eagle was named. Smith made many expeditions all over South Africa, collecting specimens. After his return to Britain in 1837 he published a magnificent series of papers entitled *Illustrations of the Zoology of South Africa* (available in facsimile reprints). His two main interests were bird and reptiles, especially snakes. He described over 90 species of birds, including five owls found in Kenya: the African Grass Owl *Tyto capensis*, African Marsh

Owl *Asio capensis*, Cape Eagle-Owl *Bubo capensis*, African Wood Owl and African Barred Owlet *Glaucidium capense*, and some five-star birds like the Rufous-breasted Sparrowhawk *Accipiter rufiventris*, Mountain Buzzard *Buteo oreophilus* and African Hobby *Falco cuvierii*; the list goes on and on. The African Broadbill is named for him, *Smithornis capensis*, and he also described the Green Mamba *Dendroaspis angusticeps* and the Boomslang *Dispholidus typus* six times as detailed elsewhere in this book in the reptile chapter.

We met a number of East Africa's early collectors in the plant chapter; those who also collected birds include Baron Karl Klaus von der Decken, Johan Hildebrandt, Gustav Fischer and Sir Harry Johnston and Arthur Donaldson Smith. All have Kenyan birds named after them. The reverend Thomas Wakefield was a Methodist missionary, at Ribe near Mombasa, between 1861 and 1888. As well as being a man of the cloth Wakefield was an explorer and a naturalist. He collected plants and sent a big collection of bird skins to Richard Bowdler Sharpe at the British Museum of Natural History. Wakefield is commemorated by a coastal subspecies of green pigeon. Kenya's first legislation specifically protecting wildlife, Queen's Regulations for the Preservation of Game, enacted by the commissioner Sir Arthur Hardinge in 1900, specifically protected five groups of birds; ostriches, owls, the Secretarybird, vultures and oxpeckers. And the man who really put Kenya on the map where ornithology was concerned was Sir Frederick Jackson.

Jackson was born in Yorkshire but in later years lived in Norfolk, in England. He came to Kenya in 1884 and spent the next 20 years exploring, shooting and getting involved in various military adventures. He gained a reputation as a notable sportsman and expert on big game shooting. The Kenyan naturalist Ian Parker considers Jackson the true founder of the Kenya Game Department. It was between 1884 and 1903 that Jackson collected most of the animals that bear his name. He probably has more species and subspecies of Kenyan birds named after him than anyone else; fifteen taxa, including Jackson's Francolin *Francolinus jacksoni*, Jackson's Widowbird *Euplectes jacksoni*, Jackson's Hornbill *Tockus jacksoni* and Jackson's Golden-backed Weaver *Ploceus jacksoni*. A string of reptiles also bear his name (Jackson's Tree Snake *Thrasops jacksoni*, Jackson's Chameleon *Chameleo Triocerus jacksoni*, Jackson's Centipede-eating Snake *Aparallactus jacksoni* etc.), as does a catfish (*Amphilius jacksonii*, Marbled Mountain Catfish). Jackson's adventures from this time are commemorated in his posthumously published book *Early Days in East Africa* (1930). Curiously, he hardly mentions the species that are named after him; there are only a few pages on birds in the entire book, although there are some nice descriptions of people and places and some memorable thumbnail sketches of Kenyan 'characters' of the time. But Jackson's *magnum opus*, also published after his death, was his three-volume work *The Birds of Kenya Colony and the Uganda Protectorate*. This appeared in 1938 after its completion by the prominent British Museum ornithologist William Sclater; it represents the original baseline data of Kenyan ornithology. In 1907 Jackson was appointed Lieutenant Governor for the East African Protectorate and in 1911 he was promoted Governor and Commander in Chief of Uganda, a post he held until 1917. Jackson was originally stimulated to go to Africa by Sir Henry Rider Haggard, an old friend from Norfolk. Haggard was later an administrator in South Africa, but made his name as the author of blood-and-thunder adventure novels like *She* (who must be obeyed) and *King Solomon's Mines*. It is said that the hero of this latter book and several others, Allan Quatermain (portrayed on the screen by Sean Connery, Stewart Granger and Richard Chamberlain, among others), was based on Jackson.

One very colourful character associated with ornithology was Colonel Richard Meinertzhagen, a man who had a remarkable career as a soldier, spy, adventurer, big-game hunter and – it subsequently turned out – downright rogue. He was based in Kenya between 1902 and 1906 and was known for being good with a gun. In her foreword to the reprint of his book about his Kenya days, *Kenya Diary*, Elspeth Huxley (1983) described him bluntly as a killer who killed for pleasure. He shot much of what he saw and his book is full of hunting stories; and yet there are also many prescient observations about the future of Kenya, and some excellent photographs. Some of the Colonel's material did make its way to the Natural History Museum in London, including the first specimens of the Giant Forest Hog, from Kabwuren, in western Kenya, which the curator Oldfield Thomas described as *Hylochoerus meinertzhageni*. Meinertzhagen was withdrawn from Kenya in 1906 after various courts of enquiry decided he had damaged the reputation of the British in Kenya following an incident when a Nandi Orkoiyot or religious leader was shot dead at Kaidparak Hill along with more than twenty of his followers, after Meinertzhagen and his men gunned their way out of what they said was an ambush. Meinertzhagen came back to Africa and was active in the First World War in pursuit of Von-Lettow Vorbeck, and after his retirement he frequently returned to collect birds. He died with a fine reputation as an ornithologist; a subspecies of broadbill is named after him, and several Palaearctic birds. He described a few species himself, including a new snowfinch. But after his death his legend began to unravel. His major work, *The Birds of Arabia*, is largely based on an unpublished manuscript by a man called George Bates, who gets little credit. Meinertzhagen claims in some of his books to have observed birds in countries where, it was later established, he had never been, such as the Cape Verde Islands, and it emerged that while researching the bird collection at various museums, Meinertzhagen had been in the habit of stealing specimens. He stole two Chinese specimens of Blyth's Kingfisher *Alcedo hercules*, from the Natural History Museum and the American Museum of Natural History in New York, and then presented them to the Natural History Museum as having been collected by himself in Burma. His deception was unravelled by the ornithologist Robert Prys-Jones, using clues from the way the birds had been stuffed. (Bird taxidermists have

Jackson's Francolin, montane species and important in Kikuyu folklore. (Laura Spawls)

their own signature skills, in the style and arrangement of their stitching.) The sad thing is that because of his duplicity suspicion hangs over the provenance of all of his specimens – many of which are undoubtedly genuine.

Whilst in Kenya, Meinertzhagen met both Frederick Jackson and his nephew, Geoffrey Archer, whom he described in *Kenya Diary* (1957) as 'an enormous lad and mad keen on birds'. In Uganda, Archer was sent by his uncle to collect birds in the west of the country at Lake Albert. Later he went on to become Sir Geoffrey Archer, KCMG, a famous ornithologist, Governor of Somaliland, Uganda and Governor-General of the Sudan. The Kenyan town of Archer's Post is named after him. Archer, with Eva Godman, published a massive four-volume book, nearly 1,600 pages long, entitled *The Birds of British Somaliland and the Gulf of Aden*, (1937–61) illustrated by the talented Scottish artist Archibald Thorburn.

Anyone familiar with the natural history of Kenya will know the name Van Someren. Four Van Somerens have been associated with ornithology in East Africa. Robert Abraham Logan Van Someren was a doctor with the Uganda medical service and ornithology was his hobby. He began collecting there in 1906. His son Vernon Donald Van Someren, a zoologist and ichthyologist, came to Kenya in the 1930s and in 1958 published a delightful book *A Bird Watcher in Kenya*, illustrated by his own superb black-and-white photographs. He also wrote a book on trout in Kenya, and co-authored papers with the famous ichthyologist Peter Whitehead of the British Museum. Vernon rose to become director of the East African Fisheries Research Organisation, in the curiously named Ministry of Forest Development, Game and Fisheries. Robert's brother, Victor Gurner Logan Van Someren was both doctor and dentist and was appointed government medical officer in Kenya in 1911. The two brothers published a little book entitled *Studies of Birdlife in Uganda*. Victor was soon finding specimens throughout Kenya. He was an all-round naturalist who collected mammals, birds, insects, plants and even a few reptiles. He sent 300,000 insects to the Natural History Museum in London, and 16,000 butterflies to the American Museum of Natural History. Victor published many scientific papers, including a major revision of the butterflies of the genus *Charaxes*, and a 500-page book *Days with Birds*. He painted more than 2,000 Kenyan birds, and some of his paintings were published in a book on weavers. His son, Gurner Robert Cunningham-Van Someren, known by all as 'Chum', was appointed curator of ornithology at the National Museum in the 1960s. A reclusive man, Chum had no radio or television in his house and claimed not to know how to use a telephone.

Ornithology in Kenya between 1946 and the mid-1960s was dominated by John G. Williams. Born in Cardiff, Williams had trained with the National Museum of Wales as an ornithologist and taxidermist, before serving with the RAF in the Second World War. In 1945 Williams and his wife, Pippa, married in Cairo and then came to Kenya, where they asked Louis Leakey for a job. Leakey got Williams to prepare some skins and mounted specimens, and then appointed him. Williams' first expedition was with C. J. P. Ionides, the famous hunter and herpetologist, to the Uganda side of the Virunga volcanoes. Ionides shot a gorilla and Williams did the taxidermy; for many years the big ape was on display in a forest diorama at the Kenya National Museum. Williams was curator of birds at the National Museum between 1946 and 1966. He published few papers, save a major revision of sunbirds, but he did produce four books; a guide to the birds, later updated, a guide to East Africa's national parks and a guide to Africa's butterflies, this latter illustrated with his own paintings. After leaving the museum, he worked as a safari consultant and guide, before returning to Britain in 1978. He then produced two books on orchids, of Europe and North America. There is more about Williams in the literature section. His place at the museum was taken by Alec Forbes-Watson, an ex-military man with great experience in dry north-east Africa and Socotra. He has a swift named after him.

Kori Bustard painted by Archibald Thorburn, from Archer and Godman's work on the Somali bird fauna. (Stephen Spawls)

Long-tailed Nightjar, legend says you must not damage its eggs. (Stephen Spawls)

Another towering figure in the world of ornithology who made his home in Kenya was Leslie Brown. Born in 1917 in India, of Scottish stock, Brown came to Kenya from Nigeria in 1946. He held the posts of deputy director of agriculture and then director, but by 1963 was so well known as an ornithologist and ecologist that he left his job and became a consultant. A big, strong, bearded man, Brown did not suffer fools gladly, and anyone who got in his way, or the way of the information he wanted to impart to the world, be they askari or publisher, was liable to get a severe tongue-lashing. Brown loved birds of prey, in particular eagles, and his publications include *African Birds of Prey*, *British Birds of Prey* (a remarkable work considering its author was resident in Kenya) and *Eagles, Falcons and Hawks of the World*, with Dean Amadon. Brown was a field man, par excellence; he believed that discoveries were to be made in the wild, not in the laboratory. His work on the location he called 'Eagle Hill' near Embu is unsurpassed, and all significant works on African birds of prey cite the data he gathered there. For the big picture on African raptors, Brown's *African Birds of Prey* is excellent, despite being published more than 40 years ago – rigidly scientific material presented in a clear and easily read style. His work and reports on Ethiopia highlighted the urgent need for environmental action in that ravaged land, a call which sadly has not been heeded. He was co-author, with Emil Urban and Ken Newman, of the first volume of the major work *The Birds of Africa*, which has now run to seven volumes.

Brown was something of a polymath; he published a book on East African coasts and reefs and a natural history of Africa, although his facts were occasionally wrong. He once conducted an angry debate with James Ashe, curator of Nairobi Snake Park, on the mechanisms of venom spitting in spitting cobras. Brown said that the jet was assisted with a blast of air, Ashe said it was muscles only that propelled the venom. Ashe was right, of course, he knew his snakes, and he called Brown a dabbler, which enraged the Scotsman. Anyone wanting something of the flavour of the man should read his book *Encounters with Nature*. It was written in 1977, while Brown was held under clubhouse arrest in Somalia, after the plane he was travelling developed a compass error and strayed off its intended route across Ethiopia; Brown's ire at Somalis and Somalia are well-developed in the preface. The book has chapters on nightjars, pelicans, tigers and his work on Eagle Hill; the chapter on how Brown found the first baby Lesser Flamingos on Lake Natron is a simply magnificent, riveting adventure story, all the finer for being true. Brown nearly lost his feet; in fact he almost lost his life, trying to reach the colony. He attempted to walk out there across a mud flat and kept getting stuck; he became exhausted and dehydrated, and when he finally gave up and turned back, he almost couldn't walk because the trona had got into his boots and caused frightful septic sores. The man's enthusiasm for natural history and ornithology shine through. He was completely unsentimental, one might say a necessary prerequisite for a zoologist. He kept a Leopard in Nigeria, but as it grew he decided it would be impractical to release it. So he simply shot it. Brown died of a heart attack at his home in Karen in 1980; tragically his only son was killed in a car crash three months later.

Hardworking and professional ornithologists now dominate the Kenyan ornithology scene, museum scientists like Peter Njoroge, Hendry Ndithia, Philista Malaki and Titus Imboma. There are also a hard core of professional birdwatchers who earn a living in the safari business, venerable experts like Don Turner, who has lived in Kenya since 1959, Chege Wa Kariuki, Brian Finch, and Terry Stevenson. One expert who has bridged the divide between professional ornithology and birdwatching is Dale Zimmerman, who has combined a university career with that of a museum ornithologist and is also a birdwatcher with a life list of more than 5,000 species.

BIRDS AND HUMANS

Humans are closely connected to the world of birds. For a start, we eat them in large numbers. In some countries even little passerine birds are trapped or shot and eaten, but in most nations of the world chickens, ducks, turkeys, guinea fowl and geese, and a few other species, including ostriches, are served up on our tables. There are more than 30 million chickens in Kenya; their eggs and meat provide protein to the common people, and free alarm calls before dawn. The cockerel, *jogoo* in Swahili, has potent significance in Kenya, it is the symbol of the political party KANU (Kenya African National Union); many people simply refer to the party itself as *jogoo*. In parts of Africa, chickens are often used as sacrificial animals, and in South African standard folk treatment for snakebite is to split a live chicken open with a panga and tie it onto the wound; it is believed to suck the venom out. Birds are kept in huge numbers as pets, for display or to carry messages; they provide companionship and beauty; they are useful. In China, in the great cities where little wildlife is seen, many people stay in touch with nature with a caged songbird. The detached observer watching the masses engaging in tai-chi in the early morning while the birds sing from their cages must balance the benefits to the individual against the unkindness of imprisoning a wild creature that can fly. There is a massive trade in wild and captive-bred birds. Birds are used for hunting; falconry is several thousand years old, originally the pursuit of nobles in the Middle East and Europe, although it never caught on in sub-Saharan Africa. Birds appear in religious texts, they have inspired art and music. The English composer Ralph Vaughan Williams' 'The lark ascending' was voted the nation's most popular piece of music by listeners to the BBC. Birds are involved in pollination of a few significant plants. Birds were used to carry messages; early seafarers used birds as a kind of compass. The House Crow *Corvus splendens*, now naturalised on the Kenyan coast, was used for this purpose. Ships would take a cage of crows on board, and if the mariners were lost and out of sight of land, they would release a crow; the direction of its flight showed the way to the nearest land – a kind of flying compass. Birds clean up carrion, and in the old days in Kenya this might include human corpses, which were often put out for scavengers. Bird guano is used as fertiliser. Feathers are used by humanity for adorning themselves; in parts of Africa ostrich eggshells have been (and still are) used as water containers and for jewellery. In sub-Saharan Africa there is an astonishing symbiotic relationship between honey hunters and the Greater Honeyguide *Indicator indicator*, which leads humans to wild bees' nests; legend says that if you fail to give the bird its share of wax and honey when you raid the nest, next time the bird will lead you onto a Lion or a deadly snake. There are many superstitions about birds, especially birds of the night. Pigeons have worked on assembly lines, in the United

Brimstone Canary. (Stephen Spawls)

States Turkey Vultures *Cathartes aura* have been trained to locate pipeline leaks with their acute sense of smell. The Atlantic Canary *Serinus canaria* was used for the purpose of detecting deadly gases in coal mines. Humans enjoy watching birds. Birds are significant in attracting a certain kind of tourist ('birders' and 'twitchers'), and of course add to the general enjoyment of those who come to Kenya to see its wild places and wildlife; it is estimated that more than 400 specialist birdwatchers come to Kenya every year just for its birds, bringing in more than 50 million shillings, 600,000 US dollars, although this is small beer compared to South Africa, where bird tourism is reckoned to bring in more than 20 million dollars each year. And there is a downside; birds are notorious crop raiders, they can damage buildings, birds fly into aircraft and can cause crashes, and birds can carry nasty zoonoses, diseases that pass from birds to humans (including the notorious bird flu).

Birds continue to be heavily used in trade. It is estimate that more than 5 million birds are traded internationally per year, a trade worth in excess of 80 million dollars. Wild birds used to be traded from Kenya and Tanzania, and without doubt brought in foreign exchange, but at the expense of much suffering and loss of biodiversity. Kenya banned the trade in live animals in 1978, and at present this trade is also (theoretically) banned in Tanzania. It has been estimated that up to 60 per cent of wild-caught birds die before they reach a pet owner. With the improvements in husbandry, and a growing awareness of the ethical treatment of wild animals, one hopes that those who are involved in the bird trade are able to breed in captivity sufficient specimens to satisfy the demand, and provide the birds with big enough aviaries at least to fly a few metres. The poet William Blake wrote, 'A robin redbreast in a cage, puts all heaven in a rage …' Birds belong under the sky, not in boxes.

At one time, there was a big trade in feathers as fashion items and many were supplied from ostriches. The only flightless birds in Kenya are the two species of ostrich, which until recently were regarded as a single species. Ostriches are also the world's heaviest birds, weighing up to 150 kg, and the unfertilised egg of an ostrich is the world's biggest cell. Ostriches are unusual birds: they have the largest eye on any terrestrial vertebrate and the male ostrich also has a grooved penis, which is unusual among birds, although ducks and flamingos also possess

Red-billed Firefinch. (Stephen Spawls)

one. Ostrich feathers are unique in that the barbs are of equal length on either side of the shaft; consequently the ancient Egyptians regarded the feather as a symbol of justice and truth. According to the Book of the Dead, after death your heart would be weighed against an ostrich feather, overseen by the god Thoth, who had the head of an ibis. If your heart was light you went to paradise, if heavy your heart was seized by a frightful beast. Ostriches were farmed in Kenya in the early years of the nineteenth century; there were farms around Machakos and Ulu and in the rift valley. Setting-up costs were low; the licence only cost 10 rupees and the birds themselves were free, if you had the wit to catch them, or find eggs or freshly-hatched chicks. But maintenance was a headache. Philip Blayney Percival lost many of his birds to Lions and hyenas. One of Kenya's first professional hunters, the Australian Leslie Tarlton, gained fame as a man who could round up one hundred cock ostriches at a time. Lord Delamere farmed ostriches, and imported a pedigree cock from Somalia. Like many enterprises in Kenya in those early colonial days, money was made initially but good times didn't last. It was a risky job herding the birds, many herdsmen had legs broken by the kick of the birds, and feathers were often stolen – they were valuable and easy to conceal. Strangely though, the trade was eventually destroyed by Henry Ford. When women began riding in open-topped cars in the years before the First World War, they had to discard their ostrich-plumed hats. The feather business collapsed, although a few farms continued, selling leather and meat and feathers for high-class feather dusters. The ban on trade in wildlife products after 1978 closed the remaining bird farms, but in 2003 Kenya Wildlife Service dropped its ban on selling ostrich products and an ostrich farm now exists near Kitengela, selling meat to the restaurant trade.

Birds can cause humans some problems. Bird strikes have become an increasingly significant risk in aviation; they cost aircraft companies close on a billion pounds per year. Fortunately, although there have been some horrible crashes, there have been relatively few deaths. Most bird strikes happen on take-off or landing, or during low flight, where there is little recovery room if the aircraft loses speed, although an aircraft once hit a vulture over the Ivory Coast at the astonishing altitude of 11,300m. One of the most serious strikes occurred in Ethiopia

Saddle-billed Stork, Tsavo West, the yellow eyes tell you it's a female. (Stephen Spawls)

in 1988, as a Boeing 737 took off from Bahar Dar airport, beside Lake Tana. A number of pigeons were sucked into both engines, one lost power immediately and the other cut out as the pilot tried to get back to the runway; the plane crashed and 31 of the 105 people on board were killed. Michael Grzimek, who was researching in the Serengeti with his father, the famous zoologist Bernard Grzimek, was killed after his plane hit a vulture. Two awful crashes occurred in quick succession in the Maasai Mara in 1991 and 1992. In the first, a Piper Navajo aircraft hit a White-headed Vulture *Trigonoceps occipitalis* which entered the cockpit; the plane crashed and nine passengers and the crew died; then in 1992 a Cessna 401 hit a Marabou Stork and damaged the wing, and once again all on board perished. The Israeli Air Force has a lot of trouble with bird strikes, as Israel lies on the major north–south migratory flyway. But most bird strikes don't result in damage to the aircraft. In places, birds have been used to try to prevent bird strikes; trained falconers employed by the airport patrol the runways with Lanner *Falco biarmicus* and Peregrine Falcons, which chase other birds away.

Humans are rarely at direct physical risk from birds, although a few people have received serious injuries from the attacks of the angry parents when they have approached too near the nest of a bird of prey. The Zimbabwean raptor expert Peter Steyn records that the skull of a child was found in a Crowned Eagle's nest. A few people have been kicked to death by ostriches and cassowaries. Humans can acquire a handful of diseases from birds. Much attention was recently focused on so-called 'bird flu', or avian influenza, caused by a virus sometimes called H5N1. A number of human cases occurred in Asia, and a few in Africa. Opinion is divided. It has been suggested that this virus 'is likely to mutate', which indeed it might, viruses mutate relatively rapidly. But the assumption is being made that the virus is likely to mutate into something much more deadly, and this is unwarranted. Other bird diseases transmissible to humans include psittacosis, carried by parrots and turkeys, and caused by the bacterium *Chlamydophila psittaci*. It can be treated with antibiotics. A few other diseases can be caught from handling birds, such as avian tuberculosis and salmonellosis, but in reality few people are at risk from bird-borne diseases. Kenya's only real bird problems are caused by the Red-billed Quelea *Quelea quelea*, or Dioch.

Farmers keep a wary eye out for flocks of quelea. (Stephen Spawls)

The bird in question is a small weaver, weighing about 20g, and when not breeding it looks quite nondescript, although it has a rather heavy pale red bill; it is a seed-eater. In the breeding season the males develop a rosy flush and a black face mask. They look innocuous but they are Africa's greatest avian pest. For a good reason. They are the most numerous bird in the world, with an estimated population of 10 billion. They occur throughout the savannas of sub-Saharan Africa. Breeding colonies in West Africa have contained over 500,000 nests. Because the birds breed in such numbers predators are rarely able to make much of an impact; however, the ornithologist Oscar Wambuguh describes a massacre in Tsavo East, not far north of Lugard's falls. He observed an estimated 500–800 Marabou Storks and around 1,000 Steppe Eagles *Aquila nipalensis* raiding a quelea nesting colony. One startled Marabou vomited up 22 quelea nestlings. Although the diet of quelea mostly consists of non-agricultural plants such as wild rice, wild sorghum and the grass *Echinocloa colona*, when these are not available they take to raiding crops. A single flock consisting of several million birds can eat 3–4 tonnes of food per day. Since quelea readily eat unripe grain, constant vigilance is needed to protect crops in vulnerable areas, which tend to be arid regions where a crop has been coaxed to grow, often by irrigation. The farmers find they have to spend weeks in the field, with watchers perched on high platforms to warn of the approach of a flock, which can literally darken the sky. If a flock comes, the whole village rushes out, firing guns if they have them, banging drums and shouting; and even then it is likely that some of the grain is going to be lost. A South African farmer once fired a shotgun into a flock, and both barrels of No. 9 shot brought down more than 600 birds. An astonishing video on YouTube shows a massive flock of quelea in Tsavo East, desperate to drink, scaring elephants away from a waterhole. Quelea are the one bird in Kenya that you can freely kill without licence, and at present they are a major menace to the ordinary citizen who farms in dry country. Over 180 million have been killed by pest controllers, they have been attacked with guns, poison and explosives; it seems to have made little difference to their numbers. African farmers are going to be fighting quelea for a good while yet. The University of Cape Town's website has a bibliography of more than 1,800 scientific papers devoted to this pesky little bird. Thorough but slightly dated coverage of bird/human conflict is found in *The Problems of Birds as Pests* (Murton & Wright 1968).

Augur Buzzard, you'll have a good day if you see it on the right. (Stephen Spawls)

There are many legends about birds, especially night birds, for obvious reasons. As gregarious primates we are scared of the dark, and thus by extension any animal that operates in darkness. Owls are often birds of ill-omen, one landing on your house can mean bad luck and in parts of Africa it is a deadly insult to suggest that someone eats owls. In the forests in Ghana, there are people who are said to be able to change into owls. The deep hoot of an eagle owl in the dead of night is frightening, as is the call of some flufftails. Owls are also reputed to be very wise, but this may not be the case; the noted South African animal husbandrist Dave Morgan had a tame Spotted Eagle Owl *Bubo africanus* that he called 'Fifteen Watts', because it was so dim. The little brown Scops Owl *Otus pembaensis* confined to Pemba Island is believed by the inhabitants to give live birth, not to lay eggs. The nightjar is a bird to be respected; legends in northern Tanzania say that if you crush a nightjar egg, bad luck will follow. Nightjars' eyes reflect light brilliantly (you can sometimes pick up a nightjar at night by approaching one with a torch, keeping the light in the bird's eyes, provided you don't let your hand get in front of the torch). There is a legend that as a nightjar passed a blacksmith's forge, a spark alighted in its eye, on the return journey the other eye was also affected, and hence it has eyes like tiny sparks. In Europe, nightjars are reputed to suck milk from goats (their generic name, *Caprimulgus*, means literally 'to milk she-goats'), although this legend isn't known in Africa.

The Augur was a roman official who foretold future events by omens derived from the actions of birds, and this is how the Augur Buzzard got its name; if it is seen on the right when you set out, you will have a good day, if it appears on the left, things will turn out badly. In dry areas, the same belief attaches to the Bateleur. The WaKamba and Sanye people have a similar belief about the call of the woodpecker; if it's on the right things will be fine, on the left is so-so, and if it calls from behind, you had better go home. In high central Kenya, many older Kikuyu people believe that if you are about to undertake a mission of some importance, you are advised to go out at dawn and find some francolin in a glade; the direction they run in will give an indication of the outcome of your mission. The British naturalist Mark Cocker has been working for 10 years on a project 'Birds and People', gathering stories about just such legends.

THE LITERATURE

Sir Frederick Jackson's 1938 book was the first comprehensive work on East African birds. This was followed in 1955 by the two-volume work, *Birds of Eastern and North-eastern Africa*, by the grandly named Cyril Winthrop Mackworth-Praed, an ornithologist whose other talents including shooting for Britain in the Olympics (he won gold medals) and who visited Africa regularly on self-financed ornithological safaris, and Captain Claude Grant, a decorated soldier and professional ornithologist at the Natural History Museum who had also worked in Tanzania. Praed and Grant actually produced three series of books, as well as the East African ones: there was also *Birds of the Southern Third of Africa* and *Birds of West Central and Western Africa*. For some years, Praed and Grant's book was the only reasonably comprehensive work on Kenyan birds; most species were illustrated in colour, and some in black and white; there was a map for each species (very rare in early guides), brief synonymy, and notes on the call, nest and eggs, which even very modern guides do not have.

In 1963, John Williams' book on eastern African birds, *A Field Guide to the Birds of East and Central Africa*, was published by William Collins. Four years later, Collins put out a guide to the East African national parks, also by Williams. This book gave data on the parks (maps, habitat, roads, faunal lists) and described mammals and some birds, it had thirteen plates showing mammals and nineteen plates showing the so-called 'rare birds' of eastern Africa. In 1980, Williams put together an upgraded version of the first bird book, with plates by Norman Arlott, with 660 species illustrated in colour. They were the first reasonably thorough books to cover Kenya's birds, they were tough, portable and nicely produced, as one might expect with Collins' field guides, easily slipped into a rucksack. These were the first useful field books for those of us who started birding in Kenya in the 1960s. But Williams' books had their defects. Adrian Lewis, a British geologist who, with Derek Pomeroy, wrote the *Bird Atlas of Kenya* (Balkema 1989), a superb although very expensive book, with maps and data essential for the professional ornithologist, was quite scathing about Williams. It was his opinion that 'the market has … rather tragically been cornered by the … John Williams Guides … I have met exceedingly few people experienced in Kenyan birds who consider these to be good field guides.' The first Williams book described about 460 species, rather briefly, there were no maps and the plates were a mixture of colour (180 species) and black and white (280), with the more vividly coloured birds like parrots, kingfishers and rollers in colour, and the duller birds (where colour really would be useful) shown in black and white. The standard of illustrations, by Williams and Rena Fennessy is best described as poor to adequate, although Rena Fennessy was actually a very talented artist. A number of species were mentioned (but not illustrated) under the heading 'allied species'. No Palaearctic migrants appeared.

Between the 1960s and 1990s, these books were really all that were available to East African birders, although in the 1960s Shell Oil produced a little soft-backed, large-format book showing about 140 common East African birds in colour, grouped according to habitat (coast, acacia woodlands, mountain, semi-desert etc.). This was a super little book to have in your pack or car; it not only showed the birds but the first glimmerings of zoogeographic groupings. Some good scientific material was becoming available; three volumes of *The Birds of Africa* (variously edited by C. Hilary Fry, Leslie Brown, Stuart Keith and Emil K. Urban and published by Academic Press) had appeared, and the East African Natural History Society produced a really useful book/checklist, *Birds of East Africa*, in 1980. Edited by Peter Britton, but written by such experts as Don Turner, Graeme Backhurst, Tony Diamond, John Gerhart and Clive Mann, among others, it lists all species with accurate notes on their distribution. It has some nice monochrome habitat pictures but only four plates showing some twelve unusual East African bird species. So Kenya's birders were still waiting for a comprehensive field guide, with all the species in colour and with maps. And then, like buses, three came along quite close together.

The first, a Collins illustrated checklist, *Birds of Eastern Africa*, was written by a Dutch ex-forester called Ber van Perlo. Virtually no East African ornithologist had heard of him, he had no publishing history, scientific or otherwise. This book appeared in 1995 on the Kenyan ornithological scene like a bolt from the blue. Every species was described briefly, with its size, habitat and call. All were illustrated in colour, and for some, like the white-eyes, five or six colour versions were shown. There was a small but reasonable map for every species, covering not just

East Africa but Ethiopia, Djibouti, Somalia and Socotra. And it was compact enough to be very easily portable and usable in the field. It had a few shortcomings; the map wasn't next to the picture so you had to leaf to the back and remember the bird's number, the pictures were small, the maps were simplistic, and the 'jizz' (in the English sense, meaning the general appearance) of some species wasn't quite right, indicating that they had been drawn from a skin. Some of the names used are strange – it's been said that you can tell birdwatchers who started with van Perlo because they call Paradise Flycatchers 'Paradise Monarchs', and talk about the Blacksmith 'Lapwing', instead of Plover – but for the first time, a decent-sized field book showed all the Kenyan species in colour, and where they lived.

Close on its heels, in 1996, came *Birds of Kenya and Northern Tanzania*. This one was written by East African ornithologists that everybody knew – Dale Zimmerman, Don Turner and David Pearson. This is a fairly expensive and heavy book, but it is absolutely superb for Kenya, with big lovely plates, the jizz and colour just about perfect, even for really tough creatures like cisticolas, greenbuls and brownbuls. The notes are short but thorough (although nobody since Mackworth-Praed and Grant has done anything on eggs and nests). The maps are good, although they don't show much detail save the two big rivers and Lake Turkana, but the team that produced this book had massive Kenya experience and it shows; a soft-back version is available for field work.

But no one had yet produced what you might call a 'classic' field guide, which is one with the text, the bird and the map all on the same double-page spread. One finally came along in 2002, a Helm Field Guide entitled *Birds of East Africa*, by Terry Stevenson, 30-year veteran of ornithological safaris in Kenya, and John Fanshawe, of Birdlife International. This is as good as Zimmerman *et al.* in the field, and has two advantages: it covers all of East Africa and all the information is grouped, as mentioned. It is heavy, however. The maps lack internal detail, but if you want to watch birds in Kenya, you need either this or Zimmerman in your car. If you are backpacking, then van Perlo might be better. Fanshawe and Stevenson, along with Nigel Redman, have also produced another Helm Guide, *Birds of the Horn of Africa* (2009), similar in format and equally good.

In 1982 Ray Moore's handy guide *Where to Watch Birds in Kenya* was published by TransAfrica in Nairobi. Although hard to find, it is worth having if you have a vehicle and want to spend time birdwatching independently in Kenya. Although slightly dated, Moore's descriptions of how to reach the 110-plus sites described in this book are accurate, as are his lists of resident species. The slight downside to this book was that some of the favoured sites – for example the Sabaki River Mouth – began to attract birdwatchers and consequently armed robbers, so the independent traveller should seek security advice. South African publishers have produced a plethora of books on the birds of that region, and some of them contain data that is useful in Kenya; for example the constantly-updated *Roberts' Birds of Southern Africa* (e.g. Maclean, 1985) contains much information on vital statistics, including weights and wingspans. In 2003 Struik publishers in South Africa produced *Birds of Africa: South of the Sahara*, a 750-page paperback showing all 2,100-plus species known from the area, by Ian Sinclair and Peter Ryan; a second edition appeared in 2011 (New Holland Publishers). This is quite a good book – the whole lot is in there – and the map, text and picture are all in the same place. For a keen birder travelling widely in Africa and constrained by weight, this would be the book to take. Where it is unsurpassed is in its maps, you can immediately get a handle on where the species is found Africa-wide and thus its zoogeography, which adds to the picture. Interestingly, Sinclair and Ryan offer both South African and East African names for a number of species (Verraux's/Black Eagle *Aquila verreauxii*, Magpie/Long-tailed Shrike *Urolestes melanoleucus*, Green/Red-billed Wood-hoopoe *Phoeniculus purpureus*). For the visitor to Kenya, Zimmerman *et al.* or Stevenson and Fanshawe are a better bet in that the maps are more detailed and the plumage/jizz shown is more accurate for Kenyan forms. It is often said, the more local a guide book, the better it is. A recent review of Sinclair and Ryan by top Kenyan bird guide Brian Finch took the publishers to task for the low quality of much of the artwork. For conservationists, *Important Bird Areas of Kenya* by Leon Bennun and Peter Njoroge, published in 1999 by the East African Natural History Society, is indispensible. It contains data on 60 important bird sites throughout the country; each site gets a description, a list of significant birds and other flora and fauna and conservation issues. The opening essays, the references and appendices are goldmines of conservation information. This really is *the* book for anyone concerned with the protection of Kenya's birds (indeed, all of its flora and fauna) and their habitats, and it is enlivened with lovely drawings (including the cover) by Edwin Selempo, a silver-level professional safari guide and enthusiastic all-round naturalist.

In the future, of course, field work and birdwatching is going to be transformed by the electronic age. Indeed, it is already happening. In a recent issue of the *BBC Wildlife* magazine that remarkable birdwatcher and comedian Bill Oddie describes a birdwatching expedition he was on in a South American forest. A rare bird was spotted; the group leader indicated its position in the canopy with a laser pointer aimed below the bird. A member of the group with a lens the size of a bazooka took a picture, it was blown up on the screen and an identification was made (and maybe one day, species recognition technology, like face recognition technology, will check the picture for you against a database and do the identification). A Smartphone was used to find the call of the bird, which was played and down the bird came to find its rival. But at least birdwatchers will still have to go out and find them, until a virtual forest is created.

Two books give the 'big picture' on African ornithology: Reg Moreau's *The Bird Faunas of Africa and Its Islands* (1966), as mentioned earlier, and Professor Gordon Maclean's *Ornithology for Africa* (published by the University of Natal Press in 1990). The latter's nine chapters cover all aspects of ornithology from the African perspective. Those who want to understand birds as living organisms within the African landscape, as opposed to just ticks on a list, should read it.

Scientific data on Kenya's birds can be found in a range of scholarly journals. A mass of information is available in *Scopus*, an East African ornithology journal, published for 30 or more years by the East African Natural History Society. That august society has also published a lot of important ornithological papers in the society's journal. For a number of years the National Museum also produced a slim but useful journal entitled *Kenya Birds*. Since 1994 the African Bird Club has been producing two scholarly issues per year on the birds of the continent, with much relevant to Kenya. The journal cover is always a piece of beautiful bird artwork, the website is user-friendly with free downloadable checklists for all African countries, and subsidised rates for African members, this is a thoroughly laudable organisation. Much good material relevant to Kenya is published in the *Journal of African Ecology*, in *Ostrich* (the South-African-based journal of African ornithology) and in *Ibis*, the quarterly journal of the British Ornithologists' Union. Other journals with regular ornithological input include *Auk, Science, Condor, Journal of Avian Biology, Emu*, etc. Ornithological research has increased in quality and quantity in recent years, after a period in the doldrums, and some major molecular studies are being conducted, but a complication is that many researchers find birds easy subjects to work on, both in the field and in the laboratory, and it's hard to keep pace with relevant material … but that probably goes for any field of scientific endeavour.

Chapter 8

The Reptiles

Meeta empur nemejo nanu eedo kidong'oe ('there is no gecko that does not claim to possess the longest tail')
– Maasai proverb

Asiye shoka, mwepuka nyoka ('if you have no axe, avoid the snake') – Swahili proverb, meaning don't argue with someone cleverer than you

This chapter focuses on Kenya's reptiles. Reptiles are vertebrates; most lay eggs and are ectotherms, gaining much of their heat energy from outside their body. Those who study reptiles (and amphibians) are called herpetologists, which derives from the Greek *herpes*, meaning 'to creep'. Kenya has a diverse reptile fauna, with about 260 species: roughly 125 species of snake, close to 120 lizards, one crocodile, one worm lizard and fourteen chelonians (tortoises, turtles and sea turtles). Kenya has its share of dangerous snakes: the fauna includes 31 dangerous species, of which 19 are known to have caused human fatalities. The Puff Adder *Bitis arietans*, which is widespread throughout Kenya, is Africa's most dangerous snake and Kenya's Puff Adders are the largest on the continent, reaching nearly 2m in length. The Kenyan reptile fauna includes animals typical of the great central Africa forest, true desert species and species endemic to Kenya, but most Kenyan reptiles are savanna dwellers. Kenya has six species of endemic snake and thirteen species of endemic lizard. Three of the endemic snakes are vipers, confined to high central Kenya, and nine of the endemic lizards are chameleons, with small ranges in hill country. Several of the endemic species have never been photographed alive.

The chapter starts by looking at the origins of reptiles and at fossil reptiles in Kenya, before moving on to give an inventory of Kenya's reptile fauna with some notes on classification and details of the zoogeography of Kenya's reptiles. There are also sections on reptile conservation, on the history of herpetology in Kenya and on the interaction between reptiles and humans. As usual, the chapter concludes with a review of the literature. Each section is self-contained and can be read separately.

REPTILES, ORIGINS AND FOSSIL REPTILES IN KENYA

There are six main extant 'groups' of reptiles: snakes, lizards, tortoises and turtles, crocodiles, worm lizards and tuataras, plus a number of extinct groups like dinosaurs. However, reptiles are not all descended from a single ancestor. It is sometimes easier to define reptiles in terms of what they don't have (fur, fins, larval stages) rather than what they have. They get their heat from outside (although the Leatherback Sea Turtle *Dermochelys coriacea* has its own internal heating mechanism). Reptiles are much maligned; they are seen as slow-moving, slimy, solitary and mostly venomous. Most of these generalisations are inaccurate. Reptiles have a dry, waterproof skin. There are no venomous lizards in Africa, and only about 25 per cent of Kenya's snakes are dangerously venomous. Most reptiles are fast moving and efficient animals, and although their body heat is derived from the outside, they do like to be warm. In highland Kenya the occasional Puff Adder may be found torpid on a road at dawn, caught out while prowling by a sudden fall in temperature. They have efficient sense organs. All reptiles have middle and inner ears, used for balance and hearing. Crocodiles and some lizards have a true outer ear (which can be closed, for example crocodiles close their ears when they dive) and excellent hearing. All snakes lack external ears and a tympanum (hence the snake charmer's cobra cannot hear music, it responds to the movement of the charmer and his flute) but snakes can perceive some low frequency sounds in the air; their lungs may play a part. In addition, the stapes bone in the middle ear rests against the quadrate bone, part of the jaw assembly, and hence snakes can detect vibrations through their jaws. The Zimbabwean herpetologist Richard Isemonger, observing a group of Mozambique Spitting Cobras *Naja mossambica* that lived in a termitarium, noticed that they could detect the vibrations of walkers on a nearby path more than 50m away whereas cyclists could get closer, 40m or so, before the snakes fled. This is the reason why walkers in the bush, especially heavy-footed ones, see few snakes; the snakes feel the person coming and take cover. Some pedestrians in Africa know this, and not only deliberately walk heavily, but bang the ground with a stick. All reptiles have eyes, save some blind snakes, where the eye has either been lost or reduced to light-sensitive spots under the skin. Some snakes, for example the rattlesnakes and pit vipers, can

The only worm lizard known from Kenya, the Voi Wedge-snouted Worm Lizard Geocalamus acutus. *(Stephen Spawls)*

The Reptiles

Kenya Montane Viper Montatheris hindii, *small endemic viper from the Aberdare moorlands. (Stephen Spawls)*

detect infrared radiation by means of heat-sensitive pits between the eye and nostril. They are so sensitive that they can detect temperature differences of as little as 0.05°C, and hence can strike warm-blooded prey in total darkness. African pythons have such pits along the lip scales.

Reptiles have a good sense of smell. Many nocturnal reptiles stalk their prey and also find potential mates using chemical cues; the snake's forked tongue picks up particles and transfers them to the Jacobsen's organ in the roof of the mouth. The different concentrations on the two tongue tips tell the snake which direction the prey or potential mate is in. Reptiles were the first vertebrates to move wholly onto land, and there internal fertilisation is mandatory. In the water, fish and amphibians can simply move next to one another and dump eggs and sperm into the water, knowing they will meet, but on land gravity is significant. So females must keep their eggs inside until the male introduces sperm and fertilisation occurs. Later, eggs are laid. Reptiles don't have a larval stage, unlike amphibians; the hatchlings leaving the egg look just like their parents. All crocodiles and chelonians lay eggs; Nile Crocodiles *Crocodylus niloticus* have been known to lay 95 eggs in a single clutch, but sea turtles may produce 200 eggs in one clutch. Most lizards lay between five and fifteen eggs, although a Kenyan Flap-necked Chameleon *Chamaeleo dilepis* once laid 88 eggs. A few snakes and lizards give live birth; in Kenya the live-bearers include the vipers (except night adders), sand boas and some high-altitude species like the Common Slug-eater *Duberria lutrix*, mole snakes *Pseudaspis cana* and some montane chameleons and skinks. The females retain the eggs in their bodies until close to hatching; one obvious benefit of this is that the female can then bask and obtain warmth for the eggs. On the Aberdares, females of the tiny Kenya Montane Viper bask frequently while pregnant, so if you are lucky enough to see one basking it is most likely to be a female. Montane Vipers usually have two, or at most three young, but a pregnant female Puff Adder from Mukogodo Ranch, near Nanyuki, 1.57m long, weighed 7.7kg and gave birth to 147 young. This was surpassed by a Kenyan Puff Adder in a Czech zoo that had 156 young, a world snake record. Some lizards and snakes have developed a placenta, similar in function (though not origin) to that of mammals, which allows food to be transferred from the mother to the developing young. A few reptiles show some parental care – building nests, guarding the young – but no reptile feeds its young.

Reptiles are an old group. The oldest reptile fossils seem to be about 310 million years old. There was a tremendous radiation of reptiles during the late Palaeozoic and Mesozoic eras, 250 million to 60 million years ago. Tortoises and turtles split off before 240 million years ago. The earliest snake fossil, *Lapparentophis*, from the Sahara, is about 110 million years old, the earliest lizard is older than that at 160 million years. Reptilian evolution continues, in highland Kenya there has been a remarkable radiation of small chameleons.

Some interesting reptile fossils are known from Kenya. A lot of ancient crocodiles have been found in the Lake Turkana basin, at Lothagam, the lower Omo and Koobi Fora. Specimens are also known from Rusinga Island and Ombo near Lake Victoria. At one time five species co-existed around Lake Turkana. Two of these are still extant: the Nile Crocodile and the Slender-snouted Crocodile *Crocodylus cataphractus*, which in East Africa today is only known from Lake Tanganyika. A few fossil lizards are known from Kenya; mostly from the early Miocene of the Lake Victoria Islands. They include plated and monitor lizards, fossil chameleons and worm lizards or amphisbaenians. Tortoise and turtle records from Kenya include a soft-shelled turtle of the family Trionychidae, described from the Miocene beds at Karungu in western Kenya, as well as from the Lake Turkana basin, and a species named *Impregnochelys pachytectis* from Rusinga Island. Lothagam has yielded six species of chelonian, including a massive land tortoise. Relatively few snake fossils are known from Kenya, but a fossil python was found on Rusinga Island and a 6-million-year-old fossil python vertebrae was found at Lemudong'o, near Narok.

INVENTORY AND CLASSIFICATION OF KENYA'S REPTILE FAUNA

There are about 300 species of turtle worldwide and fifteen forms are found in Kenya. Five species of sea turtle have been recorded off the Kenya coast and there are four species of land tortoise (few compared to southern Africa, where there are ten endemic species). The strangest is the Pancake Tortoise *Malacochersus tornieri*, flat with a flexible shell. Kenya also has four hinged terrapins, the Helmeted Terrapin and the Nile Soft-shelled Turtle *Trionyx triunguis*. Tortoises take a while to warm up in the sun, so they are interesting indicators of altitude. In Kenya 1,600m is about the upper limit for a tortoise. They can be kept at higher altitudes, but they cannot breed there as the soil is too cold to incubate the eggs.

Lizards are the most numerous, the most obvious and in many ways the most diverse group of living reptiles. No one can visit Kenya without seeing a few lizards, whereas even experts may have to search hard to see some snakes. Find some sun-warmed rocks, or a well-lit tree trunk, or turn over a rock, and you will find a lizard. There are around 4,500 known species of lizard and about 120 species are known from Kenya. The Komodo Dragon *Varanus komodoensis* of South-east Asia, a monitor, is the world's biggest lizard, reaching 160kg. Kenya's largest lizard is the water-loving Nile Monitor, which reaches 2.7m. Kenya's smallest lizards are the dwarf geckoes, reaching 6cm, jolly little tree-dwellers, some with bright yellow and black heads. Unlike tortoises, lizards occur throughout the range of habitats, from sea level up to 3,500m on Mt Kenya and the Aberdares. Up there on the cold moorlands, which freeze at night, there are three skinks, a lacertid (Alpine-meadow Lizard *Adolfus alleni*), a grass lizard and Von Hoehnel's Chameleon *Chamaeleo hoehnelii*. Despite local legends, no Kenyan lizards are venomous. Many people are scared of lizards in Kenya, particularly of chameleons, but agamas and geckoes as well. But lizards should be viewed as friends because they eat a large quantity of insect pests.

Kenya's lizards belong to the following families: geckoes, skinks, typical lizards or lacertids, grass lizards, monitors, agamas, chameleons, plated lizards and girdled lizards. In general, the number of lizard species in any area depends upon the rainfall and the altitude – the warmer and wetter it is, the higher the number of lizards. At Watamu, at sea level, some 26 species of lizard are known, Nairobi at 1,600m has seventeen species, and Limuru at 2,300m has just six. But other factors must also enter into it because Voi, at just under 600m altitude, with lower rainfall than Watamu, has 33 species, and on the top of the Aberdare moorlands, at over 3,000m, there are six species. Voi benefits from rocky hills and large tracts of sun-warmed open sandy soil, where some specialist lizards live and the Aberdares have undisturbed open grassland.

The Reptiles

Our longest lizard: Nile Monitor Varanus niloticus *at Mzima Springs. (Stephen Spawls)*

Kenya's smallest lizard, a hatchling Forest Dwarf Gecko Lygodactylus gutturalis *from the Mara woodlands. (Stephen Spawls)*

Female Jackson's Chameleon Chameleo Trioceros jacksonii *from Nairobi. (Stephen Spawls)*

Geckoes are the most numerous family of lizards in Kenya, with just under 40 species. Geckoes are mostly night lizards, so are not found at high altitude where nights are cold; Nairobi is about the upper limit for nocturnal geckoes. Night geckoes often have amazing eyes, with what appears to be several pupils. They have huge cones (colour receptors) and a visual sensitivity more than 200 times greater than humans; they can detect colour at night in very low light. There are about 25 species of skink in Kenya; these are smooth-bodied, shiny, diurnal or burrowing lizards, most with relatively small limbs. The Western Serpentiform Skink has only tiny limb buds, and that inhabitant of Tsavo, Percival's Legless Skink *Acontias percivali* (named after Arthur Blayney Percival), has no legs at all. Kenya's most visible lizard is the Striped Skink *Trachylepis striata*, a species often found in and around towns and villages, sometimes in big numbers. Chameleons are creatures of legend, and many people are scared of them; although they are totally harmless they are associated with magic. They, and monitor lizards, are the only types of Kenyan lizard that cannot shed their tail if seized. Just over twenty species of chameleon are found in Kenya; there are about 115 species in Africa. Nine small species are endemic to Kenya and live on isolated hills and mountains. Three species of tiny pygmy chameleons are known from Kenya. When handled they defend themselves by vibrating violently, by exhaling puffs of air; the resulting buzz feels like a minor electric shock.

About twelve species of agama occur in Kenya; the family is found throughout Africa, southern Europe, Asia and Australia. Diurnal and often highly visible lizards, the agamas can be split into those brightly coloured species that live in structured colonies, with a male and many females, and those who live in pairs and are small and drabber. In far western Kenya, a vivid and recently described species of the rocks is Finch's Agama *Agama finchi*. The head, neck and tail are bright red, the body indigo; it was named after the noted Kenyan ornithologist Brian Finch, who discovered it. Most agamas are known by the Swahili name *mjusi kaffiri*, literally 'the unbelieving lizard', as those praying at the mosque noticed that these lizards bobbed their heads up and down, like the faithful at prayer, but it was believed that this was a deliberate mockery of the faithful. Some people erroneously believe that big agamas are venomous.

Kenya has twelve species of lacertid, or 'old world lizard'. Most live in semi-desert but some can tolerate high cold country and in the coastal forests the Green Keel-bellied Lizard creeps around the branches. The remaining

A tiny chameleon that vibrates when handled, the Kenya Pygmy Chameleon Rieppeleon kerstenii. *(Stephen Spawls)*

Coastal gem, Green Keel-bellied Lizard Gastropholis prasina *in the Arabuko-Sokoke Forest. (Stephen Spawls)*

Extinct crocodile from Lake Turkana, the long-nosed Euthecodon. *(Stephen Spawls)*

families have only a few representatives in Kenya: there are three plated lizards, two girdled lizards and one grass lizard, a slim inhabitant of the high grassland with minute stick-like limb buds. The tally is completed with two big monitor lizards: the Nile Monitor and the Savanna Monitor *Varanus albigularis*. The lodge in the Shimba Hills is one of the best places in Kenya to see Nile Monitors; a group are resident in the dam below and can be watched throughout the day, lounging about, swimming and hunting. It has to be said, they are remarkably fat, due to handouts from the kindly lodge staff. The Savanna Monitor is a scavenger, eating anything it can find out there in the dry country; it will eat carrion and tackle large snakes. It is often present in big numbers; a small kopje on the western Tharaka Plain had twelve Savanna Monitors living on it.

Of other reptiles, worm lizards or amphisbaenians, a strange group of burrowing animals that look like earthworms, are the least known of Kenya's reptiles. A single species is found in Kenya, in the sands of Tsavo. Twenty-three living species of crocodile are known worldwide, of which three species are known from Africa. Only one species, the biggest, the Nile Crocodile, is found in Kenya. The Nile Crocodile is Kenya's largest predator; large adults can be almost 6m long and weigh 1000kg. It is estimated the Nile Crocodiles kill more than 100 people per year in Kenya. Human–crocodile conflict has existed for a long time in Africa; fossil evidence from Olduvai Gorge in Tanzania indicates that our ancestors were sometimes attacked by crocodiles.

There are between 120 and 130 species of snake known from Kenya at present; a total of about 3,000 species are known worldwide. How many occur in any discrete locality varies, as with the lizards. Species numbers decrease with altitude; Watamu, at sea level, has some 45 species, and thus is Kenya's most 'snake-infested spot', but bear in mind that a number of professional herpetologists have worked there and thus boosted the list. Voi at 600m has 39, Sultan Hamud at 1,200m has 27, Nairobi at 1,600m has 23, Limuru at 2,300m has nine and Kiandongoro at 2,900m on the Aberdares has two species alone. However, rainfall and vegetation are also significant. At Wajir, altitude 200m, based on the figures above we would expect around 40 species of snake, but it only has 17. Wajir has less than 250mm annual rainfall and is thus technically desert and so there are none of the specialist frog-eating snakes, like night adders or white-lips, there. But nowhere in Kenya compares with the 63 species of snake

Kenya Sand Boa Eryx colubrinus, *popular with Blayney Percival's fellow soldiers. (Stephen Spawls)*

found around Mt Nlonako, in Cameroon. Even higher numbers are known from localities in the Amazon forest of Peru, but the South American forest is larger and has persisted much longer than the African forests.

Kenya's biggest snake is the Rock Python *Python sebae*, growing to around 6m, although stories of much bigger ones exist. Some experts have split the species into two, but this is controversial. Big pythons are dangerous, although actual documented cases of people being killed are rare; pythons prefer antelope. Humans are not suitable prey for Rock Pythons, being too wide across the shoulders, although small children might be at risk. Kenya's snakes also include a small boa, the blunt-headed fat-tailed Kenya Sand Boa, which like the giant boas of the Americas, gives live birth to up to twenty young. Sand boas are vividly coloured; the ones from red soil areas are bright orange with chocolate-brown blotches. Unusually, Somali naturalists who often know a great deal about the local snake fauna, mistakenly believe that sand boas are deadly. Arthur Blayney Percival had a good story about sand boas. He was attached to the intelligence department during the First World War and found himself stationed at Voi. Many of the British 'Tommies' were interested in wildlife, and one of the men brought a sand boa for Percival to identify (it is a common species in Tsavo). The soldier wanted to preserve his sand boa in spirits to take home, so Percival gave him a bottle and some preserving alcohol. A steady stream of soldiers then turned up with more sand boas that they 'wanted to take home'. As Percival drolly remarked, 'if Great Britain does not now possess an adequate stock of preserved Sand Boas, lay not the deficiency to my charge'. But the alcohol ran short, and when Percival asked an officer about more, that soldier smiled and said 'ask the next chap if he doesn't think it's a pity to waste good spirits on a dead snake when there's a live man about'. Sand boas remain common at Voi, however.

Kenya has nine or ten species of blind snake and seven or eight species of worm snake, small and rather ugly snakes. They are sometimes called 'two-headed snakes', as the tail is as short and blunt as the head, and the snake can move both ways. They live underground and eat invertebrates, mostly eggs and larvae of ants and termites, quite an unusual diet for a snake. Often they actually live in termite hills, and the females never come out, but the males emerge during heavy storms and make their way to another termitarium, dispersing genetic material.

Legendary strangler, the harmless Worm Snake Leptotyphlops scutifrons. *(Stephen Spawls)*

Kenya's smallest snake is probably Drewes' Worm Snake *Leptotyphlops drewesi* (which reaches just 14.3cm).

About 65 species of colubrid snake occur in Kenya. Colubrids are what you might call regular or (to use an earthy colloquialism) 'bog-standard' snakes, found worldwide, with obvious heads and necks and nine big scales on the top of their heads. They have either solid teeth, and are technically harmless, or have grooved fangs set towards the back of the upper jaw and are venomous; when they bite, venom runs down the grooves (these latter animals are sometimes called back-fanged or rear-fanged snakes, although some professional herpetologists don't like these terms).

Worryingly, recent work has revealed that many so-called harmless snakes actually have quite toxic saliva, and although their teeth have no canals or grooves, if they chew on you, you can get some painful local effects.

The position of the fangs on back-fanged snakes, some distance back on the jawbone, meant that it was hard for the snake to get them in without seizing and chewing. In addition, the venom of rear-fanged snakes was initially regarded as not being very toxic, and many early snake books indicated that if you were bitten by one of them you might get some swelling and a headache, but nothing more – until a big rear-fanged tree snake called a Boomslang (Afrikaans for 'tree snake') killed several people in South Africa. Although Boomslangs are back-fanged, they have long fangs which they can embed without having to chew, and a highly toxic but slow-acting venom that causes internal bleeding. Victims who are badly bitten by one die after a few days from internal haemorrhaging. Once the toxicity of the venom was recognised, it was realised that several people who had died of what was attributed to blood poisoning had in fact been bitten by Boomslangs a few days prior to their deaths. The Boomslang is an interesting snake. Sir Andrew Smith, the Scottish doctor and naturalist whose work is mentioned in Chapter 7 on birds, described the Boomslang no fewer than six different times, for a very interesting reason: each was a different colour. Smith described his six new species in the genus *Bucephalus*, a dramatic name meaning 'ox-head' which classical scholars will recognise as the name of Alexander the Great's horse. Unfortunately, this name turned out to have already been used for a genus of flatworm, and hence the snake's name had to be changed to the next generic

Stay away, a Boomslang inflates its throat in a threat display. (Stephen Spawls)

name that the species had been described under, *Dispholidus*; (which means, prosaically, 'split scale' as their body scales are strongly keeled), the Boomslang's full scientific name is *Dispholidus typus*. No African snake shows as many colour phases: you can find olive, grey, brown, green, yellow, black/white striped and red Boomslangs. There are also green Boomslangs with black speckling, black ones with yellow speckling, dull blue-grey ones, and the juveniles are a totally different colour, they have a brown, blue-spotted vertebral stripe, a bright yellow or orange throat patch, an apple-green eye pupil and a brown head. In general, green Boomslangs are males and brown are females, but not always. In Kenya, a curious straw yellow form with white spots is found in dry country, as is a brown form with a red head.

Boomslangs occur throughout sub-Saharan Africa, anywhere where there are enough trees or bushes for them to live, and chameleons and birds for them to eat. Despite their great variety in colour, they are not too difficult to identify; their body scales are strongly keeled and they have a big short egg-shaped head with a very large eye, the pupil of which isn't perfectly round but elongate at the front, giving the snake binocular vision, which is very rare in snakes. Boomslangs can see motionless objects, a useful skill if you eat chameleons. If angry, they inflate the anterior third of the body and can look very threatening; a big angry specimen (they reach 1.85m) is an impressive snake. And yet, there are virtually no cases on record of an innocent victim being bitten by a Boomslang; they are shy and alert snakes who avoid confrontation and move away if approached. Almost all known bites by this snake (and there are several every year on the Kenya coast) are suffered by incompetent snake handlers, although Sanda Ashe records a bite at Watamu where a child put his hand into a bird's nest and was bitten by a juvenile Boomslang. There are few wild snakes that, if you put your hand onto them, will not bite. Little Boomslang antivenom is produced; there is one manufacturer in South Africa who will supply the liquid or freeze-dried product to treat those who have been bitten.

Since the discovery of how dangerous the Boomslang could be, a few other deadly rear-fanged snakes have been identified. Four belong to the genus *Thelotornis*, the vine or twig snakes (some people call them bird snakes).

How rear fangs snakes kill their prey, a Tiger Snake Telescopus semiannulatus *envenomates a Striped Skink.* (Stephen Spawls)

Inoffensive but deadly, the Vine Snake Thelotornis mossambicanus *has binocular vision, few other snakes do.* (Stephen Spawls)

Jackson Iha, professional snake collector from Gede, freehandles deadly Boomslangs. (Stephen Spawls)

These are slim grey-brown snakes, often with greenish heads. At one time it was believed that only a single species existed, in both the savannas and forests of Africa, but now the genus has been split into four species, two of which occur in Kenya, along the coastal plain and in the Taita Hills. Vine snakes are astonishingly thin, with a curious keyhole-shaped pupil (they also have binocular vision and eat chameleons), and rely on their camouflage. They are also non-aggressive, although there is a legend in parts of southern Africa that they can shoot through people like an arrow. A Tanzanian game warden, Eric Locke, was killed in 1953 by a vine snake that he had deliberately provoked until it bit him; he haemorrhaged from every internal organ. The well-known German herpetologist Robert Mertens was also killed in 1975 by a vine snake, 18 days after being bitten. With black humour, Mertens wrote in his diary on his last day, 'a singularly appropriate end for a herpetologist'; he was 81. No antivenom is produced for vine snakes, although a bite could theoretically be treated by continuing transfusions and coagulation factors. But again, only snake handlers are at risk; as with the Boomslang, there are hardly any cases where the victim was unaware of the snake's presence.

There are a couple more large and thus potentially dangerous rear-fanged snakes in Kenya. They are the Red-spotted Beaked Snake *Rhamphiophis rubropunctatus* of the dry country, which grows to 2.5m, and Blanding's Tree Snake *Boiga blandingii*, a huge 3m snake of Kenya's western forests and points west; one of only a handful of nocturnal tree snakes. The curator of Nairobi Snake Park, Jackson Iha, was bitten by a big Blanding's Tree Snake but only suffered some local swelling. Other interesting rear-fanged snakes in Kenya include eight species of sand snake, fast-moving striped snakes, and the vividly banded Tiger Snake. Kenya's harmless snakes, snakes with no fangs at all, include the well-known Brown House Snake *Lamprophis fuliginosus*, wolf snakes and green snakes of the genus *Philothamnus*, often mistaken for Green Mambas. Kenya also has two file snakes; these weird snakes eat other snakes, including highly venomous species, to the venom of which they are immune. A form of file snake found in Tanzania has a white vertebral stripe and local legend holds that if you see this stripe, you will become blind.

Forest Cobra Naja melanoleuca: *fast, intelligent and highly venomous. (Stephen Spawls)*

Kenya has eleven species of elapid, snakes with short venom fangs set at the front of the upper jaw. Although two of these are garter snakes, small banded creatures that are not particularly venomous, the rest of the Kenyan elapids are spectacularly dangerous – not snakes to be trifled with. There are five cobras: the Forest Cobra, Egyptian Cobra *Naja haje*, Ashe's Spitting Cobra, the Black-necked Spitting Cobra *Naja nigricollis* and the Red Spitting Cobra *Naja pallida*. The first three in this list are huge; they reach lengths over 2.5m and have powerful venoms. They are amongst the most difficult snakes in the world to try to capture alive. The Egyptian Cobra, found in central Kenya from Naivasha eastwards to Mtito Andei and south to the border, is equally venomous. Ashe's Spitting Cobra, recently described (it was concealed within the Black-necked Spitting Cobra group), is a massive snake, growing to 2.7m long. It lives in the dry country of northern and eastern Kenya and the coast. It was named after James Ashe, the charismatic snake man who curated Nairobi Snake Park in the 1960s and

Africa's biggest spitting cobra, recently named for James Ashe, Naja ashei. *(Stephen Spawls)*

then set up the BioKen herpetological research centre at Watamu. Like the Black-necked Spitting Cobra and the smaller Red Spitting Cobra, it can squirt venom from its fangs, to a distance of 2 or 3 metres. The blast from the two fangs comes out like a squirt from a water pistol, and as the cobra spits the reaction force pushes its head back, so the spray travels upwards. This is a purely defensive reaction, probably designed against predatory birds but equally effective against humans. If the poison spray lands on your skin, it doesn't do any harm (it's just a cocktail of proteins; you could drink it), but if it gets into your eyes is stings frightfully and is absorbed into the blood vessels and membranes of the eye, causing paralysis of the optic nerve; temporary blindness can result, infection and corneal damage. You must wash it out with copious quantities of water and use antibiotic eye ointment. Spitting venom is a logical defence: it keeps the enemy at arm's length, because if a poisonous snake gets too close to a large animal, the encounter is all too likely to end in the death of one of the two. It isn't a method of prey capture. Spitting cobras don't spit at their prey, they bite it.

One unusual elapid found in the forests of western Kenya is Gold's Tree Cobra *Pseudohaje goldii*. The little we know of its habits came from a professional snake collector in Kakamega, Luka Matekwa, who found specimens in squirrel traps and in riverine forest. Gold's Tree Cobra has what is arguably the most toxic venom of any African snake but there are no recorded bite cases. Proof if any were needed that arboreal snakes tend to stay out of the way of humanity.

There is no more notorious snake in Africa than the mamba, in particular the Black Mamba *Dendroaspis polylepis* (which is not black; it is named for the black coloration inside its mouth … theoretically). As always, the legend is somewhat bigger than the animal, although mambas are big snakes. There are four species of mamba in Africa, the Black, Eastern Green, Western Green *Dendroaspis viridis* and Jameson's. Three occur in Kenya: the Green Mamba along the coast, with a couple of inland records in the Nyambeni Hills and around Kibwezi, Jameson's in the forests of the far west, and the Black Mamba in most of the low altitude savanna and coast. Jameson's mamba *Dendroaspis jamesoni* and the Green Mamba are retiring snakes, they like to live in trees and are unlikely to be seen unless you are looking hard for them, although they may be present in numbers. In the thickets around Gede there are estimated to be 200–300 Green Mambas per square kilometre. But few people see them; they hide in the thick stuff.

Above: Common, secretive and dangerous: Green Mamba Dendroaspis angusticeps *from Gede. (Stephen Spawls)*

Left: Black Mamba, not black, not aggressive, but big and highly venomous. (Stephen Spawls)

The Black Mamba is a big snake, although exactly how big is disputed. Jameson's Mamba and the Green Mamba reach about 2.5m, but Black Mambas grow to more than 3m. There are stories of them reaching 4.6m in South Africa, but there are no actual measurable museum specimens that big. The largest we have measured was about 3.2m. This still makes it the second largest venomous snake in the world; the largest being the King Cobra, which may reach over 5.5m, although the South American Bushmaster *Lachesis muta* is also reputed to grow to 3.2m. But big snakes seem to shrink when you look for hard evidence. Dr Rick Shine, a thoroughly professional Australian herpetologist, was asked by a reporter in 2004 to comment on a captured South-east Asian Reticulated Python *Python reticulatus* which was purported to be '49 feet' (15m) long. 'Retics', as python enthusiasts know them, actually are the world's largest snakes, and reach at least 8.5m, possibly more, but not 15. When measured,

Mt Kenya Bush Viper, restricted range endemic snake, only discovered in 1967. (Stephen Spawls)

the '49-footer' actually turned out to be nearer 20 feet. Shine quipped that, as all herpetologists know, giant snakes shrink by up to 50 per cent or even more when you put them in a cage and come along with a tape measure. The gullible journalist reported this as though it was a biological fact!

Black Mambas are not actually black, but olive-green, grey, brown, khaki or yellow-brown; specimens from very dry areas are often heavily speckled with black. But the inside of the mouth is a rich purple-black, and when a Black Mamba gets cross, it opens its mouth, hisses and shows that black lining. It is a warning best heeded. The venom is a powerful neurotoxin mixed with a cardiotoxin; a bad bite can paralyse your lungs in an hour or two, and death follows soon. Treatment of Black Mamba bites has often required 100ml of antivenom, ten ampoules of the precious stuff, and even then the victim has not always survived. The stories about mambas – how they attack cars, or chase horse riders, or take revenge for a murdered mate – are untrue, but they can move pretty fast, they climb well and they can be truculent if threatened, rearing up and spreading a narrow hood (they are evolved tree cobras, after all). If you do encounter one, it is best to admire it from afar. But they are not aggressive in the true sense of the word. We were once sent a series of photographs, taken in the Maasai Mara National Reserve, by a scientist working there. A snake had begun coming into his tent in the afternoon and resting there, on a bunk bed, where his young son slept. He wanted to know what it was; it seemed quite friendly, so he said. The photographs showed a decent-sized Black Mamba.

Kenya has thirteen species of viper; four night adders, three bush vipers, Montane Viper, North-east African Carpet Viper, Kenya Horned Viper, Rhinoceros Viper, Puff Adder and Gaboon Viper. All vipers are dangerous: they have long, hinged poison fangs at the front of the upper jaw; the fangs are so long that if they didn't fold back, they would stick out of the snake's mouth when it was shut. The bright yellow and black Mt Kenya Bush Viper *Atheris desaixi*, known from the forests of eastern Mt Kenya and the Nyambene Hills, is endemic and was discovered in 1967. The Kenya Montane Viper, an unusual snake, is also endemic and probably has the smallest range of any snake in the world; it is known only from the moorlands of the Aberdares and Mt Kenya, above the timberline. As you might guess, it is not nocturnal!

Rhinoceros Viper Bitis nasicornis, *Kakamega, a deep-forest snake of central African affinities. (Stephen Spawls)*

Concealed Puff Adder. The essence of animal camouflage: if you don't look carefully, you won't see it. (Stephen Spawls)

Sporadically scattered across northern Kenya are populations of the North-east African Carpet Viper *Echis pyramidum*, a spirited little snake, which bites a lot of people, but rarely causes death, although in other parts of Africa carpet vipers are highly venomous. In those bits of northern Kenya where they occur, carpet vipers can be very common. Jonathan Leakey's collecting team found 7,000 of these snakes in 4 months near Moille Hill.

The genus *Bitis* contains some of Africa's most deadly vipers. Four occur in Kenya. One is small and hasn't killed anyone, the Kenya Horned Viper *Bitis worthingtoni*. It is an attractive little grey, pale-striped snake that is endemic to Kenya. It has a small sharp horn above each eye, the function of which is unknown and it occurs along the high central rift valley, from the northern Kedong Valley through the Kinangop Plateau and Naivasha north to Nakuru, and up out of the rift to Eldoret. It is much in demand for the international pet trade and often illegally exported. Two large vipers occur in the forests of western Kenya, they are the Gaboon Viper *Bitis gabonica*, a fantastically marked fat viper with deadly venom, and the Rhinoceros Viper, with equally bizarre markings and a big horn on its nose. Their spectacular patterns actually serve as camouflage when the snake is resting in leaf litter. The Gaboon Viper is widespread in thick woodland and forest of sub-Saharan Africa; it reaches nearly 2m in length. It is a surprisingly good-natured snake and rarely bites, although when it does so, it is a medical emergency. The Rhinoceros Viper is nearly as vivid, but is smaller, rarely more than 1.2m. It is also more of a true forest snake than the Gaboon Viper, which will tolerate woodland. The final member of this big trio is the Puff Adder, which, in terms of the problems it causes to the ordinary rural dweller, is Kenya's most dangerous snake. Vividly marked with chevrons down the back, various shades of brown, grey and orange, and a fat triangular head side-loaded with big venom glands and long fangs, this snake is bad news for farmers. It occurs throughout Kenya, apart from thick forest and above 2,500m. It can reach 1.85m in length, and Puff Adders near that size occur in most of low northern and eastern Kenya; at higher altitude they tend to be smaller. It is dangerous for three reasons, firstly it is fairly common throughout much of its range, secondly it strikes very quickly, it has long fangs and a potent tissue-destroying venom, and thirdly it is well camouflaged and tends to remain motionless when approached; people don't see it and tread on it. Death is unusual in properly managed cases, but permanent damage often results due to the tissue-destroying effects of the venom. But for the rural dweller who lives a long way from the nearest clinic, a Puff Adder bite can be fatal, and until the rural poor all have decent footwear, use a torch at night and farm with long-handled tools, Puff Adder bite is going to remain a major hazard in Kenya.

ZOOGEOGRAPHY OF KENYA'S REPTILES

Reptiles are excellent indicators of zoogeography, because, unlike birds, they are rarely prone to vagrancy. However, the occasional translocation does happen, especially with animals like tortoises and chameleons, which young people may have captured for a pet and then released at home. From time to time a vagrant snake has turned up in Nairobi, usually near the museum or snake park, an escapee. One morning a Forest Cobra was found meandering down the path outside the park; fortunately located by staff before it was seen by the general public. Forest Cobras are not found anywhere near Nairobi. One of the snake park's most spectacular escapes occurred when a consignment of Eastern Diamond-backed Rattlesnakes *Crotalus adamanteus*, that had arrived from the United States, as part of an exchange scheme, managed to force their way out of their box. Three were found very quickly, as they hadn't got out of the holding room. The next day, one was found outside the park. A few weeks later, the Snake Park got a call from the Boulevard Hotel, across the river, where the manager said there was a big snake in the car park. The curator and a snake handler went down quickly, and there, under a car, was one of the diamond-backs. When it had been bagged, a watching American visitor came up and asked 'What sort of snake was that?' 'It was a Puff Adder,' muttered the director (who had better remain nameless). 'Strange,' said the visitor, 'it looked just like one of our rattlesnakes.' The last escapee diamond-back, incidentally, was never found. Hopefully it didn't make its way to the Mathare Valley. No alien reptile, as yet, seems to have established itself in Kenya

Above: Taita Hills endemic Blade-horned Chameleon Kinyongia boehmei. *(Stephen Spawls)*

Left: Kenya Horned Viper; tiny endemic from the high central rift valley. (Stephen Spawls)

Most of Kenya's reptiles are Afrotropical, true African species, and associated with the savannas. A few truly Palaearctic species, animals of Asia and Europe, actually reach the semi-desert of northern Kenya, like the Northeast African Carpet Viper *Echis pyramidum*. There are also some reptile genera of Palaearctic affinities that are represented by Kenyan species, like the Kenya Sand Boa. There is a distinct reptilian forest fauna within Kenya, which seems initially surprising considering the small area of closed forest. A number of Central and West African forest animals enter Kenya from the west. Examples include the Rhinoceros Viper and the Red-flanked Skink *Lepidothyris hinkeli*. However, the major assemblages of Kenyan reptiles are associated with savanna and woodland, and fit fairly nicely in White's (1983) vegetation zones. Zambezian savanna species include the Mole

Hybrid sand snake, several species of which interbreed along the coast. (Stephen Spawls)

Snake and the Leopard Tortoise, Sudanian species include the Slender Chameleon *Chamaeleo gracilis* and the big and truculent (but harmless) Hook-nosed Snake *Scaphiophis albopunctatus*. Typical animals of the dry savanna Somali–Maasai region include the Pancake Tortoise, the Speckled Sand Snake and the Red Spitting Cobra. This is an interesting habitat; it has little rain, and not only acts as a conduit for dry country animals to move in from the north, but has generated its own reptilian fauna, with many species of half-toed or *Hemidactylus* geckoes. Thus Kenya has forest, desert and savanna reptiles. Some of these forms became isolated in the eastern portions of the East African region, in coastal and higher elevation forests. There are also a handful of reptiles that inhabit the high, cold Afromontane areas, and a number of these are endemic. This represents a typical evolutionary scenario, of rapid speciation of small animals in isolated pockets. Kenya's three endemic vipers are all Afromontane specialists: the Montane Viper on the Aberdares, the Kenya Horned Viper in the high central rift valley, and the Mt Kenya Bush Viper in the forests of Mt Kenya. A group of small chameleons, of the Side-Striped Chameleon *Chamaeleo bitaeniatus*, complex also occur in the high country. In the last 10 years some four new species have been described, from isolated mountains in the north and north-west.

Kenya's final zoogeographic region is the East African coastal mosaic, a mixture of high-rainfall woodland, thicket and moist savanna extending along the East Africa coast and inland in some areas, from southern Somalia down to South Africa. Typical Kenyan reptiles of the region include the Green Keel-bellied Lizard, the Green Mamba and the Rufous Egg-eater, and some species have used the thick country as a conduit to move up from the south, creatures like the Savanna Vine Snake *Thelotornis mossambicanus*.

Kenya's distinctive reptile fauna is more varied than in any other African country. This is a consequence of tectonic events and climatic changes, and possibly human activity. Within the last 30,000 years East Africa has been subjected to dramatic changes in rainfall, temperature and forest size. Before this, the formation of huge mountain ranges and giant valleys, barriers to range expansion, took place. And in recent times, people have modified the vegetation, in particular by burning and cutting of forest, and overgrazing. A diverse land, with a very diverse reptile faunas, arose.

REPTILE CONSERVATION

If there are difficulties with reptile conservation, it is largely because of the way the public perceive them. We started this chapter by stating that reptiles are much-maligned. To many people, reptiles are small and insignificant compared to great creatures like the Lion or Black Rhinoceros, but to others reptiles are simply a dangerous nuisance; farmers in Ukambani might well wish that all Puff Adders were dead and gone. But reptiles do have their place in the scheme of things, and as more and more of us live in cities, we should come appreciate the entirety of wildlife in wild places, reptiles included.

Reptiles have certain advantages where their conservation is concerned. For a start, there is (as yet) no great demand on reptiles for their body parts, although such a trade does exist in the Far East (it was such a trade that led to the loss of our rhinos). Certain species, such as the Kenya Horned Viper; are in demand for the pet trade, and a small number are exported illegally, such unusual endemics do need to be monitored. Many reptiles are relatively small and secretive and are not quite as vulnerable to habitat loss and exploitation as rhinoceroses and Wild Dogs. Many reptile species can survive on farmland and in suburban areas where larger animals cannot. In addition, many of Kenya's reptiles occur widely in arid and savanna regions, and are thus not under any pressure. The forest animals are more vulnerable; there are few big forests in East Africa and most of these are threatened by development. However, many of our forest reptiles are widespread in the great central African forests; and although it would be tragic if animals like the Green Bush Viper disappeared from Kenya due to loss of its habitat, there is at least a reservoir of these animals elsewhere.

Kenya's vulnerable reptiles are the endemic species, especially those with limited ranges. Many live in tiny habitats, which are often forested. With population increases, these forests are vulnerable, for their timber and potential agricultural use. At present, our conservation priorities should be to identify such habitats, document their fauna and flora and implement protective strategies. A commendable start in this process has been made in Tanzania by the conservation group Frontier Tanzania, an organisation documenting the flora and fauna of the Tanzanian coastal forests. However, no such activity has been started in Kenya. The usual problems accrue: lack of funds and the low profile of reptiles.

Camouflage: a Bark Snake Hemirhagerrhis hildebrandti *creeps up a Shimoni Tree.*
(James Ashe)

In some ways the future is bright for Kenya's reptiles. The country has an active tourist industry that employs many people and brings in outside money, thus the government has a financial interest in protecting wild places. There is a large, well-developed and protected national park and reserve system, helped to some extent by the forest reserves. Most of the big ecosystems have a large national park associated with them; for example Tsavo and the Mara. Key smaller national parks and reserves for the protection of vulnerable reptiles in Kenya are Sibiloi, Mt Elgon, Kakamega, Marsabit, Aberdares, Mt Kenya, Hell's Gate, Arabuko-Sokoke Forest, Shimba Hills and the Chyulu extension in Tsavo West. However, there are a number of vulnerable areas that are unprotected and yet contain key reptile species; they include the Tana River delta, mid-altitude forests around Mt Kenya and the Nyambeni Hills, the Mau and the Taita Hills. Attention needs to be given to reptile surveys in these areas, and some thought to protection. For example, the spectacular small endemic Mt Kenya Bush Viper does not occur (as far as is known) within any protected area. It is forest-dependent, in forests that are under great developmental pressure and it would be tragic if it were to become extinct.

There are also a number of areas that have never been herpetologically surveyed, and may well contain interesting species. These include most of south-western Kenya, isolated high hills such as Mt Kulal and Endau, the coastal woodland between Lamu and the Somali border and the country north-east of Lake Turkana. No one really knows what reptiles are in these areas.

THE HISTORY OF HERPETOLOGY IN KENYA

We have already met a number of adventurers who were significant in the Western exploration of Africa's reptile fauna, people like Sir Andrew Smith, Gustav Fischer, Johan Hildebrandt, Sir Harry Johnson and Arthur Donaldson Smith. Donaldson Smith collected, in Ethiopia, what is probably Africa's rarest snake, the Beautiful Sand Snake *Psammophis pulcher* (beautiful is actually its name, not an epithet, it is the literal translation of *pulcher*). This little finely striped gem was long known from four specimens only. George Boulenger described the species from Donaldson Smith's single specimen, from the upper Webi Shebelli River in 1895. A second specimen was found in Somalia (without further details), in 1961 Jonathan Leakey's team found one at Voi and a fourth specimen was collected near Mwingi, in Ukambani, in 1972. Astonishingly, as we went to press, the safari guide Sean Flatt collected not one but two specimens, at Bisanadi on the Tana River.

As with ornithology, much sterling work in East African herpetology was done by the Germans in the nineteenth and early twentieth century. Eduard Rüppell explored Egypt, the Sudan and Ethiopia in the 1820s and 1830s, and his name is preserved in the Kenyan Rüppell's Agama *Agama rüppelli*. Sadly, in the last 25 years of his life he became totally antisocial; the rugged individualism that had served him well in Africa let him down and his colleagues shunned him. Dr Wilhelm Peters (who must never be confused with the villainous and brutal Karl Peters, founder of the German East African Company) was a German zoologist who explored Angola and then Mozambique in the 1840s, took his huge collection back to Berlin and wrote a long series of excellent papers. He is commemorated in the names of Peters' Writhing Skink *Lygosoma afrum*, Peters' Sand Lizard *Pseuderemias striata* and several frogs. Based at the University of Berlin, Gustav Tornier didn't do fieldwork in Africa but received many specimens from the German colonies, including German East Africa (now mainland Tanzania), and in 1896 published a major work on the reptiles and amphibians of East Africa. He is remembered in the scientific name of the Pancake Tortoise *Malacochersus tornieri*. Oskar Boettger was an unsalaried herpetologist at the Senckenberg Museum in Frankfurt who didn't visit Africa but published widely on specimens from the continent. An interesting man, in the period when he did his best work, between 1876 and 1894, he was an agoraphobic who could not leave home; the museum specimens were brought to him. Several Kenyan frogs are named for him and he described the Striped Bark Snake *Hemirhagerrhis kelleri* and the Somali–Maasai Clawed Gecko *Holodactylus africanus*, inhabitants of Kenya's dry country. We must also mention poor Richard Sternfeld, who was based at the Senckenberg Museum and published a series of excellent papers on the reptiles of the

Rufous Beaked Snake Rhamphiophis rostratus *basking, Kajiado; the open holes indicate the termitarium isn't being used by termites. (Stephen Spawls)*

ex-German colonies, including Tanzania; he described the Montane Egg-eater *Dasypeltis atra* and the Mt Kenya Side-striped Chameleon *Chamaeleo schubotzi*. Sternfeld was Jewish and although he had served Germany in the First World War, he was arrested by the Nazis in 1943 and murdered in a concentration camp; one of his fellow prisoners wrote to the German herpetologist Robert Mertens (the man who was killed by a vine snake) that Sternfeld 'loved his fatherland … and felt himself to be entirely German'.

One of the leading figures of African herpetology, Arthur Loveridge, was born in Wales in 1891. He was fascinated by wildlife, in particular snakes, from an early age. In 1914, purely as a result of a fortuitous acquaintanceship with a civil engineer who was working in East Africa, Loveridge was invited to apply for the post of curator at the museum being set up by the 'East Africa and Uganda Natural History Society' in Nairobi – the society's first paid employee. Having carefully briefed himself on whales, which he felt was the one zoological subject he was uninformed on, he nervously attended an interview in London, only to find it was a foregone conclusion; the committee simply asked him to sign on the dotted line. Shortly afterwards he found himself in Kenya, the first curator of what was to become the Corydon Museum. When Loveridge arrived, the Nairobi Museum, as it was then, was a small stone building at the corner of what are now Muindi Mbingu Street and University Way. It then moved to near where the Serena Hotel now stands, finally moving to the present site in 1930. The new building was named the Coryndon Memorial Museum (often shortened to the Coryndon Museum) after Sir Robert Coryndon, charismatic adventurer and Governor of Kenya between 1922 and 1925. After Coryndon died in 1925, Lady Coryndon established a fund to build a better museum in memory of her husband; the government offered matching funds, and the museum was built between 1928 and 1929. It is now the country's flagship museum, and was renamed the National Museum of Kenya in 1964.

Interestingly Loveridge never took a formal scientific degree. A careful organiser and classifier, he started work on the collections, but was interrupted by the war. In 1915, Loveridge joined the East African Mounted Rifles,

Moment of life – a Spotted Bush Snake Philothamnus semivariegatus *slides out of the egg. (Stephen Spawls)*

and after training found himself at Kajiado. He spent much of the war fighting in Tanzania, finishing in northern Mozambique. To the frequent amusement of his fellow troopers, and the resigned tolerance of the top brass, Loveridge combined his soldiering with museum collecting to a remarkable degree. As described in his first of his four works of autobiography, *Many Happy Days I've Squandered*, published in 1944, he sometimes had to dodge sniper fire while collecting specimens. While digging in at Longido, he was constantly summoned to catch small snakes and lizards turned up by the soldiers, prompting his exasperated sergeant-major to ask, 'Is there a war on or is this a blooming British Museum expedition?'

Returning to Nairobi after the war, he resumed his work. James Ashe, curator of Nairobi Snake Park from 1964 to 1970, told an interesting story of Loveridge's time in Nairobi. Loveridge was a teetotaller and a devout Christian, a member of the Plymouth Brethren. A group of civil servants, including the Governor, Major-General Edward Northey, were talking with Loveridge when Northey used the expression 'For God's sake.' Loveridge turned to the Governor and said, 'Would you mind leaving the building, Sir Edward. I'll have no taking of the Lord's name in my museum.' Once outside, one of Northey's breathless acolytes said, 'What are you going to do about that, Sir Edward?' 'Nothing,' Northey replied. 'We have chosen the right man as curator.'

Loveridge left Nairobi in 1921, to take up the post of assistant Game Warden of Tanganyika, which put him in the field, where he wanted to be. However, in 1924 he moved to the Museum of Comparative Zoology, at Harvard, in the United States, where he was based for the remainder of his working life, curating the collection there. He retired in 1957 to St Helena, where he died in 1980. While at Harvard he made five year-long expeditions to eastern Africa, between 1926 and 1949. Loveridge wrote over 200 scientific papers, the bulk on African reptiles, including major revisions of more than fifteen African reptile groups, as well as four books of autobiography.

In 1940 Louis Leakey was appointed honorary curator of the National Museum, after a personality clash between V. G. L. Van Someren and Peter Bally. Leakey took dynamic action: he raised funds and he opened the museum to all by reducing fees for ordinary people, in the face of some frightful racist attitudes amongst the European trustees and Nairobi residents, who said that 'Africans were smelly and Asians overscented'. Leakey also took a broad interest in all aspects of natural history; this included exhibiting some live snakes and building up a spirit collection. In 1943, he asked C. J. P. Ionides to collect snakes for him. It was a propitious move.

Constantine John Philip Ionides, known to his family as Bobby and to his friends as 'Iodine', was a legendary figure in East African herpetology. British, but of Greek descent, an ex-Indian army soldier, he had joined the Tanganyika Game Department in 1940. A keen naturalist, his passionate early interest was hunting He spent all of his free time and money in pursuit of rare and unusual mammals, from northern Sudan to Malawi, often undergoing many days of desperate hardship to obtain his quarry. Iodine was a sporting hunter of the old school, scorning such practices as following his quarry by vehicle or shooting over a bait. He allowed months for each hunt, scouted, tracked and shot; when stalking a Scimitar-horned Oryx *Oryx dammah* he crawled 800 yards over hot sand, giving himself septic sores which took months to heal. He hunted specimens for the National Museum; for 40-plus years his animals were displayed in the dioramas in the main hall, and included a Mountain Gorilla, a Bongo shot on the Mau, a Yellow-backed Duiker *Cephalophus silvicultor* and a White Rhino.

Iodine took on Leakey's request with enthusiasm, he taught himself how to catch and handle snakes, and the Coryndon Museum (as it was then) benefited from his specimens. In 1956, he had to give up trophy hunting after suffering from major circulation problems in his legs; he then caught venomous species commercially for zoos and serum manufacturers. Between 1956 and 1961 he collected nearly a thousand Green Mambas and 700 Gaboon Vipers within 10 miles of Newala, in southern Tanzania, where he lived. He also sent specimens to herpetologists such as Captain Charles Pitman, ex-game warden of Uganda, Arthur Loveridge and James Battersby at the Natural History Museum in London. But Iodine continued to collect for the museum, many of his specimens are preserved there, and he was actually on a collecting trip to Lake Baringo in 1968 when he collapsed, and died shortly afterwards in Nairobi Hospital. There are two books about Ionides. The first, *Snake Man* by Alan Wykes (1960), is very decently written but Ionides detested it, feeling it sensationalised him, which was anathema to this shy and modest man. Many of the stories in this book are untrue or exaggerated, including the opening one where a Green Mamba descends at night into a hut and kills all but one of the occupants (Green Mambas are diurnal and non-aggressive). The second, *Life with Ionides* by Margaret Lane (1963), Ionides was noncommittal about, save feeling that it portrayed him as an unclean character. The best book is his autobiography, *A Hunter's Story* (1965), published in softback as *Mambas and Man-eaters* (1966). Much of the writing was actually done by Dennis Holman, who has written some excellent books on Kenya. Ionides himself felt that this book gave exactly the right impression and Holman kindly insisted that the hardcover version went out under Ionides' name alone.

In 1959 funds were raised by the museum to start a snake park, a modest building with exhibition cages based around a large central pit. In later years further pits, a large aviary and an aquarium were added. The snakes that had been on display in glass tanks in the museum were moved to the snake park, and Leakey's eldest son, Jonathan, was appointed curator of the Nairobi Snake Park, as it was known, in 1961. It was the start of an important phase in Kenyan herpetology. Jonathan only stayed for two years before moving to Baringo to start his own commercial reptile farm, but in the meantime he trained a number of snake-collecting teams. Some of the field men he trained then continued to collect for the museum. The depth of the spirit collection is a monument to the skill of these men; and they should be remembered. Luka Matekwa collected in Khayega, Kakamega, and much of what is known on the snakes of western Kenya is based on his material. Mutui Mutisya did his collecting at Ngomeni, near Kitui, and was the man who found the second Kenyan specimen of the Beautiful Sand Snake; his son still works for the museum. Jackson Iha was based at Watamu, and greatly improved our understanding of the coast reptile fauna; after working for the museum he started several small snake parks along the coast. At Chuka, Humphrey Macheru collected the first specimens of the Mt Kenya Bush Viper, he passed his specimens to Frank DeSaix, a United States Peace Corps volunteer. The new snake was described as *Atheris desaixi*, but in all honesty it should have been called *Atheris macherui*.

The Reptiles

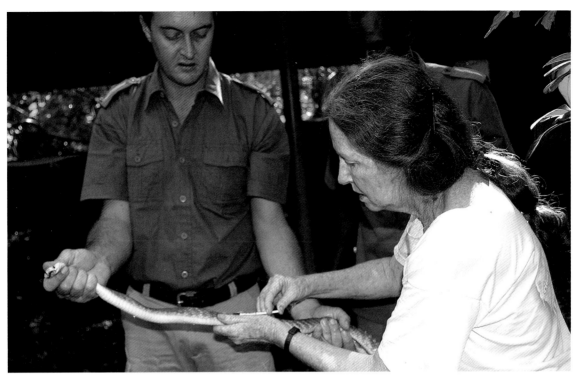

Sanda Ashe and Royjan Taylor treat a Black Mamba at Bio-Ken research Centre, Watamu. (Stephen Spawls)

After Jonathan Leakey's departure, the curatorship of the snake park was passed to Cecil Webb, an Englishman and famous animal collector who had been responsible for restocking London Zoo and Whipsnade Zoo after the Second World War. Webb and his wife, Daisy, had retired to Kenya, but Webb sadly died in 1964. The museum authorities then asked James Ashe to take over. It was an inspired choice.

Ashe was an ex-paratrooper and mining engineer with South American experience. A dynamic bearded man who at the time was running the engineering side at the Athi River cement factory, he maintained a large snake collection at home and this had featured in the *East African Standard* newspaper. Ashe quickly raised the profile of the snake park, to the point where it became a regular feature on the tourist circuit. He instituted a weekly demonstration of snake venom extraction, known as 'snake milking'. This was an exciting event, and sometimes involved the striking abilities of a huge Egyptian Cobra, with Ashe being the target. The show attracted some high-profile visitors to the snake park, including the actor James Stewart and Paul McCartney of the Beatles. Ashe encouraged people to collect reptiles for the museum, and the park acted as a centre for East Africa's leading herpetologists and naturalists; visitors behind the scenes could find themselves drinking tea with the likes of Joy Adamson and Dian Fossey. A number of keen schoolboys began to visit the park, including both authors. Ashe knew the potential of developing such a group; he called them his 'youth wingers'. He was also a remarkable and witty raconteur, with some memorable stories.

Ashe and his wife, Sanda, left Kenya in 1970, to work with Bill York's Lion Country Safari Park in Florida, but returned to Kenya in 1980, and set up Bio-Ken, a small research centre and snake park at Watamu. Bio-Ken became well known for always having stocks of antivenom, even when the coast hospitals had run out. What was less well known was that, if the victim couldn't pay for the stocks, the Ashes waived the cost. They also stocked Boomslang antivenom, which the coast hospitals found difficult to get. Ashe died in 2004, but Sanda Ashe continues the work at Watamu; a good number of people along the Kenya coast today wouldn't be alive if it weren't for the efforts of James and Sanda Ashe.

Dr Patrick Kinyatta Malonza, senior herpetologist at Kenya's national museum. (Stephen Spawls)

After Ashe left the snake park, a journalist, Peter Nares was appointed director, with Jackson Iha as curator. Both had worked as field men for Jonathan Leakey, and they concentrated on showing local snakes. At one time more than 40 species of Kenyan snake were on live display at the park. After Nares left in 1972, Iha went to Malindi to start a snake park and the Nairobi park was directed by Nicholas Odhiambo, who later left to become a chief in western Kenya. Since then the park has been run from the museum. Between 1964 and 1995, the herpetologist at the National Museum was Alex Duff-MacKay. MacKay, a Kenyan born in Mombasa in 1939, was a forest entomologist at Muguga before joining the museum. His major interest was amphibians, but he published very little, seeing himself as a behind-the-scenes museum man whose business was to curate. However, he did write a major paper on the 7,000 carpet vipers collected at Moille Hill, a short book (with his wife, Joy) on Kenya's venomous snakes and an annotated checklist of the frogs of Kenya. He also trained Damaris Rotich, who became curator after his retirement. At present, Dr Patrick Malonza heads the department of herpetology, ably assisted by a team of professional herpetologists that include Drs Beryl Bwong and Victor Wasonga. They have described a number of new species and are surveying the herpetologically unexplored areas of Kenya, particularly the forested hills.

REPTILES AND HUMANS IN KENYA

Reptiles are significant to the average rural dweller in Kenya. Lizards eat insects, crocodiles may lurk in the rivers and there are venomous snakes. Sea turtles and their eggs were once eaten. The fishermen on Kenya's northern coast had an ingenious way of catching turtles; they used a tethered remora or sucker fish *Remora remora*, and when a turtle was seen the remora was released on a line. It would attach itself to the turtle, and was then pulled upwards, slowly, until the turtle was close, then it was harpooned. In theory, turtles and their eggs are now protected along the Kenya coast.

Not the view you want if you're swimming, a 5m Nile Crocodile at water level. (Stephen Spawls)

Not all crocodiles are dangerous. Only two species, the Nile Crocodile and the Estuarine or Saltwater Crocodile *Crocodylus porosus* of South-east Asia and Australia, regularly take humans. To a big hungry Tana crocodile the creature on the bank looking after their goats is just a thin-skinned delicious primate. No one knows how many people fall victim each year to crocodiles in Kenya, but in the 1960s it was estimated that more than 100 people per year were killed on the Tana River. The Athi–Galana–Sabaki river also has a bad reputation for crocodile attacks, as does the northern Uaso Nyiro. There is a general belief that in Kenya river crocodiles are more dangerous than lake crocodiles, and this may be due to that fact that in lakes, where the water is usually clear, the crocodile can see their preferred prey of fish, but in rivers, which are muddy during the rainy season, the fish are invisible and the crocodiles switch to taking drinking animals, which can easily be seen. Lakes Baringo and Turkana both have plenty of crocodiles but are usually regarded as safe places to swim, although fatal attacks are known from Lake Turkana. In 2002, a high profile attack occurred in Lake Chala, when a British teenager was killed by a crocodile; prior to that date the crocodiles there were thought to be non-aggressive. On the Tana, it's a different matter. In some areas, such as downstream from Garissa, people collected water with a container on a long pole, not daring to approach the water's edge. For stock to drink safely, a corral of sunken posts may be used to cordon off an area of the water. George Adamson described in *Bwana Game* (1968) how he saw the posts being shaken by frustrated crocodiles as the cattle drank. The safety rules are clear: take care when you approach the water; don't sit near it, as crocs will rush out and grab a victim; if you have to use a river to gather water or wash, pick a shallow place where a crocodile will find it difficult to approach undetected; and never, ever approach the water at night.

Venomous snakes are a hazard in most rural areas of Kenya. But it is worth remembering that not all snakes are dangerous, of Kenya's 125 species of snake, some 31 are dangerous, and of these 19 are known to have caused fatalities. However, five species alone – the Puff Adder, the Carpet Viper, Black-necked and Ashe's Spitting Cobras and the Small-scaled Burrowing Asp – cause the majority of bites. Accurate data on the actual numbers of snakebites is hard to gather; in the countryside people very often visit a local healer first, to get a potion or two

Partially buried Carpet Viper, Wajir, a major cause of snakebite in northern Kenya. (Stephen Spawls)

and only go to hospital if things start to get worse, by then they are more likely to die or suffer permanent injury. The clinic or hospital may be many miles away; there may be no transport other than walking and no medical professional or serum available when they do get there. Moreover, people are often hazy about what snake bit them, particularly since the bites often happen at night, or in thick vegetation; typical scenarios are that someone goes outside the house at night barefoot or with inadequate footwear and treads on a snake, which bites and disappears in the darkness; a snake gets into a building at night and someone sleeping at ground level rolls on it; or a farmer reaches down to clear vegetation and is bitten on the arm. In the West African savanna some reliable but rather shocking data have been gathered on snakebite. It was estimated that in the Sahel there could be as many as 23,000 deaths per year. Most of these were due to three main species, the West African Carpet Viper *Echis ocellatus*, the Puff Adder and the Black-necked Spitting Cobra. One study in Kenya indicated bite rates that were much lower, with countrywide rates around 14 per 100,000 population per year, but varied from 2 to nearly 70, depending on the locality. This still translates to about 4,500 bites per year, with yearly death rates around 150 per year, a 3 per cent mortality. But such data can only be gathered at organised hospitals and clinics; the actual death rate is almost certainly much higher in remote areas.

The snakes that send a shiver down most people's spines, the mambas, are actually rarely involved. Mambas are diurnal; they are alert and tend to move away when people approach. To the rural dweller, the hazardous snake is the one that is likely to be moving around on the ground in the two hours after dark (Puff Adder, carpet viper, Spitting Cobras, Burrowing Asp), hidden in a pile of leaves or grass during the day (Puff Adder, carpet viper) or living in a hole near a chicken run or houses where rats and toads might be found (cobras). High-profile recent bite cases in Kenya involved a Red Spitting Cobra (got into the bed in a lodge banda in Tsavo at night and the victim rolled on it), Puff Adders (one got into a tent at night in Samburu and bit the victim on the head, another at Baringo was trodden on by a camper when she went outside to urinate at night),

Obvious Puff Adder, the culprit in most of Kenya's snakebite accidents. (Stephen Spawls)

burrowing asp (entered a Malindi hotel room under a door at night and the victim trod on it when he got up) and, surprisingly, Boomslang (a visitor at a Bamburi hotel tried to catch it, under impression it was harmless; he told the doctor it was a 'harmless tree viper'). During the last 30 years, James and Sanda Ashe gathered data on snakebite on the coast and in some other parts of Kenya. According to Sanda Ashe, bites around Nanyuki are mostly due to Puff Adders, but on the coast the main culprits were burrowing asps (about 50 per cent), Puff Adders (about 25 per cent) and cobras, both Ashe's and Forest. (12 per cent). Also implicated but in small numbers were Green Mambas, Black Mambas, Velvety-green Night Adders *Causus resimus* and a few others.

The prevention of snakebite in Kenya lies with raising living standards. The health worker trying to prevent snakebite will be aware of the impracticality of telling the rural poor to always use a torch at night, not to use short farming tools and always wear strong footwear. But the better-off should be able to do those things. Keeping your home and its vicinity

Sharp end of a Puff Adder, the long folding fangs fully erected. (Stephen Spawls)

African Rock Python from Kitui, Africa's biggest snake. (Stephen Spawls)

clear of suitable hiding places for snakes, getting beds up off the floor and making sure that doors fit snugly are also good precautions. And always look where you are going.

There are a great many legends about snakes in Kenya. A persistent story is that of the deadly crested snake; a myth known all over Africa and also in the Caribbean. Sometimes the snake is said to crow, like a cockerel, and its crest is described as being like the fleshy wattles of a chicken. The basis of this story might be the Black Mamba, which is big, deadly, occasionally truculent and sometimes fails to shed the skin on its neck, which builds up to a mat. Some of the weirder night sounds have been attributed to this snake, including the calls of various owls, tree hyraxes and flufftails. Another fabulous snake has a light; it travels at night with the light in its mouth and puts it down to attract animals that it wants to eat. Another creature of myth is the snake – usually big and black – that guards a spring; one such snake was reputed to live in Hell's Gate. In parts of Africa, people are believed to be able to turn themselves into snakes, and bite people who have crossed them.

Worm snakes and blind snakes are feared in some areas of highland Kenya. They are thought to be able to spring into your nostrils and somehow strangle you, a legend that probably arose after someone died and worms were seen in their throat. There are also stories that safari ants live in symbiosis with blind snakes, and gain some sort of benefit from them. Certainly blind snakes have been seen moving either beside or along a trail of safari ants and not being attacked; they have also been found inside the ant's nest. At the other end of the snake size scale, legends of gigantic snakes exist in parts of Kenya, often near lakes where the huge serpent is said to hide; these are easily explicable as simply being pythons, seen under frightening conditions. Such huge snakes are believed to have eaten people. There is no authenticated case anywhere in Africa of an adult human being swallowed by a snake, although the professional hunter Alan Tarlton did find a human baby inside a Kenyan Rock Python that he shot, and the Anaconda of South America has occasionally eaten

The Reptiles

Male Jackson's Chameleon, a miniature Triceratops. (Stephen Spawls)

people (small ones). The size of big snakes, especially if coiled, or in water, is notoriously hard to judge. The American archaeologist Hyatt Verrill encountered an Anaconda basking on a rock in Guyana, South America, and asked his companions to judge its length. Their estimates ranged from 20 to 60 feet. Verrill then shot the snake and measured it; it was 19 feet and 6 inches. Black Mambas are also believed to grow to great lengths; you hear stories of mambas crossing roads 'with their heads two metres up in the air and their body still covering the width of the road'. It's unlikely, although a snake crossing a road at high speed can seem bigger than it is. There is a story that if you kill one snake, you must burn or otherwise destroy it, or its mate will come and kill you in revenge. This is a myth; snakes do not have partners, they meet briefly in the mating season but that's all. Some snakes don't seem to have an understanding of death; live snakes have been seen attempting to mate with dead ones.

Snakes can move at speed. Meinertzhagen claimed to have timed a Black Mamba moving at 7 miles an hour on the Serengeti Plains (the ones in Kenya, between the Taita Hills and Kilimanjaro, not the more famous ones in Tanzania). One of us (S. S.) estimated the speed of a frightened mamba at Materi, on the Tharaka Plain, at around 10 miles an hour. But the story of how a traveller on horseback in Africa was chased by a mamba and bitten seems unlikely, as does the story of how a motorist who had run over a mamba was pursued by the snake, which jumped into the car and bit the occupants.

Lizards are feared in parts of Kenya, despite the fact they are all harmless. There are many legends about the humble (and totally innocuous) chameleon. Some people think chameleons are venomous but it is more commonly believed that they are evil, or some sort of bad luck will come your way if you touch one. There seems to be a general legend that the chameleon was sent by God to carry a message that man would not die, but it went slowly and God thought it hadn't ever arrived, so he sent another messenger (a toad, or a hare), which got there first but scrambled the message to the effect that man would die. Arthur Loveridge records that young Kikuyu men would kill Jackson's Chameleons and exhibit the horns to the young women they wanted to impress, to show they were brave. In parts of Kenya, geckoes are feared or disliked. They may move about in mosques during prayers, showing what is thought to be disrespect. It is also believed that if they walk on an unsuspecting person, or their droppings fall on them, illness will result, usually leprosy. You can see the connection between that disease and the way geckoes shed their skin. As we mentioned earlier, similar legends attach to agamas.

THE LITERATURE

Baseline information on Kenya's reptiles can be obtained from www.reptile-database.org, which lists all the world's reptiles. A sound overview of the reptiles and their origins is provided by the undergraduate textbook *Herpetology* by George Zug, Laurie Vitt and Janalee Caldwell, published by Academic Press in 2001, and updated in 2008. There are few books covering all of Africa, but we might mention *The Dangerous Snakes of Africa* by Stephen Spawls and Bill Branch, published in 1995 by Blandford, *Venomous Snakes of Africa*, by Maik Dobiey and Gernot Vogel (mostly just coloured photos and maps), published by Chimaira in 2007, *Tortoises, Terrapins and Turtles of Africa* by Bill Branch (Struik Publishers 2008) and Colin Tilbury's *Chameleons of Africa*, published by Chimaira in 2010 with over 800 beautiful colour illustrations. Tony Phelps's *Old World Vipers*, published by Chimaira in 2010, covers all of Africa's vipers, with distribution maps for all and beautiful pictures of virtually all species. Many excellent books cover the reptiles of the southern African subcontinent, particularly those by Don Broadley, Johan Marais and Bill Branch. Regional material for the reptiles of eastern Africa is very variable, but includes H. W. Parker's 'The snakes of Somaliland and the Socotra Islands' (1949) and 'Lizards of British Somaliland' (1942), both published as scientific papers, and a major work by Karl Patterson Schmidt (the man who was killed by a Boomslang) and Gladwyn Kingsley Noble, *Contribution to the Herpetology of the Belgian Congo*. This last is actually three huge papers, 600 or so pages, on the reptiles and amphibians of the Congo, originally published between 1919 and 1930, but republished together with an updated taxonomy by the Society for the Study of Amphibians and Reptiles in 1998. Also useful is Malcolm Largen and Stephen Spawls' *Amphibians and Reptiles of Ethiopia and Eritrea*, published by Chimaira in 2010.

There are few specific books on Kenya's reptiles, and those comprehensive works that exist are mostly scientific papers, which are not available to the public save through specialised libraries and museums. In addition, many of the books that have been published are now out of print and consequently only available from specialist natural history or herpetological book dealers. The major work covering Kenya's reptiles at present is *A Field Guide to the Reptiles of East Africa*, by Spawls, Howell, Drewes and Ashe (A and C Black, reprinted 2008); this book has a description and map for every species, and a photograph for almost all, although it is looking slightly dated due to name changes and a few new species described in the last five years. A shorter version of this, including common amphibians was published as *A Pocket Guide to the Reptiles and Amphibians of East Africa*, by Spawls, Howell and Drewes (A and C Black 2006). Other books dealing with East African reptiles that are still in print include Bill Branch's *A Photographic Guide to the Snakes and Other Reptiles and Amphibians of East Africa* (Struik 2005), which is an inexpensive handy little book with simple maps. The photographs are excellent, Branch is a professional herpetologist and probably Africa's most talented herpetological photographer. However, many of the actual specimens illustrated were not photographed in Kenya, and thus do not always look like their Kenya representatives, and a number of prominent Kenyan species are not included, including Nairobi's most common snake, Battersby's Green Snake.

Copies of Norman Hedges' *Reptiles and Amphibians of East Africa* are usually available for sale in Nairobi. Published by the Kenya Literature Bureau (1983), this book has simplistic descriptions of a number of common East Africa species (54 snakes, 24 lizards, two crocodiles, six chelonians, sixteen frogs) illustrated by photographs, some good, some very poor. Many of the scientific names are well out of date. Hedges was an engineer and

Vivid Yellow-headed Dwarf Gecko Lygodactylus luteopicturatus, *found on the south coast. (Stephen Spawls)*

Right: Common in the highlands and often mistaken for a Green Mamba, the harmless Battersby's Green Snake Philothamnus battersbyi. *(Stephen Spawls)*

enthusiastic amateur herpetologist; while extracting a gecko from a rock crack at Kibwezi in 1973 he was stung by some paper wasps and tragically died within 15 minutes. Also available in Nairobi is Hugh Skinner's *Snakes and Us*, published by the East African Literature Bureau (1974) and reprinted many times, never with corrections. This a fairly dreadful book, although it has a few good pictures. A small handy book is Alex MacKay's *Poisonous Snakes of East Africa and the Treatment of Their Bites* (Innes-May publicity, 1985), listing all the dangerous East African snakes, with utilitarian descriptions, illustrated by line drawings by his wife, Joy.

The original major work on East African snakes, with much material relevant to Kenya, but now long out of print, is Captain Charles Pitman's *A Guide to the Snakes of Uganda*, a comprehensive scholarly description of all the

Ugandan snakes. This was first published in serial form in the *Uganda Journal* (1936–7), containing 23 beautiful coloured plates. Five hundred extra copies were bound into books, published in 1938, but half the stock, sent to a London warehouse, was destroyed by a bomb in the Blitz. Copies of this book are now worth several thousand dollars. A single revised edition was published in 1974 by Wheldon and Wesley; copies of this edition are worth several hundred pounds per copy. Pitman himself was an ex-Indian army officer, who saw service in the First World War, farmed in Kenya and was appointed chief game warden of the Uganda Protectorate from 1925 until 1949, save a brief stint as acting game warden of Northern Rhodesia (now Zambia) and Director of Security Intelligence in Uganda from 1941 to 1946. Pitman wrote two books about his time in Uganda, *A Game Warden among His Charges* (1931) and *A Game Warden Takes Stock* (1942). He acquired a reputation for ruthless efficiency with his elephant control programme, and he came in for a lot of criticism from Alistair Graham in his seminal book about conservation, *The Gardeners of Eden* (George Allen 1973), of which almost a whole chapter ('Valkyrie') is devoted to Pitman. Graham himself did major work on Kenya's crocodiles and his superbly illustrated book *Eyelids of Morning*, written with the controversial photographer Peter Beard, is a gem. It was published by the New York Graphic Society in 1974 and reissued by Chronicle Books in 1990.

The first complete listing of East African reptiles was Arthur Loveridge's 1957 *Checklist of the Reptiles and Amphibians of East Africa* (published as a Bulletin of the Museum of Comparative Zoology at Harvard, vol. 117, no. 2). This is still very useful; it gives basic distribution and all the synonyms. Loveridge produced over 200 major papers on African herpetology, including over a dozen thorough revisions of African reptile genera and families. He retired to St Helena in 1957 but continued to publish; his last paper appeared in 1979, a year before his death.

Some useful regional papers include Don Broadley and Kim Howell's 'A checklist of the reptiles of Tanzania with synoptic keys' (*Syntarsus* No. 1, 1991, pages 1–70), Stephen Spawls' 'A checklist of the snakes of Kenya' (*Journal of the East African Natural History Society*, Vol. 31, 1978) and 'An annotated checklist of the lizards of Kenya', by Stephen Spawls and Damaris Rotich (*Journal of the East African Natural History Society*, Vol. 86, 1997, pages 61–83). A single journal is devoted to African herpetology; it is called (as you might expect) the *African Journal of Herpetology*, published by the Herpetological Association of Africa.

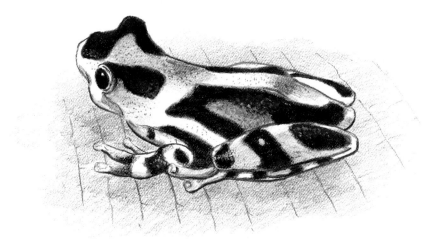

Chapter 9

The Amphibians

These foul and loathsome animals are abhorrent because of their cold body, pale colour ... filthy skin, fierce aspect, calculating eye, offensive smell, harsh voice ... and terrible venom; and so their Creator has not exerted his powers to make many of them. – Linnaeus, 1758, describing amphibians

Churamiti marididadi – scientific name of a bright red and yellow tree toad described in 2002 from the Ukaguru Hills in Tanzania; the first time a new species of animal has been described with a Swahili name, it means 'fancy tree frog'

Nairobi ... the home of frogs innumerable – Colonel Ewart Grogan, in his Foreword to *Farming and Planting in British East Africa*, 1917

The word 'amphibian' is derived from the Greek term *amphibios*, meaning two kinds of life, and refers to the fact that many amphibians live in water and must have water or moist places to breed in. Amphibians are vertebrates (although most frogs have only eight vertebrae, less than any other vertebrates); most lay eggs and gain their heat energy from their surroundings (ectotherms). Virtually all are carnivores, eating mostly insects when adult (one Brazilian frog eats fruit). There are three orders of amphibians: frogs and toads (Anura, meaning 'tailless ones'), salamanders and newts (Urodela, meaning 'tailed ones') and the worm-like caecilians (Apoda, meaning 'legless ones'). Amphibian-like creatures were the first vertebrates to colonise the land, 370 million years ago; to do this, they needed limbs, to overcome gravity. The oldest fossils of these orders are around 160–200 million years old. About 6,600 species of amphibians are known worldwide: roughly 570 salamanders and newts, 190 caecilians and more than 5,800 frogs. Kenya has about 100 species of frog and five species of caecilian. No salamanders and newts are known from sub-Saharan Africa.

This chapter first describes the origins, classification and adaptations of amphibians and then details the inventory, biology, zoogeography and ecology of Kenya's amphibians. A further section looks at amphibian conservation and the global decline in amphibians, and the chapter concludes with a discussion of the literature and a view of the history and current status of the study of amphibians. As usual each section is self-contained and can be read separately.

THE ORIGINS, CLASSIFICATION AND ADAPTATIONS OF AMPHIBIANS

As you might have guessed from the opening quotation, Linnaeus had a low opinion of amphibians. But few smell, and those that do smell largely pleasant. Their voices are sometimes harsh, but many are melodic, evocative of the African night. All have some degree of toxic skin secretions, but they are no danger to humanity, unless you choose to lick or bite them. Few look threatening and many, especially tree frogs, are appealing animals. And there are a fair number of them; they are a reasonably successful group. Good zoologist that he was, Linnaeus was misinformed about amphibians.

The wet, large-eyed, gulping frog that needs water to live in is how many of us picture amphibians; few people know exactly what a caecilian is. They are strange, elongate, limbless amphibians, five species of which are known from Kenya. They are usually found in moist forest soils and leaf litter, but one genus is known from the floodplains of large rivers. However, it is possible to see them on a road or track in the forest after a heavy rain when they have been flooded from their burrows. At first glance they can be mistaken for large earthworms but a closer inspection reveals eyes (though these are covered with skin) and a mouth. The body is marked by conspicuous rings, or annuli. They can move rapidly through leaf litter or soil. If held they feel slippery and squirm vigorously, producing copious mucus from glands under the skin. Like frogs, they eat invertebrates.

Frogs we know. Many of us will have kept tadpoles and watched them change into frogs. We may have experienced the nocturnal boom of toads from a nearby pool during the rainy season. At one time a standard school biology practical involved dissecting a frog, finding out what a living thing was like inside. Both authors had this experience. The humble *chura* is an important component of Kenya's fauna. Frogs feature in a number of legends; one involves the frog, rather like the chameleon, of which a similar story is told, taking a message to humans from God. The original message was that humans would not die, but the frog thought that humans would spoil the land and consequently changed the message, stating that humans would die. Another legend from Embu suggests that frogs lost their buttocks from constantly jumping in a competition. Although they need water or very moist soil to breed, frogs have radiated throughout the Kenyan landscape in a remarkable way, from coastal lowland to montane moorland and forest, from the wettest swamps to arid deserts. In dry country they live an astonishing life, adapted to breed rapidly in temporary pools that appear after rainstorms. As the pools form, the male frogs dig themselves out from under the soil, move to the water's edge and start calling, and their calls identify them. Females are attracted; other males come to join the fun. Mating occurs, with the male on the female's back, grasping her, a position known as amplexus. Females release eggs, the males release sperm that flows over them, fertilisation occurs. The eggs develop, tadpoles appear and metamorphose into froglets. They eat and grow like crazy, putting on weight before the pool dries out. Very few of Kenya's frogs actually live in rivers. As the land dries up, they bury themselves, and spend many months or even longer aestivating, buried in the soil or deep in vegetation, waiting for rain. The sudden appearance of hundreds of frogs after a big storm has prompted the legend that they have dropped from the sky (they have, of course, simply dug themselves out from underground, stimulated by the patter of raindrop)s. Frogs have also proved to be environmental indicators, their porous skins rendering them sensitive to pollution. They may be warning us of risks to our planet. They also eat a great many insect pests; they are the farmer's friends. There is more to frogs than meets the eye.

Anchieta's Rocket Frog Ptychadena anchietae, *a champion jumper. (Stephen Spawls)*

Amphibians form the second smallest class of vertebrate animals (mammals are the smallest), with approximately 5,800 to 6,600 living species divided into three distinct orders. The uncertainty about the species numbers is a reflection of both current and vigorous debate on the status of a number of 'species', and new discoveries. New species are being found all the time, particularly in countries with intensive research efforts, like Madagascar. The global number of species has gone from about 4,500 in 1992 to 5,500 in 2003, and is now thought to be close to 6,600. Controversy about amphibian species status is not new. Some amphibian species are hard to define. Unlike reptiles, frogs don't have scales that can be counted, and many types not only change colour as they grow, but have differing colour forms within the same population. In the Argus Reed Frog *Hyperolius argus*, found along the Kenya coast, males are yellow or green, females are rich brown with many bright cream or yellow blotches (Argus was a mythical Greek giant with 100 hundred eyes).

Intraspecific variation among some of Kenya's frogs, particularly the reed frogs, has caused problems with identification. Recently, the rise of molecular taxonomy, the use of frog calls to designate species and the changing concept of the 'evolutionary species' has caused a number of widespread species that vary across their range to be described as 'species-complexes'; subpopulations have then been split into 'good evolutionary species'. This is controversial stuff, and the status of some East African frogs – whether they are full true species, or subspecies, or 'forms' – has caused near-violent disagreement among frog experts. In southern Africa occurs the strange-looking genus called rain frogs *Breviceps*; they resemble meatballs to which the chef has added little legs and eyes. A new species of rain frog was recently described in Tanzania; it is identical to another species from the same area in every respect save its call. Some think that it is just a new call, not a new species. A helpful and, one would have thought, fairly innocuous website, 'The Amphibian Tree of Life', developed by a team of eminent herpetologists, listing the world's frogs grouped on a mixture of traditional taxonomy, molecular taxonomy and cladistics prompted a simply ferocious response from another eminent herpetologist, who called the entire system 'fatally flawed'. You can read about it in the literature section at the end of the chapter.

Female Argus Reed Frog, bright jewel of the coastal strip. (Robert Drewes)

The Gymnophiona (caecilians) and Anura (frogs) are widespread in tropical Africa, but the Caudata (newts and salamanders) are not found in sub-Saharan Africa. Modern amphibians are the descendants of the first terrestrial vertebrates, which appeared about 370 million years ago during the Devonian period. Some are highly specialised, particularly the frogs. The earliest amphibians may have evolved from a group of air-breathing, lobe-finned bony fish that lived in warm Devonian swamps. It seems likely that these fish sometimes dragged themselves onto land, possibly to avoid low oxygen concentrations in shrinking pools, perhaps to escape predators in the water, or to eat the insects and other invertebrates that had already colonised the land.

Fossil evidence shows that by the Triassic period (225 million to 190 million years ago) a rapid burst of evolution had produced a lot of different amphibians, some of enormous size (*Mastodonsaurus*, a huge amphibian with a head like a crocodile is estimated to have reached a length of 4m). But they didn't survive to dominate the world. Perhaps the niche that they occupied was taken by crocodiles; the huge ancient amphibians could not compete and died out, leading to a reduction in size and significance of the surviving amphibians. A second theory is that the huge amphibians began to reproduce in their larval stages, retained gills and thus were stuck with life in the water, a process called neoteny. A bold move, although ultimately unsuccessful, as no really large amphibians survive. This may indicate just how important it was to develop internal fertilisation and the shelled egg. Reptiles did that, and these two crucial developments meant they were no longer tied to water. They did dominate the earth. Although these archaic groups of huge amphibians all became extinct, their ranks must have included the direct ancestors of our living amphibians. It is generally believed that all modern amphibians (the lissamphibia, 'liss' means smooth) had a common ancestor, because they share many characters, but there is no fossil that proves it. All three modern orders first appear as fossils during late Triassic and early Jurassic times (about 200 million to 160 million years ago). The earliest known frog fossil is over 200 million years old, from the Triassic of Madagascar; some say this means that frogs originated in Gondwana, but the next frog fossil, 180 million years old, was from Arizona.

Male Argus Reed Frog, considerably duller. (Robert Drewes)

No fossil amphibians of any great age are known from Kenya, although a few frog fossils of still extant species (*Xenopus*, clawed frogs, and *Ptychadena*, rocket frogs) are known from bed one (over a million years old) at Olduvai Gorge, according to Mary Leakey, and the Brazilian herpetologist Ana Maria Baez has described a fossil pipoid frog (in the same group as *Xenopus*, flat aquatic frogs) from Mahenge in Tanzania. Frogs are generally delicate animals and do not fossilise as well as mammals. But fossil frogs may be waiting to be discovered. Recently a huge fossil frog 70 million years old from Madagascar was described, and out of respect for its huge size (over 40cm long and weighing, the collectors estimated, more than 4.5kg) it was named *Beelzebufo*, literally the 'toad from hell'.

The modern amphibians are ectotherms, their heat comes from outside, which means that (unlike endothermic birds and mammals) the body temperature is always very close to that of their surroundings. When this becomes too low for them to remain active, they become torpid and hibernate, in which state they are able to survive cold conditions for a long time. They tolerate low temperatures better than reptiles, thus they are more abundant at high latitudes and altitudes. There are eight amphibians in Alaska but no lizards or snakes; Sweden has two newts and ten frogs but only three lizards and three snakes. Amphibians have adapted well to life on land; they are now found in nearly every terrestrial habitat, from Arctic tundra to some of the driest deserts on earth and from sea level to altitudes over 5,000m on the Tibetan Plateau. This has been achieved despite constraints imposed by their inherited dependence on water and the fact that many of these habitats are inherently hostile to their existence. Amphibians have, to some extent, got out of water, but they still need it. We all do, of course, no living organisms on earth can survive without water (except viruses, but some scientists are prepared to define viruses as non-living). Mammals, birds and reptiles have developed internal fertilisation for life on land; amphibians largely survive without it. But if you have to have water for mating, where you can live – or more specifically, where you can mate – is limited. And amphibians have two other problems: their skin isn't waterproof (with a few odd exceptions) and their eggs don't have a shell.

There is an advantage from having a skin that is highly permeable to water and must be kept moist. It can absorb oxygen by diffusion; some frogs get half their oxygen via the skin. But it has to stay wet. And thus, once on land, water flows outwards, to keep the frog moist. Dehydration and death by evaporation from the skin poses a constant threat to a frog out of water, and this is why most frogs only come out at night, when it is cool and damp and rates of evaporation are low. Walk in the Arabuko-Sokoke forest by day, in the wet season, you might see a little frog or two. But come at night and you can be deafened by frogs.

Frogs that live in the dry savanna and semi-desert have also evolved a variety of adaptive features to help minimise evaporation, or to allow rapid rehydration following periods of drought. Put your average tree frog on an acacia branch beside a Kakuma waterhole and it will dry out in a few hours and die. Water escapes from their skin as quickly as from open water. Even in the cool of night, some frogs are known to lose over 25 per cent of body weight in the form of water while prowling. They rehydrate later. The eccentric English frog-hunter Martin Pickersgill once had some rare reed frogs escape from a bag in his Zambian hotel room. He had the clever idea of putting down a soaking wet cloth on the floor and by the morning, the frogs had made their way there; they can sense water and without somewhere wet to get to, they would have died. But foam-nest tree frogs, those agile inhabitant of the dry country, have a way of dealing with dehydration. They waterproof their skin, tuck in their limbs to reduce skin surface area, and turn white to reflect heat away, excrete solid uric acid; they control the little water they do lose to keep their body temperature down and their blood can carry oxygen at high temperatures. That is a marvel of adaptation; these are frogs that can live in semi-desert, waterproof frogs.

It is not the only highly adapted Kenyan frog. As the proverb says, 'necessity is the mother of invention' and Kenyan frogs have had to cope with the drying out of the land. (A number, no doubt, have also disappeared, but we will never know about them.) In Kenya, there are many colour varieties of the Common Reed Frog *Hyperolius viridiflavis*. They largely live in savanna and rest on plants during the day, so they are at risk of drying out. They too have some remarkable adaptations. The juveniles, which have a bigger surface-area-to-volume ratio than the adults and are thus at greater risk, have a skin that is more resistant to water loss than that of adults. They also remain inactive all day, limbs tucked in. They can slow their metabolic rate, and if the temperature gets high they release water stored in their bladder through the skin to cool themselves by evaporation. As with the Foam-nest Frogs, they can turn white. This reflective colour comes from their diet; waste food is converted to substances called purines which form platelets in the pigment cells of the skin, parallel to the surface and these reflect heat away. The inside of their legs is bright red; this was believed to be a startle colour, as it is only visible when the frog jumps. It may be so. But this red skin is also ultra-thin and rich in blood vessels, so it rapidly takes up water; a few drops of rain or dew falling on the back of the animal runs down to the thighs and belly and rehydrates the frog.

Amphibians have more diverse reproductive strategies than any other group of vertebrates and these are often related to the fact that their eggs, unlike those of reptiles, are surrounded only by a gelatinous capsule and have no protective shell. Such eggs must always be deposited in moist places, if fatal loss of water is to be avoided. And the eggs then turn into tadpoles, which largely have to live in water. No other vertebrates have a larval stage in their life cycle. Frogs in general lay eggs in water, which turn into tadpoles; a lot of eggs are laid and usually only a few survive. The eggs may sink or float, be free or attached to vegetation. The reed frogs can lay up to 700 eggs per clutch. Some species lay their eggs in open water but eggs can also be deposited in pockets of water in plants, tree holes or other little reservoirs. Bunty's Dwarf Toad *Mertensophryne micranotis*, of the Kenya coast lays eggs in snail shells or coconut shells. It also, rather astonishingly, has internal fertilisation. (Its English name honours the redoubtable Alice 'Bunty' Grandison, for many years the curator of herpetology at the Natural History Museum in London; Bunty spent some time collecting in Kenya.) Caecilians either give live birth or lay eggs in moist soil, the young hatch into miniature versions of the adult, without a larval stage.

Some frogs have most unusual reproductive strategies. The Common Squeaker *Arthroleptis stenodactylus*, a strange little frog of Kenyan forests and wet savanna, lays eggs in burrows or under leaf litter and the eggs develop into adults, there is no tadpole stage. The snout-burrowers of the genus *Hemisus* lay eggs in small holes they have dug near water, and when the hole floods the tadpoles swim to the pool. Some species of the leaf-folding frogs,

Foam-nest Frog Chiromantis xerampelina *turns white to reflect away heat. (Stephen Spawls)*

The Blunt-faced Snout Burrower, Hemisus. *(Michael McLaren)*

Afrixalus, lay their eggs on leaves above water and glue the leaf edges together, either to prevent desiccation or to hide them from predators. The African tree frogs, genus *Leptopelis*, bury their eggs inside holes which then flood; the tadpoles emerge and make their way across land to the nearest pool. Perhaps the most remarkable protective system is that used by the foam-nest frogs, and explains their name. A male positions himself on a branch above a pool and starts to call (only male frogs are vocal). Females arrive, the male clasps the female, and she produces a cloacal fluid that she beats up into foam with her back legs. Often other males arrive. The female produces eggs and the male releases sperm over them, the fertilised eggs drop into the foam. The foam dries quickly on the outside but inside it remains moist, leaving a white foam nest hanging above the water, looking like pale candy floss. These nests can often be seen above temporary pools during the rainy season in low country Kenya. The eggs turn into tadpoles inside, and after several days the bottom of the nest gives way (probably due to enzymes produced by the tadpoles) and the tadpoles fall into the water. The last act in this drama sometimes occurs when a wicked little robber waiting nearby, the Spiny Leaf-folding Frog *Afrixalus fornasinii*, which has been attracted by the calls of the foam-nest frogs, dives into the drying foam nest and eats the eggs.

KENYA'S AMPHIBIAN FAUNA: THE INVENTORY, BIOLOGY, ZOOGEOGRAPHY AND ECOLOGY

About 830 species of amphibian are known from sub-Saharan Africa, which is quite a rich fauna (although more than 1,800 are known from South America). Kenya has five species of caecilian and just fewer than 100 species of frog and as always these numbers depend on whose authority you accept. There are almost certainly a number yet to be discovered and named (this is probably true for most African countries). Regionally, Kenya has quite a rich amphibian fauna, but it is lower than it might have been on account of the large areas of semi-desert and desert in northern and eastern Kenya, where few amphibians are found. Uganda has about 70 species of amphibian, Ethiopia only 63 (despite being over twice the size of Kenya). Tanzania has more than 140 species of frog (and nine caecilians), a reflection of its larger size and huge number of endemics associated with the Eastern Arc Mountains. Only about 35 species of frog are known from Somalia, as it is a highly arid country, although the lowest amphibian species count in all of Africa is Libya, with only five taxa. Central and West Africa has the continent's richest frog fauna. The Democratic Republic of the Congo has around 210 species and Cameroon just over 200, including 51 endemics. Research around Mt Nlonako, in Cameroon, indicates that more than 90 species of frog are known from the area, and other national parks in West and Central Africa have yielded lists of 35–55 frog species. Nowhere in Kenya approaches this sort of species richness. Those high values for West and Central Africa are indicative of an area with lots of rainfall that has not lost and regained its forest in the last few million years, i.e. it was a refugium and was stable, permitting speciation.

As we saw with reptiles, there is a general trend in Kenya for species numbers to decrease with altitude, but this is also affected by rainfall, the availability of open water and the presence of forest. Some Kenyan amphibian values are as follows: Arabuko-Sokoke Forest 24 species, Taita Hills 26 species, Kakamega Forest 25 species, Voi 23 species, Nairobi National Park seventeen species, Meru National Park thirteen species, Limuru nine species, Wajir seven species and the Aberdare moorlands seven species. If you consider the Arabuko-Sokoke Forest, Nairobi, Limuru and the Aberdares, a clear decline in numbers with increasing altitude is seen. Kakamega is at similar altitude to Nairobi, but its increased species richness is due to Central African forest forms, Wajir is very low because it is a very dry area with no permanent water. A study at a waterhole near Bushwackers Camp, Kibwezi, found ten species active around the pond at night.

About 160 species of caecilian are known worldwide, with a disjunct distribution in the tropics. They are found in tropical Africa, South America, the Seychelles and South-east Asia, arguing a Gondwana origin. Two families occur in East Africa, one in Kenya, the Caeciliidae, tailless caecilians. Essentially, caecilians look like worms (see page 309), and they live in the same sort of places that worms do. If you pick up a caecilian, you will find it is

Robber of Foam Nest Frog eggs, the Spiny Leaf-folding Frog, Afrixalus fornasinii. *(Robert Drewes)*

slimy, and as it struggles to escape it produces more and more slime. The Kenyan species are large for caecilians, reaching 30–40cm, with a fairly thick body and rounded head and tail. The cloaca opens transversely. Three of the five Kenyan species are endemic. The Mud-dwelling Caecilian *Schistometopum gregorii*, occurs on the Tana River Delta floodplain and also in coastal Tanzania; it is black or brown. The three Kenyan endemic caecilians all belong to the genus *Boulengerula*: one lives in the Taita Hills, one on Sagalla Hill near Voi and the third in Ngaia Forest near Meru. The fifth, also in *Boulengerula*, lives at Changamwe and in the Shimba Hills and is also found in Malawi (but oddly, not in Tanzania, at least not yet). The species found in the Taita Hills (*Boulengerula taitanus*) is probably the easiest to find, should you desire to see a representative of an ancient group; they can be located by digging in moist soil in shambas, although the most effective way is to go into one of the remaining forest patches, like Ngangao, and look under or in a fallen log.

Some twelve families of frog (out of a world total of 42, recently elevated from 32, although there is much debate on the status of some families) occur in Kenya: they are the squeakers (family Arthroleptidae), toads (Bufonidae), snout-burrowers (Hemisotidae), tree frogs (Hyperoliidae), narrow-mouthed frogs (Microhylidae), clawed frogs (Pipidae), regular frogs (Ranidae), rocket frogs (Ptychadenidae), puddle frogs (Phrynobatrachidae), bullfrogs (Pyxicephalidae), torrent frogs (Petropedetidae) and foam-nest frogs (Rhacophoridae). The largest family is the tree frog family (Hyperoliidae), with about 30 species found in Kenya. Three of the families are represented in Kenya by only two species in each: the squeakers, snout-burrowers and clawed frogs.

The Hyperoliidae, the African tree frogs, are particularly interesting. It is a large group, with eleven genera, of which three are found in Kenya: leaf-folding or spiny reed frogs (*Afrixalus*), running frogs or kassinas (*Kassina*) and reed frogs (*Hyperolius*). Sadly Kenya misses out on the tree frogs of the genus *Phlyctimantis*, although they are known from Uganda and Tanzania, as they rejoice in the onomatopoeic common name of 'wot-wot', from their advertisement call. The frogs of the Hyperoliidae family are mostly small, but many are brightly coloured, some astonishingly so, with vivid reds, yellows and greens. They are climbing frogs (although a few species have reverted to ground life). Some tree frogs are difficult to identify. Some of the reed frogs have simply huge ranges, others have minute ones. The little green Sharp-nosed Reed Frog *Hyperolius nasutus*, occurs virtually throughout the savannas of tropical Africa (and some experts are in the process of splitting it). But David Sheldrick's Reed Frog *Hyperolius sheldricki*, is known only from Aruba Lodge and the eastern Galana River; the Maasai Reed Frog *Hyperolius orkarkarri*, is known only from Gong Rock and Galen's Drift in the Serengeti. These frogs are small animals, with short lifespans, and the climatic changes that have taken place in East Africa have isolated, both temporarily and permanently, many populations. Such little creatures have many enemies, in particular birds, snakes, crocodiles and fish, so their survivability depends on being hidden. These are all pressures that can cause changes in populations, particularly with a limited gene pool; this is evolution in operation.

A classic example is the so-called 'common reed frogs', of the *Hyperolius viridiflavis/glandicolor* species complex. Although these frogs share a common ancestor, even the finest experts cannot agree on the scientific name, and representatives of the group occur virtually throughout the savannas of sub-Saharan Africa. At least six quite different colour forms occur in Kenya, including a pale, vividly red-legged form at Nairobi, a yellow-spotted bright green form in the west and a grey and yellow speckled one in Tsavo. Despite strenuous efforts, no herpetologist has yet managed to make a key that will infallibly identify the frogs of this genus; there is just too much variation. Alan Channing and Kim Howell's book *Amphibians of East Africa*, (2006) has keys for every genus – except *Hyperolius*. Even those professionals were defeated by the reed frogs. The Danish expert on African treefrogs, Arne Schiotz, reckons there are at least 30 varieties within the *viridiflavis/glandicolor* group, but suggests that to give them all species status would cause taxonomic chaos. What is particularly interesting is that, to the naturalist in the field, these frogs with their bright colours and distinctive calls can be reasonably easily grouped. But if you put them into a jar of alcohol the colours and calls are gone. Anyone who wants to make sense of the species needs to spend time with them in the field. But these gorgeous little frogs illustrate something profound about evolution – that it isn't a static thing. Taxonomy is our human attempt to create order; we want to know what groups exist, give them a name and say where they live. With a population that breeds rapidly and are tied to surface water in a dry land with a changing climate, isolation is going to occur, small gene pools are going to drive changes and these changes may lead to sexual incompatibility between subsets of the population – hence speciation. We may not be able to pin all the populations down with names, but it's possible to have fun trying (if you have the funds!).

The Common Reed Frog has another remarkable trick up its sleeve. Some females have been known to change into males. In a laboratory study involving a variety from northern Tanzania, brown with yellow speckling, seven females changed sex. They laid eggs, but then began to sit on leaves in the position that males adopt when they are about to call. At first they could not produce any sounds (only male frogs have vocal sacs) but they then developed vocal sacs and began to produce calls and fight with other males. Sex change is a useful strategy in populations with not enough males; a number of fish show this trait, especially coral reef species. If the dominant male is killed, his lead female becomes a male.

Some eighteen or so species of *Hyperolius* occur in Kenya, out of about 140 (at present) in sub-Saharan Africa. This is a lot, although it pales beside the South American genus *Eleutherodactylus*, rain or tink frogs, of which some 700 species are known; the tink frogs are the most speciose genus of vertebrates on earth. But *Hyperolius* are pretty diverse. One species, the Mountain Reed Frog *Hyperolius montanus*, is endemic and specimens can be found on top of the Aberdares, at over 3,000m altitude. It is also found at lower altitude, at Maua in the Nyambene Hills, at Molo and on the Mau; in those places it is active by night, but nobody knows what happens on the top of the Aberdares. It freezes up there, and you wonder if this frog becomes active by day. No other reed frog is diurnal. They can supercool; you can find them inactive on plant stems at dawn beside frozen ponds.

Some six species of small pretty leaf-folding frogs, *Afrixalus*, are known from Kenya. Frogs of this genus have a curious diamond-shaped pupil, but otherwise look quite similar to reed frogs. *Afrixalus* are often striped and are creatures of moist country. Their calls (where known) are a mixture of buzzes, cricks, creaks and clacks; the call of that robber of foam-nest frog eggs, the Spiny Leaf-folding Frog, has been likened – perhaps appropriately – to that of a tiny machine gun. Curiously, *Afrixalus* are virtually absent from the central highlands; all Kenyan forms

Green form of the Common Reed Frog. (Stephen Spawls)

Vividly coloured, infinitely variable and able to change its sex, Northern Tanzanian form of the Common Reed Frog, Hyperolius viridiflavis. *(Robert Drewes)*

The Kikuyu seer Mugo wa Kibiru prophesied the arrival of strangers with skin the colour of the Banana Frog; meaning the endemic Mountain Reed Frog Hyperolius montanus *(Stephen Spawls)*

occur on the coast or in the west. There are three species of kassinas or running frogs *Kassina*, in the Hyperoliidae. As their common name suggests, these frogs don't jump. The Senegal Kassina *Kassina senegalensis*, nicely striped, occurs throughout the country, save the mountains, and has a beautiful call, a rising 'quoip', sounding like a drop of water falling into an upturned bell. The skin of the Red-legged Running Frog, (*Kassina maculata*) which is found on the coast, produces highly toxic secretions that lower the blood pressure, elevate the heart rate and stimulate flow of pancreatic juice; they combine to make anything that eats this frog violently sick. Predators thus learn to avoid it, although too late for the first victim (but his friends benefit). The third species of running frog, the Somali or Spotted Kassina *Kassina maculifer* is known in Kenya only from Wajir. The first Kenyan specimens were found by Bob Drewes in a rainwater tank at the house of John Miskell, the expert on Somali birds, who was teaching in Wajir at the time.

The family Arthroleptidae is represented in Kenya by two species of squeaker, *Arthroleptis*, little forest frogs, and six species of tree frogs of the genus *Leptopelis*. These tree frogs look something like the reed frogs but are larger, with a vertical pupil. The one species that lives in central Kenya is called Bocage's Burrowing Tree Frog *Leptopelis bocagii* which, as its name reveals, does not live in trees! It is quite common around Nairobi and may be found on wet nights, hopping across roads. *Leptopelis* are rarely found above 2,000m altitude in most of Africa, save in Ethiopia.

Everybody knows toads: warty brownish amphibians with lovely golden eyes. They live more on land than water and can produce a white poisonous skin secretion from the parotid glands behind their eyes. Toads are familiar characters in folk tales and children's stories: the bumbling Mr Toad of Toad Hall plays a major part in Kenneth Grahame's children's classic *The Wind in the Willows* (1908). In some African legends the toad plays the part of a reliable but slow character. Over 400 species are known, of which some 230-plus were originally placed in the genus *Bufo*. This is – or was – a worldwide genus, found on every continent and absent only from Madagascar and New Zealand. In 2006, a team lead by the American herpetologist Darrel Frost split the genus dramatically, as part of their major project, 'The Amphibian Tree of Life', and all Kenya's toads were assigned to a new genus,

Senegal Kassina Frog, its melodious 'quoip' adds magic to the night. (Robert Drewes)

Dull toad from the dry land, the Desert Toad, Bufo xeros. *(Stephen Spawls)*

Amietophrynus, named after Jean-Louis Amiet, a French herpetologist and entomologist who researched and taught in Cameroon for a number of years and described more than 30 species of frog. Some herpetologists are not happy with the name change, however.

African toads tend to have deep booming calls, and have kept many people awake at night. Several toad species are quite amiable about urbanisation; if you find an amphibian in a garden in the suburbs of Nairobi, Eldoret, Mombasa or Nakuru, it's quite likely to be a toad. If there's a dripping tap outside, odds are that you might find a toad there if you look after dark. Kenya has about fifteen species of toad, including the Kisolo Toad *Amietophrynus kisoloensis*, the males of which turn bright yellow in the breeding season, the Nairobi Toad, *Amietophrynus nairobiensis* (or *Bufo nairobiensis*, take your pick) which was described in 1932 by Arthur Loveridge and has never been photographed alive, and the Lake Turkana Toad *Amietophrynus turkanae*, found only there. Several Kenya toads were originally regarded as subspecies of the Leopard or Square–marked Toad, *Bufo regularis*, and indeed they all look similar – medium-sized warty brown toads, with a body 8–10cm long, the back marked with darker paired blotches on either side of the spine. The group is now seen as a species-complex, which it probably is, with discrete populations accumulating changes.

The family Ranidae are known as common frogs – and common some of them are. There used to be about 36 forms in Kenya, in the single family. Now Frost and his team have split them into four families. The family Pyxicephalidae includes Kenya's biggest frog, the African Bullfrog *Pyxicephalus adspersus*. It can weigh over a kilogram (although Africa's biggest frog, the West African Goliath Frog *Conraua goliath*, can weigh a mighty 3kg). The generic name *Pyxicephalus* means 'box-head' and the English name comes from their call, a deep low that sounds like a bull. Its distribution in Kenya is poorly understood. It lives and breeds in temporary pools, and can aestivate for a long time; in southern Africa specimens have been found emerging in a heavy storm, having not been seen for 7 years. They are unusual frogs in several ways. On their lower jaws are two razor-sharp projections, not at all what you want embedded in your thumb. They fight, dominant males attack other males to prevent them mating with nearby females. The adult male shows parental care, digging channels to allow tadpoles in a drying pond to reach another. The newly metamorphosed young will eat anything that moves, including others of their own kind. Adults have been known to eat fair-sized snakes; one ate seventeen young Rinkhals *Hemachatus haemachatus*, a South African spitting cobra. A smaller species of the same genus, the Edible Bullfrog *Pyxicephalus edulis* occurs along the Kenya coast and in Tsavo, and is eaten by some people. Another 'new' family split out of the Ranidae, the Ptychadenidae, includes the rocket frogs, *Ptychadena*. They have sharp noses and ridged backs (some authors call them ridged frogs, and some, grass frogs). Eight species occur in Kenya. These frogs are tremendous jumpers, able to cover 2m or more in a single leap. They often occur alongside rivers and lakes and can startle walkers by suddenly making a huge leap from under their feet. They really do 'go like rockets', sharp nose forward, low trajectory and everything streamlined, so unlike the reed and tree frogs, which loop upward with limbs out, ready to grab. At the annual Calaveras County jumping frog competition in California, (a competition started as the result of a famous Mark Twain story), the record was once held by a rocket frog entered by Jonathan Leakey, and the all-comers record is still held by a Sharp-nosed Rocket Frog *Ptychadena oxyrhynchus*, which jumped 9.83m in three leaps! The family Phrynobatrachidae includes the puddle frogs, *Phrynobatrachus*. These are little brown, rather warty frogs, very rarely larger than 4cm in length, although a 'giant' species with a body 5cm long, endemic to Kenya, was recently described from the Irangi Forest on the south-eastern slopes of Mt Kenya. About 10 species are known in Kenya; there are probably more. These little frogs will call during the day during the rainy season; the snoring sound you might hear as you approach a flooded grassland or reed bed during the day will be due to these little frogs. Over 85 species are known from sub-Saharan Africa, and some are very difficult to identify. DNA analysis helps but the exact identity of Africa's puddle frogs is tying up some bright herpetological minds and funding at the moment. (One might wonder if it's worth it.) The family Petropedetidae contains one of Kenya's most enigmatic (and possibly now extinct) frogs. In 1934, at around 7,200 feet altitude on the Koitobos River on the eastern slopes of Mt Elgon, Dr C. A. du Toit of Stellenbosch University collected a little brown torrent frog; it was later described by Arthur Loveridge as *Arthroleptides dutoiti*, some authorities call it *Petropedetes dutoiti*. It has not been seen since 1962, although several groups have carefully searched for it.

Vivid Garman's Toad Amietophrynus garmani *from Naivasha. (Stephen Spawls)*

Krefft's Warty Frog, Eastern Arc endemic that just reaches Kenya in the Taita Hills. (Robert Drewes)

Red-banded Rubber Frog, with venomous skin secretions and a enchanting trilling call. (Robert Drewes)

Other Kenyan frogs include three species of foam-nest frog, a couple of weird warty frogs of the genus *Callulina*, from the Taita Hills (which resemble E.T.), two snout-burrowers, species of the genus *Hemisus*, little frogs with pointed snouts that dig holes head-forwards (most frogs, if they dig at all, do so with their hind legs), and the Red-banded Rubber Frog *Phrynomantis bifasciatus*. This vividly marked frog of the moist savannas has toxic skin secretions that, if they get into a cut or onto grazed skin, can cause unpleasant symptoms including elevated heart rate, difficulties in breathing and nausea. A coast resident found one of these frogs in their shoe, but only noticed it after walking some metres; the venom entered abraded skin on his foot and caused intense irritation. These frogs have an astonishing call – a long melodious trill like an electronic warning buzzer.

We have saved to the end what may be Africa's most famous genus of frog, the clawed frogs *Xenopus*. The name means 'strange foot'; they have black claws on three of the five toes, used for defence and for tearing at large prey. Clawed frogs are smooth-skinned, flattened frogs with long powerful legs, strongly webbed feet and tiny eyes on top of their heads. On their flanks are stitch-like sensory organs that can detect pressure differences. As you might guess, they spend most of their lives in water, although they sometimes emerge en masse and move to fresh pools; a group of more than 1,000 were observed near Iten, at the edge of the Kerio Valley, moving across country on a wet season night.

Clawed frogs belong to the family Pipidae, and representatives of this family live in Africa and South America only. Since frogs cannot disperse across salt water, this argues that the family originated on Gondwana and was separated when the land masses moved apart. This led to an interesting scientific spat about the use of the molecular clock, as initial calculations based on their differing DNA indicated that the split between the South American and African pipids occurred less than 80 million years ago, that is after the continents separated … which would mean that they then swam from Africa to South America (or vice versa), an unlikely proposition. Later work with the DNA indicated a much earlier split, fortunately; cynics may recognise that

The Amphibians

The strange warty Maasai Reed Frog, Hyperolius orkarkarri, *described from Gong Rock in the Serengeti. (Robert Drewes)*

famous scientific principle, that your results might as well fit in with the irrefutable evidence! But *Xenopus* are famous for another reason: the pregnancy test. In the 1930s it was discovered that if a few drops of urine from a pregnant woman was injected under the skin of the female of the Common Clawed Frog *Xenopus laevis*, the frog began laying eggs. For more than 20 years, this was the standard test for pregnancy, until immunoassay tests became available. Many hospital imported clawed frogs from Africa, and it was soon found that it was an excellent laboratory test animal, usable for a variety of experiments, as it was hardy, could live under a variety of conditions and bred easily. It remains so and, unfortunately, it has managed to escape and survive to form feral breeding colonies in a number of places, including such cold and wet locations as Wales. In 2004, the South African zoologist Che Weldon indicated that the species may also have been responsible for the spread worldwide of the dangerous chytrid fungus.

The patterns of distribution of Kenya's frogs indicate zoogeographical affinities essentially similar to those of the other vertebrates, and, as always, closely tied to Frank White's (1983) botanical regions. The largest group of frogs, more than 30 species, are members of the Zambezian savanna fauna, centred on the savannas of southern, south-eastern and eastern Africa. This group includes the African Bullfrog, Red-banded Rubber Frog, Southern Foam-nest Frog, and Bocage's Burrowing Frog. There is also within Kenya a large group of Pan-African frogs, found virtually throughout the savannas of West, Central, East and Southern Africa; members include the Senegal Kassina, Sharp-nosed Reed Frog, Guinea Snout-burrower and the huge Groove-crowned Bullfrog. Another reasonably large group are those frogs typical of the East African coastal mosaic, coming up from Mozambique; Kenyan examples include the Argus Reed Frog, Tinker Reed Frog, Red-legged Kassina, Ornate Tree Frog, Edible Bullfrog and Bunty's Dwarf Toad. There is a small group of forest frogs, associated with the great Central Africa forests, Guinea–Congolian species, which just reach Kenya in the west. These include the Forest White-lipped Frog *Amnirana albolabris*, and Osorio's Leaf-folding Frog. But this is a small group, unlike the forest snakes, of which a fair number occur in western Kenya, like the Rhinoceros Viper and two green bush vipers. Likewise, there are a lot of Somali–Maasai reptiles in Kenya, as that arid landscape suits them, but only a handful of Somali–Maasai frogs, examples being Keller's Foam-nest Frog *Chiromantis kelleri* the Somali Kassina and the Big-eared Frog (sadly there is no Noddy's Frog) *Hildebrandtia macrotympanum*. A couple of toads, the Desert Toad *Bufo xeros*, and the Square-marked Toad, associated with the desert north, reach northern Kenya. There are a few Kenyan frogs that originate from Sudan savanna (Balfour's Reed Frog *Hyperolius balfouri* and Steindachner's Toad *Bufo steindachneri*), and in the Taita Hills is a frog typical of the Eastern Arc mountains, Krefft's Warty Frog *Callulina kreffti*.

There is an interesting group of endemic amphibians in Kenya. The high land of eastern Africa, along the rift valleys, has a number of endemic species. Amphibians are quite tolerant of high cold conditions provided they are wet, unlike reptiles. East African montane endemics include the Kisolo Toad and the Kerinyaga Toad *Amietophrynus kerinyagae*. Endemics confined to Kenya include 14 frogs and three caecilians. Ten of the endemic frogs live in the central highlands. Some have fairly large ranges, for example De Witte's River Frog *Afrana wittei* is known from the Mau, Aberdares and Mt Kenya, the Silver-bladdered Reed Frog *Hyperolius cystocandicans*, occurs in the Nyambene Hills, on the high land between Mt Kenya and the Aberdares, south to Limuru. But Du Toit's Torrent Frog is only on Mt Elgon and the Kinangop Puddle Frog *Phrynobatrachus kinangopensis* is only on the Kinangop plateau. The other endemics occupy a disparate range of localities: Mackay's Tree Frog *Leptopelis mackayi* is in the Kakamega forest, David Sheldrick's Reed Frog is in Tsavo East, and the Shimba Hills Reed Frog *Hyperolius rubrovermiculatus*, the Lake Turkana Toad and the Nairobi Toad are named for where they live. The three endemic caecilians live on Mt Sagalla, the Taita Hills and the Ngaia Forest below the Nyambene Hills.

AMPHIBIAN CONSERVATION AND THE GLOBAL DECLINE IN AMPHIBIANS

Over the last twenty years, amphibians have increasingly been seen as bellwethers of environmental deterioration, as they are sensitive to changes in their surroundings. This is because they have constantly wet skin. Dissolved gases, liquids and some dissolved solids enter into their bodies by both diffusion (fluid flow from high to low concentrations) and active transport (movement against the concentration gradient). If something nasty is dissolved in the water they live in, or is present in the air and dissolves in water, the frog has a chance of absorbing it, which makes it sensitive to pollution; other land vertebrates don't absorb very much through the skin. In addition, amphibians' constantly wet skin and relatively warm body provide ideal breeding conditions for fungi and bacteria.

Many amphibians also need wetlands to survive, and wetlands are anathema to developers and farmers; you can't build anything on or very near to a wetland, or grow much except rice. Wetlands also encourage insects, particularly mosquitoes. As humanity needs more and more living space, food, energy and possessions, wetlands are drained, land is impounded, buildings are put up and the air and water become polluted. The presence or absence of amphibians, and changes in their populations, may be useful tools to measure how we are damaging our planet. And amphibians are under threat. In the last 25 years, some nine species of frog have become extinct, and another 50-plus haven't been seen and may be extinct. Three of the most spectacular are the Golden Toad of the cloud forests of Costa Rica, not seen since 1989, and the two species of Australian gastric brooding frogs, which were unique among amphibians in that the eggs were brooded and tadpoles raised in the mother's stomach. The cause of amphibian declines and extinctions is much debated, and has spawned several books on the topic.

Extinction in Our Times: Global Amphibian Decline, by herpetologists James Collins and Martha Crump (2009), put virtually all the blame on the chytrid fungus. This fungus *Batrachochytrium dendrobatidis*, was first identified on the skis of dead and dying frogs in Australia in 1993, but more recent research has shown that it has been around for a good deal longer than that. As we mentioned, the researcher Che Weldon believes it was spread across the world by exports of clawed toads from Africa for the laboratory trade, although not all scientists

Undescribed riches; an unknown caecilian recently found in the Ngaia Forest, near Meru. (Stephen Spawls)

Endemic caecilian from Ngangao Forest, Taita Hills, Boulengerula taitanus. *(Stephen Spawls)*

Kenya's most vivid amphibian, the Shimba Hills Reed Frog, Hyperolius rubrovermiculatus. *(Stephen Spawls)*

accept this. There seems to be no effective treatment of the fungus, and no one knows exactly how it kills, although affected animals lose keratin from their skin and become disorientated. The problem has given rise to a conservation initiative called 'Amphibian Ark', in which zoos take in frogs and set up a breeding colony, isolated from the fungus, until hopefully someone finds a cure. A desperate measure; frogs belong in wetlands. But things may not be so bad. A team of German and Scandinavian herpetologists recently sampled 860 frogs from eleven sites across Kenya, from the Shimba Hills to the Kakamega Forest. The chytrid fungus was found at all sites. On average 30 per cent of frogs were affected, a total of 271, but it varied a lot, from 6 per cent infected in the Shimba Hills to over 70 per cent on the Aberdare moorland and near that value at Nyahururu. Among species, the figures varied strangely: 40 per cent of reed frogs were infected, but none of the *Leptopelis* tree frogs. But no dead or dying frogs were found and no declines have been reported; so it seems that Kenyan frogs can live with the fungus … so far. A dismissive review of Collins and Crumps' book in *Nature*, by herpetologists Alan Pounds and Karen Masters (2009), indicates that disease shifts are down more to global warming and environmental problems than to chytrid fungus. They state that amphibian declines are part of a chorus of canaries, all indicating that the earth's life-support system is in trouble. This is, at present, a very contentious topic; it has to be said, however, that most field scientists agree with the proposition.

Some monitoring of the effect of global warming is being carried out by teams from the department of herpetology at the National Museum, but little money is available. Monitoring of amphibians is important. Most of our amphibians like wet places. The amphibians of Wajir and Amboseli can probably cope with a bit of drying out, while those of Kakamega or the Cherangani Hills may not be able to (not that anyone knows what amphibians are in the Cherangani Hills). The extensive wetlands set up for rice growing in parts of Kenya are probably good news for frogs, provided that the farmers don't use herbicides like atrazine, which sterilises some male frogs and turns others into females. But the damming of rivers is a complicated business. A permanent reservoir of fairly static water is created, but this is liable to pollution, and it floods suitable habitat, as happened in China with the Three Gorges Dam and the Chinese giant salamander. Holding back rivers has its risks. A frightful amphibian extinction took place in Tanzania quite recently. In 1996, Kim Howell, professor of zoology at Dar es Salaam, and Dave Moyer, a Tanzanian zoologist, found a beautiful little orange-yellow toad in a unique wetland at the base of the Kihansi Falls where the river comes down from the Udzungwa Mountains. It was recognised as a new species and described by a team lead by Dr John ('JC') Poynton, high priest of African frog studies. They named the frog *Nectophrynoides asperginis*, the Kihansi Spray Toad. Kim Howell estimated that there were about 8,000 of these toads, all living in a tiny area kept wet by spray from the falls … and nowhere else, not even the surrounding moist woodland. Alas, a dam was being built above the falls, which reduced the amount of water coming down by more than 80 per cent. A sprinkler system was put in place to replace the spray from the falls, in 2003, but not long after that the population crashed and the toad was extinct in the wild by 2009. It was suspected that the toads died of chytrid fungus, brought in on the boots of conservationists, but no one is sure. A number of Kihansi Spray Toads, fortunately, had been taken to zoos in the United States and it is hoped that in the not-too-distant future they will be reintroduced. So far, no such extinction has taken place in Kenya … that we know of.

There may yet be little colonies of some unknown frog or caecilian, nestled away in Kenya's hilltop forests. One thing is for sure: if the developers get there first, the odds of survival of a creature with very precise habitat requirements are going to be reduced. That is why it is important to survey Kenya's vulnerable wild places, especially those with water. As humanity takes more of the land, what happens to its original inhabitants? Kenya has two major eastward-flowing rivers. Water is extracted from them for agriculture, which, in itself, is a good thing. But if the flow of water is reduced, it will reach fewer areas downstream when it floods. What is happening to the amphibians that certainly bred in those areas? Early herpetologist found many wonderful reptiles and amphibians in the Tana Delta but some have never been seen since. Do we know what's happening to them? The problem might be complicated by the fact that large areas of the delta are now being leased to foreign countries. One big problem in Kenya is that huge areas have never been initially surveyed, let alone monitored. The Athi River is fed by the Nairobi River but we do not know what industrial chemicals are making their way downriver, or what is happening to the small rivers that feed our rift valley lakes, or Lake Victoria, or what is the effect of pesticide run-off into rivers. In parts of the world (although not as yet in Kenya) there have been increases in numbers of deformed amphibians, often with extra limbs. Chemical pollution has been suggested as a cause, along with increasing ultra-violet radiation, predation and parasites.

Whatever the external factors affecting Kenya's amphibians, the crucial challenge at the moment is discovery. We need to know what is there, and then numbers can be monitored. Kenya has a good network of protected areas, but for only four – Arabuko-Sokoke, Meru National Park, the Aberdares and Kakamega Forest – is there anything even approaching a comprehensive list of amphibian species. It isn't difficult to survey amphibians and Kenya's scientists shouldn't have to plead for money to document the country's richness of biodiversity. You can't conserve without knowing what you've got.

THE LITERATURE AND HISTORY OF THE STUDY OF AMPHIBIANS

Amphibians are a major vertebrate class, and consequently have generated a fair amount of scientific literature. Solid background material is provided in William Duellman and Linda Trueb's *Biology of Amphibians* (The Johns Hopkins University Press, 1986 but reprinted 1994). Excellent coverage of all aspects of amphibians and their biology, including all African groups, is found in a sound herpetological textbook, *Herpetology* by George Zug, Laurie Vitt and Janalee Caldwell (Academic Press 2001). Both these books are pitched at undergraduate and postgraduate level. Taxonomists who want to know the most up-to-date details on all the world's amphibians and their status should consult Darrel Frost *et al.*'s classic paper 'The Amphibian Tree of Life', published as Bulletin No. 297 of the American Museum of Natural History in 2006, and available also as a free online reference as Amphibian Species of the World: an Online Version, continuously updated at http://research.amnh.org/herpetology/amphibia/index.php/. All African frogs are listed here. The original paper is not without its critics, as we mentioned. John Wiens, a herpetologist at Stony Brook University, in a caustic review of 'The Amphibian Tree of Life' in the *Quarterly Review of Biology* (vol. 82, March 2007), called it 'a disaster', suggested that herpetologist 'simply ignore the study' and implied that it was fatally flawed, largely on the basis of some of the molecular taxonomy used. The authors then refuted Wiens' arguments in the journal *Cladistics* (vol. 24, 2007). It was strong stuff; among other things Frost's team suggested that Wiens had deliberately held back details of a publication he had authored supporting the opposition's findings.

Amphibian Conservation, edited by Raymond Semlitsch (Smithsonian Books 2003) is a wide-ranging scholarly summary of the global decline in amphibians. Useful websites include amphibiaweb.org, which is user-friendly and easily searched; you can choose a country and get a listing of all amphibians known from that country, or choose a genus, and get all species, and so on. Unfortunately some of the African material seems to have been drawn up by someone without field experience and, among other things, the same species is listed under both old and new names. Also very useful is Dr Breda Zimkus's website, africanamphibians.lifedesks.org, which is constantly updated by a team of professionals who really know African amphibians.

For Kenya, the single most important book is *Amphibians of East Africa* by Alan Channing and Kim M. Howell, published by the Cornell University Press in 2006. This 418-page comprehensive guide to the amphibians of Kenya, Tanzania and Uganda describes all of East Africa's 194 frogs and nine caecilians; 146 (just over 70 per cent of the total) are illustrated by colour photographs and there is a distribution map for each species. Most of the book consists of species descriptions but it includes a fascinating essay on the history of amphibian studies in East Africa and a ten-page section on conservation. The authors are hard-core veterans of the African amphibian world. Kim Howell has worked at the University of Dar es Salaam for more than 30 years and is currently professor of zoology there. Alan Channing, from the University of the Western Cape in South Africa, has been publishing on African amphibians for more than 35 years. This book is slightly biased towards Tanzania, and the maps in it are somewhat simplistic. For example De Witte's River Frog *Afrana wittei*, for example, is said to be 'from the Kenya highlands', and there is a single dot on the map; in fact this species is known from the Mau, Aberdares and Mt Kenya. However, there is no better comprehensive amphibian text for Kenya. It is a bit heavy for the field (nearly 1kg) and the pictures are bound together in a block at the centre, but there is no other book that gives complete coverage.

Anyone who has talked to naturalists knows that what they want in a field guide is good colour illustrations. Channing and Howell's book fits this criterion, although some of the pictures are rather small and dark. Two other books show a good selection of colour photographs of Kenyan amphibians. One is *A Pocket Guide to the Reptiles and Amphibians of East Africa* by Stephen Spawls, Kim Howell and Bob Drewes (published by A and C Black in 2006), which has colour photographs and distribution maps for 85 prominent East African amphibians and descriptions of 110 more. The other is Arne Schiotz's *Treefrogs of Africa* (Edition Chimaira 1999), which covers all the frogs that get into trees; thus including the foam-nest tree frogs, leaf-folding frogs, reed frogs, kassinas and tree frogs. Schiotz is a Danish zoologist; he was the director of the National Aquarium in Denmark for over 30 years and did his PhD on African treefrogs. He spent over three years on fieldwork in West and East Africa; including several extended safaris, with Alex Duff-MacKay, around the East African frog hotspots between 1968 and 1971. Schiotz has a good crack at trying to clarify the confusing taxonomic situation with tree frogs and their wildly varying colour patterns; these patterns (as explained earlier) are so variable that they make it almost impossible to construct a key to the treefrogs. This book is an expansion of Schiotz's earlier papers, and one such paper is the 232-page 'Treefrogs of Eastern Africa', which Schiotz published in 1975 (in the journal *Steenstrupia* at Copenhagen); this is a scholarly work with accurate maps, descriptions, sonograms, many black-and-white and some colour pictures.

An excellent book on the amphibians of an important area of Tanzania and Kenya, entitled *Field Guide to the Amphibians of the Eastern Arc Mountains and Coastal Forests of Tanzania and Kenya* has just been published by Camerapix in Nairobi. The authors are Elizabeth Harper, John Measey, David A. Patrick, Michele Menegon and James R. Vonesh; it is in English and Kiswahili, with Kiswahili translation by Imani Swilla. Two other books (both mentioned in the previous chapter) with some Kenyan frogs in colour are Norman Hedges' *Reptiles and Amphibians of East Africa* (Kenya Literature Bureau 1983), showing sixteen common species, and Bill Branch's *Snakes and Other Reptiles and Amphibians of East Africa* (Struik 2005), which illustrates 42 frogs and two caecilians in colour, with distribution maps, although these maps are somewhat erratic. Another somewhat strange book on East African frogs is Martin Pickersgill's *Frog Search* (Chimaira 2007). This is an expensive, 574-page account of the frogs collected by the author, an eccentric self-funded Englishman, in southern and eastern Africa over six years. Written in a chatty and occasionally strangely informal style, with some odd interpretations of biological terminology, it does include 160 excellent colour plates showing a wide range of Kenyan frogs. Pickersgill has done some good work on a difficult taxonomic group and published a handful of papers. However, in the late 1990s Pickersgill attracted a certain amount of controversy by collecting a number of new species and then offering to name them after anyone who contributed money to his collecting fund. This practice wasn't uncommon in the past, and great expeditions often named new species after those who funded them, but nowadays the description of new taxa is expected to be totally free from any sort of commercial taint. The newspapers said he was going to use the money raised to hunt for the tokoloshe, a mythical small monster from southern Africa.

Some good regional material exists. The Belgian herpetologists Gaston-François de Witte and Raymond Laurent did major work on frogs in the Democratic Republic of the Congo. Based at the Musée Royale d'Afrique Centrale at Tervuren outside Brussels, De Witte carried out seven major expeditions to the Congo between 1924 and 1957, and collected in excess of 170,000 herpetological specimens. He authored massive works on the herpetology of three major national parks: Garamba, Upemba and Albert. Laurent was based in the Congo and Rwanda between 1947 and 1961; he was in Katanga when it seceded – a dangerous place at a dangerous time. Laurent wrote major systematic reviews of eight genera of African frogs. Southern Africa has produced some very sound frog books, including Alan Channing's *Amphibians of Central and Southern Africa* (2001), *South African Frogs* (1979) by Neville Passmore and Vincent Carruthers, *Amphibians of Malawi* (1967) by Margaret Stewart and *Amphibians of Zimbabwe* (1989) by Angelo Lambiris.

The scientific literature on Kenyan frogs and caecilians is woefully lacking. Channing and Howell's East Africa book, as described above, contains a thoroughly comprehensive, sixteen-page listing of all relevant papers on East African amphibians but the bulk of these papers, tellingly, do not concern Kenya. Most of them fall into four main groups: (a) taxonomic studies of whole families or genera, (b) laboratory research done outside East Africa but on East African frogs, (c) studies made within Tanzania (especially based around the Eastern Arc Mountains), (d) works by French and Belgian scientists around the Albertine Rift. Major works with relevance to Kenya include Bob Drewes' 70-page tree frog revision, 'A phylogenetic analysis of the treefrogs of Africa, Madagascar and the Seychelles'. This was published in 1984, as issue 139 of the occasional papers of the California Academy of Sciences. Drewes and Jean-Luc Perret also described, in 2000, a new species of giant montane puddle frog from the Aberdares foothills. Drewes has also published a nice article (with colour photographs) on the frogs of the Arabuko-Sokoke forest, in *SWARA* (the East African Wildlife Society magazine) (vol. 20, no. 2); following on from this Drewes also has an excellent website at http://research.calacademy.org/research/herpetology/frogs/list.html, with photographs and sounds of the frogs of this unique coastal forest. The startling paper mentioned above, about how reed frogs changed sex, was published in 1989 in *Copeia* (vol. 4, pages 1024–9) by Thomas Grafe and Karl Linsenmair. Arthur Loveridge's *Checklist of the Reptiles and Amphibians of East Africa* (published in 1957 as a Bulletin of the Museum of Comparative Zoology at Harvard, vol. 117, no. 2), as mentioned in the previous chapter, remains an excellent summary; the synonymies are particularly useful.

But a surprising situation exists concerning amphibian studies within Kenya, in that up to the end of the last century, hardly any had been made. In some ways, this is quite astonishing. The National Museum and Nairobi Snake Park was, and remains, a centre of local expertise, with a full-time professional curator and museum herpetologists; other vertebrate zoologists based there have published widely. In addition, the University of Nairobi has been training zoologists for over half a century, and amphibians are an easily studied group of vertebrates; they can be found and collected without too much difficulty during the rainy season. Kenya Wildlife Services have a staff herpetologist, and a number of qualified researchers, plus control of a swathe of conservation areas that, as yet, do not have any sort of decent faunal checklists. And yet the number of substantial amphibian papers produced between 1950 and 2010 by anyone associated with these institutes can be counted on the fingers of two hands. Alex MacKay was based at the museum for 31 years, and was curator of herpetology for much of that time, and yet he described only two amphibians, both in joint papers. One was a new tree frog from Tsavo, which was named after the charismatic warden David Sheldrick, *Hyperolius sheldricki* (A. Duff-MacKay and Arne Schiotz, 'A new *Hyperolius* from Kenya', published in 1971 in the *Journal of the East African Natural History Society*, vol. 19, no. 128, pages 1–3). The other was a desert-dwelling toad, *Bufo xeros* (M. Tandy, J. Tandy, R. Keith and A. Duff-MacKay, 'A new species of *Bufo* from Africa's dry savannas', published in 1976 in the Pierce-Sellard series of the Texas Memorial Museum, vol. 24, pages 1–20). In 1980 MacKay also produced a 44-page summative paper on Kenya's amphibians. Entitled 'Conservation Status Report number 1: Amphibia', it is a listing of all the frogs and caecilians known from Kenya then, with brief notes on their range, habitat and conservation status. This is quite a useful paper – MacKay knew a lot about amphibians – but only a few hundred copies were produced, on a duplicating machine within the museum and distributed locally. MacKay's paper is so little known that Alan Channing and Kim Howell were unaware of its existence when they published their book.

There have been a few field studies of amphibians in Kenya. In 1979, a paper by R. and M. Bowker (in the journal *Copeia*, vol. 2, pages 278–85) described the breeding activities of ten species of Kenyan frog near Kibwezi, and in a paper in *Tropical Zoology* in 1996 Bob Drewes and Ronald Altig described how the Spiny Leaf-folding Frog raided the Foam-nest Tree Frog's foam ball to eat the eggs ('Anuran egg predation and heterocannibalism in a breeding community of East African frogs', vol. 9, pages 333–47). James Hebrard, Geoffery Maloiy and Daniel Alliangana published a nice paper on the habitat and diet of the Taita Hills Caecilian, in the 1992 *Journal of Herpetology* (vol. 26, pages 513–15). In 2005, a short paper on the distribution patterns of amphibians from the Kakamega forest was published by Susanne Schick, Michael Veith and Stefan Lotters in the *African Journal of Herpetology* (vol. 54, no. 2, pages 185–90). The same team, plus Jorn Kohler and Beryl Bwong, published a description of a new species of tree frog from Kakamega forest, in 2006 ('A new species of arboreal *Leptopelis* from the forests of western Kenya', *Herpetological Journal*, vol. 16, pages 183–9). They named the new frog after Alex MacKay. A paper on the chytrid fungus in Kenyan frogs was authored by Kielgast, Rodder, Veith and Lotters in 2010 in the journal *Animal Conservation* (vol. 13 (Suppl. 1), pages 36–43). In 2007 Victor Wasonga, Afework Bekele, Stefan Lotters and Mundanthra Balakrishnan published an ecological study entitled 'Amphibian abundance and diversity in Meru National Park', in the *Africa Journal of Ecology* (vol. 45, no. 1, pages 55–61). They gathered data on the amphibian community of the savanna grassland and woodland. The July 2006 edition of *SWARA* contains a four-page article by the well-known safari guide Brian Finch on East Africa frogs, with some nice pictures. A herpetological survey team lead by Michael Cheptumo analysed the herpetofauna of the Kora National Reserve in 1986 (published as 'Survey of the reptiles and amphibians of Kora National Reserve', in Malcolm Coe's book *Kora: An Ecological Inventory of the Kora National Reserve*, Kenya, Royal Geographical Society, London, pp. 235–9), but they recorded only five amphibian species, whereas most savanna habitats should have at least twenty.

This lack of research is really a very strange situation, although one must bear in mind that in a country like Kenya research scientists have greater priorities than the lives and taxonomy of amphibians. And, of course, money is tight and work needs to be prioritised. And yet the usefulness of amphibians as environmental indicators should commend them to zoological researchers. One can look at the bright side. There is tremendous potential for future work, and Kenya is training young and enthusiastic zoologists The national museum teams include amphibian enthusiasts and experts like Beryl Bwong, Vincent Muchai, Patrick Malonza and Victor Wasonga. In Tanzania, Dr Charles Msuya is doing sterling work, and was the first Tanzanian to describe a new frog. But at the same time, one is reminded of Reg Moreau's prescient comment (mentioned again in Chapter 13 on conservation) about African biologists scrabbling in the ruins of their flora and fauna. Much of Kenya's landscape is under threat, particularly the moist forests. Beautiful amphibians live there; many may not even have been discovered yet and could become extinct before we are aware of their existence. Kenya's frogs should be enjoyed by all. We would urge you to go out at night and see them – but take all sensible precautions. A good torch, waterproof shoes and some insect repellent are all that is needed, but you may need to take care in wild places and will want the assistance of a local expert. But don't be put off; there is delicate beauty and wondrous sounds to be experienced on the margins of Kenya's freshwater ponds in the rainy season, and a night with frogs at a waterhole in the Arabuko-Sokoke forest is an unforgettable experience.

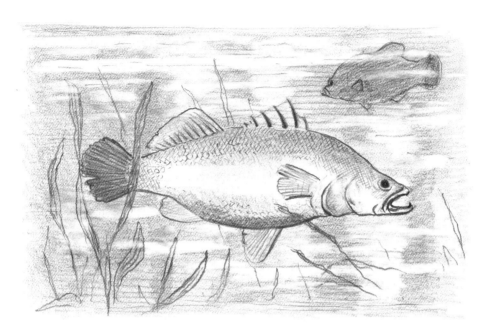

Chapter 10

The Freshwater Fishes

Hata furu humtuma sato / fulu bende oro ngege ('even the furu gets help from the tilapia') – Swahili/Dholuo proverb, meaning that the young should get some help from the old

3rd Fisherman: … I marvel how the fishes live in the sea.
1st Fisherman: Why, as men do a-land; the great ones eat up the little ones
– William Shakespeare, *Pericles* II:i

The non-biological term 'fish' includes the hagfish (which have no vertebrae) and five groups of living vertebrates: the lampreys, cartilaginous fish (sharks, rays and allies), ray-finned fish (regular fish), Coelacanths and lungfishes. Lampreys and hagfish are jawless, but are not found in Kenya. This chapter looks at Kenya's freshwater fish, which include ray-finned fish and lungfishes. Around 32,000 species of fish are known worldwide, of which more than 12,000 are freshwater fish; this is surprising considering that fresh water makes up less than 0.009 per cent of all water on earth. Kenya's fish fauna includes 206 freshwater species, excluding the Lake Victoria cichlids (mouthbrooders). Some authorities state that more than 700 species of cichlids inhabit Lake Victoria, an astonishing evolutionary event. The number of cichlid species in Kenyan waters is unknown.

The chapter will look first at the biology and origins of fish, then at fossil fish in Kenya and what they mean. Kenya's fish fauna, the catchment areas, zoogeography and inventory, are described, followed by a section on the story of the Lake Victoria cichlids, entitled 'Evolution in action'. The chapter concludes with a look at some of the prominent people in the study of Kenya's fishes, and reviews the important literature and research on fish. The sections are self-contained and may be read separately.

FISH BIOLOGY AND ORIGINS

Fish are the oldest, most abundant and yet perhaps the least known group of vertebrates. We all know something about fish, but mostly as an item of diet, a fighter on a line or pets in a tank. Most of us have eaten fish at one time or another, although some peoples in Kenya have traditionally avoided fish in their diet, particularly the pastoralists. Fish are streamlined vertebrates that live in water, although a few have adapted to spend time on land. Most have external fertilisation, with the male and female releasing eggs and sperm in close proximity, although a few have internal fertilisation, and give live birth. Most fish breathe by means of gills (save the lungfishes); as water flows across the four or five pairs of gills, oxygen dissolved in the water diffuses into the bloodstream. Bony fish have a skeleton that largely consists of a skull, backbone and ribs. They move with the assistance of fins but their pelvic fins are not directly connected with the vertebral column; likewise their pectoral fins are connected to a bony plate in the body wall, not articulated to the spine. The sharks and relatives have a skeleton made of cartilage. Fish are ectotherms and they are at the temperature of the water surrounding them. The term bony fish includes all 'regular' or ray-finned fish, plus Coelacanths and lungfishes.

Fish were the first vertebrates with jaws and a mineralised skeleton. The earliest known fossil ray-finned fish was found in Russia and is about 420 million years old. Kenya's oldest fish fossils are over 230 million years old.

Evidence indicates that after the evolution of bony fish, the Sarcopterygii, (Coelacanths and lungfishes), appeared. The lungfishes evolved lungs, weird fins and nostrils that connect to the mouth; in bony fish these nostrils end in a blind sac. Six species of lungfish are known: four are African, of which three are found in Kenya. The other two occur in South America and Australia, indicating a Gondwana origin. Lungfishes are curious fish; they can breathe air and they survive when a water body dries out by aestivating in the mud. Lungfish have bony fins, but at least one African form 'walks' on the lake bottoms using its pelvic fins. Lungfishes lay their eggs in a nest, and the Marbled Lungfish *Protopterus aethiopicus*, found in Kenya, can be over 2m long. Hugh Copley, one of Kenya's fish wardens, noted that its flesh is red, like beef, and excellent eating; he also observed that when a lungfish comes up to breathe and take a gulp of air, the gasp is audible some distance away. In Swahili it is called *mamba*, but this word is also used for the crocodile (but not the snake mamba).

The Coelacanths were originally known only as fossil fish. They have curious lobe-shaped bony fins, and were believed to have become extinct 75 million years ago; many fossils were discovered in the early twentieth century. How a live Coelacanth was trawled off the South African coast in 1938 and described by J. L. B. Smith is one of science's most exciting stories. This was the discovery of a living fossil. The next one was found 14 years later, off the Comoros. Since then, a number of these great blue fish have been found off the eastern and south-east African coast. Kenya's first Coelacanth was caught east of Malindi in 2001, and a second species of this ancient fish was described from Indonesia in 1999.

Fittingly, ray-finned fish appear to have originated in fresh water, and spread to the sea. The first bony fishes to appear, judging by the fossils, seem to have been the Chondrostei, large thin fish with a lot of cartilage. The group includes the sturgeons – source of caviar – and the bichirs, weird elongate fish with many dorsal fins. Two species of bichir occur in Lake Turkana, where they spend the day in reedy shallow water. The majority of modern bony fish are placed in the Neopterygii; the name means 'new fins' as these fish had the standard tail, dorsal, paired pectoral and paired pelvic fins and could move relatively fast. Most living fish today are placed in a division of this group, the teleosts (the name means 'complete bone'), fish that have tail lobes of roughly equal size and can protrude their jaw outwards from the mouth. Most of Kenya's diverse freshwater fish are teleosts.

FOSSIL FISH IN KENYA AND WHAT THEY MEAN

Kenya's earliest fossil fish are from the Maji Ya Chumvi beds, near the coast. The fish remains within this bed are of early to middle Triassic age (250 million to 230 million years old) and belong to genera that went extinct in the Triassic, *Boreosomus* and *Australosomus*. Identical fish are known from fossils from the same age horizon in western Madagascar, indicating that that island was joined to Kenya at that time. However, evidence also indicates that Kenya's present fish fauna came into existence after separation; Madagascar and Kenya do not share a single indigenous freshwater fish species nowadays.

Not a lot of fossil fish are known from Kenya's Mesozoic (200 million to 65 million years ago) rocks. It's a different matter for the last 65 million years. Parts of some fish preserve well, in particular teeth, bones and fin spines. Data on African fossil fish from the Cenozoic, the last 65 million years, was gathered in a long but elegant paper, 'The freshwater fish of Neogene Africa', by the Canadian expert on fossil fish Kathlyn Stewart, and published in the appropriately named journal *Fish and Fisheries* in 2001. Stewart looked at data from more than 40 fossil sites throughout Africa. Kenyan sites examined included Lake Turkana and its environs, Kadam, Rusinga Island, Maboko, Kirimun, Karungu, Loperot and the Tugen Hills. The big picture that emerged is that about 25 million years ago, in the Miocene epoch, a number of prominent fish species, including members of the Nile Perch genus (*Lates*), African lungfish (*Protopterus*), bichirs (*Polypterus*), some cichlids (tilapia and friends) and squeakers (*Synodontis*), were widely distributed across north-eastern and eastern Africa. Some scientists describe this as a 'Pan-African' fauna; others call it a Nilotic or Soudanian fauna. What it does indicate is that at various times more than 25 million years ago, connections existed between the major waterways of Africa and fish moved between them.

We have certain data that support this. In a paper entitled 'The drainage of Africa since the Cretaceous', Oxford geographer Andrew Goudie (2005) indicates that the present drainage patterns of Africa are relatively

The Sabaki Tilapia Oreochromis spilurus. *(Stephen Spawls)*

young and that in the past the Nile, as we know it, did not exist. Twenty-four million years ago a river complex flowed westwards across Egypt, and 16 million years ago a huge river, the 'Late Quena', flowed from eastern Egypt down into Central Africa. As we saw in Chapter 4, more than 30 million years ago Lake Victoria was not there and huge rivers flowed westwards across East Africa to join the Congo River system. The big picture is that Africa was a plateau and there was ample opportunity for the fish faunas to get mixed up. The larger species (which take longer to reproduce and thus speciate more slowly) had huge distributions. The later upwarping created the present discrete catchment areas.

Like reptiles, sub-Saharan African fish genera tend to be confined to Africa and found nowhere much else save Asia. The prominent species mentioned above are largely endemic to Africa although the Nile Perch is believed to be a marine immigrant that entered African rivers (other members of the genus occur in the Indian and Pacific Oceans). Some freshwater fish may have entered Africa from Asia. Big catfish of the family Clariidae are present as fossils from a lot of African sites, nearly all in the east and north-east. Living catfish of this family nowadays occur across Africa and Asia, and it has been suggested that they entered Africa from Asia. But they must have come more than 20 million years ago as fossils of that age are known from Kirimun, near Maralal, and Ngorora, in the Tugen Hills. Cichlids nowadays occur in Africa, Central and South America and parts of Asia and are probably of Gondwanan origin, although this is uncertain. Fossil cichlids are known from Europe. Until the wide acceptance of continental drift and the initial close association of Africa and South America, the distribution of cichlids provided a frightful headache for zoogeographers. An excellent book on zoogeography was published in 1957, however its author, Philip Darlington of Harvard University, was slightly sceptical of continental drift, and his explanation of how cichlids might have got from Africa to South America (or vice versa) via bizarre 'land bridges' or around the edges of the northern continents, is a masterpiece of convoluted logic. Sometimes the simplest explanation really is the best.

The uplift of the land along the present rift valley had a dramatic effect upon the fish fauna of eastern Africa. One of the most obvious consequences was the creation of the rift valley lakes and Lake Victoria, and the resulting isolation of the fish faunas of East Africa. Rivers no longer flowed to the west to join the great Central African whirlpool. Those streams rising on the west of the spine of high country between Mt Elgon and the Mau ran into Lake Victoria and outflow from that system went to the Mediterranean Sea, down the Murchison Falls, which prevents fish moving upstream to reach Lake Victoria. Kenya's rift valley lakes became isolated basins; rivers ran into them but not out. A river that had run through Lake Turkana disappeared, although it left behind Pan-African or Nilotic fish that occur nowhere else in Kenya, like the bichirs, the tigerfish *Hydrocynus forskahlii* (known as 'lokel' by the Turkana) and the bizarre pufferfish *Tetraodon lineatus* (and the Nile Perch *Lates niloticus*, later introduced into Lake Victoria). The eastward flowing rivers like the Tana, the Athi–Galana system and the Northern Uaso Nyiro were left with a handful of species. Some speciation took place, particularly in Lake Turkana, but the increasing aridity of eastern Africa (see Chapter 4) meant that fish populations in the rivers had a hard time of it. The rivers often dried out and their fish fauna remains impoverished; we shall return to this point in the next section. For a variety of fish, you need deep permanent water with a variety of habits, conditions notably absent from most of Kenya. A number of fish families, widely distributed elsewhere in East and Central Africa, are absent from the eastward-flowing rivers; they include the labyrinth fishes (family Anabantidae, curious fish that can breathe air), the mountain catfishes (family Amphiliidae) and the bichirs. A final piece of proof is that the East Africa fossil deposits that contain fish and were laid down prior to 5.5 million years ago are largely river sediments, while those subsequent to 5.5 million years ago are mostly lake deposits. The big rivers had gone. And many of the fish in these later deposits are species confined to eastern Africa. So the uplift of East Africa in the last 5 million years created a discrete and impoverished river fish fauna. But in the bigger freshwater lakes it was a different matter, as we shall see.

KENYA'S FISH FAUNA: THE CATCHMENT AREAS, ZOOGEOGRAPHY AND INVENTORY

Kenya has only two lengthy permanent rivers, the Athi–Galana and the Tana, both running east from Mt Kenya and its southern foothills. There are a number of smaller rivers, and some lakes, large and small, but much of Kenya has no permanent water. This is very significant in terms of fish species numbers; freshwater fish need fresh water.

Kenya has essentially five riverine drainage or catchment basins (mentioned in Chapter 4, and see Figure 4(2)), largely influenced by the Gregory Rift Valley. This bisects the country from south to north. The drainage basins east of the rift run eastwards, those to the west run westwards, and those within it drain into rift valley lakes. The five basins are: the Lake Victoria basin, which drains westwards, the Rift Valley, which drains from the slopes downwards, and the Athi River, the Tana River and the Northern Uaso Nyiro. These last three all drain eastwards. The Lake Victoria basin is essentially all the land west of the western wall of the Gregory Rift, that is, all the rivers that run westwards off the Mau Escarpment and the high land between there and the Cherangani Hills. This is quite a rich system, with eight major rivers and about 56 species of fish. The rift valley is a somewhat complex drainage system, all internal; it includes Lake Turkana and the three big (but seasonal) rivers running into it (Turkwell, Kerio, Suguta) and all the smaller rift valley lakes and their feeders. Lake Turkana and its feeder rivers have some 58 species, not that there is much in the feeders; the Kerio River has no more than two or three species. The central lakes (Baringo, Bogoria, Nakuru, Elmenteita, Naivasha, Magadi) have small feeder rivers associated with them, but an extremely impoverished fish fauna (no fish occurred in Lakes Elmenteita and Nakuru in the nineteenth century). They probably had a much larger fish fauna in the past, when the huge superlakes existed in the central rift, but all six have only a handful of natural species now.

The northern Uaso Nyiro is the largest and driest catchment area. The Uaso Nyiro (and its feeder, the Uaso Narok) are usually permanent in the upper reaches but the Uaso Nyiro has often dried up, and is going to continue doing so, a situation not helped by extraction of its waters on the Laikipia Plateau for agriculture. A number of small seasonal rivers enter the Uaso Nyiro from the north, and there is a large permanent river on Kenya's far north-eastern border, the Daua. This system has only 24 recorded fish species, although it does include lungfish, elephant-snout fish and some barbus.

The Tana River is Kenya's largest river; it is permanent, and is fed from the eastern side of Mt Kenya and the Aberdares. There are five dams on the Tana, which had had an effect on its fish fauna. Some 47 species of fish are known from the Tana system, thirteen of which occur in this river alone. There is a blind river within the Tana catchment, the eastward-flowing Tiva, but nothing is known of its fish fauna although almost certainly some killifish and barbus are there. South of the Tana is the Athi River catchment area, fed by the rivers of the southern Aberdares and the high land south and west of Nairobi; the watershed lies between the Chania and the Ruiru Rivers. The Athi River is known by three names: the Athi in its upper reaches, the Galana further down, the Sabaki as it approaches the sea. Around 35 fish species are known from Athi system. One of these is an undoubted endemic, the little Hildebrandt's Elephant-snout Fish *Mormyrus hildebrandti*, named after Johan Hildebrandt.

Some authorities consider the Pangani drainage as a separate catchment area; it is a small area that consists essentially of some small streams running off the southern and south-eastern flanks of Mt Kilimanjaro, and thence into Tanzania, but this drainage also includes Lake Jipe, and Lake Chala, on the border, is suspected to have an underground connection with the Pangani system. Fourteen fish species are known from the Pangani system.

It should be noted that these three river systems also contain a number of marine species that can tolerate freshwater and move up rivers (these were not included in the totals above). The most notorious of these is the Zambezi or Bull Shark *Carcharhinus leucas*, a big and dangerous shark that often enters freshwater, and has penetrated hundreds of miles up the Zambezi River.

Kenya's freshwater fish are rather poorly known, although a decent checklist, by Seegers, De Vos and Okeyo has been published in the *Journal of the East African Natural History Society*. This paper lists 206 species of fish as occurring in Kenya, of which 35 are endemic. However, that total does not include the Lake Victoria cichlids. The cichlids (family Cichlidae) are a group of 'typical' fish (in shape); most of them protect their young by allowing them to hide in their mouths, and are thus sometimes called 'mouthbrooders'. There is evidence (as detailed in the next section) that they can speciate very rapidly, and the boundaries that separate certain species are both slight and flexible. The following discussion does not refer, unless stated otherwise, to the Lake Victoria cichlids, which will be discussed in the following section.

Comparable species figures for other nearby countries and their rationale are as follows. Somalia has only 57 freshwater fish species, of which three are endemic, but the country has no freshwater lakes and only two permanent rivers. There will have been more fish there in the past; the aridification of north-eastern Africa slaughtered them, as it did Kenya's eastern river fish. One hundred and thirty-six species are known from the Sudan, but only three of these are endemic. Ethiopia has 175 species, of which 41 are endemic; this is a reflection of a rich variety of rivers and lakes at high altitude. Uganda has around 200 species, not including Lake Victoria cichlids. All these figures pale however, when compared to the Democratic Republic of the Congo, which has 1,086 species of freshwater fish, a reflection of a huge, ancient and stable river system.

The structure of Africa, and subsequently its fish fauna, has changed a lot in the last 30 million years. The paucity of fish fossils makes it difficult to construct zoogeographic regions for old Africa, although increasing fossil data may change this. A zoogeographical analysis based on modern fish in Africa is very different to that for land vertebrates, for obvious reasons. Fish are confined to water, and thus their zoogeography is based around river and lake systems. Africa is regarded as having seven main ichthyofaunal regions: the Maghreb (north Africa west of the Nile), Nilo-Sudan (all of west and north-east Africa), Lower Guinea (Cameroon south to the Congo basin), Congo (catchment area of the Congo River), Quanza (Angolan coastal rivers), East Coast (land south-east of Lake Victoria to the sea) and southern Africa (everywhere south of and including the Zambezi and Cunene catchments). Kenya lies astride the Nilo-Sudan and East Coast regions, the dividing line between them running from south of Lake Turkana to the coast south of the Sabaki River mouth.

Kenya's freshwater fish belong in 38 families, out of a total of more than 80 families worldwide; the 200-plus fish species known from Kenya belong to at least 88 genera. The largest family is the Cyprinidae (carps, chubs and barbus), with 50 species, and includes nine or ten species of *Labeo* and more than 30 species in the genus '*Barbus*', although there are classification difficulties with these. Most ichthyologists (fish experts) in Kenya believe that the African species of *Barbus* should be placed in another genus and are not the same as the European *Barbus*. The common fish that you can see at the tank in Mzima Springs, which was a *Barbus*, *Barbus oxyrhynchus*, the Pangani Barb, has been moved to another genus, *Labeobarbus*. The Cyprinidae might be called 'typical' fish. Other large families include the cichlids, with 28 species (excluding those of Lake Victoria) and the mormyrids or elephant fish, also called stonebashers, many of which have, as you might guess, curious elongate snouts; fifteen species of this family occur in Kenya. They are unusual fish. They have a relatively large brain – comparable to humans in terms of the ratio of brain size to body size – and a pair of electrical organs in their tail sets up a continuous electric field around the fish, which may be used to detect their prey or enemies and are used in communication.

Other families well represented in Kenya include the Mochokidae, the squeakers and suckermouths, with fifteen species or so in Kenya. Squeakers are smallish fish, often spotted, with pointed noses and strong locking spines on their elongate dorsal and pectoral fins; these spines fossilise well and hence the ancient distribution of these spiny fish is quite well known. Squeakers are not beloved by anglers as they can often cause injury with their sharp, strong spines. In Graham Bell-Cross and John Minshull's book *The Fishes of Zimbabwe* (a useful book, with much material relevant to Kenya and good pictures), the authors uncompromisingly advise that the best way to get a hook out of a squeaker's mouth without a painful stab is to beat it to a pulp and remove the hook from what is left of the mouth! Another speciose fish family in Kenya is the Alestidae, or characins, with twelve Kenyan species that include the little robbers and tetras, but also the savagely-toothed, hard-fighting tigerfish. And then there are the Aplocheilidae, the family of killifishes. They don't actually kill anything; the name comes

Ripon Falls Barb, Barbus altianalis. *(Stephen Spawls)*

The Pangani Barb, Labeobarbus oxyrhynchus, *this one appears to have encountered a Mzima Springs Crocodile. (Stephen Spawls)*

Lake Victoria Squeaker, Synodontis victoriae, *a fish with a wicked spine. (Stephen Spawls)*

from a Dutch word *kuil* meaning a ditch or puddle, and refers to where they live. The twelve Kenyan members of the family are all small fish belonging to the genus *Nothobranchius*. Many are brightly coloured. They are highly unusual in several ways. They probably have the shortest lifespan of any vertebrate; a Mozambiquan species may only live between 3 and 6 months. They inhabit seasonal pools and have a simply astonishing lifestyle. For part of the year they exist only as desiccation-resistant eggs in the dried mud of what was a seasonal pool. When rain falls and the pool reforms the eggs hatch, the fish grow rapidly, spawn and die; the eggs survive in the mud and the cycle continues. In dry times, the entire Boji plain killifish population (*Nothobranchius bojiensis*), exist only as eggs.

Kenya's remaining 32 fish families are represented by less than ten species per family. Some of them are very interesting, despite belonging to small families. Kenya's biggest river fish is the Sharptooth Catfish, *Clarius gariepinus*. Sharptooth Catfish can weigh up to 70kg, and are not fussy eaters, consuming anything that moves, provided they can get it in their mouths. Safari guide Rick Mathews caught one in the Mara River that had eaten a Puff Adder and an Ethiopian specimen had swallowed a fully grown Cattle Egret. These catfish can survive for a long period in thick mud, due to their ability to breathe air. They usually breed in flooded grasslands, and the female may lay up to half a million eggs. The eggs hatch in 36 hours and the fry remain in the flooded areas, to avoid predation, until they are forced back to the river by falling water levels. Although they have a reputation as a sluggish fish, when hooked, they will actually jump in an attempt to throw the hook. A 30kg fish is always going to put up a fair fight, on account of its weight if nothing else. Some anglers say that the best bait for a catfish is either beef heart or blue-mottled soap!

Some interesting endemic fish are found in Kenya, but the status of some is debated taxonomically. Although fish do have taxonomic characters that can be measured and documented very precisely, like size and number of spines, scales on the head, colour, size dimensions and so on, there is a lack of data and specimens of many species, and the life colours (which the fish start to lose as soon as they are dead) are often missing from descriptions. But the major complicating factor in Kenya is that fish in separate river systems (classic examples being those species that occur in the northern Uaso Nyiro, the Tana and the Athi–Galana) are isolated. This gives separate gene pools, and the

The Butter Catfish, Schilbe intermedius. *(Stephen Spawls)*

Kenya's biggest river fish, the Sharptooth Catfish. (Stephen Spawls)

existence of man-made barriers likes dams complicates the situation. And many of these animals were members of species that were split by the rift valley uplift. With isolation comes changes in the population, more significant in a small gene pool. The opportunity for exchange of genetic material between the populations is small; not many fish can get from one river system to another. It's not impossible, of course. Some fish, like catfish, which can breathe air on account of a modified gill chamber, the labyrinth organ, can move across land when it is raining heavily. The Athi and Galana systems are only a few kilometres apart just north of Ol Donyo Sabuk; a catfish could wriggle across the watershed. A fish that could tolerate salt water could make its way from the Tana Delta to the Sabaki mouth. Any fish in the Lake Victoria drainage rivers could theoretically get washed into the lake and then make its way up another river. So exchange of genetic material is possible. But in general, changes in landscape and increase in aridity has had two major effects upon Kenya's fish fauna. Firstly, the creation of the rift valley has split the fish fauna into three groups – east of the rift, in the rift and west of the rift – and there is virtually no overlap between them. Secondly, the aridity of the rift valley and the north means that really heavy rain and subsequent large-scale flooding of the type that unites catchment areas is very uncommon, so the river and lake fish of those places are largely isolated. This creates species, of course, provided that the habitat is large and varied. But in Kenya's eastern rivers and shallow lakes it isn't, so you are left with an impoverished but changing fish fauna.

This is manifest in the endemics and the taxonomic debate. Lake Turkana has ten endemic species, including several cichlids, a barbus, the 'Dwarf Lake Turkana Robber' *Brycinus minutus*, and a sardine *Chelaethiops bibie*. These seem to be good species. The Lake Victoria basin rivers have five endemics. The Tana and the Athi river system have three endemics each, but they are mostly of doubtful status, meaning that it is not certain whether or not they are discrete species or just variants of some other form. For example the Nairobi Labeo *Labeo trigliceps*, an Athi River putative endemic, is regarded by some as just a variant of the Redeye Labeo *Labeo cylindricus*, known from most of the rivers of east and northern Kenya; two labeos from the Tana are undescribed and are probably new species, but their status as yet is uncertain. Many of Kenya's other water bodies have one or two endemics; there is one in the Turkwell, two in the Nzoia, two in the northern Uaso Nyiro (one of which *Labeo percivali*, is named after Arthur Blayney Percival), one in the Suguta, one at Mzima Springs and so on. This is indicative of small isolated water bodies, a situation that gives rise to speciation but the limited habitat means only a few specialised fish can make a success of life there. The conservation implications are obvious: Kenya's little water systems need (a) investigating to see what is there, (b) protecting to prevent them drying up or being poisoned and (c) monitoring. As always, the time for action is now.

In general, freshwater fish occur in Kenya from sea level to above 2,500m. As with most other vertebrates groups, the higher you go, the fewer the species, although fish are good – much better than amphibians and reptiles, for example – at tolerating the cold. There are a lot of fish in Antarctic waters; fish are the oldest vertebrate group and their enzymes have evolved to cope with low temperature extremes. But the recent uplift of the Kenya highlands has had a curious effect. Many of the rivers coming down from the highlands have waterfalls, and the water up there is cold. None of the lowland warm river species made it onto the montane moorland. Instead, that niche is occupied by an alien species, trout. Three species of trout have been introduced to Kenya Rainbow *Oncorhynchus mykiss*, Brown *Salmo trutta* and American Brook Trout *Salvelinus fontinalis*. Fertilised eggs and fingerlings of Brown and Rainbow were placed in the rivers of Mt Kenya, the Aberdares, Mt Elgon and the Cherangani Hills, in the cold waters of the moorland streams, from 1910 onwards. However, the initial introduction of trout was in 1905, by Colonel Ewart Grogan; the famous Kenya settler who walked from the Cape to Cairo to win the hand of the girl he wanted to marry. The American Brook Trout were released in some of the glacial lakes of Mt Kenya, although it is doubtful if they still survive. No other fish were on the moorlands; the only indigenous fish that ever gets that high is the African Mottled Eel *Anguilla bengalensis*, which has been recorded at 3,000m in the Ragati River on the south-western slopes of Mt Kenya (having made its way from the sea right up the Tana system). Introduction of alien species is usually bad news for indigenous creatures, but in this case the trout on Kenya's montane moorlands seem to have done no harm, and have provided Kenya's sport fisherman with what many aficionados regard as the

Redbreast Tilapia, Tilapia rendalli, *an introduced species that has somehow spread throughout Kenya. (Stephen Spawls)*

purest form of sport fishing. pursuit of trout with a fly and a casting rod. Two surviving highland inns, the Pig and Whistle at Meru and the Izaak Walton at Embu, were set up to provide hostelries for trout fishermen. The Izaak Walton is named after the famous English fisherman of that name, who published his lovely book *The Compleat Angler* in 1653, and said 'is it not an art? ... to deceive a trout with an artificial fly?' The only possible losers, according to Rupert Watson, an expert on trout in Kenya, are the Cape Clawless Otters. Otters and trout both eat freshwater crabs, and the trout were apparently more successful than the otters at finding them.

Other alien fish from outside the continent introduced in Kenya include the Common Carp *Cyprinus carpio*, the Grass Carp *Ctenopharyngodon idella*, the Crucian Carp *Carassius carassius*, the Mosquito Fish *Gambusia affinis* (widely introduced for obvious reasons), the Guppy *Poecilia reticulata*, the Black Bass *Micropterus salmoides* (into Lake Naivasha), the European Eel *Anguilla anguilla* and the Bluegill *Lepomis macrochirus*. Introductions of African fish into Kenya waters include the Three-spot Tilapia *Oreochromis andersonii*, from Botswana into Nairobi dam, Mozambique Tilapia *Oreochromis mossambicus* into the Ramisi River and the Korogwe Tilapia *Oreochromis korogwe* into Lake Challa. Introductions of Kenyan fish into waters where they normally didn't occur include the Nile Perch, from Lake Turkana into Lake Victoria (where it thrived) and Lake Naivasha (where it didn't), and a lot of tilapia translocations, for example the Redbelly Tilapia *Tilapia zilli* into Lakes Victoria and Naivasha, the Athi River Tilapia *Oreochromis spilurus* into Lake Naivasha (Naivasha has had an alarming number of alien fish dumped into it) and the Lake Magadi Tilapia *Alcolapia grahami* into Lake Nakuru (three times!). No one has any real idea of the effects of these introductions; they might be beneficial, neutral or downright harmful. At present just about all we have for Kenya's fish populations are checklists of species present; little quantitative work – other than a bit in Lake Victoria and the associated Lake Kanyaboli – has been done on fish populations, let alone population changes, so the effects of introductions (as well as things like climate change, pollution, riverside agriculture and chemical run-off) are undocumented. Since fish are significant members of the biological community, and provide food for a great many Kenyans, this is work that really does need doing.

EVOLUTION IN ACTION: THE STORY OF THE LAKE VICTORIA CICHLIDS

There are three great lakes in Central Africa: Lake Victoria, 67,500km^2 in area, Lake Tanganyika, 33,000km^2, and Lake Malawi, 22,500km^2. All are associated with the formation of the rift valley and in geological terms all are relatively young, although their ages are debated; a debate complicated by the fact that parts of them have been full and other bits dry. For example, during the last 2 million years the level of Lake Tanganyika fell and turned it into three 'sub-lakes', the Kigoma Basin, the Kalemie Basin and the East Marungu Basin. During this time forest snakes got across from the western forest and established themselves in the woodland of the central eastern shore. However, general opinion is that Lake Tanganyika is the oldest, at least 9–12 million years old. Lake Malawi is younger, although opinion varies, some authors say 4–5 million years, others suggest as little as 1 million. All scientists agree that Lake Victoria is the youngest of the three.

There is violent scientific debate about just how young it is, however, and it is connected with the fact that Lake Victoria is shallow – nowhere deeper than 69m. The other lakes are far deeper: Lake Tanganyika is the world's second-deepest lake, 1,470m, and Lake Malawi is just over 700m deep. Both these lakes are in valley troughs, while Lake Victoria was formed, as detailed in Chapter 4, by the flooding of savanna when the shoulders of the Albertine Rift rose up. A lake existed there around 20 million years ago, as evinced by Miocene freshwater sediments; the geologist Jim Wayland called it 'Lake Karunga'. However, the uplift that eventually created Lake Victoria started 10 million years ago, although general opinion is that the basin began to fill about 800,000 years before the present. What happened since then is contentious. One thing is certain, Lake Victoria as we now know it formed about 100,000 years ago, and it shrank dramatically between 18,000 and 14,000 years ago. American geographers Tom Johnson and Curt Stager are certain that the lake actually dried right out. Their proof is that cores obtained by drilling in the lake bed all show the presence of a paleosol, an ancient soil that the sun shone upon. Other scientists dispute that the lake totally dried out, claiming that parts of the lake remained wet; a string of smaller lakes may have existed. The reason for such intense debate over this moot point is something biologically exciting: the greatest and possibly fastest evolutionary speciation of vertebrates ever known, namely the explosive diversification of cichlid fish.

Cichlids (pronounced 'sick-lids'), are bony small to medium-sized fish that look like perch. They are loosely known as tilapia which comes from the Setswana word for fish, *tlhapi*. The smaller, brighter-coloured species form a tribe, the haplochromines, a slightly unfortunate name as haplochromine means 'one colour'. They occur in South America and Africa, with a few species in Asia. This distribution suggests they are of Gondwanan origin, but Africa split from South America 120 million years ago and the oldest unequivocal cichlid fossils, from Mahenge in Tanzania, are about 45 million years old. Some cichlids are tolerant of salty or brackish water but most live in fresh water. They are unusual for three reasons.

Firstly, the females of a great many species keep the eggs in their mouths until they hatch and sometimes longer, a clever protective technique that has lead to these cichlids being known as mouthbrooders. This has lead to a further unusual behavioural adaptation. When most fish mate, fertilisation is external: the female releases huge quantities of eggs into the water, the male hovers close by and releases seminal fluid, known technically as milt; the gametes meet in the water and fertilisation occurs. Many don't meet and are lost, hence the large numbers of eggs. The female mouthbrooder, however, produces small quantities of eggs and then turns quickly to take them into her mouth. She then has to get them fertilised. Males of many mouthbrooders have vivid pale spots on the anal fin, known as egg-dummies or egg-spots. As the female releases the eggs, the male positions himself at eye level, and the female sees his egg-spots. She mistakes them for eggs, and lunges towards them to swallow the eggs. With perfect timing the male's reproductive organ releases the milt, which drifts into the female's mouth; thus the eggs are fertilised.

The second reason cichlids are unusual is that they possess pharyngeal jaws, a second set of jaws within the throat, which are useful for handling various types of food. And thirdly, they seem able to speciate very rapidly,

Lovely little Haplochromine, the Lake Victoria mouthbrooder, Pseudocrenilabrus victoriae. *(Stephen Spawls)*

and this might be connected with the pharyngeal jaws, enabling the use of a range of foods. They don't live only in lakes, of course; one species is confined to the swamps of Amboseli, another is found only in the warm water of Buffalo Springs, a third tolerates the hot water of springs on the shore of Lake Magadi where water temperatures reach 42.8°C. But it is in the big lakes that they have done remarkable things.

The Natural History Museum in London has a long and noble history of work on cichlids. The first African lake cichlids were described in 1864 by Albert Günther, keeper of zoology at the museum, from dried skins brought back from David Livingstone's expedition to Lake Malawi, and in the early twentieth century George Boulenger, also of the museum, produced a monumental 'Catalogue of the Freshwater Fishes of Africa' (1909). In the 1930s, the remarkable and talented ichthyologist Ethelwynn Trewavas stared work at the museum on the natural history of the cichlids, and published major works on those in Lake Malawi. But the really important early work on cichlids, in Lake Victoria, was done by Humphry Greenwood, a fisheries office in Jinja, Uganda, in the late 1940s and early 1950s. Greenwood realised that the tilapia, as they were then known, that thronged the lake shore were of many different species; he began publishing on them in 1940 and thus commenced 40-plus years' work on cichlids. Subsequent work indicates at present that there are about 700 cichlid species in Lake Victoria, a similar number in Lake Malawi and – surprisingly, seeing it is the oldest lake – some 280 in Lake Tanganyika. These figures are also contentious, depending upon whether you are a taxonomic splitter or lumper, but these figures are astonishing whichever way you look at them. As an indicator of why cichlids are so unusual, consider the lungfish. That probably entered Lake Victoria not long after it was formed, but in the subsequent 100,000 (or maybe 15,000) years it hasn't changed one little bit. It remains a single species. But a cichlid – or maybe a few – entered the lake and has split into many hundreds of species in quite a short time. It was how this occurred that ichthyology has to explain.

Much work was done in Tanzanian waters around the Mwanza Gulf between 1977 and 1992 by HEST (Haplochrominis Ecology Survey Team), a team of Dutch ichthyologists based at the University of Leiden. Their research enabled the little cichlids, the Haplochromines, locally known as 'furu' (or fulu) to be split into feeding groups, depending upon what they ate and how they ate it. It was here that the pharyngeal jaws really came into play. A superb book, published in 1996 by Dutch fish expert Tijs Goldschmidt, entitled *Darwin's Dreampond*, tells the story of the Lake Victoria furu. The feeding groups included fish classified as mud biters, algae-scrapers, leaf-choppers, snail-crushers, snail shellers, zooplankton eaters, insect-eaters, prawn-eaters, fish-eaters, paedophages (which sucked out and ate the offspring of other furu, by ramming their head into the mouth of a mouthbrooding female), scale scrapers, cleaners … the list goes on. Goldschmidt observed several fish with missing eyes, and speculated, although he couldn't prove it, that one species of furu might live by eating the eyes off other fish, an eye-biter. And why not – two Lake Malawi species live by biting scales off other cichlids; one species specialises in biting off scales on the right-hand side of their victims, the other by biting scales off the left-hand side. The Dutch team located more than 150 new species in Lake Tanganyika, so many that team members became exhausted and not all were formally named. Goldschmidt describes how he came across a purple fish with a black face, an angry male, and let him go, because he simply couldn't face discovering yet another new species. The colours of the male furu are simply astonishing, vivid 'day-glo' colours – reds, yellows, blues, greens, pinks, matching the sort of colours you see on coral reef fish; bright things to attract females. There are disadvantages in being so bright; Ethelwynn Trewavas examined the furu she found in the gut of a cormorant and they were nearly all brightly coloured males. The sort of colours that commend you to the ladies also make you visible to a fish-eating bird.

Studies in the 1990s established that the cichlids of eastern Africa were monophyletic, that is they had a single common ancestor (Humphry Greenwood thought they were polyphyletic, i.e. had several ancestors). In the journal *Science*, in 2003, a team of Belgian and German fish experts put forward their suggestions for the diversity, in a paper entitled 'Origin of the superflock of cichlid fishes from Lake Victoria'. The team, consisting of Jos Snoeks, Eric Verheyen, Axel Meyer and Walter Salzburger, eminent fish men all, looked at DNA from the mitochondria of fish from Lakes Victoria, Kivu, Edward, Albert and George. Within a cell, there are two sources of DNA, the nucleus and the mitochondria. The mitochondria are energy-producing bodies. The existence of two lots of DNA in cells led the brilliant and controversial American biologist Lynn Margulis to propose that mitochondria started life as independent prokaryotes, and were incorporated into the bodies of larger organisms. The mtDNA, as mitochondrial DNA is known for short, is often more suitable for taxonomic work, as it doesn't get scrambled in meiois, the production of gametes. The fish men worked with haplotypes, lengths of DNA, or genes that could be identified and isolated from the fish. With some elegant haplotype network diagrams, what they found was that the cichlids of Lake Victoria were all derived from the older Lake Kivu, which has only fifteen cichlid species (and Kivu itself was seeded from the rivers of Tanzania). Representatives of these fifteen species came down into Lake Victoria roughly some 100,000 years ago, in two waves, and radiated explosively into some 700 species. To understand how they do that you have to remember that fish live in three dimensions. Moving into a new 3-D waterscape, in a new lake without major predatory fish, some headed for the muddy bits, others for the sandbanks, others for the margins of submerged rocks. Some went deep, some stayed in shallow water, some on the clear sandy shorelines, others in the reeds. Some started to feed on insects, some scraped algae and some attacked and ate their fellow cichlids. In time, subgroups formed isolated populations, began to interbreed, accumulated changes and eventually looked different from nearby friends.

The pioneering paper by the European ichthyologists was picked up on by the famous zoologist Richard Dawkins, who devoted a long section in his book *The Ancestor's Tale* (2004) to their discoveries. But the original paper was followed a year or so later by a cross response from three geographers, led by Curt Stager, (2004) who had done the work on analysing the Lake Victoria cores. Stager and colleagues had no problem with the cichlid radiation, but were convinced that the lake had totally dried out between 14,000 and 18,000 years ago … and even if it hadn't, the geographers were certain that what refugia, or small lakes and rivers were left, would have harboured only a few cichlid species. They stated that their investigations of the lake floor showed no signs of ancient river channels. This was, in effect, an even more exciting evolutionary scenario: 700 species appearing over 15,000 years, rather than 100,000. But the ichthyologists came quickly back on this, (Verheyen 2004) pointing

The Dembea Stone Lapper Garra dembeensis, *named because it eats algae off stones. (Stephen Spawls)*

out, sensibly enough, that no one could be certain that a few shallow lakes did not exist in the Lake Victoria basin while most of it was dry. Exactly this situation exists in the Kenyan rift valley today, with some small shallow lakes, and some of them do have cichlids.

So it looks like a fact: in 100,000 years – or less – the most astonishing evolutionary speciation effect ever known took place in Lake Victoria, with a handful of pioneer species moving into the new lake and radiating into 700 or more. Some other bits of data support these conclusions. Verheyen and his team found that, despite the huge radiation of species in Lake Victoria, the 700 species there, as a whole, were less genetically diverse than the twenty-odd species found in Lake Kivu. This means that the Kivu species flock was older, because evolution is driven by mutation. A mutation is an unexpected change in a genetic sequence, and is like changing a bit of code in a computer program – it may have a significant effect, either beneficial or negative, or an insignificant one. It can be caused by chemicals, by penetrating rays like gamma, by viruses or by bits of DNA actually moving. But the crucial thing about mutations is that they take place at fairly regular intervals, and this is the basis of a measuring technique known as the molecular clock. The 'clock' makes use of various biological molecules, including DNA. Once two organisms have diverged in the past into two separate evolutionary lines, mutations cause their DNA to start accumulating differences. If you can work out how often changes take place, and how different the DNA of two related organisms is, you can work out – theoretically – how long in the past they had a common ancestor. So when you look at a group of individuals of a particular species, or a group of related species, the amount of genetic diversity amongst the specimens indicates how old they are. The fact that people from areas 20km apart in Kenya are more genetically diverse than a Spaniard and a Chinese person, indicates that humanity has been in Africa a lot longer than in the rest of the world.

In addition, the differences between species in Lake Victoria were often confined to minor characters, like a change in male colour. Two cichlids found in Tanzanian waters of Lake Victoria go by the scientific names

Pundamilia pundamilia and *Pundamilia nyererei*. *Pundamilia* is the Swahili word for zebra; it means literally 'striped donkey'. The second species was named for one of East Africa's most famous sons, the first president of Tanzania, Julius Nyerere. These two cichlid species look very similar; their females look virtually identical but the males of *P. pundamilia* are metallic blue, the males of *P. nyererei* are vivid red and live in slightly deeper water. The females of the two species select their males by colour. But if you put males and females of both species in a tank and illuminate them by monochromatic light, so the females cannot tell the colour of 'their' males, then they will mate with males of either species, and the hybrid young are not sterile. In other words, the isolating mechanisms that keeps them apart in clear water, and makes them into separate species, breaks down. They obviously have a recent common ancestor and the genetic differences between them doesn't prevent them interbreeding and producing fertile offspring. With two different species, the genetic distance between them dictates what will happen if they try to interbreed. Two very closely related species, as in this case, may produce fertile offspring. Two species slightly further apart (i.e. their common ancestor was probably further back in time) may be able to interbreed but their offspring may have genetic problems. For example, the mating of a male donkey and female horse will produce a mule; which can survive but cannot breed; male mules are sterile. If the two species are even further apart, then development of the embryo will not take place. The two sets of chromosomes, from the male and female, do not match. The interbreeding of the two *Pundamilia* species argues that they are very closely related; the DNA evidence from certain specific sequences in many species confirms that. The flock of cichlids in Lake Victoria have not changed genetically very much. They are young, and this means that colour and morphological changes – size, shape, type of jaw – can occur rapidly. This leads to an important point in taxonomic studies; that both the appearance and the genes matter.

There remains the question of why there are many fewer cichlid species in Lake Tanganyika, even though they are so much older, than the Lake Victoria ones. This might be connected with predation; for example there is a highly specialised aquatic cobra in Lake Tanganyika, the Banded or Storm's Water Cobra *Naja* (formerly *Boulengerina*) *annulata stormsi*, which specialises in eating cichlids. It might also be due to habitat variety; the cichlids cannot exploit Lake Tanganyika's great depths whereas Lake Victoria is shallow, its depth and area fluctuate, and cichlids are shallow-water fish. But this Lake Victoria variation might also be due to the dynamics of the situation. Rapid diversification has taken place in a short time. There are many species out there. As time goes by, some species/populations might not be viable and gradually die out while others may become more diverse. An interesting example to illustrate this process can be observed in the rise of the mobile or cell phone. In the early days many firms made and sold them. As time passed, some firms were more successful, and the smaller firms could not compete and were swallowed up. Nowadays only a handful of big firms make mobile phones. The same evolution may take place with the Lake Victoria cichlids. One hundred thousand years from now, species numbers might be smaller and more stable.

Another factor has also been significant in the story of Lake Victoria's fish life in the past 50 years – the introduction of a huge predator into this 'dreampond' of happy fish, and its effects. Nile Perch, which can weigh up to 200kg, were probably introduced into Lake Victoria in the 1950s. The actual details of this stocking are uncertain. Previously unknown above the Murchison Falls, they were definitely introduced into Lake Kyoga in 1955, by the Ugandan fisheries officer John Stoneman. In terms of providing protein, the introduction had a dramatic effect: the Lake Kyoga commercial catch rose over 14 years from 4,500 tonnes to nearly 49,000 tonnes. This episode was documented in his book *The Shamba Raiders*, by Major Bruce Kinloch, a highly decorated soldier who became chief game warden of Uganda after Captain Charles Pitman. Kinloch saw the introduction purely in terms of human benefit – Ugandans got more fish and foreign exchange; the effect of the Nile Perch on the original fish inhabitants of Lake Kyoga is not known. But it was a different matter in Lake Victoria.

Popular belief is that Nile Perch were released in Lake Victoria for the first time in the 1950s, or possibly the 1960s, but this later date is almost certainly wrong, because Humphry Greenwood found a Nile Perch vertebra in the gut of a Lake Victoria lungfish in 1957. The stories also indicates that there was a lot of debate about the

Innocuous looking baby Nile Perch. (Stephen Spawls)

wisdom of the introduction, with fisheries experts suggesting that the furu were useless as a commercial fish so if the Nile Perch did eat them, it would improve the edible fish stocks; but that Nile Perch co-existed with cichlids in many other lakes and didn't decimate them anyway; and that Nile Perch, or at least their ancestors, lived in the area, so their introduction shouldn't do any harm. It is true that a Miocene fossil species of *Lates* (the genus of the Nile Perch) was known from the Lake Victoria shore, but the contradictions in the above arguments will be obvious. Conservationists, especially those at the East African Fisheries Research Organisation (EAFRO) at Jinja, didn't like the idea. They argued that introducing the Nile Perch was too much of a leap in the dark, that thinking it would just eat 'useless' fish was unwarranted, and in any case, who was to say that the bony little furu were 'useless'. As in any food chain, the further up you go, the less efficient it becomes. In terms of human consumption, tilapia grazing on water plants are an efficient way of producing protein, but the fish that eats tilapia uses most of the energy it gets from them in living, and only a small amount of the total energy (around 10 per cent) is actually available to those who eat the perch.

Even while the arguments were going on, Nile Perch were released into Lake Victoria, both from Lake Albert and Lake Turkana. As a result of the environmental upheaval that was caused, no one initially wanted to take responsibility. But in 1986 a Kenyan fisheries officer, John Offula Amaras, in a letter to *The East Africa Standard*, claimed that in 1954 he single-handedly stocked Lake Victoria with Nile Perch from Lake Albert, after getting clearance from Alec Anderson, a senior Ugandan fisheries officer.

Slow to get going, the first Nile Perch was recorded in Kenyan waters in 1966, the first from Tanzanian waters in 1978. The perch took until the late 1970s to become statistically significant in commercial catch figures. Initially, catches in Lake Victoria were largely of native tilapia, of two species: Graham's Tilapia, *Oreochromis esculentus*, and the Victoria Tilapia *Oreochromis variabilis*. A few other fish were also regularly caught, like the Sharp-spined Sewu or Sudan Catfish *Bagrus docmac*, the Ningu or Victoria Labeo *Labeo victorianus* and the Marbled Lungfish. But between 1940 and 1960, pressure grew on fish stocks as increasing numbers of boats

Mega-mouthed fish eater, Nile Perch caught at Eliye Springs, Lake Turkana. (Robert Drewes)

and fishermen joined the hunt, cheaper and more durable boats, tough artificial-fibre nets and better outboard motors became available. By the 1960s, the tilapia stocks were becoming exhausted, and pressure began to increase on the big catfishes and the lungfish. By the 1970s, Lake Victoria showed signs of what is called a 'fishing down process', whereby as bigger species disappear, more and more of the smaller, faster-growing species are caught. This can give the illusion of sustainable fishing – the total catch weight stays more or less stable – but suddenly one runs out of useful species and the fishery collapses.

The fishing-down model was then temporarily reversed in the 1970s by the appearance in catches of many large Nile Perch and the Nile Tilapia *Oreochromis niloticus*, which was introduced in 1957. By the 1980s the Nile Perch made up 50–80 per cent by weight of the catch in some areas. The remainder were largely Nile Tilapia and the Lake Victoria Sardine *Rastrineobola argentea*, locally known by the DhoLuo name 'omena', or 'dagaa' in Swahili; a spicy little fish that is dried and eaten. And at the same time, the numbers of furu were declining. In the 1990s in Tanzania a number of species of these bright

Nile Tilapia, introduced into Lake Victoria and now the most common cichlid there. (Stephen Spawls)

Lake Baringo fisherman on his ambatch boat with a Catfish and a Lungfish. (Tim Spawls)

little cichlids seemed to have disappeared; the HEST team were only seeing 60 or so species where before they had observed close to 130 species. The Nile Perch were eating them, and it was feared that many species were either extinct, or about to become so. This discovery struck a chord with many environmentalists. A dramatic and well-received documentary film, *Darwin's Nightmare*, covered the demise of the furu while touching on other contemporary problems of the developing world, such as AIDS, the selling of luxury food to the West by poor African countries, and the flow of Western aid and guns into the developing world. The big picture seemed to indicate a mass extinction had taken place, as a result of human stupidity.

However, nature had an ace up her sleeve. The furu were indeed under threat, but, hidden in the often cloudy waters, secret back creeks and hidden marshes, they were fighting back. There were places to hide, to get away from the big-mouthed predator that stalked the open water. It's a crucial consideration in conservation; the bigger and more varied the dispersal areas, the better the chances of survival. In the United States, a little fish called the Devil's Hole Pupfish *Cyprinodon diabolis* lives in a tiny limestone sink hole; several times human intervention has saved the population but it lives on the edge of extinction. But Lake Victoria is huge, and there are places of safety. A paper in the journal *Bioscience* in 2003, by the Ugandan fish expert John Balirwa and a bevy of other ichthyologists indicated that not only were the Nile Perch declining and the size of mature adults decreasing, but that the furu were reappearing. In addition, new species were observed and it was speculated that they were the offspring of two species that were previously thought to be different. It seems that many of the furu manage to hide out in areas where the Nile Perch couldn't find them, such as shallow reedy waters. As one might expect with so many newly emerged species, the furu are flexible in evolutionary terms. And very big water bodies are hard to document as much is hidden. Balirwa and his colleagues suggest that the key to keeping the furu happy in Lake Victoria is to over-exploit the Nile Perch. So we should certainly keep on eating those delicious fish!

FISH PEOPLE, FISH LITERATURE AND FISH RESEARCH

Traditionally, many people relied on fishing in certain areas of Kenya, on the coast, along the major rivers and around the lakes. The Il-Tiamus or Njemps people fished for tilapia on Lake Baringo, the El Molo lived on the southern shores of Lake Turkana and fished for tilapia and Nile Perch; the Pokomo along the Tana River practised agriculture and fishing. A major Luo community along the shores of Lake Victoria farmed and fished; we know a great deal about this from the researches of the Luo historian Paul Opondo, and his thesis is available as free download here; http://uir.unisa.ac.za/bitstream/handle/10500/4301/thesis_opondo_p.pdf?sequence=1. The fishing communities developed their skill over generations, knowledge gained as they moved up the Nile and into the lake country of Central Africa. Fishing communities were clan-based, on particular sandy beaches, and there was both organisation and a clear division of labour; the men fished, made and maintained the boats, the women collected, split, gutted and dried or smoked the fish and sold the excess. The 'home' beach, known as the 'wath', was administered by the clan elders. The elders were skilled in a pragmatic type of ecological control; fishing was banned and the beach closed during the weeding season, during heavy rains and when fish were spawning; this was a community in harmony with the environment. Fishing took place using fish traps, fibre nets and spears made from local material. Hand and drag nets were often made of papyrus and this useful plant also provided fibrous wood for smoking the catch. The fishermen might use rafts of lashed logs or canoes; initially dugout canoes made from a single big tree trunk were made and lasted 5–10 years. Later, the most prized craft became the big Ssese canoe. As its name suggests, the Ssese canoe originates from the Ssese Islands on Lake Victoria's western shore in Uganda; with the use of joined pieces of wood these canoes were no longer limited by the size of the parent tree and some really large ones were made. Even today one can see Ssese canoes used on the lakeshore for ferry work as well as for transport. Later, the arrival of improved nets from abroad, metal hooks, metal and fibreglass boats, outboard engines and the bicycle (enabling fish to be carried inland) transformed the fishing community and commercial activities began in earnest. Fish are significant to both Kenya's nutritional needs and its foreign exchange; some 200,000 tonnes of edible fish are landed yearly in Kenya's Lake Victoria towns. More than 40 per cent of this is exported. Commercial fishing is worth over 100 million US dollars to Kenya and employs close on a million people.

The industry needs monitoring; a rather appalling recent development in the Kenyan waters of Lake Victoria involves poisoning fish by simply dumping commercial insecticide into it. The fish thus killed are then sold for eating! The Kenyan entomologist Dino Martins found that in bays where pesticides were used, only one or two species of dragonfly remained, compared with twenty-plus species in unpolluted bays.

The colonial authorities were aware of the potential of Lake Victoria – after all, they had built a railway there – and in 1927 they asked two British ichthyologists, Michael Graham (after whom both Graham's Tilapia *Oreochromis esculentus* and the Lake Magadi Tilapia *Alcolapia grahami* are named) and E. B. Worthington, to report on the situation. They spent a year at Lake Victoria, produced a book, *The Victoria Nyanza and Its Fisheries* (1929) and suggested the construction of a fisheries research institute at Jinja; this was built in 1947. Worthington then returned to Kenya in 1930 and made an expedition to the rift valley lakes and Uganda; during this expedition he collected the first specimens of the Kenya Horned Viper, at Lake Naivasha. H. W. Parker at the Natural History Museum named it *Bitis worthingtoni* in his honour. In 1917, the colonial authorities began selling fish licences, and from 1926 onwards official fish wardens (coastal and inland) and fisheries officers were appointed. The first was Captain R. E. Dent, the man who saw the fabled 'spotted lions' on the Mt Kenya moorlands (for more about this, see the mammals chapter). Dent was fish warden from 1926 to 1937. He was followed by Hugh Copley, who was in post from 1937 to 1948 and then was an acting game warden on and off for a number of years. During his time two government fish farms were started: in 1948, the Sagana Fish Farm for freshwater species and the Kiganjo Fish Farm for trout. Copley was followed as fish warden by Major D. F. Smith. Trout fishermen around Meru in the late 1950s may remember Smith's active policing of their activities, and woe betide any fishermen found with anything other than a permitted fly. One fisherman was prosecuted when detected with a grasshopper on his bare hook.

The Freshwater Fishes

Drying omena (Lake Victoria sardines) on Rusinga Island. (Stephen Spawls)

Copley himself wrote several books, on mammals, the seashore and fishes. His work includes a really useful paper, 'A short account of the freshwater fishes of Kenya', published in the *Journal of the East Africa and Uganda Natural History Society* in1941; this has some handy line drawings of common fish, details of the various water bodies, descriptions of the fish and a checklist with local names. There is some interesting material within this paper; Copley was aware that Lake Turkana contains many Nile fish, found nowhere else in Kenya, and that Lake Victoria had dried out within the last 15,000 years, although the refilling and subsequent cichlid radiation he attributed to a 'pluvial', as was believed in the 1940s. Copley followed this paper in 1958 with a slim volume entitled *Common Freshwater Fishes of East Africa*, although it is really all about Kenya and Lake Victoria. It has a handful of black and white photos and 89 line drawings; chapters on the history of fishes, what a fish is, and fish distribution and migration are followed by a key and descriptions of some 199 East African fish. Although valuable, the names and data in this little book are very dated, but no popular account of Kenya's fish has been published since. This may seem a surprising state of affairs, but when one considers how popular birds are, and the fact that the first fully comprehensive full-colour Kenyan bird book didn't appear until 1995, perhaps the absence of a similar book for the far less popular fish is unsurprising. Not many people come to Kenya to see fish, although some visitors do go to Lake Malawi to see the cichlids. There may be potential there. In the meantime, the ichthyologists at the National Museum are working on a book but until that happens, the best information on Kenyan fish can be found from the 'fishbase' website, www.fishbase.org, a superb site, easily searched, that will categorise fish for you, for example all of Kenya's fish, all of Kenya's introduced fish, all of Kenya's threatened fish, and so on.

Early scientific work on Kenya's fishes begins with George Boulenger. He was a Belgian who commenced work at the British Museum in 1881, became a naturalised Briton and rose to the post of 'first-class assistant'. Although primarily a herpetologist, who did sterling work on snakes, Boulenger also described over a thousand new species of fish. Between 1909 and 1916 he produced the massive, near two-thousand-pages, four-volume *Catalogue of the Freshwater Fishes of Africa in the British Museum*. This monumental and still relevant work is available as a free download using this link http://biodiversitylibrary.org/bibliography/8869 – (but check you have lots of disc space!) from the Biodiversity Library. It is worth reading; the original descriptions of a great many African species are there, along with superb pictures, taken from a range of scholarly publications or specially prepared for the catalogue. Another Natural History Museum scientist interested in African fish was the redoubtable Ethelwynn Trewavas, who was taken on by Tate Reagan, himself a distinguished ichthyologist, in 1931. Trewavas's main interest was cichlids, particularly those of Lake Malawi, but she also collected in Kenya in the 1980s, at the age of 80. As a zoologist who was keenly interested in the natural distribution of African fish; her caustic comment that people were 'moving tilapia around Africa in a slap-happy way' seems particularly pertinent today in view of problem introductions.

Trewavas was one of the first Western scientists to document the cichlid radiation, in Lake Malawi, but it was Humphry Greenwood who did the early important work on those fish in Lake Victoria. Greenwood was born in Cornwall, in England, but raised in South Africa. He studied zoology at the University of the Witwatersrand and then became a colonial fisheries research student at the EAFRO headquarters in Jinja, Uganda, and rose to become its research officer in 1951. In 1959 he was appointed as a principal scientific officer at the Natural History Museum in London where he continued his work on the Lake Victoria fish. He published many papers but his two major works were *The Fishes of Uganda*, published in 1966 and his 1974 paper 'The cichlid fishes of Lake Victoria, East Africa: the biology and evolution of a species flock'. By all accounts Greenwood was a charming and witty man, but he was reputed to have a ferocious temper. Many of his students were scared stiff of him; he was so exacting, hypercritical of any deviation from the scientifically precise. His death was tinged by tragedy; he died of a stroke which felled him while working late in the laboratory; if a night watchman had found him he might have survived. One of his four daughters, Pippa, is famous in Britain as a regular presenter on the BBC *Gardeners' Question Time* radio programme.

Some good work on Kenya's not-so-glamorous river fish was done by a very glamorous man, Peter J. P. Whitehead. Born in Kenya in 1930, Whitehead served in the Royal Artillery and then took a degree at Cambridge. Returning to Kenya in 1953, he worked as a fisheries officer for EAFRO; with a particular interest in the fish of the eastward-running rivers. His publications during this time, some 30-odd papers on Kenyan fish, include one with Vernon Van Someren on eels in Kenya rivers, several papers on fish of the Athi and Tana and the description of a number of new fish species, including the Athi Sardine *Neobola fluviatilis*, Van Someren's Suckermouth *Chiloglanis somereni*, the Tana Squeaker *Synodontis serpentis*, and (with Humphry Greenwood) the Tana Churchill *Petrocephalus catostoma tanaensis*. In 1961 Whitehead was appointed as a principle science officer at the Natural History Museum. Whitehead was an angry man; his twin brother, Rowland, was born first and inherited the family title of Baronet; this caused lasting resentment. Thin, bearded and energetic, Whitehead was twice married but had a reputation as a ladies' man; he travelled each day to work on the train from Oxford and seemed to be able to regularly pick up attractive waitresses on the train's dining car. He wrote a massive, scholarly work on clupeoid fishes of the world (the clupeids are herrings and sardines). He discovered a lost Mozart manuscript while researching a collection of papers in a Polish library. Tragically, he was diagnosed with a brain tumour in 1991 and went off to have a final fling. In his wonderful book about the machinations of the Natural History Museum, *Dry Store Room Number 1*, Richard Fortey (2008) describes how while Whitehead was away his wife tried to get into his office but the keeper of zoology locked her out; it was too full of incriminating details of his extramarital affairs. Whitehead sent the staff a final postcard from Mexico, showing him in a boat with a topless Mexican beauty.

The existence of EAFRO and the presence of Game Department fish wardens and fish farms seemed to have an inhibiting effect on the Corydon/National Museum, which originally had no department of ichthyology. One of the museum's directors in the 1960s, Dr Bob Carcasson, did a lot of work on sea fish (and wrote a book on reef fish) but this was his hobby; by profession he was an entomologist. But in 1997 a new department of ichthyology was set up at the museum by Dr Luc De Vos, a Belgian fish man known to his friends as Tuur.

He had extensive experience in Burundi, the DRC and Rwanda; his doctorate was on African catfishes and he spent time researching at the Royal Africa Museum in Tervuren, near Brussels. De Vos was a dynamic man, kind and friendly, and well-liked at the museum. During his time there he visited most of Kenya's water bodies, collecting fish. He described a number of new species and located, in the Tana River, the first Kenya specimen, and only the second specimen known to science, of the weird, huge-headed Somali Giant Catfish *Pardiglanis tarabinii*. This is a fish that looks in outline like a giant ice-cream cone, nearly a metre long; its head makes up over 60 per cent of its total body. De Vos also presided over the acquisition and preservation of Kenya's first Coelacanth. In 2003 he was not only preparing an exhibition of this primitive fish for the museum, but finalising his checklist of the freshwater fishes of Kenya, in collaboration with Lothar Seegers, a German expert, and Dr Daniel Okeyo, a scientist from western Kenya who authored a major paper on the ecology of the fish of the Athi–Galana system. Their checklist has some nice images (Seegers *et al* 2003). It lists all species (except the Lake Victoria Haplochromines) with their distributions – we found it extremely useful for this account – and the reference lists includes all major papers of relevance to Kenya's fish fauna. It was published in the *Journal of the East African Natural History Society*, vol. 92, in 2003. Shockingly, the same journal also contains De Vos's obituary (Snoeks and Akinyi-Odhiambo, 2003). He had died in Nairobi, quite unexpectedly at the age of 43, while being treated for a lung infection, as a consequence of a long-tem kidney problem. His legacy lives on; he drew into the museum some fine ichthyologists, and the department continues under the able stewardship of Dr Wanja Dorothy Nyingi and Dr Elizabeth Akinyi Odhiambo. They have 30,000 freshwater fish preserved, and have published a string of research papers; details of ongoing research can be found on the National Museum's website.

Chapter 11

The Arthropods

On no other continent in the world has the struggle between insect and man been so acute as on this immense area [Africa] – C. Gordon Hewitt, *A Review of Entomology in the British Empire*, 1916

Dudu liumalo silipe kidole ('don't extend your finger to an insect that bites') – Swahili proverb, meaning it is best not to associate with evil people

The arthropods are relatively small invertebrate animals with segmented bodies; most have six or more jointed legs and a tough external skeleton. The best known arthropods are the insects, from the Latin insecto, meaning 'to cut into sections'. Other typical arthropods include spiders, scorpions, millipedes, crustaceans and the now extinct trilobites. Arthropods are successful and abundant. More than one million species have been described, most of which are insects, but it is believed that between three times and 100 times that number remain undescribed. It is estimated that there are a billion billion (10^{18}, or one followed by eighteen zeroes) arthropods on earth. In terms of distribution, number and species diversity, arthropods are the most successful of all animal phyla. However, none of them has yet worked out how to play football or spray large numbers of humans with lethal chemicals although a flightless Kenyan Bombardier Beetle *Stenaptinus insignis*, defends itself by producing an accurately directed acrid spray that reacts to reach a temperature of 100°C. This chapter will look at arthropods, their origins and fossils, and then at Kenya's arthropod fauna, the inventory and zoogeography. There is a section on arthropods and humans and the chapter concludes with a review of the literature, the personalities and the history of invertebrate research in Kenya. As always, each section is self-contained.

ARTHROPODS, THEIR ORIGINS AND FOSSILS

The arthropods are a subdivision of the invertebrates, animals without backbones. Invertebrates make up 95 per cent of all known animals; over 1.4 million invertebrates have been described. A lot of these live in the sea (corals, jellyfish etc.) and we look at them briefly in Chapter 12. We have not discussed the 40,000 or so species of segmented and round worms or other various tiny creatures (water bears, velvet worms, flatworms, rotifers etc.). Interesting though they are, they have not been well studied in Kenya and a good general textbook on the lower invertebrates will tell you all you need to know about their general characters. But nevertheless, the remainder of the invertebrates, the arthropods, are the majority. Well over a million species are known.

Arthropods are small flying or crawling animals, members of the phylum Arthropoda. They comprise things like beetles, flies, scorpions, spiders, lobsters and millipedes. *Arthron* is Greek for 'joint' and 'arthropod' means 'jointed leg'; most arthropods have these. Some arthropods give us pleasure, for example beautiful butterflies and moths, but many are disliked. The average person is none too keen on spiders, scorpions, flies, ants, fleas, ticks, cockroaches and mosquitoes. At first blush, it looks as though arthropods are nothing but trouble. Many people have both rational and irrational phobias about arthropods; many find their appearance and behaviour disturbing as they don't move or look anything like vertebrates. That's why we call them 'creepy-crawlies'. However, arthropods are important in the web of life. Arthropods are pretty successful creatures, too; since they made it onto land, out of fresh water, they have colonised every habitat available, from sea level to the tops of high mountains. Spiders are found above 7,000m in the Himalayas and species of ground beetle occur on Mt Kenya above 4,300m. Some even live on the surface of the sea, the sea skaters, *Halobates*, although no insect lives under the sea surface; the arthropods that dominate this realm are the crustaceans.

Arthropods are abundant and important animals, ecologically, economically and medically. We depend upon bees, flies and other insects to pollinate our crops. Pawpaw is an important crop in parts of Kenya and pollen is

The Black-tailed Scorpion, Parabuthus liosoma, *its small pincers and fat tail indicate it's highly venomous. (Stephen Spawls)*

Papilio demodocus, *known as the 'Daddy Christmas' Butterfly. (Stephen Spawls)*

Edible insects that form huge swarms, Lake Flies, Chaoborus, *caught in a web. (Stephen Spawls)*

The Picasso Bug, Sphaerocoris, *beautiful and edible. (Stephen Spawls)*

Dung Beetle rolling dung, the Ancient Egyptians worshipped them. (Stephen Spawls)

carried from the male to the female tree by hawkmoths. Research by the entomologist Dino Martins (2009) has shown that if patches of natural habitat are not left standing for hawkmoths to live in, the pawpaw crop suffers. Bees supply us with honey. Insects are eaten by a number of people in Africa. Although they are not a major dietary item, flying termites often make a welcome snack at the beginning of the rainy season when food is scarce; they have a nutty flavour. They may be trapped by placing a deep bowl at the emergence exit of the mound, when the alate (winged) termites emerge in the early afternoon. The Giant Burrowing Cricket *Brachytrupes membranaceus*, is the world's loudest insect; their high intensity buzzing can be heard over 1.5 km away. In parts of Tanzania they are eaten roasted. Around Lake Victoria, the Green Grasshopper *Ruspolia knipperi* is often eaten, and specimens of the Red, Desert and Migratory Locust are also edible; apparently they taste like shrimp. In Tanzania, caterpillars and beetle grubs are occasionally eaten, and honey hunters may eat bee grubs. A most unlikely edible insect is the Lake Fly *Chaoborus edulis* (although the specific name gives a clue, *edulis* means 'eatable'), which is collected in nets or large baskets when it swarms around Lake Victoria; the masses of flies are pressed into cakes, sun-dried and eaten. Some people make a living selling butterflies; the Kipepeo butterfly-breeding project, started by Nature Kenya and now run by the National Museum, has put more than a million US dollars back into the communities around the Arabuko-Sokoke Forest. Dung beetles clear up what animals have left behind and lay their eggs in the dung balls. Dung beetles can move balls of dung 40 times their own weight. More than 40 species of dung beetle have been released in Australia, in an attempt to deal with excess cattle dung. In ancient Egypt they were respected; the pharaohs believed that a giant dung beetle rolled the sun across the sky, in the manner that an ordinary dung beetle rolls a ball. Fly larvae break down animal corpses. Many arthropods provide food for larger animals; a typical food chain in Kenya might read grass eaten by tiger beetle, tiger beetle eaten by striped grass mouse, mouse eaten by Serval Cat. If you like to see the Serval, you've got to put up with the beetle. The food web that ends with something useful to larger animals is virtually guaranteed to include arthropods lower down.

The Dark Blue Pansy, Junonia oenone, *in the Arabuko-Sokoke forest. (Stephen Spawls)*

Arthropods are also significant pests in Kenya; locusts eat crops in the field, beetles steal stored food, fleas and ticks suck blood and cause disease to humans and stock, tsetse flies, sandflies and mosquitoes spread diseases, the Siafu or Safari Ants attack poultry, people are stung and bitten by scorpions and spiders, termites chew up houses, Chiggers (or Jiggers) infect our toes. It has been estimated that in parts of Africa insects claim about 75 per cent of crops. Human beings have been in conflict with arthropods for a long time. Thomas Eisner, charismatic and eminent entomologist from Cornell University (the jacket photograph of his book *For Love of Insects*, 2003, shows him riding a bicycle backwards) put it nicely when he explained that insects are not going to inherit the earth, they already own it. Eisner suggests that we might as well make peace with the owners. And in Kenya, you can see a lot of arthropods at certain times of the year. They are worth appreciating. In the rainy season, anywhere below 1,500m altitude, you can see a museum's-worth gathering round an outside lamp after dark. You can keep them off yourself with insect repellent.

Arthropods are the most diverse group of animals; older than vertebrates; some fossil arthropods are more than 550 million years old. The groups we are going to cover here (Arachnids, Millipedes, Centipedes and Insects) share certain characteristics: all have bodies that are segmented both inside and outside, an open circulatory system, jointed legs and a hard exoskeleton. These modifications convey certain advantages and also disadvantages.

The exoskeleton is largely made of a carbohydrate polymer, a giant molecule, called chitin; if you handle a butterfly or moth, the stuff that comes off on your hands is chitin. Chitin is light and extremely tough for its volume, like a rigid plastic. Muscles are easily mounted on it. Humanity has found a use for it; chitin is used to make a strong and flexible surgical thread, used in operating theatres; it decomposes naturally after the wound or incision heals. Chitin gives protection to arthropods and explains why, if you squash a beetle, you hear a crunch. You have broken a tough plastic-like polymer. That strength gave insects the ability to colonise

Colotis danae, *the Scarlet-tip, at Aruba Lodge. (Stephen Spawls)*

land, to overcome the forces of gravity. But a chitinous exoskeleton is inflexible. It cannot grow and it stops whatever is inside from getting any bigger; it is like a suit of armour. So if an arthropod wants to get bigger it has to shed its skeleton (this is called moulting). And while the body expands and the new skeleton is secreted and hardens in the air, the arthropod is vulnerable. Moulting is a process that requires, relatively speaking, a lot of energy. Imagine if you had to lose all your bones before you could grow at all. Hence arthropods tend to grow only in spurts; on average five or six expansions or less. Then that's it, unlike animals with an internal framework —such as ourselves – where both the skeleton and the muscles hung upon it can all gradually get bigger. The open circulatory system that arthropods have, involving a fluid called hemolymph (the term blood is reserved for a closed system, where the liquid circulates in tubes) is simple. Small tubes squirt the fluid from the heart into open spaces called sinuses; it then oozes back into the heart through pores with valves. It's not very efficient.

Arthropods also have an inefficient body oxygenation system. Vertebrates have pump systems, where oxygen-rich fluid either flows through or is actively pumped in and out of a thin-walled absorption system. In most arthropods there is no pumping; oxygen drifts into the body through complex branched tubes, and diffuses into the body fluid. Diffusion works quite well but it is passive. Coupled with the inflexible exoskeleton and open circulatory system, this inefficient oxygenation system means that land arthropods can never grow very large as they cannot get enough oxygen. The only species longer than 10cm or so need to be very thin so as to have a large surface area to volume ratio. This is why we never see beetles as big as tortoises. Arthropods may be successful animals but they cannot get large, although several of the biggest arthropods in existence occur in Kenya. Not all arthropods were successful, however; the marine forms known as trilobites flourished for more than 270 million years and then, 250 million years ago, just disappear from the fossil record.

Most arthropods reproduce sexually, and lay fertilised eggs. What happens next varies. Insects undergo metamorphosis, of two types; they change in appearance. In holometabolous insects, which usually have wings, there are four stages: egg, larva or grub, pupa or cocoon, and adult. Typical examples are butterflies, beetles and flies. Hemimetabolous insects have three stages: egg, nymph and adult. Typical examples include grasshoppers, bugs and stick insects. Scorpions give live birth; their babies are charmingly known as scorplings and are white. Spiders lay eggs that hatch into tiny spiderlings. Ticks have four developmental stages: egg, larva, nymph and adult; the frightful little ticks that get onto your legs and cause intense itching in high grassland areas, often called 'pepper ticks', are actually tick larvae. Millipedes and some centipedes lay eggs, which hatch into babies with only a few pairs of legs and as they grow they moult and get more legs. Other centipedes are born with a normal complement of legs.

The arthropods comprise several groupings, of which three landforms occur in Kenya. Cheliceriformes is the technical name for the spider and scorpion group; the name comes from the Greek *cheilos*, lips, and *cheir*, arm, named for their claw-like feeding pincers or fangs. Modern cheliceriformes include some marine forms (sea spiders, horseshoe crabs) but most are arachnids, a group that includes spiders, scorpions and the parasitic ticks and mites; they usually have eight legs. Myriapods are millipedes and centipedes. Millipede means 'a thousand legs' and centipede means 'one hundred legs'. In reality most millipedes have a few hundred legs, usually under the body. Millipedes are all herbivores, eating vegetable matter. Some huge specimens of the genus *Scaphiostreptus*, up to 15cm long, black with an undulating fringe of red legs, are found in Kenya's low country; they can often be seen crossing roads in the daytime. Popularly known as 'Tanganyika trains', their Swahili name is *chongoo*, or *chongololo*. It is fascinating to watch their legs move, a living wave machine. Centipedes, *tandu* in Swahili, are a different matter; they are not innocuous beasts. Flat-bodied, they have between 30 and 60 legs, usually sticking out the side. All centipedes are carnivores, they have claws (modified mouthparts) that inject poison to paralyse prey and assist in defence. Some Kenyan centipedes are large, 15cm or so, and can deliver a very painful stab.

The 'Tanganyika Train' or Chongololo, coastal millipede of the genus Scaphiostreptus. *(Stephen Spawls)*

Insects have six legs (the name of the subphylum, Hexapoda, means 'six legs') and are the most speciose group of organisms; there are more species of insect than all the other forms of life combined.

Arthropods seem to have made it onto the land more than 400 million years ago; before that they lived in the sea. Some bizarre marine ancestors of arthropods are known from the Burgess Shale, an outcrop of Cambrian sediments 500 million years old in the Canadian Rockies in British Columbia, creatures like *Opabinia*, which had five eyes and a snout resembling the hose of a vacuum cleaner, and the bizarre *Hallucigenia*, which had spikes on its back. Between 500 million and 400 million years ago, giant water scorpions existed, some over 2.5m long. But by 400 million years ago, several arthropod groups were on land. They had an external skeleton and limbs; necessary because once you get onto land, gravity is important. Your body is no longer supported by water. Some were large; immense millipedes over 2m long are known. Huge dragonflies with wingspans over 75cm have been found in the carboniferous rocks known as coal measures, 300 million years old, in Britain and France. No such dragonflies survive today (or giant millipedes, or giant water scorpions for that matter) and this has lead to suggestions that there was more oxygen in the atmosphere in those days than is present today. Flying insects were around more than 300 million years ago; flight has contributed to insect success, particularly since the wings are an extension of the cuticle (the skin) and thus no pairs of legs have been sacrificed in order to fly. Flying mammals and birds have had to give up a pair of limbs to be able to take to the air. In the Maji ya Chumvi formation near Mombasa, Kenya's oldest arthropod fossils have been found, 250 million years old. They include a millipede, some undetermined insects and a spider. In fact, by 240 million years ago, most of today's insect families had appeared and were on land. Some fossil crustaceans, nearly 200 million years old, are known from near Dar es Salaam. A reasonable cache of fossil insects are known from the Miocene deposits, around 18 million years old, of Rusinga Island. They include grasshoppers, beetles, cockroaches, butterfly larva and even a fossil earthworm, which is not, of course, an arthropod. A fossilized nest of Stingless or Sweat Bees was also found there.

KENYA'S ARTHROPOD FAUNA: THE INVENTORY AND ZOOGEOGRAPHY

A 2006 paper on insect diversity by Bland Finlay and co-authors, in the Proceedings of the Royal Society, starts by stating simply 'The task of quantifying insect diversity is frustrated by great uncertainties'. Which is true. There are so many arthropods, both documented and undocumented, that there is considerable discussion about the numbers. For many groups, not even a rough estimate is available. There is lively debate about how much we do know. Consider for example the beetles, those chunky industrious little insects known to all of us. Beetles all belong in the order Coleoptera, which means 'sheathed wing'; *pteron* is Greek for 'wing'. Large dung beetles are Kenya's heaviest flying insects; one species, *Heliocopris dilloni*, can weigh 15g and has an 18cm wingspan. Flying dung beetles are known to have broken car windscreens. A standard zoology text book of the 1960s, Storer and Usinger's *General Zoology*, states there are 280,000 species of beetle. Neil Campbell and Jane Reece, in their superb undergraduate textbook *Biology*, published in 2008, reckon there are 350,000 beetle species. Wikipedia says there are 400,000. All may be accurate and reflect the knowledge of the time; more and more species are being found and taxonomists are splitting some species as well. Some scientists conservatively put the potential total number of beetles at one million. Others have suggested a fantastic 100 million, but more considered opinion is 2–5 million species. Come what may, there are more beetles than any other order of insects (or any other order of anything living, for that matter). This fact prompted the famous response by the polymath and evolutionary biologist J. B. S. Haldane, when asked what evolution revealed about the mind of God: 'An inordinate fondness for beetles,' was his reply. But consider, since the estimates by (presumably) reputable scientists vary by a factor of 100, all we can take from this is that there is a lot of uncertainty. Kenya is estimated to have around 12,000 species of beetle.

In the class Insecta there are around 30 or 32 orders of insect (the term 'order' describes a division between class and family). After Coleoptera, beetles, the next biggest order is Diptera (meaning 'two wings'). This includes

The beetle display at Kisumu Museum, pride of place to the Goliath Beetle Goliathus giganteus, *heaviest insect on earth. (Stephen Spawls)*

flies and mosquitoes. There are 150,000 of those flying insects worldwide. The order Hymenoptera contains the ants, bees and wasps. Hymenoptera means 'membrane wing', and not everyone realises that ants can fly; 130,000 species of Hymenoptera are known. Two other big families are the Lepidoptera ('scale wing'), which is the butterflies and moths, with about 125,000 species, and the bugs, order Hemiptera ('half-wing'). There are more than 85,000 species of bug. The other orders are smaller, mostly with fewer than 10,000 species, with the exception of the Orthoptera ('straight wings'), grasshoppers, crickets, locusts and friends. The big picture here is diversity; when we consider that there are less than 60,000 species of vertebrate, the fact that there are well over a million described species of insect tells us that they are a successful group. And so they should be; they got onto land before the vertebrates.

According to Scott Miller and the Kenyan entomologist Lucie Rogo (2001), some 100,000 species of insect have been described from sub-Saharan Africa. But of all the African countries, only Nigeria has a published insect checklist, of some 20,000 species. The only listing to date for East Africa was prepared by Richard Le Pelley, the Kenya Government entomologist; his 1959 book *Agricultural Insects of East Africa* lists some 4,000 insect species of importance to farmers. Scott and Rogo reckon there should be about 600,000 insect species in Africa in total. If one looks at land vertebrates, which may not be a valid comparison, within Kenya you can find 35 per cent of Africa's mammals (about 400 out of 1,140), over 50 per cent of Africa's birds, about 25 per cent of Africa's snakes but only 12 per cent of Africa's frogs. This would suggest that Kenya should have at least 60,000 species of insect, 10 per cent of the African total. But this is only an estimate. The cynic would say this is guesswork. Cornel Dudley, an entomologist working in Malawi, noted that 7,500 species were known from there but then suggested that the total number of insects species in Malawi may range from 129,000 to 558,000, variation so large that it seems meaningless (1996). A book that tried to cover them all would be impossibly large. The one portable book on African insects is the *Field Guide to the Insects of South Africa*, (2004) by expert entomologists Mike Picker, Charlie Griffiths and Alan Weaving. It is 444 pages long, and describes more than 1,200 South African insects, with beautiful colour photographs. And what proportion of South Africa's insects is shown here? About 2.4 per cent. Less than a fortieth. And Picker and his team reckon that those 50,000 named species represent less than half of all actual South African species. We are talking big numbers here.

We have few figures for arthropod numbers in Kenya. No one has ever surveyed a discrete area of Kenya and tried to work out the total number of species of arthropod, or even insect, that occur there. The numbers are so large, many of the creatures themselves so small and the logistics so formidable that it has seemed impracticable ... so far. Consider research elsewhere. At Monks Wood nature reserve in Cambridgeshire, England, over 40 years some 1,740 species of insect were recorded, not including thrips and aphids. Some 11,000 species are known from the UK and yet it took 40 years to get a reasonable total. The numbers are tougher in Africa. There have been some small arthropod surveys in Kenya. J. C. M. Gardner's 1957 annotated list of East African forest insects contains about 680 species. According to Gerard Dupre's checklist of African scorpions, there are 344 species of scorpion

The 'Tarantula Hawk', spider-hunting wasp Hemipepsis, *on the sticky burr* Achyranthes, *Nairobi. (Stephen Spawls)*

*Assassin bug (*Platymeris sp.*) at Hunters Lodge; its painful bite can hurt for days. (Stephen Spawls)*

The African Queen, Danaus chrysippus, *found virtually throughout the old world. (Stephen Spawls)*

on the continent, around 27 are found in Kenya. This is only about 8 per cent, but it does include a rift valley species given the generic name *Riftobuthus*! The American millipede expert Richard Hoffman states there are 212 named species of millipede in Tanzania, a further 200-plus un-named species, but suggests that the true total is probably nearer 1,500, in Tanzania alone!

In Torben Larsen's book *Butterflies of Kenya* (1996), 870 species of butterfly are described from Kenya and this is about 25 per cent of all known sub-Saharan African species, which total around 3,500. Some 90-plus species of termite are known from Kenya out of a total of about 680 in sub-Saharan Africa, this is 13 per cent. About 500 species of ant are recorded from Kenya, out of a sub-Saharan total of 1,700 species, so 29 per cent. Some 230 species of ant were recorded by Malcolm Coe's survey team in Mkomazi, in Tanzania, and at Kora National Reserve some 400-plus species of insect were recorded. A survey on the Athi Plains by netting found 214 species of insect. We have some 170 species of dragonfly and damsel fly in Kenya, out of an African total of 970, so just under 18 per cent (and 72 of them are known from the Kakamega Forest). According to Chris Borrow, a researcher at the Natural History Museum in London, 274 species of dung beetle occur in Kenya, out of an Africa total of around 2,000; this is nearly 14 per cent. Although regional lists of species are almost non-existent for Kenya, we do know that 136 species of peacock fly are found in the Kakamega forest, out of a total of roughly 5,000 species worldwide. There are over a thousand species of bee in Kenya, and they are most diverse in the dry country. And on Mt Kenya, in the forest, sixteen species of linyphiid spider (sheet weavers, or money spiders) have been found.

But bear in mind, these are largely species that are easily visible. We can all spot a dragonfly, or catch a butterfly. But at Monks Wood, several hundred species of insect were noted that were less than 3mm long. Long-term surveys of such tiny insects have never been done in Kenya. Likewise, little work has been done on the zoogeography of Kenya's arthropods, although a couple of noble efforts do exist. The German entomologist Viola Clausnitzer's survey of the Kakamega Forest dragonflies found that some 28 per cent of them were species

The Black-Tailed Skimmer, Nesciothemis farinosa, *seen at Kisumu. (Stephen Spawls)*

associated with the Guinean–Congolian rainforest, and the remainder were largely widespread species. Robert Carcasson's 'Preliminary survey of the zoogeography of African butterflies', a pioneering paper published in the then *East African Wildlife Journal* in 1964, indicated that the African butterflies could be grouped into four broad zones, which he called sylvan (= forest), open formations (essentially savanna), Malagasy and Cape. The forest sub-region included a Kenya zone, and the 'open formations' included a Sudanese zone, a Somali zone and an eastern zone. These zones are not dissimilar to Frank White's vegetation zones (see Chapter 5 on plants). In his *The Butterflies of Kenya*, Larsen takes Carcasson's work as a starting point, and expands his analysis, bearing in mind that butterflies are mobile animals. Some insects are international, like the Small Copper butterfly *Lycaena phlaeas*, and the Painted Lady, *Vanessa cardui*. In his book *Early Days in East Africa* (1930), Sir Frederick Jackson describes meeting a Painted Lady on top of Mt Elgon; the bright little insect reminded him of his English home.

Larsen groups the Kenyan butterflies in much the same way as Carcasson, essentially Pan-African, forests, savanna, montane, Cape and Malagasy species. He splits the Kenyan forest species into those occupying 'all forest zones' (which include the coastal forests), 'main forest zones', 'equatorial forest zones', 'Ugandan forest zones' and 'coastal forest zones'. A number of butterflies occur across forest and savanna, Larsen calls them 'transitional species' and his savanna zones are dived into 'all Africa', 'Sudanian, 'Somali', 'Zambezian' and 'Maasai'. A number of endemic species of butterfly occur in Kenya, over a wide range of habitats; Larsen lists them and includes those that range just outside Kenya. Ten endemic species occur in the coastal forests, six in the forests of the far west, twelve in north-eastern Kenya, seven around the Maasai Mara, thirteen species are endemic to the mountains and fourteen from a disparate range of localities that include the vicinity of Meru, Ukambani and South Kavirondo. Larsen's zoogeographical analysis is a valuable one; it matches to a large extent White's zones. It might well be applicable to other arthropod groups, although their Kenya-wide and Africa-wide distribution is, at present, not well-known enough to do anything very comprehensive. There is no atlas for any group of Kenya's land arthropods, just spot records. Richard Hoffman's zoogeographic analysis of East African millipedes indicates three faunas exist in Kenya: a West African-Congolese millipede fauna that reaches the area around Lake Victoria, a coastal fauna that exists eastwards of the 700mm rainfall isohyet and a Sahelian-type fauna that exists everywhere else in Kenya.

Little work has been done on how arthropod faunas change with altitude in Kenya. One would expect the species numbers to decrease with increasing altitude, but there is little comprehensive data on Kenya's insect faunas. Some work was done by Professor Mahadeva Mani. His 1968 book *Ecology and Biogeography of High Altitude Insects* has a chapter on the East African montane insects. He found that on Mt Kilimanjaro some 600 species of insect occurred in the high altitude forest, but only 150 or so species on the upper moorlands and around 40 species in the alpine zone, above 3,000m. It seems species numbers decrease rapidly with altitude. The Cambridge entomologist George Salt, who wrote a classic paper on the ecology of Mt Kilimanjaro, collected a spider up there at 5,500m. Many of Kenya's higher hills and mountains are effectively 'sky islands', with unique insect faunas and endemic species.

The appearance of insects in Kenya is largely dependent upon the weather. Spend a night at Voi, in the middle of the rainy season, and you'll believe that there must be several thousand species of insect there. Stand by a lamp, and you can almost guarantee that a huge dung beetle will crash into it with a bang. You might see twenty mammal species in a day on Tsavo, or even 200 species of bird, but if you know your insects you could probably find well over 500 species without difficulty. But there is no key, as yet, for you to identify them. If there are 100,000 species of insect in South Africa, there are possibly close to 150,000 in Kenya. Possibly. That means that in a big area like Tsavo National Park, there are probably over 20,000 species of insect. Probably. Let us hope that the knowledge of Kenya's arthropods increases. Because, as we are about to describe, some have considerable influence on humanity and our well-being.

ARTHROPODS AND HUMANS

Here's a hate list of Kenyan arthropods: Anopheles and Aedes mosquito, Human Flea, tsetse fly, Jigger, Tumbu Fly (Mango Worm), tick, Army Worm, sandfly, stalk borer, bedbug, Culicoides Beachfly and cockroach. All responsible for a lot of human misery and suffering, and virtually no redeeming features. If we could get rid of the lot of them, we would all probably be much happier. Many people would willingly add scorpions, Siafu or Safari Ants, spiders, paper wasps, body lice, centipedes, Nairobi Eye, Desert Locust and termites to this list, although it has to be said that some of them do have certain redeeming features. Siafu can clean up old houses. Locusts and termites are edible. Spiders eat problem insects. Termites aerate the soil and with the assistance of microbes, break down cellulose and make its nutrients available to the soil community; without termites many ecosystems would be totally different and much smaller. But there are some villains here, and humans are engaged in a never-ending duel with them.

The battle against harmful insects can be won. In a remarkable control programme, run in 1946, 1953 and 1955, the little Black Fly *Simulium neavei*, was eradicated from western Kenya. The curse of onchocerciasis or river blindness, caused by parasitic worms carried by the flies, was removed from the land. Before the programme, the Kodera area was known locally as 'the valley of the blind' and in areas around Kakamega more than 70 per cent of the population were infected. Now the disease is gone and that's a triumph for humanity. But not all such battles against the arthropods prove so decisive. One battle that we don't seem to be winning at the minute is the one against the anopheles mosquito, vector of malaria (see Figure 11(1)). Malaria and schistosomiasis (bilharzia) are the tropical world's two most devastating diseases in socio-economic terms. (There is no place to discuss bilharzia here as it is not hosted by an arthropod but a snail, which is a mollusc. This chronic disease, which weakens its victims, due to major infestations of parasitic worms, is a curse in Kenya, particularly to people living in the Lake Victoria basin, along the coastal strip and the irrigated and rice-growing areas of Eastern Province.)

The Arthropods

Shimba Hills tsetse fly, note the powerful stabbing proboscis. (Stephen Spawls)

Paper Wasp in the Shimba Hills. (Stephen Spawls)

Mating pair of Desert Locusts. (Stephen Spawls)

Malaria is a major problem in Kenya, and humanity has been battling against malaria since the start of recorded history. Both Alexander the Great and Oliver Cromwell are believed to have died of malaria, Tutankhamen may also have been killed by it. Recent high-profile victims of malaria include Cheryl Cole, who contracted the disease while on a charity expedition to Mt Kilimanjaro, and George Clooney, who caught it while in the Sudan. It is estimated that 2,000–3,000 people die *every day* in Africa from malaria. To almost every ordinary Kenyan who lives at altitudes below 1,600m, malaria is a constant, deadly hazard. The name malaria is of Italian origin, and means 'bad air', as it was originally believed to be caused by breathing the air around swampy regions. The incorrect but persuasive logic is clear; if you lived near a swamp you were likely to get malaria, if you lived on a hill you were not. The insects that bit you

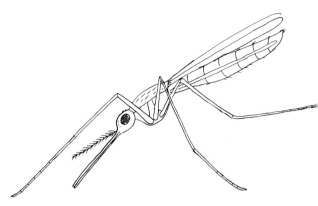

Figure 11(1) Resting Anopheles *Mosquito. The posture, with downwards pointing head, is characteristic of this genus.*

in the swampy areas were not suspected of carrying anything. It fell to Patrick Manson and Ronald Ross, over a hunded years ago, to demonstrate the role of the mosquito. There are several thousand species of mosquito and malaria is spread by the bite of the Anopheles mosquito – the name appropriately means 'useless' or 'without merit' in Greek. There are several hundred species of Anopheles mosquito and the ones that cause the most problems belong to the species complex called *Anopheles gambiae* (there are six or seven species in the complex). They can be recognised – if your eyes are sharp enough – by their characteristic resting posture, with the body angled downwards towards the head; from the side the head and antennae are below the line of the front legs and both pairs of back legs are below the wings (see Figure 11(1)) They detect their human prey by the carbon dioxide we breathe out, plus a host of other chemicals.

The disease itself is caused by a single-celled protist parasite called *Plasmodium*, which looks rather wormlike and is only a few micrometers long. There are more than 200 species of *Plasmodium*. Various animals – birds, lizards, chimpanzees – can get malaria of sorts. Four species of *Plasmodium* cause malaria in humans, of which the species known as *Plasmodium falciparum* is the nastiest. *Plasmodium vivax, P. ovale* and *P. malariae* also cause malaria; *P. vivax* can cause recurring malaria due to its ability to remain dormant in the liver. The parasites, in a form known as a sporozoite (the terminology of the gradually developing evil creature is complex) live in the female mosquito's salivary glands. When she bites, to drink blood to fuel her next batch of eggs, the parasites are injected, along with chemicals that prevent the blood coagulating and cause the itching weal that appears later. The sporozoites develop over the next few days in liver cells. The cells then burst, releasing numerous tiny parasites, merozoites, which infect the red blood cells, and turn into ring-shaped trophozoites; about a quarter the size of red cell. A skilled technician can spot these ring shapes in the blood sample of a potentially malarious patient. The red cell then bursts and releases large numbers of further merozoites, each capable of infecting a new red cell. The victim's blood is losing its oxygen-carrying capacity; in severe cases 80 per cent of the red cells can be infected. This release of huge numbers of merozoites coincides with cycles of fever. In the more benign types of malaria, infected red cells are destroyed as they pass through the spleen, but the deadly *P. falciparum* has a frightful defence against this. It produces sticky proteins on the surface of the red cells, and they clump together and stick to the walls of the vessels, so avoiding the spleen. This stickiness can cause the blockage of blood vessels in organs, and breach the blood–brain barrier. Death can follow. Kenya, and much of Africa, is engaged in a constant battle against malaria and it is sometimes said that by the time the children of low country Africa have reached the age of five they are reasonably immune to malaria; those who didn't acquire immunity are dead.

There are differing approaches to malaria. You can try to kill mosquitoes, stop them breeding, prevent bites and supply drugs to prevent the disease. A vaccine is still far away. The battle is complicated by Africa's rapidly expanding population, poverty in a number of areas, difficulty of reaching the sufferers and shortage of funding. It is a sad fact that big pharmaceutical companies find it more lucrative to produce drugs for affluent Westerners, who can pay a lot, than for poor Africans who cannot. If you have many rural poor, prophylactic drugs are rarely the best answer. Keeping a huge number of people on a drug regime is not practical. Kill the mosquitoes, supply insecticide-coated bed nets, get the children to bed before 8 o'clock in the evening, cover up bare skin, use repellent and drain stagnant water, use drugs to treat only known cases and thus prevent drug resistance; these are all answers. Well-supplied clinics with free access to all, with trained technicians who can spot malarial parasites are also essential. Local campaigns can work if the funding is available. There was a vigorous campaign against malaria in Nairobi in the 1960s and 1970s. City Council employees, spray tanks on their backs, regularly visited residential areas, and woe betide anyone if the hard-faced men of the council team came across mosquitoes breeding in the garden or yard. If the larvae were found in a rusty tin or stagnant ditch an instant fine was levied. Slow-moving streams and ponds running through the city had oil sprayed on them, grass in common areas was cut short, the eaves of dwellings and other resting places were sprayed with insecticide; garden ponds had to have a population of Mosquito Fish. In the early days of the city, there were many cases – the area was, after all, originally a swamp. In 1914 there were 20,000 people in the city and 1,400 malaria cases. in 1964 there were 300,000 people and less than a hundred cases. It was an effective campaign. But the mosquitoes are fighting back; some have become resistant to insecticides, and *Plasmodium* itself has become resistant to prophylactic drugs in some areas, particularly along the coast. Plastic bags can be a curse, with stagnant water in them providing mosquito nurseries and the larvae feeding off wind-borne maize pollen. In addition, Kenya is growing more rice, and mosquitoes love rice fields for breeding. The worst areas for malaria in Kenya are those that have reasonable rainfall and are not too high and cold: all along the coast, Ukambani, the area around the Taita Hills and the land around Lake Victoria. But there is a risk of catching it almost anywhere below 1,700m in the wet season. We need to keep fighting this awful disease.

Other mosquito-borne diseases include yellow fever, which is essentially a disease of forest monkeys that can be spread to humans via the bite of an infected mosquito, of the species *Aedes aegypti*. Yellow fever kills a lot of people in Africa, but fortunately very few in Kenya. Big vaccination programmes will hopefully eradicate it, but not everyone appreciates how widespread the disease is, or that there is no cure. The same mosquito is also responsible for the spread of dengue or breakbone fever, luckily rare in Kenya. Mosquitoes also spread filariasis, and the hideous little thread worms that cause this disease can also cause dramatic swelling of the lower half of the body, an affliction known as elephantiasis.

The battle against mosquitoes, and other insects, has been complicated by problems with insecticides. The saga of insecticides, poisoning and pollution is a continuing story that has merited many books and taxed governments. Chemicals like DDT (dichloro-diphenyl-trichloroethane) and dieldrin were once hailed as the saviour for all our insect problems, but they were only temporary solutions. They are subject to bioaccumulation, meaning that they build up in living things faster than they are lost, and thus gradually poison them. Anyone interested in how this saga started should read the classic book *Silent Spring* by the American biologist and conservationists Rachel Carson. Kenya used to produce over 90 per cent of one of the world's best environmentally friendly insecticides, pyrethrum, derived from the little white daisy-like flowers of the plant *Tanacetum cinerariifolium*. But no longer. Driving from Nairobi to Naivasha, you used to see fields of these beneficial flowers, but the development of synthetic pyrethroids coupled with mismanagement has all but killed the industry.

Eight species of the big, hard-bodied, savagely biting flies known as tsetse, of the genus *Glossina*, are found in Kenya. Tsetse means 'fly' in Setswana, the language of Botswana. They are unusual insects. Fossil tsetse flies 30 million years old are known from Colorado; nowadays tsetse occur only in Africa. They have a painful bite, as they drive in their short, tough proboscis. Swat one, and as you take your hand away the fly zooms off undamaged. Unlike most flies, they rest with their wings folded over the body. They do not lay eggs but give 'live' birth, to living larvae. There is a dramatic photograph in Bernard Grzimek's *Serengeti Shall Not Die* (1960) of a tsetse producing a larva. The larvae then burrow into the soil, pupate and emerge. Tsetse can live up to 6 years. They like to bite in the shade and if you walk in tsetse country in wide shorts or a skirt, one may fly into the shade under your clothes and give you an unpleasant nip in the wrong place. They can fly at over 40km per hour; they will chase a moving car and fly inside to bite you. Their bite forms a hard unpleasant weal that itches for literally days. But worse than all this, in some places they also carry sleeping sickness.

Also known as trypanosomiasis, sleeping sickness is an awful disease. There are two types: one affects humans, the other stock. The stock form is known as nagana, and it kills cattle readily. It is caused by trypanosomes, single-celled parasites that infest the host animal, suppress the immune system and cause anaemia. For a long time tsetse and trypanosomiasis kept the cattle herders out of the southern half of Africa. The historical distribution of pastoralists in sub-Saharan Africa was a consequence of tsetse distribution: flies = no cattle. Some remarkably wasteful techniques were later tried to eliminate the tsetse. Since the fly has to rest in the shade, and does not like flying across country, big areas of bush were simply cleared, by burning and felling. In Southern Rhodesia (now Zimbabwe) half a million wild animals were shot, in an attempt to eradicate wild hosts. Areas of bush were separated by firebreaks too wide for the fly to cross and each was successively drenched in insecticide.

Even nowadays, tsetse are still a problem. There is no vaccine against trypanosomiasis, and the only way to make an area safe for domestic stock is eradication of the flies. Modern techniques include releasing sterile males, and setting out swatches of blue cloth coated with insecticide (tsetse are attracted by blue, which is odd as there are no blue mammals). There is a theory that the zebra's stripes, far from being camouflage, serve to confuse tsetse flies and make zebra less susceptible. And, controversially, many conservationists regard tsetse as their friends, for these flies have kept humanity and their stock out of many areas that are now reservoirs of wildlife.

Human sleeping sickness is in retreat, which is good news. The number of annual reported cases dropped below 10,000 for the first time in 2009. That is still a lot of cases, although few are in Kenya, fortunately. There are two forms, carried by the two subspecies of the trypanosome. One form, *Trypanosoma brucei gambiense*, is found in West and Central Africa; its reservoir is mostly in humans and the disease it causes is slow acting. The other, *T. b. rhodesiense*, is found mostly in eastern and south-eastern Africa; its reservoir is mostly in game animals, particularly Bushbuck. The disease it causes in humans is fast-acting. The first signs of infection are a boil-like swelling which appears a few days after a bite (not an itchy weal, which usually appears the day after a bite), followed by fever and severe headaches. Eventually the nervous system is affected and this leads to the daytime drowsiness that is typical of the disease. Death follows. Oddly, there are huge areas of Africa where various species of tsetse are present and yet there is no sleeping sickness. The areas where the human disease is present are called foci; there is one in Kenya, south of the Winam Gulf on Lake Victoria. Visitors to Ruma National Park may have noticed the tsetse traps there. Various species of tsetse are widespread in Kenya, in areas with thick woody bush, water and usually at low altitude. This includes the coast, the lower Tana and Galana Rivers, the environs of the Tana south-east of Mt Kenya, along the Uaso Nyiro and Turkwell, and the Maasai Mara. Tsetse can be a nuisance in the Shimba Hills, although they don't carry the human disease there. They will just give you a bite or two when you walk down to the lovely Sheldrick Falls.

Another unpleasant flying insect is the sandfly, of the genus *Phlebotomus*. They are 2–5mm long and look like small hairy mosquitoes; the fuzz of hairs on their wings and body is clearly visible (see Figure 11(2)). They are the vectors, or carriers, of a disease called leishmaniasis, caused by a tiny single-celled organism called *Leishmania*, named after the Scottish physician William Leishman. Two varieties of leishmaniasis occur in Kenya, cutaneous leishmaniasis, which causes a septic ulcer at the bite site and is confined largely to Naivasha and Laikipia, and visceral leishmaniasis, or kala azar. This disease is relatively unknown and yet has caused considerable mortality in eastern and north-eastern Africa, in Kenya, Somalia and the Sudan. It wrecks the immune system and the victim then dies of secondary infections. Over 100,000 people died of the disease in the late 1980s in Sudan. In Kenya there are 600–1,000 cases annually, mostly in arid areas, particularly Turkana, Wajir and Mandera districts, where the sandfly is common and expanding its range. Other unpleasant biting insects in Kenya include the Tumbu or

Putzi Fly *Cordylobia anthrophaga* (the specific, second name, means 'eater of men') and the tiny black beachfly, *Culicoides*. The tumbu fly lays its eggs on clothing hung out to dry, especially if the clothing is draped over bushes; it is particularly fond of anything with the merest hint of urine or sweat. The eggs hatch out and the larvae burrow into the skin, causing a painful boil with the developing maggot inside. The black beachfly occurs along the edge of the beach, especially where there is lots of stranded seaweed. It has a sharp bite, and the little lump that results itches for days.

Kenya has a fair number of biting and stinging arthropods, creatures that don't carry diseases but can nevertheless be dangerous or very painful. They are both ground dwellers, like scorpions, spiders and centipedes, and flying insects. Let's look at the flying ones first.

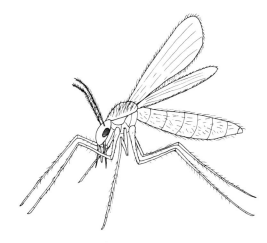

Figure 11(2) Vector of Leishmaniasis, the Phlebotomus *sandfly, like a little hairy mosquito.*

The bees and wasps of Kenya are members of the order Hymenoptera, that is ants, bees and wasps; the name means 'membrane wings'. Many are social insects, living in structured colonies and working communally, while others are solitary. The bees and wasps are in the suborder Apocrita. In those that sting, the pointed organ called the ovipositor has become modified to deliver venom; in other insects it is used for laying eggs. With the sting and wings comes social behaviour; these insects are able to defend themselves and their colonies more actively than most arthropods. It is believed that social wasps and bees are monophyletic; they are descended from a common ancestor. Many members of the suborder lay their eggs inside other living arthropods, which feed off the host as they develop. A great many of Kenya's wasps can sting but are no danger to humans, as they are totally unaggressive and are only likely to sting if physically restrained. The ones that are likely to deliver a painful sting before you actually touch them are the paper wasps and the Honeybee; insects that live in colonies and regard you as a threat if you get too near. Paper wasps, sometimes called hornets in Kenya, are big insects with a very narrow waist; they make pale-coloured nests out of chewed wood with regular, often hexagonal cells. In the bush, these nests are usually sited in a tree, and certain birds – especially the estrildine finches – are known to place their nests near paper wasp nests, knowing they will gain protection from the wasps. But the wasp nests may be in caves, in culverts and under the eaves of, or inside buildings. If you visit an outbuilding, you are quite likely to find a paper wasp nest inside or under the eaves. Don't get too close. If approached closely or molested, they will fly out and sting you and the sting is painful. However, there are rarely more than ten or fifteen wasps at a nest, and stings are unlikely to be fatal, unless you suffer an allergic (anaphylactic) reaction.

The Honeybee is arguably the most important insect to humanity. It pollinates a great many food crops. Without it, we wouldn't have cashew nuts, onions, peppers, watermelon, coconut, melon, citrus, carrots, strawberries, beans … the list continues. Nor would we have honey, or honey beer for that matter. As mentioned, more than 1,000 species of bee – both social and solitary – occur in Kenya; many are undescribed but almost all are important pollinators. Honey is culturally important in Kenya, and honey cultivation was traditionally practised in areas where tsetse flies made it difficult to keep cattle. Rock paintings in southern Africa depict honey hunters approaching wild Honeybee hives. The original Kenyan beehive simply consists of a hollowed-out log, with a wooden disc on each end, suspended in trees; you may see them in well-wooded low country, especially places like Tharaka. Swarms were not placed in them, the honey collectors simply put up lots of hives in the hope that a passing swarm would land there. Bees also nest in hollow trees and rock crevices. At the Jimba Caves near Gede one can see massive ancient bees' nests that have been there hundreds of years, the stained coral rock attesting to their great age. The Honeybee *Apis mellifera*, originated in Asia but is now found

Honeycomb of the African Honeybee. (Stephen Spawls)

worldwide, some other species in the genus are found in east Asia. Three subspecies are found in Kenya: they are *A. m. scutellata* the African Honeybee, *A. m. monticola* the Mountain Honeybee, and *A. m. yemenitica* the Yemeni Honeybee.

African Honeybees have a bad reputation. Although their venom is no more toxic than that of the gentle European Honeybee, their behaviour is very different. Professional apiarists describe African Honeybees as having a high swarming rate and a high tendency to 'abscond and depart', making them sound like convicts. They have a vigorous form of defensive behaviour. This is measured in standard tests by bouncing a black leather ball on a rubber cord in front of the hive; the observer then retreats, noting how many stings the ball receives and over what distance the bees continue to attack. European Honeybees attack in small quantities, and generally give up the attack when the observer is 5 or 6m away; the African Honeybees attack in huge numbers – often 30–40 per cent of the colony – and continue the attack up to 50m from the hive. Every year a few people die after Honeybee stings, as a result of anaphylactic shock. In Kenya, people have been killed not only by allergic reactions, but also from actual poisoning after massive attack by swarms of bees. A child was killed by a swarm of Honeybees near Adam's Arcade in Nairobi in the early 1970s; Louis Leakey was nearly killed by a swarm at Gede. It is estimated that if you are stung and receive more than ten stings per kilogram of body weight, you are at risk from death; what kills you is a protein called melittin. If you approach a hive of African Honeybees, you are likely to be stung. Anyone who has a known allergic reaction to bee or other insect stings should carry injectable epinephrine (adrenaline), preferably in pre-prepared autoinjectors.

Even elephants are afraid of Honeybees; research in Kenya found that elephants did not feed on trees that had hanging beehives. If trees showing elephant damage had hives placed in them, the elephants then left those trees alone. There is an evolutionary reason for the aggressive behaviour of African Honeybees; they have some dedicated enemies, in particular the Honey Badger or Ratel and the honey-hunting human. The sweetness of honey is something we humans crave, and our ancestors have probably been hunting honey for tens of thousands of years. In Africa a symbiotic relationship between humans and the Greater Honeyguide has evolved over the years; a relationship unpopular with Honeybees.

Many people are frightened of spiders. We all know the term arachnophobia – fear of spiders. A number of venomous spiders occur in Kenya, but there are hardly any documented cases of deaths here from spider bite. There are some very dangerous spiders elsewhere, for example the Australian Funnel-web Spiders (family *Hexathelidae*), but in general spiders are non-aggressive and reluctant to bite. A number of spiders do have neurotoxic venom, affecting the nerves, and others have necrotic venoms, which cause tissue damage. You can see pictures of unpleasant spider bites on Google Images. One of the most obvious spiders in parts of Kenya, especially the woodlands and savannas of the west, are the big orb-web-weavers of the genus *Nephila*, spindly-legged and spectacular. A few people have bumped into the huge symmetrical webs of these spiders, but their bite, although painful, is rarely dangerous. Also common in houses are the flat wall spiders of the genus *Selenops*, sometimes called flatties. In relative terms, flatties are among the world's most fast-moving creatures; many people will have seen them shoot across a wall like a stone on ice. In savanna areas you can find the huge, ground-dwelling baboon spiders of the sub-family Harpactarinae, sometimes called old-world tarantulas; they live in silk-lined tunnels. They are big and hairy. One Kenyan species, the so-called 'King Baboon Spider' *Pelinobius muticus*, is a creature of nightmares, a large beast growing up to 20cm across, but its bite, although very painful, is not deadly. The only known deadly species in Kenya belong to the genus *Latrodectus*, button spiders, one of which is an immigrant, the innocuous-looking little Black Widow *Latrodectus mactans*, with a red hourglass shape on its abdomen. Interestingly, most bites occur in outdoor toilets at night, when the spider is sat upon or hit. Records from the Lodwar hospital indicate that spider bites are not uncommon. In 2007 the death of an Irish nun was reported at Lodwar. She was bitten by a so-called 'hunting spider', but it was probably mis-identified; this name is usually applied to fast-running wolf spiders, which are not dangerous. It is not known how many species of spider occur in Kenya but it is almost certainly over a thousand, out of a world total of about 42,000 species. The National Museum entomologist Dr Charles Warui studied spiders

Selenops, the 'flattie', often on house walls. (Stephen Spawls)

Spindly-legged and spectacular, Orb-web Weaver of the genus Nephila. *(Stephen Spawls)*

on Mpala Ranch and found more than 130 species there. We might also mention the nightmarish arachnids known as solifugids, or camel spiders, which although technically harmless have a savage bite. Found in the dry country, they are often orange in colour, and hairy, with huge mandibles. If teased, they will jump at you and they can squeak. During the Second World War some British tank commanders in the Western Desert kept them as mascots; they were called 'gerrymanders'. During leisure time, the tank men matched their solifugids against each other in fights to the death, gambling on the outcome.

Kenya has only about 30 or so species of scorpion, but they cause a lot of pain and suffering to the ordinary rural dweller in low-country Kenya. At Wajir hospital in the 1970s, there were ten–fifteen admissions for scorpion sting per month. Although there were virtually no deaths, the victims were often in very great pain. Often people who arrived did not know what had bitten them; a useful rule of thumb in these cases is that if the victim is suffering intense pain that began immediately after the incident, it was almost certainly caused by a venomous arthropod – and probably a scorpion – rather than a snake bite, where intense pain usually takes a few hours to appear. Worldwide there are about 1,500 species of scorpion, of which 25 species have caused deaths. The really dangerous ones belong to the genus *Buthus* and *Parabuthus*, although one of the deadliest ones goes under the frightening English name of the Deathstalker *Leiurus quinquestriatus*. It hasn't been found in Kenya but might occur there; it's known from Ethiopia and Somalia. There are a few statistics for deaths from scorpion stings in Kenya, but in Libya and Algeria 20–100 people die every year from scorpion stings. Scorpion venom often contains neurotoxins, that interfere with nerve function, as well as enzyme inhibitors; the Israeli cardiologist Moshe Gueron found that some venoms contained cardiotoxins. Victims usually recover; an injection of local anesthetic is often helpful. Mexico leads the scorpion fatalities list, with more than 1,000 deaths per year.

Scorpions are unusual arthropods. They are cannibalistic; there is no snack a scorpion enjoys quite as much as another scorpion. They are nocturnal, spending the day concealed under a rock, in a log or in a hole, which they may dig themselves; a scorpion hole has a distinctive entrance like an inverted crescent. Under ultraviolet light scorpions fluoresce and look green; this is a good way to find them at night. They have a courtship dance during which they kiss, and they give live birth. On the underside they have comb-like chemosensory (smell-detecting) structures called pectines; no other animal has such structures. Scorpions have sensitive feet and can detect the distance and direction of an approaching insect. In general, ones with big pincers are not very venomous; they kill prey animals by squeezing them. Those with small pincers are more dangerous, as they sting their prey to death. They also cause a lot of human suffering. For those who live in Kenya's low country (scorpions are rare at altitudes above 1,500m and non-existent above 2,000m), in a wooden or mud building, and who sleep on the ground or a rickety bed, without adequate lighting, the scorpion is their enemy. Scorpions have an unpleasant habit of getting into shoes placed on the ground, or in clothing. A zoologist with an ultraviolet lamp examined brushwood huts in the town of Kakuma and found an average of two or three small scorpions in every hut. The famous flying doctor Anne Spoerry often treated scorpion sting at her clinics in Kenya's dry north. She popularized the 'electric shock' treatment, often using a power lead from the airplane's electrical system. Medical opinion is a divided on this type of treatment, but Anne Spoerry swore by it.

Bed bugs apparently were brought to Kenya by outsiders; these little pests like to suck human blood. As do lice and the Human Flea, whose scientific name is, appropriately, *Pulex irritans*, and is most common in the highlands as it detects a potential host by warmth. A horrid immigrant to Kenya from South America in the eighteenth century was the jigger, or chigger, a type of flea *Tunga penetrans*. The female drills into bare feet (usually between the toes or under the nails). She embeds there and drinks her host's blood while swelling up and producing quantities of eggs. The site may be recognized by a circular swelling with a darker centre, although some unfortunate sufferers may have masses of sores and weeping wounds from large infestations. Treatment consists of digging out the flea under sterile conditions or soaking the feet in alcohol, but many people are skilled at removing Jiggers with kei-apple thorns or needles, without rupturing the body of the pest. It was the problem of Jiggers that prompted the colonial government in several African countries to provide boots for their workers. Male Jiggers live in the sand; the females spend their lives embedded in their hosts.

Ticks suck the blood of humans and stock; these nasty small arachnids wait in the grass until a host passes by and then latch on. Tick-borne diseases suffered by stock in Kenya include East Coast fever, Nairobi sheep disease, gallsickness, heartwater and redwater. Farmers constantly battle ticks, stock must be dipped regularly. In humans, ticks can cause relapsing fever and typhus. Ticks have been around a long time; the oldest fossil ticks are 100 million years old. Classic work on African ticks was done by the Kenyan-born zoologist Jane Brotherton Walker, who published many papers on Africa ticks, despite being crippled by polio in her teens. Over 70 species of tick were known from Kenya in 1970, according to data in the *Atlas of Kenya*. There are probably many more known by now.

Another problematical group of arthropods are those who steal our food; the locusts, the Army Worms and the insects that bore into growing crops, the stalk and stem borers. These borers are the larvae of a number of innocuous-looking small and slim grass moths, several of which are immigrants from elsewhere. The moth lays its eggs in a suitable spot on the plant, and the larvae hatch out and bore into the leaves and the steam, eating their way through, causing damage. When the dry season comes, they stop developing (a process called diapause) and remain dormant in plant debris, then turn into pupa and adult moths when the rainy season arrives. The damage that these moth larvae can cause can be horrendous. In parts of dry-country Kenya up to 80 per cent of the sorghum crop has been lost; the rural communities who live in arid areas and grow sorghum for a living can ill-afford such depredations. Maize is a bigger plant and better able to stand such damage (it is the physical damage that matters, the moth larvae don't eat the actual grain or cob, they bore through the leaves and stem and disrupt the nutrient supply upwards from there) but up to 20 per cent of maize has been lost in some areas. A major paper on stem borers by Rami Kfir and co-authors in the journal *Annual Review of Entomology* points out, sadly, that the communities most at risk from these troublesome crop pests are often beyond the reach of agricultural extension teams and rarely have access to insecticides. Interestingly, planting a more attractive host grass around the crop is a very effective way of diverting the pests. Another crop raider, the 'Army Worm', is not a worm but a caterpillar of a rather ugly moth, *Spodoptera exempta*. It is a pest of small grain crops and local outbreaks can cause devastation. It gets its name from the habit of marching in huge numbers, especially at night, onto new crops.

One of the ten plagues of Egypt, mentioned in both the Quran and the Bible, was locusts. Locusts are the adult, swarm-forming phase of short-horned grasshoppers; three species are a problem in low-altitude Africa, although a fourth, the Green Milkweed Locust *Phymaetus viridipes*, can be a pest in highland areas. This latter species is a spectacular insect, with bright green, red and blue wings, and sometimes occurs in huge numbers. They are known to feed on the leaves of a dangerous plant, the Poison Arrow Tree *Acokanthera schimperi*, and may gather in thousands or more. However, the really bad locust in Kenya is the Desert Locust *Schistocerca gregaria*. Two other species, the Red Locust *Nomadracis septemfasciata*, and the Migratory Locust *Locusta migratoria*, have caused problems in Africa. The Red Locust occurs sporadically across the continent and has caused problems in Tanzania and Zambia. The Migratory Locust has a huge distribution in Africa, Asia, Australia and New Zealand but it has not caused plagues in Africa since the 1940s.

The Desert Locust is a problem insect. A swarm that entered Kenya in 1954 was over 200km^2 in area, and since it is estimated there are 50 million individuals per square kilometre, that means there were some 10 billion locusts in the swarm. Since each individual eats about 2g of food each day, that swarm would need 20,000 tonnes of food per day! Locusts are most common in arid areas, hence they are obviously a major threat to human crops, and their activities need to be carefully monitored. Desert Locusts normally occur in a belt of arid country that extends east from Mauritania across to the Sudan, thence south-east to southern Somalia and east to north-west India. This area is called the recession area; locust heartland. However, if conditions are right for breeding and many eggs hatch, the locusts undergo an interesting biological change. As they constantly bump into one another, their colour changes, they give off a pheromone that attracts them together, and a swarm is formed. They become insects with attitude. If the swarm then flies off, it can cover 100–200km per day, depending upon the wind speed. Desert Locust swarms in the past have spread south to Southern Africa, north to Spain and Russia, eastwards to the far east of India. Ethiopia has suffered particularly, losing thousands of tons of grain every year between 1949

Crematogaster ants live in the galls of the whistling thorn and defend the tree from browsers. (Stephen Spawls)

and 1962, and again in 1978. Kenya has been more fortunate, but a major swarm struck in 1962; part of that swarm reached Nairobi and locusts were found all over the city. The most recent invasion of Kenya was by a swarm that entered the north-east in 2007, originating from Yemen. Monitoring Desert Locusts takes time and money; the United Nations Food and Agricultural Organisation maintain a Desert Locust information scheme in Rome. The eccentric explorer Wilfred Thesiger, who lived for many years at Maralal in Kenya, took a job in 1945 looking for locust breeding grounds in Arabia. The purpose of this job is to spot the hoppers swarming on the ground before they fly; a job that requires intensive patrolling of the heartland. However, Thesiger's real agenda was to travel across the Empty Quarter. (He was a man who always did exactly what he wanted.)

Compared with frightful arthropods like locusts, things like Safari Ants, blister beetles and the Nairobi Eye are really just minor pests, although if you are a Kilifi farmer who has just found your entire flock of chickens slaughtered by Safari Ants, you probably don't think of them as minor. The Safari Ants, or Siafu, are members of the genus *Dorylus*; in other parts of Africa they may be called driver ants, soldier ants or army ants; more than 60 species are known. Siafu are astonishing animals; a Safari Ant colony has been picturesquely described as a machine, with the worker ants just components: they are all females, all sisters, all sterile and all blind. Siafu need forest or woodland to survive, they don't usually occur in arid country, but occur through a huge range of altitude, from the coast to the Mt Kenya and Aberdare moorlands above 3,200m. They live in underground nest chambers, which they excavate themselves. A marching column can contain over 20 million individuals. These ants are raiders, sending out columns to hunt down, kill and dismember any prey they can overpower. They also move frequently: when they have run out of food, the entire nest 'ups sticks' and moves to a fresh location. They do some good: if they enter a building they slaughter things like cockroaches. Legends attach to army ants, and they have featured in films and books; Charlton Heston (as the settler Leinengen) battled a huge ant army in the film *The Naked Jungle* and Harrison Ford takes on some frightful ants (which he calls 'Siafu') in *Indiana Jones and the Kingdom of the Crystal Skull* … but in South America!

Safari ants or siafu, the big ones are the soldiers. (Stephen Spawls)

In reality, an advancing column of Safari Ants can only move at around 20 or 30m an hour. Although the main prey of Safari Ants are earthworms, insect larvae and slugs, they will catch anything they can overpower. Most children (and adults) in highland Kenya will have inadvertently sat or stood in a Safari Ant trail and been bitten. The ants have a horrible habit of not biting until a number are well established, often under your clothing, then they all bite together. How they co-ordinate this simultaneous attack isn't known. The only course of action is to get out of their range, take your clothes off and pull the little beasts off. It is unlikely that they have actually killed anyone, but Siafu once nearly cost the professional hunter John Hunter his job. He was guiding a German baron who had a beautiful wife and was insanely jealous of her; he had hired an ex-major to stay with her at all times. The baron didn't do much actual hunting; so usually Hunter, the wife and the major went out into the field. During a lion hunt, Hunter heard the wife screaming and went to assist; he found her naked apart from her knickers, desperately trying to get a swarm of ants off. Hunter picked off the ants, and had to remove some with his knife blade; the baroness had only just got her clothes back on when the major appeared. One minute earlier and he would have found her undressed; Hunter reckoned he would have lost his licence. A bite from a big soldier ant is painful and they cling on to the death. In Kenya they have been used to suture a deep cut. The lips of the wound are squeezed together then the ant is made to bite across either side. The head is pulled off and the jaws remain locked. Siafu are a real menace to stock. If they get into a chicken coop or a rabbit hutch they will eat the unfortunate trapped animals alive. A lovebird enthusiast at Diani Beach lost his entire collection to a Siafu attack one wet night. Both of us recall sleeping, as children, on beds with their legs stood in fruit tins, into which a mixture of kerosene and water had been placed, to prevent the ants climbing onto the bed. The male ants, strangely, develop wings and fly in the rainy season. They are known in Kenya as 'sausage flies'. It took years before entomologists realised that these weird flies were actually male ants.

There are some essentially innocuous but corrosive insects in Kenya. One is the Nairobi Eye, a little beetle attractively banded in deep blue-green and orange, belonging to the genus *Paedurus*, the Rove Beetles. They may appear in gigantic numbers at night in the rainy season. If they are accidentally crushed on the skin, a caustic compound in their bodies, pederin, is liberated and causes septic blisters to form some hours later. These take a while to heal. Pederin is not actually manufactured by the little beetle but by symbiont bacteria. Brightly coloured flower-eating beetles of the family Meloidae, blister beetles, also produce a caustic chemical, cantharidin, if handled or roughly treated. Cantharidin has an unjustified reputation as an aphrodisiac; anyone wishing to try it should be aware that 10mg is a lethal dose!

Termites can be problem insects, and they have been with us a long time. Fossil termites from 165 million years ago are known. Popularly known as 'white ants', they are not ants at all, although they are also social insects. The Kenyan termites of the genus *Macrotermes* build nests (incorrectly called 'anthills') over 9m high. Over 90 species of termite are known from Kenya. It is estimated there are 700kg of termites for every person on earth. All termites were originally in the order Isoptera, but work on their DNA by the Natural History Museum entomologist Paul Eggleton in 2007 proved that termites are actually social cockroaches. Termites are now placed in the cockroach order, Blattodea. Termites are famous for two reasons: they build huge mounds and they eat wood. They embellish the landscape, wreck plantations and demolish houses. (In 1949 termites caused major damage to the Vatican City library.) Termites are the reason that most railway sleepers, telegraph poles and electricity poles in Kenya are black; they have been steeped in creosote to stop termites eating them. The other side of the coin, of course, is that termites aerate the soil and add features to the landscape. Termites are unusual insects; with the help of gut protozoa they can digest cellulose, the big carbohydrate polymer that makes up most of plants. We cannot, which is the reason we can't eat leaves.

Termites have a social hierarchy. Their termitaria are organised with division of labour and their homes, if abandoned by the original owners, provide refuges for a huge number of other animals, including blind snakes and cobras, monitors and other lizards, mongooses and hordes of insects, some birds even nest in them. An active termite hill is sealed on the outside; those that have been abandoned have open holes. The queen termite, prized

Queen termite, a machine that produces eggs every two seconds. (Stephen Spawls)

by some as food when dug out of a termitarium, has a tiny head and a huge fat white body. Essentially she is just an egg-producing machine and is immobile. She produces one egg every two seconds on average. Termites are vulnerable to desiccation and predators (the soldiers are usually blind) but protect themselves with shelter tubes of saliva and soil. At night, if a group of termites are out, they warn of impending danger by banging their heads against the ground; walkers at night in Kenya's woodlands will have heard this distinctive little sound. In the rainy season, they disperse by growing wings and emerging in the afternoon. Many animals will eat them; birds will gather and pick them off as they fly out. A Black Mamba has been seen snapping up the alates as they emerge. They are attracted to lights; after the first storms of the rainy season you are likely to find swarms of termites, or often just their wings, under an outside light; at this time the winged or alate termites may be called 'flying ants'. An Oscar-nominated documentary by the famous film-makers Alan and Joan Root, *Castles of Clay*, documents the life of termites in a Kenyan termite hill.

As we have explored, arthropods get a bad press. They are nevertheless important in the overall scheme of things. Back in 1897, the old Africa hand Sir Harry Johnston remarked in his book *British Central Africa* that 'insects … seem to have been created for an almost wholly evil purpose'. But almost every food chain and web involves arthropods. And their presence adds to the enjoyment of the wild country. Too few naturalists appreciate them. Turn any rock or log, and you will find interesting insects. Stand by an outside lamp in the rainy season and admire the diversity of small life. Put out some mashed banana and sugar in a bowl and you will attract butterflies. Alongside a lake or river you will see vivid dragonflies. Reflect upon the fact that these harmless and useful insects were long believed to have a venomous sting; in the United States they were called 'darning needles' and generations of children grew up afraid of them. If you meet a column of patrolling ants, it can be fascinating to watch their activities as the column marches and the scouts patrol. In leafy bushes there are interesting insects like praying mantises and stick insects, often brilliantly concealed among the foliage. The praying mantis is important in Khoisan mythology, a theme publicised by the South African adventurer

Violet Dropwing Dragonfly, Trithemis annulata, *dragonflies were thought to sting. (Stephen Spawls)*

Laurens van der Post. Stop to observe those little perfectly conical depressions in the dust. In the bottom are ant lions, the larval form of a big dragonfly-like insect. Many youngsters in Kenya have amused themselves dropping an ant or other unsuspecting insect into the pit; a miniature monster rises from below and seizes the unfortunate victim. In Kenya, a wildlife safari of small beasts can start in the garden.

THE LITERATURE, THE PEOPLE AND THE HISTORY OF INVERTEBRATE RESEARCH IN KENYA

The statistics are formidable. If there are 150,000 plus arthropod species in Kenya and each species merited one line in a book, the book would still be 3,000 pages long. If you had twenty pictures per page, your book would need 7,500 pages. It's no wonder that Mike Picker and his team, in the aforementioned *Insects of South Africa* (published by Struik in 2004), chose to cover only 1,200 species. Even so, it's quite a hefty field guide, although a pretty good one, and quite useful for Kenya, although the distribution maps only cover South Africa. The photographs are of the highest quality; there isn't a bad picture in the whole book. If you want to do field work on arthropods in Kenya, until something more specific appears, this is the book to get, always bearing in mind the many differences with in the Kenyan insect fauna. We might also mention Elliot Pinhey's *A Guide to the Insects of Africa*, published in 1974 by Hamlyn and illustrated by Ian Loe. Although this is a short book, aimed at the younger enthusiast, its thirteen large colour plates, showing some 60 insect species in the landscape, plus some superb line drawings of insects and their anatomy and some well-informed text, provides quite a fun introduction to Africa's insects.

There isn't a lot of summative literature on the arthropods of Kenya, or East Africa. The only book that deals with a decent selection of any order of Kenya's arthropod fauna is Torben Larsen's excellent *Butterflies of Kenya* (published by the Oxford University Press in 1996), it covers all 870 of Kenya's butterflies; each is illustrated on a colour plate. However, it is now out-of-print. Broadening our search to cover East Africa, we find Richard Le Pelley's work *Agricultural Insects of East Africa*. It is an obscure book and hard to find. Published by the East African High Commission in Nairobi in 1959, it has no illustrations. Its 300-odd pages consist of a series of lists; insects with their host plants, lists of host plants, parasites, etc. The lists are annotated with numbers, giving the original references. In 1957, J. C. M. Gardner produced a list of East African forest insects; this 48-page pamphlet was published by the East African Agriculture and Forestry Research Organisation (EAAFRO). This is quite a useful paper, as it lists just fewer than 700 insects, indicates the country in which the species is found, gives a few locality records and some natural history notes for some species. It has no illustrations, however. An excellent and handy book is Hans Schabel's *Forest Entomology in East Africa* (published by Springer in 2006), a 380-page book with lots of colour plates. It is mostly about Tanzania, as indicated by its sub-title, *Forest Insects of Tanzania*, but its nine chapters cover not just the Tanzanian forest insects but the Tanzanian environment and the history of the East African entomology. Its price is the only thing about it that isn't excellent; it costs over $200 (it's out of print but available through the Internet).

A few discrete African arthropod groups have generated some literature. There are expensive books on Africa's tiger beetles, more reasonably priced volumes on the dragonflies and damselflies of southern Africa, and a couple of works on Africa's bees, spiders and scorpions (mostly of southern Africa). There is a catalogue of the aphids of Africa, covering 244 species, a book on the mosquitoes of southern Africa and another on the sandflies of West Africa. For those enthusiastic about ticks, five volumes on the ixodid ticks of Central Africa may be purchased. African moths and butterflies, being showy and attractive, have stimulated the production of a fair number of books, but there is nothing specific – as yet – on Kenyan moths, of which there are several thousand species.

This doesn't mean that there isn't much literature on Africa's arthropods. There are many thousands of scientific papers, but mostly they deal with specific groups that are of importance to humanity, in most cases because of the agricultural and medical problems they cause. There are hundreds of papers on tsetse flies alone. Journals like *African Entomology*, the *Bulletin of Entomological Research*, *Insect science and Its Application*, *African Invertebrates*,

Taita Hills Praying Mantis, important in Khoisan mythology. (Stephen Spawls)

African Journal of Ecology, *Medicinal and Veterinary Entomology* and *Revue de Zoologie Africaine* all publish articles on African and Kenya arthropods. For a summation of the current state of African entomology, read Scott Miller and Lucie Rogo's paper, 'Challenges and opportunities in understanding and utilisation of African insect diversity', published in *Cimbebasia*, the Namibian biological journal. Scott and Rogo point out that only about 4 per cent of ecologists and 7 per cent of insect systematists are actually based in Africa. Funding is hard to get. Virtually all research on insects in tropical Africa focuses on the negative effects of insects, even though these make up less than 1 per cent of known African insect species. In the journal *Insect Science and Its Application*, between 1990 and 1995, of the 100,000 insects known from the Afrotropical region, fewer than 500 species were mentioned, and 97 per cent of the articles were about insect pests and economics. Few people regard arthropods as fun animals to research upon. However, in a noble attempt to reach the public, during the last fifteen years, *SWARA*, the magazine of the East African Wild Life Society, has published a series of articles showcasing spectacular arthropods, written and illustrated by the talented Kenyan entomologist and artist Dr Dino Martins, who has also published a string of papers and aims to finish a definitive book on Kenya's insects soon. Dr Martins offers a free insect identification service through the insect committee of Nature Kenya. You can send photos or questions to the team at insects.eanhs@gmail.com. He also has a blog, at http://dududiaries.wildlifedirect.org/

Naturalists and explorers who collected insects in East Africa in the nineteenth and early twentieth centuries include Von der Decken, who was accompanied on his second safari by entomologist Dr Otto Kersten, after whom a beautiful little ground-dwelling chameleon is named. Other insect collectors from this time include several who appear in other chapters of this book, like Sir Harry Johnston, Johan Hildebrandt, Emin Pasha and Franz Ludwig Stuhlmann; details of early insect collectors in East Africa, particularly in Tanzania, are provided in Hans Schabel's book. Most of the specimens they collected are housed in museums in London, Berlin, Vienna and Brussels. In Tanzania, a major expedition to collect insects on mountains was mounted in 1905 and 1906 by the Swedish zoologist Yngve Sjostedt, who considered 'an investigation of the insect fauna of the mountain tops of East Africa

Kenyan entomologist Dr Dino Martins with Milkweed Locust. (Dino Martins)

among the most interesting challenges of zoogeography'. His team collected 59,000 specimens of 4,300 species, of which 1,400 were new to science. Considering our state of the knowledge of Kenya's arthropod fauna, you could probably find a similar number of new species even now … if you had the funding, energy and wherewithal.

The economic factor has led to government taking a direct interest in arthropods. No country can afford to neglect agriculture, and insects have a major effect on farming. In 1908, Sir Frederick Jackson appointed J. T. Anderson to the post of Official Entomologist in the Agricultural Department of the British East Africa Protectorate, as Kenya was then, and government have employed entomologists ever since; both Richard Le Pelley and J. C. M. Gardner held the post of Senior Entomologist. Anderson's main business was dealing with problem crop insects, in particular pests of coffee; he also held the post of 'Inspector of Plant Diseases'. In 1924 he published, through the East African Standard publishing house, the remarkably titled *Bloodsucking Insects and Their Allies in the Colony and Protectorate of Kenya*. Kenya's Forest Department appointed F. G. G. Peake in 1947 as their first full-time entomologist; his remit was to identify and take action against the insect pests affecting the exotic conifer plantations in Kenya's highlands.

Few people, if asked, could name a famous entomologist; they tend not to be prominent or popular people. Some older people may know of Jean-Henri Fabre, the French entomologist who popularised the study of insects; he persuaded a group of caterpillars to follow each other around the rim of a glass for a week. Ecologists will know of E. O. Wilson, an ant specialist who has authored popular and brilliant books on ecology. Some entomologists have found fame in other fields. Charles Darwin wrote of the pure pleasure he got from collecting beetles. Sir Cyril Clarke, pioneer in research on the prevention of Rhesus disease of the newborn, was a lepidopterist. It was his work with *Papilio dardanus*, the African Mocker Swallowtail, that led him to understand the genetic basis of polymorphism and this then led to his figuring out the Rhesus blood groups. So we have to thank a Kenyan butterfly for the millions of lives saved through this discovery, for which he was very deservingly knighted. Alfred Kinsey, who did pioneering work on human sexuality, was an entomologist. The novelist Vladimir Nabokov, famed for his stylish novel of under-age sex, *Lolita*, was an entomologist who privately rated his entomological work on butterflies as highly as his literary efforts; for several years he was curator of Lepidoptera at Harvard. Theodosius Dobzhansky, the Russian geneticist who famously said 'nothing in biology makes sense except in light of evolution', did his pioneering work on flies. One of Kenya's most famous scientific sons was an entomologist – Professor Thomas Risley Odhiambo. Born in Mombasa (presumably in the evening, since the name Odhiambo signifies that), where his father was a telegraph officer, Odhiambo's first degree was in biology at Makerere. He later took the natural science tripos at Cambridge, followed by a PhD in insect physiology. After lecturing at the university in Nairobi, Odhiambo founded ICIPE, the International Centre of Insect Physiology and Ecology, out at Kasarani on the Thika road; the local area is jokingly known as 'duduville', i.e. insect town. Odhiambo was a visionary, a man who saw that pesticides were not the only solution to the problems of the African farmer, and he encouraged the use of low-cost biological and ecological tools. He also founded the department of entomology at the University of Nairobi. He started the journal *Insect Science and Its Application*, published under the auspices of ICIPE, and publication continues today, under the new masthead *International Journal of Tropical Insect Science*. Thirty-one volumes have been published. This is a journal that does good; its pages are packed with articles by dedicated entomologists aiming to assist the people of Africa in what is largely a battle with the arthropods. And the bulk of the PhD-trained scientist nurtured at ICIPE still work in Africa; this is a triumph in light of the steady brain-drain of talented people out of Africa to better-paid posts in the West. Odhiambo wasn't a man who avoided controversy. In the 1990s he clashed with the governing body of ICIPE over its future direction; he saw the institute becoming a university providing training; his opponents wanted it to concentrate on research, and they had the support of those who held the external purse strings. Tragically, Odhiambo was removed from his post by the ICIPE governors. One is reminded of Professor Bethwell Ogot's prescient remark in his autobiography, that no research and development programme can succeed in Africa if it is based on donor funding alone.

Research on insects, in connection with livestock, is also carried out at the veterinary research centre at Muguga, north-west of Nairobi; where a new institute was built in the forest there in 1949. Originally under the auspices of the East African community, in 1979 it became the National Veterinary Research Centre.

Visitors to the National Museum, or as it was in the early 1960s, the Coryndon Museum, may remember the insect hall. Filled with display cabinets, it showcased Kenya's insect fauna, with several thousand species on display. The senior entomologist at the museum at that time was Robert ('Bob') Carcasson, an ex-tank commander and farmer who had studied tropical agriculture in Italy. Carcasson was appointed to the museum in 1956; his main interest was butterflies and hawkmoths and he produced the pioneering paper we mentioned earlier on African butterfly zoogeography (Carcasson 1964). In his spare time, he published on reef fish. Controversy attaches to Carcasson's time at the museum. He had become director in 1961 and received his PhD from the University of East Africa, but it was suggested that he had the wrong attitude towards some of the talented young Kenyans that were working at the museum; the story was that he refused to let the black scientists use the main toilet. He was removed from his post in 1968, and was succeeded by Richard Leakey, who was 24 at the time. The Kenyan Mike Clifton, now a well-known safari guide, took over as entomologist, but was then replaced by a British entomologist, Dr Mark Ritchie. The insect gallery was dismantled – it was seen as old-fashioned; it probably was by modern standards – but nothing took its place, and this was still the case after Ritchie left, six years later. However, the entomology department at the museum in Nairobi is nowadays a hive of activity. The museum website states that it has over 2 million specimens, and a team of more than ten entomologists that includes the eminent Dr Richard Bagine are currently researching Kenya's arthropods; their projects and publication lists are extensive. Long may it continue.

Chapter 12

The Marine Environment

It is excusable to grow enthusiastic over the infinite number of organic beings with which the sea of the tropics teems ... the submarine grottoes, decked with a thousand beauties – Charles Darwin, *The Voyage of the Beagle*, 1839

The oceans cover more than 70 per cent of the earth's surface. Kenya's coast lies at the western edge of the Indian Ocean, the world's third largest body of water. Unlike the other large oceans, the Indian Ocean is closed on its northern side, which causes drastic seasonal changes in the winds and currents.

There are coral reefs along most of Kenya's coastline and they are of considerable age; most are fringing reefs that enclose warm shallow lagoons. Fossil coral extends inland for 10–20km along most of the coast, as well as offshore. The coral reef is the most diverse and life-rich ecosystem in the sea. The building of a coral reef can take from 7,000 to 10,000 years of work by the coral polyps. The El Nino event of 1998 caused lasting damage to 16 per cent of the world's reefs; some parts of the western Indian Ocean lost 90 per cent of their coral. Over 16,000 species of bony marine fish are known worldwide and about 750 species are known from the sea off Kenya. Of these, close to 500 species are associated with the reef.

This chapter describes Kenya's marine environment, looking at the sea and the tides, the beach, the reef and the lagoon, life in deep water and life on the reef, and exploitation and conservation. Thus, this chapter is about both the environment and the life within it. It covers a broad range of habitats and a broad range of creatures. Of course, no single chapter in a book can do justice to Kenya's marine environment and its boundary; entire books are dedicated to just small parts of this waterland. All we hope here is to cover the basics and give the naturalist a flavour of the magnificence of Kenya's coastal landscape and the astonishing variety of creatures that live there. The chapter concludes with a review of the literature, and as usual, each section is self-contained.

THE SEA AND THE TIDES

Water is dense, nearly a thousand times denser than air. A cubic metre of water weighs 1,000kg, a metric tonne, and the water molecules within it are close together. A cubic metre of air at room temperature weighs only 1.2kg and the molecules within it are far apart. Hence water heats up slowly and cools down slowly, and can also hold a lot of heat. This lends stability to the watery world. Get up at dawn on the Aberdares and the air temperature can be at zero degrees Celsius, but by early afternoon the air temperature may have risen to 13 or 14 degrees. Even at Mombasa, the air temperature can change from 22 degrees at dawn to 32 degrees by lunchtime. This never happens in the water. The water temperature one metre below the surface in Kilindini Harbour may be 24°C at dawn and it won't be any different at 2 o'clock in the afternoon. It needs a prolonged period of heating to raise water temperature. In the language of physics, we say water has a high specific heat capacity and it is this property that gives the sea its temperature stability. You can get local effects, sometimes a thin layer of hot water may overly a cooler mass; those who dive or snorkel will have seen the shimmering effect of such a layer. And as you descend, the temperature may drop suddenly in some places, but not in others. In general, however, a large water mass tends to have a stable temperature.

Kenya lies on the western shore of the Indian Ocean, which is the third largest of the world's oceans (the Pacific is the largest, followed by the Atlantic). The Indian Ocean contains 292 million cubic kilometres of water, which is about 20 per cent of the world's water; it has a surface area of 73 million square kilometres, which gives it an easily calculated average depth of exactly 4km. In reality, of course, all the world's oceans are connected, but the Indian Ocean is regarded as extending from the eastern shores of Africa to the western half of Australia, a distance of about 10,000km.

The Indian Ocean is also the world's warmest ocean, with surface temperatures over the entire ocean averaging between 22 and 28°C, and over 28 degrees in the Bay of Bengal. Its deepest point is the Java Trench, over 7,000m deep. The continental shelves (in effect, the flooded land areas) around the Indian Ocean are generally narrow, averaging 200km in width, with the exception of the western Australian coast, where the shelf extends several hundred kilometres westward.

Off the southern Kenyan coast the depth increases fairly rapidly. The 100m contour is usually less than 10km offshore along the coast from Shimoni north to Watamu, but then moves away from the shore, as the Tana Delta has created a fairly large shallow plateau. The 1000m contour is less than 30km east of Lamu, but this contour then continues more or less directly south, so between Malindi and Shimoni the 1000m contour is close to 200km offshore.

A vast current, the great ocean conveyor belt, carrying cold water northwards at depth, runs past Madagascar's east coast. Water inside this system moves very slowly; cold water down inside the current may not return to the surface for over 1,000 years. This does not directly affect Kenya. However, south of India, near the equator, this current rises (a process called upwelling), bringing cool and nutrient-rich waters to the surface. It powers an important food chain. The nutrients feed phytoplankton, which are tiny green free-floating plants, which in turn feed the tiny animals called zooplankton. The zooplankton support huge numbers of small pelagic fish (pelagic means living in or associated with open water), and are eaten themselves by big fish. Most other areas of the western Indian Ocean are nutrient-poor; hence there are few fish.

Of importance to Kenya is the south equatorial current, (which isn't actually on the equator, but 10 degrees south) which flows westwards past northern Madagascar and splits into two as it approaches the Mozambique coast, in the vicinity of Cape Delgado, at the Tanzania–Mozambique border, turning north and south. The southward flow is then called the Mozambique or Agulhas current; the northward flow is called the East African coastal current (EACC), although some authors call it the Somali current. This current flows northwards all year, right along the Kenya coast, to southern Somalia; beyond there it is affected by the season and consequent winds. The south equatorial current originates on the far side of the Indian Ocean, and explains why observers on the Kenya coast see fish that are also found in the Seychelles, and even in the western Pacific; they have been carried right across from the east in the current.

During the south-east monsoon, the Kusi (June to September, the word *kusi* is a shorthand of *kusini*, meaning south), the wind blows from south-west to north-east, at an average velocity of 9m/s. The coastal current flows north at around 4 knots and it extends right up the Somali coast to Cape Guardafui and Socotra, thence east. The Kusi, travelling a long way across the ocean, can create big waves. These thoroughly stir the water, causing a lowering of the surface temperature, as cool water from below is brought up, and thus decreases nutrient growth. Between November and March, the wind changes direction, to the north-east monsoon or *Kaskazi* (meaning either 'north' or 'cold', this is debated), which blows from north-east to south-west at an average velocity of about 5m/s, usually in the afternoon, and slows down the coastal current to about one knot. The coastal current is also forced offshore and eastwards, between Lamu and Kismayu in Somalia, becoming the Equatorial Counter Current, moving north of the Seychelles. October and April are usually relatively quiet in terms of winds.

The western Indian Ocean lies within the tropics. The surface sea temperature off the Kenya coast is rarely less than 20°C and is usually between 25 and 33°C; hence it is actually warmer than most of the Indian Ocean. The surface waters of the EACC are warmest in March, averaging 29°C, and coolest in September, average 25°C; the water is also highly saline and saturated with oxygen. This current has an interesting profile and, like the speed, it changes with the seasons. Its depth is greatest during the south-east monsoon, down to about 120m, and its temperature only falls a tiny amount between the surface (27.3°C) and 100m depth (27.1°C). But below this depth the temperature falls rapidly, inside some 20metres further it drops from 27 to 19°C. Such rapid falls in temperature over a short distance are called thermoclines. During the north-east monsoon, the current is only 85m or so deep. Further down is a mixture of static water and the slowly moving conveyor belt. As you get deeper the water gets colder; at a depth of 500m the water is only 12 or 13°C, at 1,000m it is 5–6°C. The light dies away rapidly too, with red disappearing first (it has less energy than blue); only 25 per cent of light reaches down to 10m depth, and at 100m only 0.5 per cent of the surface sunlight remains. The maximum depth to which enough sunlight penetrates to enable plants to photosynthesise, the photic zone, is never deeper than 250m and in coastal waters is usually no more than 50m. Below that, life and food webs must be animal-based. Let us leave these cold dark regions, return to the surface and talk about the tides and the waves.

The tide, that rhythmic rise and fall of the sea, is largely caused by the gravitational pull of the moon. The Kenya coast has two high tides and two low tides per day, roughly. Since the moon always travels westward, it follows that high water moves across the Indian Ocean from east to west; high tides occur in the Seychelles nearly an hour before they occur in Mombasa. The moon actually takes 24 hours and 50 minutes to go around the earth, which is why it appears 50 minutes later each day. The lunar cycle is roughly 29 and a half days, so the actual time between a high and a low tide, in theory, is 24.833 (50 minutes is 0.833 of one hour) divided by four, which gives 6.21 hours, 6 hours, 12-and-a-half minutes. So high tide is 50 minutes later each day … in theory. The tidal range on Kenya's coast is small, roughly between 3 and 3.7m between high and low water at spring tide, and in fact the tidal range for most of eastern Africa is between 2 and 3.8m, save at Beira in Mozambique where the discharge of the Zambezi can push it up to 5.7 m (although even this pales in comparison beside the world-record range in the Bay of Fundy in Canada, where the rise and fall is 17m). A number of organisms live in the inter-tidal zone, the area between low and high water (sometimes called the 'littoral zone', although strictly speaking this means anywhere near the shoreline) and the intertidal zone is where life moved onto land, or so we presume.

Across the sea, mostly from the north-east and south-east, comes the wind. It is the wind that creates waves; energy from moving air is transferred to the water. It needs a lot of energy to get a mass of liquid moving, because water is nearly a thousand times denser that air. But once it gets going, the water has a ponderous inertia that is hard to stop. One group of molecules bash into another, the bashed ones bash another, and so on. The particles themselves only move back and forward, what actually flows is energy. Watch a wave moving through water, and as it passes a floating object, you will see that object rise and fall, but it finishes up where it started. What has passed is kinetic energy, and coupled with the mass of the water, the water has a quantity called momentum. Something interesting happens as the wave approaches a beach. As it gets into shallow water, the wavelength of the wave shortens, it moves more slowly, and – since the momentum is conserved – it becomes taller. At the same time friction caused by the sea bed slows down the bottom of the wave, but the top tries to continue at the same speed. As a result, the wave starts to topple over. It may spill, running forward, or it may break, when the top of the wave plunges. The water running up the beach is called the swash, and that running back is called the backwash. Most of Kenya's beaches slope gently, and the fringing coral reef takes the brunt of the energy of the big Indian Ocean waves. The waves are largely constructive, moving the sand up the beach; the swash is more powerful than the backwash. However, if the beach is steep (and the beach profile can vary, they change all the time) and the waves break strongly, the backwash may be more powerful than the swash and this pulls heavier material back off the beach.

If the waves hit the beach at an angle, rather than perpendicularly (at right-angles to the shoreline), the sand is carried onto the beach at an angle, but gravity pulls it straight back to the water, perpendicularly. This results in debris being moved along the coast in the direction of the waves; this is called longshore drift. Material carried in water will drop as the water speed slows (you see the same with moving air, heavy objects only get taken aloft if the wind speed is high). If that material is heavy, it will drop even if the water is moving fast. If it is light, it won't drop until the water has slowed down a lot. As the water runs up the beach, it slows; as it runs down, it accelerates. This causes sorting, with larger bits of debris at the bottom of the beach, and lighter stuff further up. The bigger bits of coral rubble don't usually get onto the beach, but forms pockets of valuable habitat in the lagoons, although dangerous animals like stonefish may lurk there.

THE BEACH, THE REEF AND THE LAGOON

The shape and structure of Kenya's coastline is a consequence of its geological history. During the last 5 million years, the sea extended inland for at least 6 kilometres, virtually right along the coast; around a million years ago the sea was at least 75m higher than it is now. Then the sea level sank, or the land rose, possibly as a result of the weight of water no longer holding it down, although some geologists suspect tectonic movements. And during the last 100,000 years, the sea level dropped to at least 35m below present level before rising back up; proof of this are ancient terraces at least 35m below the present sea level. At the same time, the sea remained warm and coral flourished. So all along Kenya's coastline the land surface inland from the beach, the hinterland, consists of ancient coral, known by its quarryman's term of coral rag; rag is rock with open spaces. Here and there are sand lenses, and occasional limestone outcrops. The coral rag rests unconformably (meaning there is a gap in succession) on the much older rock below and is estimated to be at least 100m deep. The immediate offshore rock is also coral rag; nowhere along Kenya's 574km coastline does igneous or metamorphic rock outcrop on the shore. This means that the coastline of Kenya is largely an unrelenting line of sandy beaches and low coral cliffs. The smoothness of the coastline is typical of an emergent coast, formed by sinking sea levels or rising land, with few drowned valleys.

Beach material largely comes from the pounding of waves. Kenya's white sand beaches thus come from the break-up of coral fragments, calcium carbonate. In places in the north yellow silica or quartz sand is seen, brought from the interior by rivers. Quartz is a very tough material. Walk along Kenya's coast and what you see is either gently sloping sand beaches, or slightly harder coralline outcrops that have survived the pounding of the waves. Occasional low cliffs occur, bits of tougher coral rag, and sometimes the waves meet behind them, forming an island; continuing wave action wears the base, leaving a mushroom-shaped outcrop. Eventually that will fall. In places there are dunes, formed where beach sand has been blown inland and caught up in vegetation, forming a ridge that then piles up, like Shela Beach on Lamu. These dunes can be interesting places, with unusual life forms. A study of the dunes north of Malindi by Pamela Abuodha found over 170 plant species in one dune field. The beach sand tends to form the dune, as a result of the continual onshore wind, but along the Kenyan coast, fossilised sand lenses and fossil dunes exist in places, and their erosion by wind adds to the sand. If the rate of growth of vegetation is not as fast as the rate of sand piling up, a big dune can result.

The sea erodes relentlessly, and the Kenya coast is a lee shore, meaning that the wind drives the waves towards it most of the time, but the presence of the reef and the shoreward lagoons means that coastal erosion is very slow. In addition, there are few violent storms along the equator. In living memory not many of Kenya's coastal establishments have been lost to the sea. If you climb a coastal tree, what you see is a flat landscape. Dive in the sea and what you find is a gentle slope beyond the reef; there are no underwater granite spikes, like the Seychelles, or massive blue holes. The immediate hinterland, the beach and the sea just offshore slope gently, or are flat. The restless, unceasing flows of energy brought ashore by the never-ending wind and waves are absorbed on long gently sloping beaches. To many people, the beach is the place to be; somewhere you can lie in the sun and watch the waves break gently, where the only large visible animals are the ghost crabs of the genus *Ocypode* that zoom up and down the intertidal zone at speed. No one is certain if they are called ghost crabs because they are white, or because they move so fast and disappear so rapidly, or because they mostly come out at night. There are actually over 300 species of crab recorded from Kenya, including a 4kg monster, the Robber Crab *Birgus latro*; this remarkable crustacean climbs coconut trees and cuts down the nuts.

The reef along the Kenya coast is a fringing reef. Charles Darwin wrote on coral reefs, in his first monograph. He proposed three stages of reef formation: firstly a fringing reef, then the barrier reef, finally the atoll formed by a subsiding island. In Kenya as the sea level rose and fell over the last 5 million years, coral grew along the edge of the shoreline, in the shallow water, creating reefs that fringed the land; thus the reef moved both out to sea and inland. The tough limestone rock that forms the hard structure of a coral reef remains, both off and onshore. Mombasa is built on top of an ancient reef, constructed over 100,000 years ago, when the ice caps were smaller and sea level higher. The present reef is growing on top of an old one. The general pattern is that of a fringing reef, a few hundred metres offshore, forming a hard platform that the sea comes over at high tide; the reef flat. This platform is exposed to the air at low tide and hence the bits that get dry don't have living coral. The presence of the fringing reef has created, between the reef and the beach, a relatively shallow lagoon. The lagoon is rarely deeper than a few

Angry Grapsid Shore Crab in Watamu house. (Stephen Spawls)

The Marine Environment

The distant reef, the lagoon, Sacred Ibis and Great White Egret, Vipingo. (Stephen Spawls)

metres and is filled with coral outcrops and, near the beach, seagrass beds. On the seaward side of the fringing reef the seabed slopes down gently, and has coral down to a depth of around 40 metres. The seaward slope of the reef can be quite steep but is rarely high; there are no huge walls. But the fringing reef can be spectacular, nevertheless. The coral that stretches from Msambweni, near Shimoni, to Malindi is the world's longest continuous fringing reef, its length due to the fact that no big rivers enter the sea along its length and dump silt. There are gaps here and there, known as mlangos (*mlango* is Swahili for door or entrance), through which the water funnels as the tide is ebbing or flowing; they are exhilarating places to snorkel or dive, but dangerous. The lagoons are lovely places; often they are very warm, especially the shallower ones, during the day, as the sun heats the water. Coral often grows profusely. You can snorkel happily, and dangerous sea creatures like sharks virtually never get into the lagoons. The reef protects the lagoons; they are gentle places, nurseries of life.

Coral reefs don't occur on every coast. Although there are corals that occur in deep dark, cold waters – corals are found off the coasts of Norway and Scotland, for example – they generally don't produce enough calcium carbonate to make large reefs. To form large coral reefs, with overlapping species and big coralline structures, you need light, warmth, clear and shallow water. The big coral reefs of the world occur in the Indian and Pacific Oceans, 30 degrees either side of the equator, to a maximum depth of 60–70m (although there is usually very little below 40m). There are few coral reefs in the Atlantic; the only reef in West Africa is around Sao Tome. Fresh water kills corals; they absorb water by osmosis, swell up and burst. And fresh river water carries silt and mud, which darkens the water, preventing coral growth; the algae within the coral communities cannot photosynthesise. For these reasons, areas on either side of the Tana and Sabaki Deltas have no reefs and there are few in Ungwana Bay. Nor are there reefs on the Somali coast between Mogadishu and Cape Guardafui, due to the relative coldness of the sea and the absence of vegetation inland; storm runoffs cloud the water in the rainy season.

Coral itself is a fascinating and curious structure. A coral reef consists of literally millions of animals. Their soft tiny bodies, called polyps, are basically living stomachs. This stomach, or gastrovascular cavity, has a single

The beauty of the reef. (Stephen Spawls)

entrance, the other end of the polyp is cemented to a rock by calcium carbonate, or limestone, which the coral secretes at its base; forming a cuplike depression, called the calyx. It is calcium carbonate that makes up the solid structure of reefs and it is all produced by the little animals. Around the mouth/anus of the polyp are tentacles and these have stinging cells that fire like little harpoons. Coral polyps feed on small animals they can catch, from tiny fish to plankton, but also get nutrients from an unusual hitchhiker. They live in a state of mutual benefit, symbiosis, with a species of unicellular algae. Known as zooxanthellae, the algae photosynthesise, providing the coral with photosynthates (chemicals such as glucose), the coral reciprocates with carbon dioxide and waste chemicals containing nitrogen and phosphorous, which the zooxanthellae use to make proteins and other nutrients. The zooxanthellic algae is often coloured blue, green or brownish, and gives some corals their distinctive colours; it is visible as tiny coloured dots on the surface of the polyp. Several other marine organisms also have symbiotic algae, including sponges and clams.

The coral polyps are connected by a thin sheet of living tissue, the coenosarc, and thus they can share nutrients. Thus much of the surface of a coral is covered by living tissue. It is this inter-connectedness that had led some scientists to describe big reefs, like the Great Barrier Reef of north-eastern Australia as 'the world's largest living organism', the argument being that since all the polyps are connected, they are all part of a single thing. Not everyone will agree with this. Corals can actually be aggressive with each other, and if two corals are placed next to one another, the more aggressive species stings and extends filaments into its rival and starts digesting it; the zoologist Judith Lang established an actual 'pecking order' of Caribbean corals. Although the little polyps and associated algae have a relatively short life, a coral growth can last a long time, as the underlying skeleton gets bigger. Some corals have been observed growing at a rate of several centimetres per year, and corals that are estimated to be over 200 years old are known. Work by the oceanographer Dr David Obura on coral growth rates around Watamu and Malindi (1995) indicates that coral growth rates vary a lot in Kenyan waters. Staghorn Coral *Acropora formosa* may grow up to 10cm per year, while Boulder Coral *Porites* spp. may only grow a few millimetres

The gently sloping coral sand beach, Watamu. (Glenn and Karen Mathews)

per year. However, the corals at Malindi were more stressed than those at Watamu, and the stress resulted from sediment fall. This is usually caused by rivers carrying silt into the ocean, and stress is significant to corals.

If coral gets stressed it may eject the algae, and this causes what is called 'bleaching', as the colour imparted by the zooxanthellae is lost, leaving the white calcium carbonate skeleton. The coral polyps get their calcium carbonate from chemicals present naturally in seawater – calcium and carbonate ions. Coral can reproduce asexually, just making new cells, and thus a coral head grows bigger. It can also reproduce sexually, by releasing sex cells, known as gametes. With perfect timing, male and female coral gametes are both released on the same night, often at or near a full moon on a turning tide, so as to ensure the sperm and egg meet in quiet water. The simultaneous release of thousands of gametes can actually cloud the water. The male and female cells fuse, forming a zygote, and then find a suitable spot to start growing. The zygotes can drift great distances, which is a good survival technique, as they can repopulate areas where the coral has become extinct. It is by this method that coral has spread right across the Indo-Pacific, looking for a spot that is warm, well lit and nutrient-rich. We can all empathise with such desires.

Coral is an environmental indicator. It is sensitive to the amount of carbon dioxide in the atmosphere, and this is increasing. Atmospheric carbon dioxide dissolves in the sea, forming carbonic acid, so the more carbon dioxide there is, the more acid the sea gets. Increased acidity leads to a lowering of the concentration of the carbonate ion, CO_3^{2-}. This means the coral produces less calcium carbonate, so the reef doesn't grow so fast. In addition, corals die if it gets too hot (about 10 per cent of the Indian Ocean coral died in 1998 as a result of the El Nino Southern Oscillation event). They are also sensitive to pollutants in the water and to suspended particles that block out the light necessary for the zooxanthellae to photosynthesise. Our coral is a delicate bellwether of the health of both the coastline and the sea. In general, corals are protected worldwide, but the law isn't enforced in many places. About 8 per cent of the world's population are now living within 100km of a coral reef, so the reefs' over-exploitation and pollution from land-based sources has become a serious problem. In some countries – including Tanzania – dynamite fishing is practised, a frightful method of catching fish by setting off explosives underwater. It slaughters other harmless but unusable organisms and shatters the reef. In other places coral is harvested for building stone and for cement production; this is done in Kenya but with dead coral inland. If done sensitively this causes no environmental harm, but in some countries even offshore coral is dug out. Kenya still has most of its coral and luckily the Kenyan reefs are at less risk than those in South-east Asia, where it is calculated that 81 per cent of the reefs are at high or very high risk.

It is hard to convey in words the sheer beauty of the coral reef. The pleasure of snorkelling or swimming over a coral reef is like no other experience on earth. It is like being admitted to another, magical world. Gravity seems to have no effect; the observer feels as though they are flying, silent and weightless in a warm blue-green world. The colours of the inhabitants are stunning; the architecture of the living coral, unrestricted by weight, has shapes that defy the imagination. Coral reefs are the world's most diverse shallow water ecosystems; they make up only one per cent of shores but have 25 per cent of all marine life. There are reefs along about one-sixth of the world's coastlines. They are places of enchantment, and everyone deserves a chance to enjoy them. Too little attention is paid to this. If our children were given a little tuition and a chance to snorkel in a coastal lagoon, some would be converted to marine conservation for life. As we have said in other places within this book, you only want to protect what you know and love.

In the shallow water on the landward side of the lagoon you often find seagrass beds. There is no better place to see photosynthesis in action; this is living science. If you drift over a seagrass bed on a sunny day, you can see the bubbles rising in their billions. Those bubbles are oxygen. The plant is absorbing energy from the sunlight, carbon dioxide from the water, plus water itself, and turning it into glucose and oxygen. If you snorkel again near dusk, you will see that the rate of bubble production has dropped. Most divers and snorkelers splash through seagrass beds on their way out to more exciting and deeper water, and yet those beds of waving green stems are fascinating places themselves. A square metre of shallow water sand may support more than 10,000 seagrass shoots. A number of fish species, for example the Dash-and-dot Goatfish *Parupeneus barberinus*, and the Seagrass Parrotfish, *Leptoscarus vaigiensis* (the only parrotfish that doesn't change sex if necessary), prefer the seagrass beds, and the beds also act as nurseries for the fry of some reef species.

The seagrasses themselves are genuine vascular plants; about 60 species are known worldwide and some twelve recorded from the East African coast. They are not true grasses, but are flowering monocots. They serve many functions: they stabilise the seabed, oxygenate the water and start the food chain for many organisms. Dead seagrass leaves can be transported over huge distances, dead seagrass leaves have been recorded at depths of 8,000m. They decay slowly, under the influence of bacteria and fungi, and provide nutrients that start many food chains.

In the seagrass beds of a quiet lagoon you might come across a Green Turtle *Chelonia mydas* eating the grass, for unusually this turtle is largely herbivorous (and hence the alternative name for seagrass beds, turtlegrass beds). In the past, you could even see those weird sea animals Dugongs grazing in seagrass beds. There are very few Dugongs left now on the Kenya coast, although there may be a few in the quiet backwaters of the Lamu Archipelago and the Tana Delta, and the occasional specimen is seen on the south coast. Dugongs do not co-exist well with fishermen; they get hopelessly entangled in nets and tear them. Their flesh is also palatable, tasting like veal, so Kenya's Dugongs have been largely killed and often eaten. Sir Frederick Jackson had a tremendous story about Kenyan Dugongs. The Natural History Museum in London wanted two Dugong skeletons and asked Jack Haggard (Henry Rider Haggard's brother), who was the vice-consul at Lamu in 1884, if he could supply them. Haggard got the fishermen to net a couple, and when they were drowned Haggard anchored the corpses in the sea, wrapped in sacking, a couple of hundred yards from his house; he was waiting for the flesh to rot off. The Dugongs began to decompose; they swelled up and stank horribly. Jackson, who was staying with Haggard, told the vice-consul he should make some cuts in the corpses and allow them to sink, but initially Haggard wouldn't hear of it. But after a few days of the unrelenting stench (Haggard's house was to leeward), he changed his mind and went out with his small consular sword to stab the bloated corpses. At first, his sword wouldn't penetrate, but eventually it went in, unexpectedly. Haggard lost his balance, slipped forward and the Dugong then exploded, showering the unfortunate official with a blast of putrefaction. Haggard tried to stay upright, until Jackson roared at him to

Green Turtle, grazer of seagrass. (Stephen Spawls)

jump into the sea. Jackson dryly records that after his swim Haggard landed clean enough to be approachable. The innocent beasts sacrificed for science had had the last laugh.

Stingrays are sometimes found in the lagoons, especially the Blue-spotted Lagoon Ray *Taeniura lymma*. Nine species of stingray are found off Kenya's shores. Although they are non-aggressive animals, their tails are armed with sharp, serrated spines with grooves containing cytotoxic venom. Legend says that Odysseus (Ulysses) was killed by a spear tipped with a stingray sting; a stingray stab into the heart killed Australian environmentalist Steve Irwin. But a stingray is unlikely to lash at you unless you tread on it, and they are nervous animals, usually moving away if you get too close. The trick for avoiding a closer-than-desired encounter with a ray, especially in murky water, is to shuffle your feet.

The seagrass beds are also rich places for marine invertebrates; many species of small invertebrate (polychaete worms, sea lice, shelled animals, starfish, urchins and other echinoderms) live there, often in concentrations of several thousand animals per square metre (although most of these animals are very tiny). These beds are energy-rich places, although the larger animals here are at risk as they find it hard to hide; there is danger from above, the occasional Fish Eagle quarters the lagoons. A walk or a snorkel in the shallow water where the seagrass grows may reveal interesting creatures; you may see starfish, of which some 50 or so species are known from Kenya, and under small overhangs spiny sea urchins lurk; many children running about on the reefs have fallen on these black spiky echinoderms and got spines in their limbs. The old way to remove them was to bandage a slab of pawpaw on the punctures and leave it overnight; in the morning you squeezed the flesh gently and the spines popped out. Sea urchins are worth a second look if you meet some underwater, one species (appropriately of the genus *Diadema*) shows iridescent electric-blue lines on its body and has a glowing orange anus. In the shallow water you may see snake eels, which are fish but look very like small sea snakes; the half-banded *Leiuranus semicinctus* and ringed snake eels *Myrichthys colubrinus* astonishingly so. There is little risk from real sea snakes in Kenya. Only a single species is known, the distinctively coloured Yellow-bellied Sea Snake *Pelamis platurus*, black above and yellow below, with a spotted tail. It lives in the deep sea and usually only dying specimens come ashore.

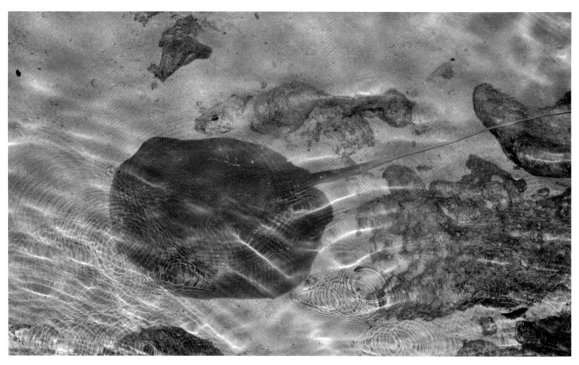

Stingray in shallow water. (Stephen Spawls)

In shallow water you may encounter the fat sausage-like sea cucumbers, or sea slugs. They are also echinoderms of the class Holothuroidea (the Phylum Echinodermata includes star fish, sea urchins, sea lilies, and sea daisies, these latter two are not flowers despite the name but animals with a radial or ring-shaped body plan; echinoderm means 'spiny skin' in Greek). They look ugly and uninteresting, but in fact they are full of surprises. Some sea cucumbers have a cathartic method of defence. If you pick one up it initially squirts water, and if further irritated it blows its guts out through a tear in its anus. The ruptured creature can survive this, but the lost material can take several weeks to grow back. Sea cucumbers have a collagen body wall, which can liquefy to allow the animal to squeeze through a narrow gap. Their skins contain a powerful toxin, holothurin; if a sea cucumber is cut in two and squeezed into a rock pool, the venom will stun or kill the fish in it. These leathery animals are a popular food and a basis of folk medicine in China, and they do actually seem to have some genuine pharmaceutical uses. The adventurer William Travis (1967), in his book about running a turtle cannery in Somalia (*The Voice of the Turtle*), describes a dangerous scam involving sea cucumbers in the 1960s. Apparently the most prized form is a reddish animal; a Japanese fishing boat crew discovered that if the common (and not so nice tasting) white sea slug was boiled in a copper pot it turned red and was thus indistinguishable – visually – from the expensive red ones. They sold their catch to Chinese food manufactures and many people were fatally poisoned. Despite this, sea slugs are still being exploited in some places; research along Sudan's Red Sea coast indicates that their numbers have been decimated.

Another interesting coastal habitat is the mangrove swamp or forest, which grows in the intertidal zone. Most trees do not tolerate salty soils, or those that are regularly flooded by salt water, but mangroves do both. The word 'mangrove' can be taken to mean any plant community that grows in the intertidal zone or near it, or just the trees that grow there, or, most narrowly, just trees of the family Rhizophoraceae. We shall use it here to mean the trees and the forest, or swamp. There are scattered mangrove forests along the length of the Kenya coast, mostly in sheltered waters, covering some 50km^2. A number on Tudor Creek and Mteza Creek were killed by oil pollution, due to tankers cleaning out oil residues. At Mida Creek there are more than 1,000 mangrove trees per hectare, and there are some magnificent mangrove forests on the Tana Delta and between Lamu and Kiwayu. They are a

Mangroves at Mida Creek. (Glenn and Karen Mathews)

fascinating habitat. Over 50 species of mangrove tree are known world-wide, of which around ten occur in Kenya. They are adapted for living in salt water in several ways. They have aerial roots (called pneumatophores) that stick up above water level and absorb and store gases from the air, so the gas supply continues even while the roots are submerged; the aerial roots also absorb nutrients. They control water loss by closing the leaf stomata; some take in salt water and secrete the excess salt, others prevent the absorption of salt ions and thus get only fresh water. Some have floating water-resistant seeds; others drop seeds in the mud. When they grow, initially a tap root is put down, and aerial roots then grow out of the trunk. Once they are strong enough the tap root dies and the tree stands on a framework of aerial roots, looking as though it is sitting on a big spider. The aerial roots are slim but tough and consequently the waves that fall on them create no great force on the tree; they can withstand the crashing forces of the sea. It is a clever design. Mangroves demonstrate the ability of the flora to thrive almost anywhere.

Mangrove swamps are useful. They give protection to soft shorelines, absorbing wave energy and preventing erosion. The root clusters provide quiet shallow water areas that act as nurseries and stable habitats for a number of marine creatures, including fiddler crabs and Mudskippers *Periophthalmus sobrinus*. Some birds nest there. And mangrove wood is hard and is termite-resistant, due to its high tannin content. For thousands of years, the mangroves of the East African coast represented the nearest supply of easily obtainable wood to the carpenters, boatwrights and builders of the Arabian Peninsula, and they harvested there. Poles of two main sizes, known as *mazio* (thin) and *boriti* (medium), were taken for use in construction. Mangrove wood was also used for manufacturing dhows, which were – in the early days – held together with wooden pins, no metal was used. If you wanted a dhow built in the Lamu Archipelago, you took a gamble, because you had to pay in advance for the entire boat, and the story was that, during the launching, there was a risk that the boat would fall off the launching cradle and smash to smithereens … but there were no refunds. A study by the botanist James Gitundu Kairo and colleagues, at Mida Creek, found that the trade in *boriti* and *mazio* mangrove poles for construction along the Kenya coast meant that the heavier trunks and branches were left on the trees. This had the unfortunate effect of cutting the light from the smaller saplings, which didn't thrive. In addition, as you might expect, the non-useful species became more common, but nevertheless the use was still sustainable. However, the demand on mangroves has decreased, and their future in Kenya looks reasonable, although in other places in the eastern Indian Ocean many mangrove forests have been felled so that shrimp farms may be set up.

LIFE IN DEEP WATER

Deep water can be found close to the Kenya coast. The 1,000m contour line is less than 25km southeast of Lamu; a few kilometres further out and the water is 2,000m deep. These are the deeps, no plants live down there in the pelagic zone.

The deep blue seems empty. On or near a coral reef, you are never out of sight of abundant life, but swim out over the abyss and often there seems to be nothing there. It can be scary, but it is an experience to savour, drifting out over the deep. You look down, and the azure deepens into midnight blue. There are organisms there, of course. Life in the ocean starts with plankton. These are small drifting organisms that live in the deep sea, divided into the phytoplankton, tiny plants, and the zooplankton, tiny animals; some marine scientists include bacteria in the group, the bacterioplankton. Most marine food webs start with the phytoplankton, which photosynthesise. Because they cannot function without light, they live in the photic zone, within 150m of the surface. They are eaten by the zooplankton, the little animals of the sea. Many zooplankton are single-celled creatures smaller than 1mm long, but some are large, for example jellyfish; included in the group are the krill, tiny shrimp-like organisms that feed some whales. Unlike their plant cousins, which are tied to the surface and light, the zooplankton may sink to great depths, often during the day and then rise back to the surface at night (a form of behaviour that confused the British Navy in the past, when the sonar appeared to indicate the sea was strangely shallow; the sound was being reflected off a layer of marine organisms). This behaviour is called diel vertical migration; they may go down a thousand metres. And the zooplankton is eaten by the nekton, a nice term that mean all actively swimming sea animals (as opposed to the drifters); the nekton range from tiny crustaceans to the Blue Whale *Balaenoptera musculus* (the biggest animal that ever lived).

The mass of plankton is less in the warm waters off Kenya's coast than in colder seas. This may seem strange – surely more warmth and sunlight should produce more life. But marine food webs start with phytoplankton, and they require nutrients. Such nutrients are usually picked up from the sea floor, so to get a rich supply of minerals like phosphorous, nitrogen and other elements, the water needs to pass along the sea floor and then well up, or be brought to the surface by wind-fuelled mixing of the water column. Upwelling does occur far off the eastern coast of Somalia, but Kenya doesn't benefit directly from it. Some areas, for example near the mouths of the big rivers like the Tana, benefit in a limited way from nutrients in the river discharge; 2–4 weeks after the surge there is an increase in zooplankton and a year of heavy rainfall may be followed several years later by a sudden increase in fish catch. But the overall lack of nutrients, coupled with the fact that there is nowhere to hide, means that the deep water areas off Kenya's coast are impoverished, and only 200 or so fish species are known from Kenya's pelagic zones. You don't get the huge swarms of fish like Herrings, Pilchards or Anchovies, although these fish do exist in limited numbers. The open sea is a hostile place to live exposed, like an open plain, but in three dimensions. There isn't anywhere to hide. The bottom is a long way away. The coral reef is like the river running through savanna, providing a nutrient-rich haven with lots of cover. Many organisms can flourish there. But open water is another matter. You avoid trouble in open water by being very small and living in huge gangs, or becoming large and swimming very fast.

There are some exciting fish, and other marine animals, in the deep sea off Kenya's coast. By virtue of their size and speed they are international species, found in tropical oceans worldwide. The endemic sea fish live in or near the reef. The deep blue water boys are worldwide species. More than 20 species of whale and dolphin have been recorded within 20km of the Kenyan shore, including the Killer Whale *Orcinus orca*, the Blue Whale, the Common *Delphinus delphis* and Bottle-nosed Dolphin *Tursiops aduncus* and the Sperm Whale *Physeter macrocephalus*. A number of sport fish, the big pelagic fast-swimming hunters of the sea, are recorded off Kenya's

The Marine Environment

Orangespine Unicornfish Naso lituratus. *(Stephen Spawls)*

Whitemargin Unicornfish Naso annulatus *in the deep blue. (Laura Spawls)*

Zebra Shark Stegastoma fasciatum, *harmless unless molested. (Stephen Spawls)*

shores. They include the Wahoo *Acanthocybium solandri* (which can swim at 75km per hour), several types of tuna (Yellowfin *Thunnus albacares*, Bluefin *Thunnus thynnus*, Bigeye *Thunnus obesus* and Skipjack *Katsuwonus pelamis*) and those superb fighting billfish; the Indo-Pacific Sailfish *Istiophorus platypterus*, Swordfish *Xiphias gladius* and the Black *Makaira indica*, Striped *Tetrapturus audax* and Blue Marlin *Makaira mazara*. All delight the big-game fisherman, and bring money into Kenya as visitors come to fish for them. They have a distinctive shape, with sleek powerful bodies and short fins that act as stabilisers; the force that pushes these pelagic fish along comes from the tail. If hooked, a billfish like a sailfish or marlin may demonstrate the sheer driving power of its tail by getting up out of the water and moving across the surface using only their tail, a technique known as tail-walking. There are few flabby, medium-sized, slow-moving fish in the open ocean, because there is nowhere to hide; what you find here are the high-speed hunters. There is no excitement in fishing quite like the shriek of line off the reel as one of these big fast fish takes a bait. The adrenalin-pumping excitement that comes as you battle with rod and line against the spirited speed merchant does not have to end with the death of the fish; most skippers will tag and release the fish, with honour satisfied on both sides.

The largest fish known off Kenya's shore is the Whale Shark *Rhincodon typus*, which reaches 12m length and weighs 12 tonnes, although there are anecdotal reports of larger specimens. Despite its huge size it is totally harmless. About 70 species of cartilaginous fish (sharks and rays) are known from Kenya's coast, out of a worldwide total of 750 species. Over 50 species of shark are recorded. Most are harmless, but a few are dangerous. Shark attacks have taken place around Mombasa Island, and Bull Sharks (also known as Zambezi Sharks) have been identified with near-certainty as being responsible for these. Bull Sharks *Carcharhinus leucas*, are big stocky sharks, growing to close on 4m and 300kg. Bull Sharks are probably responsible for the majority of near-shore shark attacks worldwide; they occur throughout the tropical seas and have penetrated long distances up the Zambezi and Ganges Rivers. In 1959, apparently, an elephant attempting to swim to one of the islands of the Lamu Archipelago was attacked by sharks and largely eaten, although precise details of this attack are hard to find.

Tiger Sharks *Galeocerdo cuvier* occur off the Kenya coast. Some big examples – up to 450kg – have been caught, but they do not seem to have the dangerous reputation they have in places like Hawaii, where several attacks take place each year. This is probably because Tiger Sharks prefer deep water during the day, and only come close inshore at night. People do not swim at night on the Kenya coast. In parts of the Caribbean, swimming in the sea at night is a recognised way of committing suicide. In the final analysis, there is always some danger from sharks if you are in the water, just as there is danger from Lions if you are on foot in wild country, and the longer you spend there, the greater the risk But we need to keep things in balance. Worldwide, there are usually only between 50 and 80 documented shark attacks on humans per year, of which around 20 per cent or less are fatal. Of course, some attacks may not be documented. But if one takes the usual sensible precautions, like not swimming at twilight or night, avoiding places where offal is dumped, not swimming alone in deep water, being aware of one's surroundings and never molesting a shark, even if small, there should be virtually no risk of shark attack. There is also a remarkable thrill – for those who want to risk it – of being in the water with a shark; the grace and beauty of these ancient predators is a sight to be savoured.

LIFE ON THE REEF

A lot of organisms live on and around coral reefs; they are the world's most productive ecosystems, stimulated by warmth, light and the kinetic energy of the water (although it must be mentioned that they tend to recycle their own products). Reefs are in shallow water, and, once established, have plenty of hiding places and nutrients. Along with rainforests, coral reefs are the most diverse natural communities on earth; both share a high species diversity, with organisms that have very precise niches and clear-cut relationships. And, unlike rainforests, the fauna of the coral reef is visible. In a big rainforest you can see very little vertebrate life without spending a lot of time looking. But on a coral reef you may be able to see 20 or 30 species of vertebrate just

Emperor Angelfish Pomacanthus imperator. *(Stephen Spawls)*

by turning your head. And what vertebrates too! No other animals display the kaleidoscope of colours that reef fish display, fish like the Picasso Triggerfish, the Emperor Angelfish or the Powderblue Surgeonfish, each an explosion of primary colours. Great flickering flocks of blue Chromis dance around the coral heads. The venomous but placid Scorpionfish, astonishingly arrayed in stripes of orange, white and black, with its absurdly long, feather-like fins, hovers in a recess; Moorish Idols drift by. Here and there are big flocks of squirrelfish, also called soldierfish, vifuvu in Swahili, and the oddly-masked Racoon Butterflyfish *Chaetodon lunula*. There is much speculation about the purpose of such garish colours in reef fish. Several reasons have been suggested: that they attract mates, they repel rivals, that they conceal the owner from predators. This last theory was put forward by that fine evolutionist Alfred Russel Wallace, the man who stimulated Charles Darwin to get publishing. But none of the theories seems very satisfactory. As with plants (see Chapter 5), the wavelengths of the electromagnetic spectrum that are useful to us for vision are pretty similar to those used by fish; it is the usual compromise between having waves of such high energy they cause damage and those of such low energy they can't stimulate a receptor. But we do know that a number of fish can see ultraviolet, which we cannot. Recognition of your fellow fish seems a promising possibility, but there might not be a single reason.

Close to 500 species of fish are known from Kenya's coral reefs, and not everyone knows that most of them can change sex quite easily; in fact more species of reef fish change sex than do not. Some populations are led by a dominant female, and when she dies, her lead male becomes female; in other species the male is dominant and when he dies his top female becomes male. Exactly how they do it isn't well known, but it's a trick that few other vertebrates can perform.

On a good day on the reef, the enthusiast will identify 50 or 60 species of fish, or more. The ornithologist Leslie Brown was a keen snorkeler, and he once saw 106 species of reef fish on the Big Three Reef at the entrance of Mida Creek. Brown identified a total of 180 species in the Mida Creek area. Ken Bock, the South African plant pathologist who made Diani his home, recorded a total of 195 species of fish in the Diani lagoon, and his

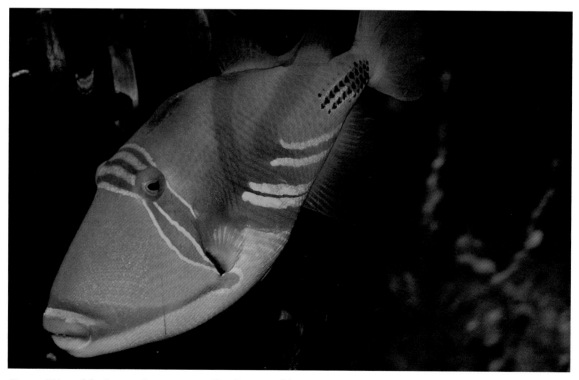

Picasso Triggerfish Rhinecanthus aculeatus. *(Stephen Spawls)*

The Marine Environment

Powderblue Surgeonfish or Blue Tang Acanthurus leucosternon. *(Stephen Spawls)*

Indo-Pacific Sergeant Abudefduf vaigiensis, *common along Kenya's reefs. (Laura Spawls)*

Respected by Muslim fishermen, the Moorish Idol Zanclus cornutus. *(Stephen Spawls)*

Coachman Heniochus acuminatus, *easily confusd with the Moorish Idol. (Stephen Spawls)*

Venomous but not aggressive, the Scorpionfish Pterois miles. *(Stephen Spawls)*

major paper on them was published in the *Journal of the East African Natural History Society* in 1996 (although sadly Bock himself died before it was published; the same issue contains his obituary). Look at the reef itself, with the many-coloured corals; the Staghorns, the Mushroom Coral, the Brain Coral, the great flat Acropora corals, like tables, and the huge 'coral heads' of the genus *Porites*; amazingly each coral head is either all male or all female. Kenya has a big diversity of corals; more than 200 species of hard coral (ones that have calcium carbonate skeletons) are known. The reef is a place of fantasy; to some it conjures up the magical landscapes painted by the British artist Roger Dean. Embedded in the coral you may see Giant Clams, with their beautifully coloured mantles waving. They need to be treated with care if irritated they snap shut, and many snorkelers may have felt a frisson of fear, imagining themselves inadvertently putting a foot into the open shell and being trapped, foot clamped inside, on a rising tide. In truth, there is little danger on Kenya's coral reefs and in its lagoons, save that of sunburn from spending too much time floating in this magical world. You might see a small shark or two – there are nurseries of Blacktip Reef Sharks in the lagoons at Watamu, and stonefish and lionfish are on the reef, and the occasional octopus, although it is unlikely to pull you down, unlike the unfortunate fisherman in Victor Hugo's novel of 1866, *The Toilers of the Sea*.

Bigger fish can be seen in some of the deeper pools inside the reef, although they are more common just outside. The underwater caves off Mida Creek are home to Giant Groupers *Epinephelus lanceolatus*, *tewa* in Swahili, big stupid-looking fish which can weigh up to 300kg and have killed humans, but only by accident. In one such incident an American diver shot a Grouper with a speargun and the huge panicked fish entangled the diver in the lanyard; he couldn't cut himself free and drowned. In deeper water you might meet two largish fish with strangely shaped lumps on their heads: the Humphead Wrasse, or Napoleonfish, which is solitary and can weigh nearly 200kg, and the Bumphead Parrotfish *Bolbometopon muricatum*, which isn't so big (up to 50kg), travels in shoals and has been recorded as eating 5 tonnes of coral per year, chewing it away with its big horny beak. On a reef, if it's quiet, you can often hear the parrotfish crunching on the coral. On the seaward slopes of the

The Humphead Wrasse or Napoleonfish Cheilinus undulatus, *one of the biggest reef fishes, reaching nearly 200kg. (Stephen Spawls)*

Blacktip Reef Shark Carcharhinus melanopterus *in Watamu lagoon. Known to mistakenly bite waders' feet. (Glenn and Karen Mathews)*

reef you may see feelers sticking out of recesses; these belong to lobsters, of the genus *Panulirus*. Delicious eating, they are secretive and hide in holes by day; their enemy is the Common Octopus. William Travis describes how he hunted lobsters in Somalia with an artificial octopus, made out of a rubber ball with rubber tube arms threaded with springy wire. The fake octopus was pushed into a likely hole and then inflated with a plastic bottle so that its arms shot out. Travis described the lobsters exploding out of every nook and cranny, producing bat-like squeaks in their panic-stricken flight.

The reef is actually home to an astonishing variety of life. Plants are limited. There are seaweeds (which are algae, without veins), some green, some brown, some red; there are over 200 species of seaweed known from Kenya's coast. There are also blue-green and green algae; in Zanzibar they cultivate it for food and sale to laboratories. But the diversity of animal life is astonishing. A phylum is a large group of organisms; for example, phylum Chordata includes all the vertebrates and some others, phylum Arthropoda, as we saw earlier, has at least a million member species. But on the reef you can see some pretty unusual phyla. There are the sponges, now in two phyla, Calcarea (calcite skeletons) and Silicea (silicon skeletons), which used to be in one phylum, Porifera. We know sponges because we use their skeletons in our bath. Several hundred species of sponge are known from Kenya; they can be red, green, yellow, blue, brown … the list goes on. The phylum Cnidaria contains the corals, jellyfish and sea anemones; all visible on the reef and all able to sting, although there are fish called anemone fish or clownfish that actually shelter in the anemone fronds; these little fish were made famous by the film *Finding Nemo*. There are comb jellies, of the phylum Ctenophora, resembling jellyfish. There are flatworms, of the phylum Platyhelminthes, and some of these are brightly coloured. Worms are diverse: worm phyla include spoon worms (phylum Echiura), peanut worms (phylum Sipuncula) and segmented worms (phylum Annelida). The phylum Mollusca covers not only the squids and the octopi, but also Kenya's huge variety of shells. These include the lovely conches, the many bivalve shells (the emblem of the Shell Oil Company is one half of a bivalve) and the venomous cone shells, which stab with a venomous dart and have killed people. Cone shells are best

The Giant Clam, Tridacna. *(Stephen Spawls)*

simply admired. There are over 50 species of those lovely organisms, cowries. The little Money Cowrie, *Cypraea moneta*, has been extensively used as currency all over Asia and Africa; in places it still is – in some West African countries the money paid to the parents of the bride always includes a few sacks of Money Cowries, and they are often used for decoration, even among people who live thousands of miles from the sea. Other strange phyla of the reef include Phoronida (horseshoe worms), Brachiopoda (lamp shells), Bryozoa (sea mats and lace corals) and Chaetognatha (arrow worms). This massive diversity points to an obvious conclusion; life began in water and the seas are the most extensive habitat on earth. Of the large organisms, there are more than 40 marine phyla and only 28 terrestrial phyla, and 90 per cent of all animal classes live in the sea.

The marine world is worth getting to know. There are dangers, of course: you risk drowning, sunburn, heat stroke, dehydration, or ear infections. In the water, sponge spicules can puncture your skin and cause irritation, you can be stung by jellyfish, stinging coral, anemones, cone shells, scorpionfish, stonefish, and stingrays. You might get slashed by coral, or a surgeonfish, stabbed by sea urchins or the Crown-of-thorns Starfish *Acanthaster planci*, bitten by Moray Eels, or attacked by a shark (extremely unlikely). But in reality, provided you follow the rules – treat everything with respect, don't touch, poke, pick up or antagonise sea life, and watch where you put your feet – a morning on Kenya's coral reefs should leave you with no scars, merely a longing for more.

EXPLOITATION AND CONSERVATION

Kenya's coast is a lovely place to live, if you don't mind the heat. It is fertile, with decent rainfall, wild areas, community and its own remarkable culture. More than 3 million people live along the coast, most of them between Lunga Lunga, near the Tanzanian border, and Mambrui, just north of Malindi. Not so many dwell around the Tana Delta. Many of the residents make a living from coastal resources. The statistics are impressive.

Ear-spot Angelfish Pomacanthus chrysurus. *(Stephen Spawls)*

Clown Triggerfish Balistoides conspicillum. *(Stephen Spawls)*

Sea Urchin Diadema, *with irridescent spots and glowing orange anus.* (Stephen Spawls)

Big Moray Eel Gymnothorax javanicus, *it can deliver a savage bite if angered. (Stephen Spawls)*

There are exclusively coastal occupations like coral rag quarrying, limestone and cement production, farming coconuts, cashew nuts, kapok, fruit and sisal, producing salt and coir rope. And tourism is a major earner. Coast hotels provide around 20,000 beds, and most coastal visitors come for the sunshine and the beautiful beaches, bringing in more than 100 million US dollars a year and giving employment to over 200,000 people. Around 50,000 people gain employment from marine fishing along the coast, of which 10,000 are fishermen, landing more than 6,000 tonnes of fish, which sells for in excess of 4 million dollars. And all those people and their families (which can mean ten or more people per earner) are dependent upon the coast remaining secure, unpolluted and not overexploited.

Interestingly, some 80 per cent of the fish landed are shallow water, reef-associated fish, caught by artisanal fishing (basket traps, fence traps, hand lines, nets of various types, spearguns and harpoons) and most are sold for local consumption. Work by Tim McClanahan, (McClanahan 1996) eminent marine biologist, indicates that the take (the MEY or Maximum Economic Yield) is pretty near the upper limit, and hence cannot be increased without detrimental effects. There is also a trade in pretty reef fish, corals and anemones for the aquarium trade. This is a big business worldwide. It is estimated that 30 million fish are traded each year, and most of these are wild-caught. Only about one per cent of marine fish in the pet trade are captive-bred. At the minute, the trade in reef fish is not particularly large in Kenya; most marine fish destined for the pet market originate in Indonesia and the Philippines. The trade is constrained by the fact that the organisms have to be kept alive and in good condition from capture until it they reach the customer's aquarium, so they cannot be harshly treated. However, the trade needs to be monitored, because some unethical practices are used, including simply pulling off coral heads to get the fish in them. In Asia many fish are caught by a horrible method: a solution of sodium cyanide is squirted into the coral, which stuns and disorientates the fish so that they can be easily picked up (and it kills the coral); to get those that don't emerge the coral is crowbarred apart. It is a technique that wrecks reefs. Reef fish are also collected by dynamiting, although this frightful practice isn't employed in Kenya … yet.

The Marine Environment

Octopus Octopus vulgaris, *enemy of the Crayfish. (Stephen Spawls)*

Statistics on offshore fisheries are hard to gather. At one time 20 per cent of the total catch in Kenyan waters was by large international vessels, trawling for plankton-eaters like sardines and mullet, and fish-eating fish like the Kingfish or Koli-koli *Caranx melampygus*, and species of tuna. The situation is now complicated by the instability in Somalia. Before the rise of piracy off the Somali coast, many international fishing boats operated there, with, it has to be said, little benefit to Somalia itself. Somalia also had its own fishing industry, but following the problems that have arisen since 1991, that industry has shrunk to almost nothing. However, the buccaneers have reduced the number of huge factory-fishing operations that were devastating Somali fish stocks. So, it seems that this cloud has a silver lining. But it's not all good news in conservation terms; an unfortunate consequence of the present situation in Somalia is that dubious waste-disposal companies have been paying Somali frontmen to be allowed to dump toxic wastes on the Somali Coast. In places, this runs off into the sea. Another spectre that looms over Kenya's coast is that created by the oil industry; a spill by an ultra-large crude oil carrier offshore could wreck the coastline for years.

All states with a coastline have absolute sovereignty up 22km (12 nautical miles) offshore and certain rights for a further 22km (the so-called 'contiguous zone'). These rights are concerned with pollution, immigration and customs/financial matters. There is also an 'exclusive economic zone' up to 370km (200 nautical miles) offshore where Kenya can exploit minerals and fishing, but the country has no jurisdiction on any ship passing through this zone (or loitering there, as it is charmingly put). But along the coastline itself, a confusing mix of organisations have jurisdiction over those who want to make use of the marine resources. Those who control bits of the coast include an alphabet soup of organisations: Kenya Wildlife Service (KWS), responsible for National Parks and other conservation areas; the Kenya Fisheries Department (KFD), which is responsible for fisheries development and management; the Coast Development Authority (CDA); the Kenya Marine and Fisheries Research Institute (KMFRI); the Ministry of Fisheries; and the National Environmental Management Authority (NEMA), whose business is to represent the government in the implementation of environmental policies. Other organisations

Porcupine Fish Diodon liturosus, *armed with inflatable spines, specimens have been found jammed in the throats of sharks. (Stephen Spawls)*

with stakeholders in the coastline include independent research organizations and donor agencies who supply funds and feel they have a right to say how those funds should be spent. There are, as always, two sides to the situation. The coast needs protecting. But the humble fisherman who just wants to make a living harvesting what he feels should be his for the taking as a citizen of Kenya, may feel that there are rather too many people out there who want to tell him what to do.

Formal protection of the near-shore flora and fauna does exist. In global terms, the Kenya coast lies within the Indian Ocean Whale Sanctuary, a huge theoretically protected areas where hunting of whales is forbidden. Kenya has five nicely positioned marine conservation areas, one up beyond Lamu, two on the north coast and two on the south, but their status is not uncontroversial. The most northerly is the Kiunga Marine Reserve, which was gazetted in 1979 and stretches from just south of the Somali border to Kiwayu. South of Ungwana Bay is the Watamu Marine National Park, lying between Malindi and Watamu. The Mombasa Marine National Park is between Shimo La Tewa (which means 'the hole of the grouper' in Swahili) and Nyali. South of Mombasa Island are the Diani-Chale Marine Park, south of Tiwi, and the contiguous Kisite Marine National Park–Mpunguti Marine National Reserve, near Shimoni. The form and amount of protection these sanctuaries offer varies. The Kiunga Marine Reserve allows traditional fishing and harvesting methods, and covers a beautiful stretch of coastline beyond Pate, but not many tourists are visiting it at the minute. The Watamu Marine National Reserve, established in 1968, extends from Malindi to just south of Mida Creek. It includes the creek itself and encloses two marine national parks: Malindi Marine National Park, offshore between Chanoni Point and Leopard Point, and Watamu Marine National Park, from Turtle Bay south to Whale Island, south-east of the entrance of Mida Creek. These were Kenya's first marine parks. No harvesting of any sort is allowed within the National Parks, traditional harvesting is allowed in the reserve. Similar rules apply to Mombasa Marine Reserve, established in 1986, and Mombasa Marine National Park, a smaller area that protects the reef on the eastern side of Mombasa

The Marine Environment

Tortoiseshell comb, made from turtle shell; one of the reasons turtles were harvested. (Stephen Spawls)

Island and the Leven Reef off Nyali Beach; the reserve itself extends from Likoni north to Shimo La Tewa. The Diani-Chale Marine Park theoretically stretches from Ukunda south to Gazi on the south coast but since its legal designation in 1995, there has been animated conflict between Kenya Wildlife Service and other stakeholders over the use of the area. There is a huge line of hotels along the beach there, and the landowners and the fishermen objected to any attempt to stop them harvesting and pay fees to be within the designated park; the situation isn't resolved and Diani-Chale is referred to at present as a management area. On the extreme southern border there is a little complex of protected areas. In 1978 the Kisite Marine National Park (east of the minute Kisite Island) and the Mpunguti Marine Reserve, around the south-eastern corner of Wasini Island and the three Mpunguti Islands, were set up. Where parks have been established, even in degraded areas, the coral and its inhabitants does seem to re-establish quite quickly.

Some international organisations are also involved in coastal conservation. In 2006, a locally managed marine area (LMMA) or community conservation area, was set up just north of Vipingo, at Kuruitu (Kuruwetu on old maps). At present, Flora and Fauna International (FFI) is trying to set up no less than seven such LMMAs in the area between Ras Jimbo, on the Tanzanian border, and Funzi Bay, conserving an area of over 12km^2. This is a commendable effort, because not only does it abut the Kisite and Mpunguti reserves, but the whole region is rich in coral reefs, including the big offshore reefs of Mwamba Cha west of Wasini Island, and around Sii Island. If it works, this will protect what is arguably Kenya's richest marine area, but – as the October 2011 FFI magazine states, laconically, 'decision-making proceeds rather slowly where there happen to be … Wazee who are particularly loquacious'. You've got to give the old men their say. FFI also promote the establishment of Beach Management Units (BMUs for short).

Marine Parks are important, and since you can oversee quite a big area with a handful of eagle-eyed scouts and some fast boats, and poaching at night on the water isn't really a practical proposition, protection of our marine resources shouldn't be too much of a problem. At the local level, though, it has to be mentioned that controlling who goes in and out of a park can be a headache. Unlike your average land-based park, where entrances can be monitored and fees charged, a marine park can be entered on foot all along the beach or by boat at a number of points. The response of the authorities has been to charge a fee to hoteliers whose establishments front the beach by the park, and this has often caused a lot of bad feeling.

But the big coastal picture is of an extremely beautiful area, with largely unspoilt white sand, sun-kissed beaches under shady coconut palms, and the potential to provide economic benefits and employment to a lot of people. But there are several threats. Coastlines are dynamic systems and are vulnerable; a single major weather, tectonic or offshore pollution event can quickly damage an entire coastline for a long period – consider the 2004 tsunami. Oil tankers may cause offshore pollution and the onshore winds tend to bring sea pollution onto the coastline. The possible effect of climate change and fluctuation in sea levels looms large. Pollution from the land, including sewage and agricultural chemicals, could be a problem, algal blooms have been observed in sheltered bays. And who knows where the continuing instability in Somalia may lead. Also of significance is the possible construction of a huge deep-water port north of Lamu, as part of China's investment in Africa. There is also the need to protect the Tana Delta, a crucial area for biodiversity. Other threats include overfishing, harvesting of shells and other marine invertebrates (it is important that visitors are discouraged from buying these), harvesting of turtle eggs and killing of turtles for food or their shells, clearing of mangroves and seagrass beds for aquaculture and salt manufacture. As yet, there seem to be no problems with introduced alien species, but this needs to be monitored. And, so far, the coast hasn't suffered too badly at the hands of foreigners willing to pay excessive amounts of money for some innocuous animal, that they may use its body parts in some sort of magical or medicinal preparation. But this could come. And, of course, there is a balance to be drawn between protecting a beautiful coastline and yet allowing those citizens who live there to benefit commercially. Some juggling is going to have to be done. Ken Bock (1978) lyrically described Kenya's coral reefs as a tranquil haven, where you can restore your battered soul. The opportunity to kick off from the shore and drift over a gravity-defying environment of warmth, vivid colours, beauty and organisms unlike any other should continue to be available for all. More prosaically, Tim McClanahan and David Obura (1996) reckon that tourism is the most benign user of the coral reef, as it earns the highest income for a fixed area and causes the smallest amount of ecological damage. The reefs need to continue to exist.

THE LITERATURE

The naturalist on Kenya's sea coast needs two books. Firstly, the *Field Guide to the Seashore of Eastern Africa and the Western Indian Ocean Islands*, which was published in 1997 but recently re-issued and revamped. Edited by Matthew Richmond, with over 450 pages and 150 colour plates, this is a simply superb guide, not just to all the plants and animals, big and small, of the East African coast, but also things like the boats, people of the coasts, how reefs are formed, coastal medicine, coastal climate, weather, seasons, tides and methods of trapping fish. The illustrations are accurate, large and clear, and everything is here, including shells, phytoplankton, fish, hard and soft corals, seabirds and mammals, sea weeds, seagrass, mangroves … the list goes on and on. Each section is authored by an expert in that field, all accompanied by thoughtful essays on the significance of that group. We have drawn extensively upon it for our material here. And because its production was extensively funded by SIDA, the Swedish International Development Cooperation Agency, who also published it, the book is available at a reasonable subsidised price. Coast scientists and naturalists should have this book.

Since only about 180 of Kenya's fish (mostly reef species, but a few pelagic) are illustrated, the diver and snorkeler will also want a good guide to the fishes. The best one at present is probably Ewald Lieske and Robert Myers' book *Coral Reef Fishes: Indo-Pacific and Caribbean*, published as a Collins Pocket Guide in 1993. It covers around 1,500 reef fishes, including a few large species that may come near reefs, and is a proper field guide with superb illustrations opposite the relevant succinct text. The notes on the Oceanic Whitetip Shark *Carcharhinus longimanus* say, briefly and chillingly, 'aggressive and dangerous'.

Two other books are also really useful in terms of local knowledge. One is Ken Bock's book, *A Guide to the Common Reef Fishes of the Western Indian Ocean and Kenya Coast*. This was published by Macmillan in 1978 and reprinted several times. About 140 species of common fish are illustrated, half by nice colour photographs and the remainder by Bock's simply superb line drawings. With three chapters on the sea, corals and the reef, nice summaries on genera and species and Swahili names, this book is a boon to the local naturalist. The second book is Leslie Brown's *East African Coasts and Reefs*. Produced in 1975 by the East African Publishing House, almost quarto size and very poorly bound (the pages fall out on the fourth or fifth opening), nevertheless this is an excellent book. Written in Brown's forthright and yet easy style, enlivened by his wry asides, dry humour and occasional blistering attack on those who didn't espouse the ethics of conservation, it has ten chapters that cover the sea, the reef, the lagoon, the creeks, the dunes, the interior, all the flora and fauna and conservation. It has sixteen pages of informative colour photographs and eight full colour plates of paintings of common fish. Brown was not just one of the region's best ornithologists (there's more about him in Chapter 7 on birds), he had travelled widely in Africa and knew a lot. This is a book that will get you up to speed on the biology of the coast and the immediate hinterland. Another old but enjoyable book is Anthony Cullen and Patrick Hemphill's *Crash Strike*, published in 1971 by the East African Publishing House. It is largely about big game fishing, but the opening chapter has a lot of useful information about the western India Ocean and the Pemba Channel. A classic early work on fishes of the western Indian Ocean was Professor J. L. B. Smith's *The Sea Fishes of Southern Africa*, originally published in 1949. The first edition sold out in three weeks; it has subsequently been updated several times.

Increasing interest in the oceans – they do, after all, cover 70 per cent of our world and are the source of a great deal of food and recreation, as well as providing transport lanes which can be freely used to cross the globe – has resulted in an increase in literature, with dozens, even hundreds of books available on marine ecology and biology. Those who want the background information on the Kenya coast and its ecology as a whole should read the relevant chapters in *East African Ecosystems and Their Conservation*, published by Oxford University Press and edited by Tim McClanahan and Truman Young. An entire section of nearly 100 pages is devoted to marine ecosystems; with a chapter by Tim McClanahan himself on oceanic ecosystems and pelagic fisheries, a chapter by McClanahan and eminent Kenyan marine biologist Dr David Obura on the coral reefs and inshore fisheries, and a chapter on intertidal wetlands by the marine biologist Dr Renison K. Ruwa. These three chapters are 'big picture' stuff, essential reading for scientists interested in the way the coast works, but note this is dense scientific material, and therefore not easy reading.

A lot of good material on the Kenya coast is published in the relevant scientific journals, including *Hydrobiologia*, *Coral Reefs* (the journal of the International Society for Reef Studies), *Journal of Hydrology*, *Journal of Experimental Marine Biology and Ecology*, *Bulletin of Marine Science*, *Estuary and Coastal Marine Science*, *ICES Journal of Marine Science*, the beautifully named journals *Sea Wind* and *Reef Encounter*, as well as in the prominent biological and conservation journals. An edition of *SWARA*, the magazine of the East African Wild Life Society (July–September 2011), was a marine special, dedicated to Kenya's Coast. It contained articles on marine conservation in Kenya, history and oceanography of the western Indian Ocean, Kenya's artisanal fishing, exploitation of sea cucumbers, lobsters, the tropical fish pet trade in Kenya, corals and managing marine resources, amongst other things. There are several articles by the professional marine biologists of CORDIO (Coastal Oceans Research and Development in the Indian Ocean), a research team based at Bamburi Beach; these include prominent Kenyan oceanographer Dr David Obura and the Australian marine scientist Dr Melita Samoilys.

Chapter 13

Conservation

The natural resources of this country – its wildlife which offers such an attraction to beautiful places … the mighty forests which guard the water catchment areas so vital to the survival of man and beast – are a priceless heritage for the future. The government of Kenya … pledges itself to conserve them for posterity with all the means at its disposal – Mzee Jomo Kenyatta, First President of the Republic of Kenya (inscription at the entrance of Nairobi National Park 1964)

This land is now ours. For 17 years the government kept us out, and refused to let us even graze our cattle here. Now we have taken it back – Villager encamped in the wreckage of the Abiata-Shalla National Park, Ethiopia, 1991

This chapter deals with the subject of conservation; it looks at what it is, at the two sides of the conservation argument and at the history of conservation in Kenya. We will talk about what is and what should be conserved, about the problems, conflicts and politics associated with conservation, and the way forward. The chapter concludes with a review of the literature and a summary timeline of landmark dates. Each section is self-contained and can be read independently.

WHAT IS CONSERVATION?

Gifford Pinchot, first chief of the United States Forest Service, defined conservation beautifully; he said it was the greatest good to the greatest number for the longest time. There are different fields of conservation. We are going to consider the conservation of the natural world: the protection and preservation of habitats and their inhabitants, particularly in Kenya. It needs to be remembered that agricultural and forestry land, wetland and water sources and even parts of the urban landscape are just as important to Kenya's continued well-being as the wilderness areas. Many of the smaller animals and plants find refuge in agricultural areas.

Wilderness conservation has two main purposes: economic and aesthetic. Is it the money or the feelgood factor that is most important? There is frequent and lively debate among the conservationists on this dilemma. Economic arguments for protecting natural habitats are straightforward. Wild places provide income from visitors, and employment to those who work in the service industry. The living organisms within wild places may be sources of useful and valuable material and chemicals for medicinal and other commercial purposes. Finally, the animals and plants can be used for food, timber and other useful products, and it may be possible to harvest these sustainably. The economic argument is significant in Kenya. Tourism brings in a sizeable chunk of revenue; over 900 million US dollars in 2010. Although agricultural exports brought in much more (over 4 billion dollars), those working in agriculture are not as well paid. The service industry, which includes tourism, employs over 2 million Kenyans. And, at the most basic level, we need animals and plants because we eat them, both domesticated and wild varieties, and plants give us oxygen.

The aesthetic argument for conserving wild places is slightly more nebulous, but essentially states that (a) we need wild places because they are good for our well-being – a yearning that the famous American ecologist E. O. Wilson has called 'biophilia', (b) we owe it to the present generation and our descendants to preserve wild places and wildlife so that all may benefit, both now and in future, and (c) that wild places have a spiritual or cultural quality, and need protecting out of respect for ourselves, our forebears and our descendants.

These are compelling arguments … to many. Logically, one would say that the question of whether to conserve or not is, to use a modern term, a no-brainer. The answer is obvious. And yet, public opinion on conservation, and what action to take, is divided. It is worth taking a look at the opposing views and the reasons they exist.

THE TWO SIDES OF CONSERVATION

The natural world is under threat. Apart from dangers such as nuclear and biological warfare, pollution, water and mineral shortages, some of which may be less dangerous than the media tell us, we are losing wild places. The population is increasing; between 1804 and 1927 the population rose from 1 to 2 billion, between 1927 and now the population has risen to 7 billion. Kenya's population has gone from about 7 million in 1960 to 40 million in 2011, a near six-fold increase in 50 years. Kenyans now live in places where they didn't in the 1960s. They need food, space, fuel and somewhere to graze their stock and this has had an effect on biodiversity. Biodiversity is the variety of life on our planet, and there are three main threats to biodiversity: habitat loss, introduced species and overexploitation. Kenya is a tropical country, and the greatest biodiversity exists in the tropics. There are more species of snake in the Arabuko-Sokoke forest than in all of Europe; the Shimba Hills has as many plant species as are known from Great Britain. But Kenya is losing habitat and species are being overexploited. In a land of great natural beauty, wild places need protection. In addition hardwood trees and usable edible species are being exploited unsustainably. Conservation is necessary.

Conservation has become a global topic, of high profile interest. There are thousands of books and dozens of scientific journals devoted purely to conservation. It has become big business, many millions of dollars are disbursed in aid of conservation. Prominent and charismatic people have lent their name to conservation efforts, people like our own sadly missed Nobel Laureate Wangari Maathai, actor Harrison Ford and Queen Noor of Jordan. Conservationists care about the environment. They want to keep our world beautiful and protect resources. Surely nobody in their right mind would want to destroy the earth's resources and its wild places so one would expect that conservationists would be admired and respected. But that is not the case. Some conservationists and their activities are distrusted and even actively disliked. A top Kenyan wildlife expert has even declared that he hates conservationists. The protection of biodiversity has become highly controversial and the debate polarised.

Downside of development; plastic rubbish despoiling the landscape. (Laura Spawls)

The reason, of course, is that the situation is complicated by various factors, moral, aesthetic and economic. It is linked to the increase in population and demand. The needs of humanity are putting pressure on untapped resources. What we should do is not clear-cut. The media sometimes give the impression that the entire field is inhabited by only two unpleasantly simplistic camps: the users and the protectors. The users, as the argument goes, want to put the needs of the poor (or simply humanity) first, and exploit the natural world; the protectors want to preserve wild places and prevent their exploitation. In reality, it is neither as simplistic nor as polarised as that, but there is an element of evangelical zeal among the more extreme supporters on both sides of the argument. And the vested interests of those who would commercially exploit the wilderness cloud the issue. This has caused polarisation; with thoroughly decent people and fanatics on both sides. In sub-Saharan Africa, the issue is further complicated. Autocratic rulers, hated by their people, have been good for conservation, protecting with an iron hand, and foreign conservationists have sucked up to them. In Ethiopia, under General Mengistu, national parks were rigorously protected. Local people were cut out. A deadly reckoning followed in 1991 when Mengistu was overthrown. At the Abiata-Shalla Lakes National Park local people entered the park, destroyed the park buildings and boats, cut down the trees, brought in their cattle and shot the wildlife. Park visitors observed the depressing sight of the severed heads of the park's antelope being offered for sale. The ordinary people had received no benefit from the park; they saw the conservation establishment as oppressive.

Foreign conservationists, wary of corruption, have sometimes bypassed state protocol and enraged the ruling government. Conservation in Africa has an image problem, seemingly being carried out by autocratic, self-righteous, well-paid white foreigners with expensive vehicles. They don't train local replacements, they keep ordinary citizens out of fertile areas. The rationale for this is often nebulous and means nothing to hungry and desperate people. And citizens of the country are not engaged. There is a telling story in Barack Obama's book *Dreams from My Father* (1995) of how he suggests visiting a national park, but his half-sister makes a connection between game parks and colonialism, asking him how many ordinary Kenyans can afford to go on safari. She asks him why land should be reserved for tourists rather than farmers, and suggest that *mzungus* (white people) are more upset about one dead elephant than they are about the deaths of a hundred Kenyan children. Torben Larsen once observed that you can watch 100 videos of African wildlife and not see a black presenter or scientist in any.

Incomes vary wildly in Kenya. The desperate poor have nothing to lose. Protecting rhinos or zebras is of no significance to those who are trying to simply stay alive. In prosperous countries, conservation problems do arise, but they often involve the well-off, trying to get their hands on public land. In many African countries, conservation problems often arise from ordinary people simply trying to make a living. A rich businessman attempting to get hold of a bit of a national park to put up buildings and make a profit is easily condemned by all; a group of desperate villagers encamped on a private game reserve and threatened with eviction does not present such a clear-cut issue. The rich entrepreneur caught smuggling a crate of rhino horn is unanimously perceived as a shameless villain; the landless hunter who shot one of the rhino and sold the horn to feed his children is not an unmitigated criminal. Many people in Africa are poor, and often have little hope of ever becoming rich; they visualise a life of grinding poverty. They are vulnerable to bribes, to schemes that may enrich them rapidly, or even just illegal and risky enterprises that will feed them and their family for a week. A powerful economic argument exists; that a hectare of land planted with crops produces more food, makes more money, gives homes to more people and feeds more people than a hectare given over to wildlife. It is estimated that if the Maasai-Mara National Reserve was turned over to agricultural production, the revenue would be more than seven times the present income from game viewing.

So the image of conservation isn't clear-cut. But it isn't all doom and gloom. In Kenya, with increasing prosperity, there has been a steady rise in domestic tourism and the involvement of ordinary Kenyans in conservation. More and more Kenyans are visiting their national parks and reserves and taking pleasure from wild places. With the growth of a reasonably prosperous society, the beautiful wilderness becomes important.

But how, and what, to conserve? There is, as always, debate about the nature of wilderness conservation. Should wild places be rigorously protected, with nothing within them save the natural flora and fauna (so-called 'fortress conservation'), or should the protected areas be allowed to have people living within them and harvesting the products, either sustainably or unsustainably. This debate also tends to polarise its

Once vermin, now the most-hoped for sighting, young Leopard and its kill in the Maasai Mara. (Laura Spawls)

protagonists and has hardened recently, largely as a result of population increases. In reality, if there was not much pressure on land, then most people would agree: it is desirable to preserve wild places and their inhabitants. And some things are sacrosanct. It would undoubtedly benefit the poor if Uhuru Park, or Hyde Park, or Central Park, were grubbed up and a huge swathe of low-cost housing was built there. Why don't we do it? Surely the benefits outweigh the disadvantages? But in reality, we don't do things like that because such open spaces, often steeped in history, do matter to us. Even the poor of Nairobi would object to Freedom Corner being dug up for low cost housing. Some things are sacrosanct.

Let's move up a scale. Here is a classic dilemma. If you approach Nairobi at night, in an aeroplane, you see a fascinating phenomenon. As the plane rounds the Ngong Hills and settles on its final run down to Jomo Kenyatta Airport, you are passing over populated country; little lights twinkle below, and the great gleam of Nairobi City is visible on the horizon. And then, suddenly, the land below you is dark, a stretch of open country, with not a light showing. This is Nairobi National Park, a natural wilderness area on the outskirts of a great city. Its 120km^2 contains forest, savanna and rocky gorges. It has a huge variety of wild animals and plants; big game including rhino, Leopard, Buffalo, Lion. No other African city has such a large and diverse sanctuary so near to its centre.

KENYA NATIONAL PARKS AND NATURE RESERVES

1. Nairobi Nat. Park
2. Hell's Gate Nat. Park
3. Mt Longonot Nat. Park
4. Lake Nakuru Nat. Park
5. Lake Bogoria Nat. Res.
6. Lake Kamnarok Nat. Res.
7. Maasai Mara Nat. Res.
8. Aberdare Nat. Park
9. Mt. Kenya Nat. Park
10. Ol Donyo Sabuk Nat. Park
11. Mwea Nat. Res.
12. Mt. Elgon Nat. Park
13. Rimoi Nat. Park
14. Saiwa Swamp Nat. Park
15. Kakamega Forest Nat. Res.
16. Ruma Nat. Park
17. NDere Island Nat. Park
18. Impala Nat. Park, Kisumu
19. Meru Nat. Park
20. Bisandi Nat. Res.
21. Kora Nat. Park
22. Rahole Nat. Res.
23. Mwingi Nat. Res.
24. Samburu Nat. Res.
25. Buffalo Springs Nat. Res.
26. Shaba Nat. Res.
27. Maralal Game Sanctuary
28. Laikipia Game Sanctuary
29. Marsabit Nat. Res
30. Losai Nat. Res.
31. South Turkana Nat. Park
32. Nasalot Nat. Park
33. Sibiloi Nat. Park
34. Central Island Nat. Park
35. South Island Nat. Park
36. Malka Mari Nat. Park
37. Tsavo East Nat. Park
38. South Kitui Nat. Res.
39. Tsavo West Nat. Park
40. Chyulu Hills Nat. Park
41. Amboseli Nat. Park
42. Mombasa Nat. Park & Res.
43. Shimba Hills Nat. Res.
44. Kisite Marine Nat. Park
45. Mpunguti Marine Nat. Park & Res.
46. Diani/Chale Marine Nat. Park & Res.
47. Malindi Marine Nat. Park & Res.
48. Watamu Marine Nat. Park & Rcs.
49. Kiunga Marine Nat. Res.
50. Dodori Nat. Res.
51. Boni Nat. Res.
52. Tana River Primate Res.
53. Arabuko Sokoke Forest Res.
54. Arewale Nat. Res.

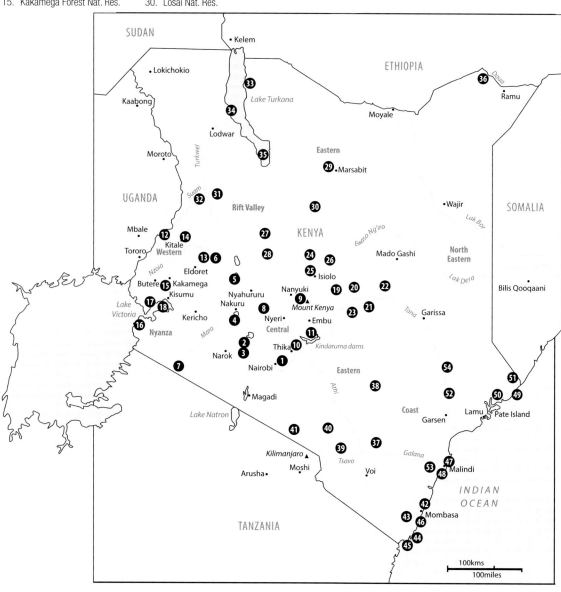

And there are those who say that this park is unique, and it must be preserved in its entirety. They believe it would be criminal to damage it, and future generations of Kenyans will curse those who might have, by negligence or otherwise, let the park be despoiled or any part of it lost. On the other side, there are those who say that such a huge area of land, lying idle on the outskirts of a huge city, is an anachronism. Only a few hundred metres away, on the slopes above Nairobi Dam, is Kibera, full of decent and yet desperately poor people. Why shouldn't they live in the park? They say that only a few rich people visit the sanctuary; that few tourists go there anyway, preferring the Maasai Mara or Lake Nakuru. They say that the land should be given over to housing, and development; it is near the city, there are plenty of other parks in areas where there is less land hunger. Who is right? Is there a right or wrong? There are two sides to every argument.

CONSERVATION AND ITS HISTORY IN KENYA

For most of its history, humanity has regarded the natural world as something to be cheerfully exploited. And for much of history, there was nothing wrong with this attitude. There were fewer of us and the resources seemed limitless. It is only in the last 200 years, and particularly the twentieth century, as we saw, that the planet's population has boomed, that we have started to think about what we are using up, and perhaps conserving things. The realisation that there was – or might be – a problem gradually sunk into the public's consciousness in the first half of the twentieth century, in Kenya particularly during the 1940s and 1950s. Debate then began about what to do, but opinions were divided on how to conserve, and they remain so, often violently. Few developments illustrate this better than the story of Africa's elephants and their ivory; a story brilliantly told in both the scientific papers of Thomas Hakansson and Esmond Bradley Martin, and the books of Ian Parker and the Douglas-Hamilton family (see the literature section at the end of this chapter). It was control of the ivory trade that started the conservation ball rolling in Kenya, as we shall see.

Ivory is a tough, durable and attractive substance. Cro-Magnon man was carving mammoth tusks, 40,000 years ago. Four thousand years ago, the Egyptians were importing ivory from Nubia. By the time that the un-named author of the Periplus of the Erythrean Sea (Africa's first travel guide, written in the first century AD) was dryly describing the opportunities for trading on the East African coast, the East African ivory trade was under way. In the fifteenth and sixteenth centuries there was widespread elephant hunting in the interior of East Africa, the lure of elephant's teeth (and gold and slaves) proved greater than the risks of penetrating into the dangerous and disease ridden hinterland. Much of the ivory harvested was exported to China, India, the Middle East and (increasingly) Europe.

By 1840, it was estimated that the elephant populations along the East African coast between Malindi and the Pangani River in Tanzania, and for a hundred miles or so inland, had been exterminated. Early explorers to the interior, like David Livingstone, described the caravans of slaves and ivory that they encountered. By 1890 elephants had largely disappeared from the entire East African coast (with the exception of the area around the mouth of the Tana River), they were also gone from the interior as far west as Mt Kilimanjaro. And no wonder; in 1848 Zanzibar exported 350 tonnes of ivory, which (at an average of 45kg per pair of tusks) represents over 7,000 elephants. In 1891 ivory prices peaked in Zanzibar at 170 Maria Theresa silver dollars per frasila (an archaic unit of 15.5kg). Much of this ivory now went to Europe and the United States, to make combs, toothpicks, piano keys and billiard balls, which was why Colonel Richard Meinertzhagen made his famous remark, to the effect that he could not understand why elephants needed to be killed so that creatures little more intelligent should play billiards with balls made from its teeth.

In Kenya today, Tsavo National Park is one of the best places to see elephants, especially big elephants. In 1898 Lieutenant-Colonel Colonel Patterson was trying to build a railway bridge across the Tsavo River, in the heart of what is now the park, and also trying to deal with man-eating Lions (see Chapter 6 for the full story). But in Patterson's best-selling book *The Man-eaters of Tsavo* (1907), he does not mention elephants in Tsavo. Presumably he saw none there. Sir Charles Eliot, who was governor of Kenya from 1900 to 1904, never saw an elephant in Kenya. There were elephants left in the interior. For a number of years, professional ivory hunters had made a living by shooting elephants, in a hinterland that was beyond anyone's control; famous elephant hunters like Arthur Neumann were shooting lots around Mt Kenya, and W. D. 'Karamoja' Bell was hunting up in north-west Kenya. But the elephants of eastern Kenya had been shot out, as was noted by the colonial authorities.

In 1895, the British Government then took over control of the territory. One of the first actions of the colonial government was to outlaw commercial ivory hunting, in 1896. No more elephant hunting, except under licence. For good measure, they also declared the existence of 'Kenia District Game Reserve', extending from Nairobi north and east to Habaswein. A Game Department was established in 1899 and in 1900, two huge reserves, the 'Northern Game Reserve' (Mt Kenya north to Lake Turkana) and 'Southern Game Reserve' (Nairobi south to the border, east to Tsavo) were proclaimed. Arthur Blayney Percival was given the job as the first 'Ranger for Game Preservation' in 1901, but had only six assistant scouts. In 1907, at the suggestion of Sir Frederick Jackson, a 'Game Rangers Department' was formed; the Kenyan wildlife expert Ian Parker suggests that this was the real starting point for the Game Department. At the time, the laws still permitted ordinary Kenyans to hunt game for food, but this permission was rescinded after 1909. In 1910, Richard Bowen Woosnam was appointed the country's first 'Game Warden'. Woosnam was a scientist, an entomologist, but under him the Game Department's core business was controlling the poaching and sale of ivory. Sadly, in 1915 he was killed in action at Gallipoli. After the war, Blayney Percival was reappointed as 'Game Ranger'; his problem was the depredations of wildlife upon agriculture. A major step forward resulted from the appointment, as Game Warden, in 1924 of Captain Archibald 'Archie' Ritchie.

Born in Dublin, Ritchie had attended Oxford where he read zoology. A tall, imposing man with a trim moustache, Ritchie fought with distinction in the First World War (initially in the Foreign Legion) and ran the Game Department in a military manner; he was in charge from 1924 to 1949. The emphasis at the Game Department at the time was on game control, the hunting of crop raiders, as well as controlling poaching. In addition to his Kenyan game scouts, Ritchie and his team of six European wardens appointed a number (between 80 and 90) of 'Honorary Wardens', who assisted with curbing poaching and shooting or dissuading problem animals; these wardens included prominent grandees like the American pioneer Sir Northrup McMillan, and professional hunters like Alan Tarlton and Denys Finch Hatton. The animals they controlled were not just the big problematic animals like elephants and Buffalo, but also what Ritchie termed 'vermin'. This was a category that included not only such animals as baboons, warthogs, bushpigs and porcupines, but also Leopards; it is an odd thought that those beautifully spotted cats, nowadays so often the highlight on a day's game viewing, were regarded as vermin, to be shot on sight. Even Lions were gratuitously shot. In the late 1920s Ritchie employed John Hunter to put down lions in Maasai country, as they were causing problems with stock; Hunter shot 80. He and his team also killed more than 1,000 rhino in Ukambani. Ritchie's wardens included Captain Robert Whittet, in charge of the Northern Frontier, the eccentric Tom Oulton, a British Israelite who always slept with his bed pointing due north and was in charge of 'game and vermin control', a late appointee, one George Adamson, and the man who caught more poachers than anyone before or since, G. C. MacArthur. MacArthur was an interesting man; as well as being a championship heavyweight boxer, he had been a police detective. MacArthur, who was based at the coast, went after the Tsavo elephant and rhino poachers with a vengeance. In one year, 1931, he and his team of fifteen game scouts apprehended and got convicted some 477 poachers. The steady flow of ivory and rhino horn eastwards to Somalia was staunched, at least for a while. MacArthur retired after the Second World War and started Mac's Inn at Mtito Andei, now Tsavo Inn.

In the late 1930s, the writer Elspeth Huxley travelled extensively in northern Kenya, and noted with horror the paucity of game there. She wrote an indignant letter to her cousin by marriage, the famous biologist Julian Huxley. He organised it to be forwarded to the Colonial Office and then to Government House in Nairobi. The Game Department at the time existed on a shoestring budget, but the government were stung by the criticism. Funds were found, and Ritchie hired George Adamson to be game warden of the northern frontier. It was a job

Rescued baby Elephant at Dame Daphne Sheldrick's orphanage. (Stephen Spawls)

that George took on with gusto, and the chapter describing his adventures there, in his book *Bwana Game* (1968), is one of the most fascinating of the book.

At the time Kenya was generating revenue of a sort from its wildlife. In the early part of the twentieth century, many of the foreign visitors who came to Kenya came to hunt big game. The idea that visitors might come to simply watch and photograph wildlife was as yet unheard of (although the America couple Osa and Martin Johnson spent four years based at Lake Paradise on Mount Marsabit, taking photographs of northern Kenya, in the 1930s; Osa's (1941) book is a gem). A brotherhood of professional hunters emerged in Nairobi, men who guided the rich foreign visitors on their hunting safaris and found trophies for them to shoot. Among their number were Richard ('R J') Cunninghame, Charles Cottar, Bill Judd, Philip ('Phil') Percival (younger brother of Arthur Blayney Percival and immortalised as 'Pop' in Ernest Hemingway's *The Green Hills of Africa*) and Alan Black. Their clients included men like the ex-American president Teddy Roosevelt; it has been said that Roosevelt really started the fashion. Other high-profile clients included Winston Churchill, several members of the British Royal family, including the Duke of Connaught, the Duke and Duchess of York (later to become King George VI and Queen Elizabeth the Queen Mother) and the Prince of Wales (who, as King Edward VIII, would later give up the throne), Baron Rothschild and Ernest Hemingway. At that time, hunting was seen as a manly and honourable pursuit; the stigma that attaches nowadays to hunting mammals (but curiously not to hunting fish) did not exist. On 12 April 1934 a group of hunters met at Nairobi's famous Norfolk Hotel and formed the East African Professional Hunters Association. Archie Ritchie was present at the meeting. Although it may seem odd that a team of professionals whose business was assisting their clients to shoot wildlife could actually assist in protection, the hunters proved to be some of conservation's most able friends. A stringent set of laws was drawn up, that ensured sporting conduct. No shooting was allowed at night, or near the vehicle; females and young animals could not be shot. Automatic weapons were not allowed. Only animals for which you had obtained a licence could be shot, and numbers tightly controlled. Wounded animals were to be followed and shot, not left. The country was divided into hunting blocks, and a professional wanting to hunt in a particular area must book that block and no more than two parties could be in the same block. Since the clients were mostly after trophy animals, a small take of usually elderly males did not prove detrimental to the entire population. The presence in remote areas of a group of well-informed and armed professionals who knew the land tended to deter poachers, and the relatively large fees charged were a benefit to Kenya's exchequer. The comings and goings of the famous on sporting safaris generated publicity for the country. However, the idea that Kenya's wildlife and wild places might need active protection seemed not to have occurred to the colonial government in the 1930s and it required the dynamic enthusiasm of one man to change that. That man was Mervyn Cowie, and if any one person might be singled out as the father and founder of the beautiful national parks that grace Kenya today, that person would be Cowie.

Cowie was born in Nairobi in 1909, but studied accountancy in Britain. On returning to Kenya in 1932, he was disturbed at the disappearance of game. Cowie thought that if national parks could be established, it would generate animal tourism and the money raised would pay for the parks. But the colonial government was not interested; indeed they had diametrically opposite ideas. A despatch from the Secretary of State for the Colonies, William Ormsby-Gore (later Lord Harlech) suggested that national parks should be established, but for the purpose of shooting wildlife! Cowie was in despair, but he came up with a remarkable ruse. With the connivance of George Kinnear, the then editor of the *East African Standard*, Cowie wrote a disingenuous letter to the paper under the nom de plume 'Old Settler' and suggested that (a) all wild animals in Kenya should be shot by machine gun or poisoned, (b) all the land should be turned over to farming. To Cowie's initial horror, the Farmer's Association at Nanyuki enthusiastically supported his idea, but then the backlash began. Letters poured in, and most of the writers recoiled from 'Old Settler's' suggestions; some threatened to shoot him! The *Standard* editor kept the pot boiling with a playful chastisement of those who advocated violence, and support for wildlife protection grew in momentum. A big public meeting was held and then a game policy committee was formed. Initially Cowie was not on it, because the acting governor at the time, Sir Walter Harragin, thought Cowie would be 'a damned nuisance'. Cowie confronted the governor in person, and was added to the committee. At the same time, Cowie had been cultivating a pride of Lions in the 'Nairobi Commonage', the area that would become Nairobi National Park, and taking important people (including an ex-governor) to see 'his' pride; he gave them names like 'Makora' (rogue), Lulu, Dickens and Jones. Although the area suffered from being a military firing range during the Second World War, in 1945 the then governor, Sir Philip Mitchell, proclaimed the National Parks Ordinance, and on Christmas Eve 1946 the park was formally proclaimed, with Ken Beaton appointed the first warden. According to Cowie, that night the Lions roared and he and Ken Beaton 'jumped over the moon'. As well he might: it was the start of what is arguably the best, and most important, cluster of national parks in Africa. The initial entry fee was one shilling; the entrance was near the present 'Carnivore' Restaurant. Visitors to the park were required to stay within their cars and not drive at night, unpopular innovations at the time.

Other national parks and reserves followed. Tsavo, the Mara, Amboseli and Marsabit were proclaimed in 1948, Mt Kenya National Park in 1949 and the Aberdares in 1950. The first marine national parks were gazetted in 1968. Now Kenya has some 29 national parks and 25 national reserves, covering nearly 37,000km^2, although Tsavo is by far the largest of them all, at nearly 20,000km^2. This means that over 6 per cent of the country lies within a park or reserve. The roll of protected areas also includes about 270 forest reserves covering some 16,000km^2 and close on 50 private sanctuaries covering 12,000km^2. This means that biodiversity is formally protected, to a greater or lesser extent, on about 65,000km^2 – over 11 per cent of Kenya's land area. It is a record to be proud of.

Some remarkable and charismatic park wardens were appointed. Cowie had been put in charge, and as well as Ken Beaton, he appointed David Sheldrick and Bill Woodley, young men who had been raised in Kenya and who proceeded to knock Tsavo into shape; Woodley went on to do the same in the mountain parks. Many of the wardens were ex-military men, often of British extraction, and they oversaw the national parks in an honest but militaristic fashion. For an interesting insight on this period, it is worth reading Alistair Graham's 1973 book *The Gardeners of Eden*. The parks were run with a firm and paternalistic hand; roads were built, dams were constructed,

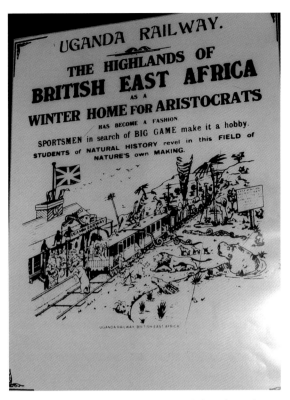

Lake Paradise, Mount Marsabit: where Osa and Martin Johnson laid the foundations for photographic safaris. (Robert Drewes)

Poster showing how East Africa appealed in the early 20th Century. (Stephen Spawls)

signposts were put up. Graham has a amusing and perspicacious dig at the plethora of signboards that littered some of the parks; signs that largely forbade the public to do anything – the prevailing attitude of many of the wardens was that the public were there under sufferance. In Nairobi National Park you met signboards during the first 5 miles that forbade you to stand up in your car, forbade you to travel over 30 miles an hour, travel after dusk … it went on and on. But by and large, the 1950s and early 1960s were a time of consolidation for the parks, with steady development, improved roads and consequently increasing tourist numbers. Kenya began to be seen as an international tourist destination for the average citizen, not just an exclusive departure for the well-heeled.

There can have been no more important event in the early days of the parks than the visit in 1952 by the then Princess Elizabeth and Prince Philip to Treetops Hotel, in the Aberdare National Park. Many know the story that she ascended the wooden steps into the famous hotel a Princess and emerged as a Queen, her father, King George VI, having died in the night.

The first lodges were built in the 1960s, oddly in the face of some stiff opposition. Many of the park administrators felt that lodges would somehow despoil the parks, and dilute the pristine wildernesses they were in charge of. It is a point of view that emerges from time to time in conservation, where those in charge feel that their crusade for what they perceive as the greater good of humanity leads them to forget the existence of the common people who, in the end, have to benefit. One warden received a grant to improve the approach road to 'his' park and develop the main road within the park; he spent the money instead on building dams. When quizzed about this, his response was to suggest that improving the approach road 'would encourage riff-raff to visit the park'; by 'riff-raff', he meant ordinary Kenyans. It needs to be remembered that in many cases people were actually living within what became the park; they were moved on and not always pleasantly. In Cowie's autobiography *Fly Vulture*

(1961), he describes how he struggled to move the Somali communities out of Nairobi National Park, where they had lived for some time. One brave old lady, enraged at having been forced out of her home and having to forcibly sell her cattle, was due to be paid a sum equivalent to 2,000 pounds. As Cowie put the money into her hands she contemptuously spat at him and cast the banknotes to the winds.

But the lodges were built, and thus ordinary people were able to experience the exquisite pleasure of spending the night in wild places. They could drive in the afternoon watching the wildlife, get back at dusk, maybe see game drinking in the night at a lodge waterhole, sleep with the sounds of the bush around them and rise and drive again at dawn in the awakening wild world. There is no pleasure like this. Some magnificent lodges were constructed, the first in 1962 was Kilaguni, in Tsavo West, with its stunning view over the waterhole towards the Chyulu Hills. This was followed by Keekorok, in the Mara. Nowadays, there are over 100 lodges and tented camps in the Mara area alone; it can feel crowded at times, and yet, if you pick your spot, you can still find solitude.

In the 1950s and 1960s, Kenya also benefited from the publicity gained by some remarkable natural historians and the animals they attached themselves to. One of them was Joy Adamson. George Adamson was actually her third husband. Born Friederike Victoria Gessner, she was first married in Austria to Ritter Viktor von Klarwill, then to the botanist Peter Bally, who persuaded her to change her name to Joy. She married George in 1944, and in 1956 began raising Elsa, a Lion cub whose mother George had shot while pursuing a man-eater. Joy's book *Born Free*, describing the raising of Elsa, was published by the Harvill Press in 1960, and was an instant best-seller. When Elsa died in 1961 of a tick-borne fever, it was headline news in the *East African Standard*. Joy went on to write twelve further books, and raise a succession of Lions, Leopards and Cheetahs. The film of *Born Free*, starring Virginia McKenna and Bill Travers as the Adamsons, was chosen for the Royal Command Performance in London in 1966 and set Travers and McKenna themselves on a lifelong crusade to better the lives of captive wild animals, a calling that their son, Will, continues to this day. Joy became an international celebrity; and travelled widely, speaking to enthralled audiences worldwide. Tragically she was murdered in early 1980 at her camp in the Shaba National Reserve. George Adamson wrote two books himself, but he too died violently. He was shot by Somali bandits in Kora National Reserve in August 1989, at the age of 83. Typically he was in action as he died, charging the robbers down with his Landrover, pistol in his hand, full of the quiet courage that typified his life.

In the 1960s, Louis Leakey encouraged three young women to study the great primates; in Africa these were Dian Fossey, who studied gorillas in the volcano country of the Albertine Rift Valley, and Jane Goodall, who worked with chimpanzees in Gombe Stream Game Reserve on the shores of Lake Tanganyika. Another prominent wildlife man of the time was Dr Iain Douglas-Hamilton, who did pioneering work with the elephants of Lake Manyara in Tanzania. Other wildlife 'characters' of the time in East Africa included Armand and Michaela Denis, wildlife film-makers who lived in Nairobi, and the wildlife cameraman Alan Root and his wife Joan. And at the same time, Hollywood came to East Africa, with adventure films like *Hatari*, starring John Wayne. The professional hunters and guides operated out of Nairobi, and took high profile clients like Bing Crosby and the writer Robert Ruark on safari.

It cannot be denied that these larger-than-life adventurers and their activities heightened awareness about East Africa, and stimulated people to visit, bringing their money with them, and this must be chalked up to their credit. However, the activities and lifestyles of some of the naturalists were controversial – in the case of Joy Adamson, surprisingly so. The unsanitised version of her life, as well as George's, is told in a very readable near-500 page 1993 book, *The Great Safari*, by Adrian House, who was an editor at Collins. Joy had many affairs after she married George, including one with Sir Billy Collins, her publisher. She often had a curiously cavalier attitude towards the animals she was supposed to love and support. She owned a leopard-skin coat and a colobus-monkey shawl, she shot an elephant, and once left a baby leopard in a bag in a car in Nairobi's hot sun, where it suffocated. She had a reputation of being high-handed with her Kenyan staff, and when her manuscripts arrived in London, her editor, Marjorie Villiers, had to go through them and remove the abusive references to ordinary Kenyans. It appears that she was murdered by one of her Kenyan staff, although even now this isn't certain. Opinion is equally divided on the merits of the work she did with the big cats; and whether her writings actually contribute

Conservation

On the balcony of the legendary Treetops Hotel. (Stephen Spawls)

Professional hunter briefing clients before the walk to Treetops Hotel. (Stephen Spawls)

409

towards carnivore biology or are simply whimsy. And, of course, there is much debate about whether or not her rehabilitation efforts served any real purpose. Her attempts to give freedom to her animals are easy to understand; anyone with a heart who has seen a great wild cat caged and miserable, unable to run and hunt, would wish to give it freedom. But her animals, and George's, hurt and killed several people. George's right-hand man, Stanley, was killed by Boy, one of George's Lions, and Mark Jenkins, the son of the Warden of Meru National Park, Peter Jenkins, was also mauled by Boy. George got into conflict with the explorer Wilfred Thesiger, who suggested that George's rehabilitation attempts were simply creating man-eaters. But Thesiger was arguing from an entrenched viewpoint; he had shot more than 70 Lions himself.

However, George was a well-liked man, respected by his staff and colleagues; he was one of nature's gentlemen and no one has a bad word for him. His two books are simply superb, *Bwana Game* in particular catches the spirit of northern Kenya. Likewise people such as Iain Douglas-Hamilton, the zoologist and elephant expert, who has dedicated his life to protecting the African Elephant and its environment; Douglas-Hamilton is a hard-bitten professional zoologist whose popular writings and scientific papers are thoroughly sound. But others are not so well regarded. Dian Fossey is a case in point. Her work was popularised by the film *Gorillas in the Mist*, with Sigourney Weaver starring as Fossey (the film was actually made in the Aberdares). There is no doubt that Fossey raised the profile of the gorilla, and raised funds; these may have ultimately benefited gorillas as a whole. She was well-liked by the Western media but she had some strange attitudes. Initially she had no zoological training, although she later took a doctorate. She hated the idea of tourists visiting the gorillas, although it is tourism that has proved the saviour of the Mountain Gorillas; the funds it brings in pays for the running of the parks to some extent. She shot at Rwandans and visitors alike. Fossey also disliked the local people, unless they 'knew their place', i.e. as subordinates. She was often hideously cruel to people whom she believed to be poachers, even to their children; in many countries she would have been jailed for such actions. In her diaries she called the local people 'wogs' and 'things', and describes wanting to kill the people who had poached gorillas; all this is carefully documented in Farley Mowat's excellent book *Woman in the Mists* (1987). Fossey alienated many Rwandans. In the end she was murdered, in 1985, although no one is certain by whom. The Rwandan hunter charged with her murder hanged himself in prison.

In the late 1960s some problems emerged in Kenya with the national parks. The young republic, independent since 1963, wanted jobs to be made available for Kenyans, and some people felt that entrenched expatriates were too slow in training successors and making way for them. Many of the 'old school' wardens, strangely, distrusted scientists, although one would imagine that the first prerequisite for running a conservation sanctuary was biological training. But no, the expatriate wardens were unhappy at the idea of young Kenyan scientists moving into their jobs, and indicated strongly that their potential replacements should have been trained in military-type establishments and skills. To the Kenyan biologists, this was anathema; just a ruse to keep them out of top posts. Mervyn Cowie lost his job in 1966, and was replaced by Dr Perez Olindo. Some of the wardens, who had spent a long time in their parks, were moved; David Sheldrick was transferred to Nairobi, to a job involving planning. Sheldrick wasn't happy, and sadly died of a heart attack the following year. And yet, a number of extremely competent men came to the fore around this time. Apart from Perez Olindo himself, who oversaw a major expansion of the parks, took Meru and Amboseli into the fold (and had been known to fearlessly rugby-tackle poachers), one thinks of men like Sam Ng'ethe, who was head of the capture unit and warden of Nairobi National Park, and Joe Kioko, the flying warden who managed to stop the Maasai spearing rhino in Amboseli; later he held the post of director of Kenya Wildlife Service.

Poaching became a major problem in the late 1970s and 1980s, and to some extent it was home grown. There was also insecurity in the parks: rangers trying to protect the wildlife often found themselves in gun battles, some died, and some foreign visitors were robbed and shot dead. This was a major problem for a country with a tourism-based economy. It's a matter of perception. You are probably at greater risk of being murdered in London or New York than in Kenya, but your city is familiar. The reports in the press of robbings and shootings in the Mara, Meru and Tsavo parks scared visitors off. Holidays were cancelled and tourists stayed away. Also at this time

Harassed Cheetah tries to eat its kill, Maasai Mara. (Stephen Spawls)

a lot of elephants were also being poached. The elephants of Tsavo have had mixed fortunes. At the beginning of the twentieth century there were few there, but their numbers gradually increased, particularly in the 1950s. This was a tribute to David Sheldrick; he and his men had arrested and jailed most of the men who hunted elephant with the bow. But soon there were so many elephants that large areas of the park's *Commiphora* woodland were destroyed. In 1968 there were reckoned to be 40,000 elephants in the park. A cull of 300 animals was done for scientific purposes, but it was proposed to extend it. Sheldrick didn't like the idea, and the team who had the contract to cull the elephants had an efficient but brutal way of dealing with them. Since shooting a few out of a herd leaves the remainder traumatised and often leaderless, the culling team's technique involved surrounding the herd and then shooting – often with automatic weapons – all of them. It is probably the best way of reducing numbers, but it is horrible to see. Elephant are sentient and social animals; watching a herd being shot down, first the big matriarchs, then the younger ones and finally the babies is a sickening experience. There was also violent disagreement as to how many elephant were present and how many needed to be shot … if any. In the end, the big cull did not go ahead. And in 1971 a drought came and huge numbers of elephants died – between 9,000 and 15,000 depending on whose figures you believe.

After Sheldrick was transferred away, poaching in Tsavo changed. Originally, it was mostly local people, but then Somalis moved in from the effectively unpoliced lands out east of the park. And national park staff were also involved. Politicians were in control of the poaching, at the highest level; at one point government vehicles arrived at Voi and took away a huge stock of Tsavo ivory that had been collected from dead elephants; the tusks were never seen again or accounted for. Elephant experts like Patrick Hamilton and Joyce Poole noticed that there were carcasses everywhere, hardly any old elephants were left in the park, and the herds had an inordinate fear of green 4x4 vehicles. The tuskers were being poached by the very people paid to protect them. In Meru National Park, a showcase group of semi-tame White Rhino, imported years earlier from South Africa (no White Rhino occur naturally in Kenya), were gunned down by poachers. Tsavo lost most of its Black Rhino; in fact rhino were

Problems with drought, baby Elephant struggles to reach the water, Voi, 2009. (Stephen Spawls)

decimated throughout the land. If you drove for a morning in Tsavo East in the early 1970s, you would come across five or six Black Rhino without having to look hard for them. By the end of the 1980s there were none left. Elephant numbers, estimated at over 200,000 in Kenya in the 1970s, dropped to around 20,000 by the end of the 1980s.

Now that the situation is improving (although not yet perfect) it is possible to take a detached look at Kenya's problems in the 1970s and 1980s. The happy early days of freedom following Independence were followed by difficult times of adjustment, as invariably occurs after an autocratic government comes to an end. Many patriotic Kenyans were saddened to realise, in the following decades, that they had swapped the colonial government for the equally harsh regime of free world economics and Western markets, and that not everyone in government and its agencies was incorruptible. This is a theme perfectly explored in Ngugi wa Thiong'o's magisterial novel *Devil on the Cross* (1987). In the absence of a group of foreigners in government, whom all could dislike, there was no clear-cut enemy. In-fighting began, people favoured their own tribe. Many people in authority used their office to enrich themselves; elephant ivory, rhino horn and spotted cat skins were valuable. There was also administrative manoeuvring; in 1976 the Game Department (which was widely regarded as having become very corrupt) was merged with the smaller National Parks (not so corrupt) into the new Department of Wildlife and Conservation Management, under the Ministry of Tourism, which was also perceived to be a hotbed of corruption. Theoretically, one streamlined department could cover all wildlife business. But there was too much graft, including poaching and bribery. The *mzungu* wardens may have been anachronistic in their outlook, but they were enthusiastic about their work and largely seen as incorruptible. As wildlife disappeared, Kenya's image abroad suffered. In 1977, all sport hunting was banned, and many people approved of this, but it did little to curb poaching. Indeed, some said it made the situation worse as the presence of the expert – and armed – professional hunters in remote areas had tended to deter poachers.

Things began looking up in May 1989. Cometh the hour, cometh the man. Richard Leakey was appointed head of the Department of Wildlife and Conservation Management, which then shed its old name and image and became the Kenya Wildlife Service, KWS. In a brilliant gesture, Leakey arranged for President Moi to set fire to 12 tonnes of ivory, in Nairobi National Park, in July 1989. The event generated a vast amount of publicity; pictures flashed around the world. Many people thought it was a mistake, a waste of money, wildlife experts in southern Africa sniggered and said it was a publicity stunt. But in reality the event sent out a signal; that Kenya's conservationists held in contempt those who wore jewellery made of ivory or used ivory artefacts. Subsequent burnings have tried to drive home this message. Leakey also found money to get the parks' vehicles moving, he kept on fund raising, relentlessly, he sacked corrupt and incompetent workers and he took a close personal interest in the welfare of his team. In the face of attacks on his staff and park visitors, Leakey got his men equipped with assault rifles. Kenya managed to get African Elephants put onto CITES Appendix 1 (CITES stands for Convention on International Trade in Endangered Species of Wild Fauna and Flora; Appendix 1 species cannot be traded) at a meeting in Switzerland, after an initial setback in Gaborone, Botswana. The price of ivory declined steeply afterwards, although it is creeping back up again. Leakey cared about his employees. One of us once asked a ranger how life was under a subsequent director; the scout pointed to his boot, which had a flapping sole, and said, 'Under Leakey, this wouldn't have happened. I'd have been given a new pair or this would have been fixed.' Leakey's story can be read in his second volume of autobiography, *Wildlife Wars*, grandly sub-titled *My Battle to Save Kenya's Elephants* (Leakey & Morell 2001).

Not everything went smoothly after Leakey took charge. Just after the ivory burning, George Adamson was shot dead in Kora, attracting world-wide negative publicity. Some of Leakey's comments to the press about how his well-equipped teams would deal with poachers lead to the press stating that KWS had a 'shoot-to-kill' policy. For a time, the code of discipline that allowed KWS rangers to take use their weapons freely when in contact with poachers was

What brought the early hunters to Kenya, big ivory in Tsavo East. (Stephen Spawls)

withdrawn, although it was later reinstated. And in June 1993, the light plane that Leakey was piloting ran out of power and went down. The exact cause of the crash remains unknown. Leakey survived but subsequently had both his lower legs amputated. Leakey had to undergo a further ordeal when KWS was investigated by a government team for irregularities. Leakey felt that they were trying to find evidence that he had broken the law. In 1994, he resigned; subsequently he was a politician and then a civil servant, before accepting an academic post in the United States and running Wildlife Direct, a charitable conservation organisation.

The job of director of KWS became something of a political hot potato. After Leakey, a string of people occupied the post, including the ex-chairman of the East African Wildlife Society, Dr Nehemiah Rotich, the hard-bitten ex-warden Joe Kioko, Amboseli expert David 'Jonah' Western and veterinarian, Dr Evans Mukolwe. No one was sure where things were going. But since 2005 the post has been held by a steely, down-to-earth and honest professional, Dr Julius Kipng'etich, who has brought stability to the organisation.

Ivory burning site, Nairobi National Park. (Stephen Spawls)

The unbeatable views: elephants in the swamp at Amboseli. (Stephen Spawls)

WHAT IS AND WHAT SHOULD BE CONSERVED?

As mentioned, there are four main types of protected area in Kenya: national parks, national reserves, forest reserves and private protected areas. A few other types of protected area, like national monuments, sacred forests and city parks, exist. Eleven percent of the land has some form of protection. The national parks are run by KWS, which is a state corporation. In general, nobody lives in national parks. National reserves are run by county councils. Although the same principles apply – conserving biodiversity – the reserves sometimes permit people to live within them. The forests are protected by the recently established Kenya Forest Service, originally the forestry department; a curious raft of legislation relates to the actual forests themselves. The private protected areas are under the control of their owners, although they are obviously constrained by the laws of the land, for example hunting of game is not allowed.

Kenya's protected areas cover a wide range of habitats. The only large region of Kenya conspicuously lacking any sort of protected areas is the north; in the north-east there are no forest reserves and only one national park, at Malka Mari; in the central north there are no reserves anywhere north of Marsabit save Sibiloi, and the northwest is without a single sanctuary anywhere west of Lake Turkana.

This does not mean that all of Kenya's biodiversity is conserved, however. It is hard to conserve if you don't know what you've got, and as has been mentioned in different contexts in several previous chapters, lack of data means that in Kenya we don't know exactly what we've got so it's hard to say what the strategy should be. Do we need more protected areas? One can take the realistic view that humanity is eventually going to take over a large proportion of the habitable land. In that case, it would be sensible to try to preserve at least part of each and every ecosystem. Evidence shows that it is best to try to protect habitats rather than individual species. But this sort of conservation is complicated in Kenya, *inter alia*, by the huge variety of habitats within the country; variety brings both advantages and disadvantages. This book has illustrated that Kenya has a huge range of differing habitats and many endemics; one thinks of the Boji Plain Killifish, the Mt Kenya Bush Viper, the Ngaia Forest Caecilian. All live in areas with no formal protection. And why are White Rhinos carefully conserved in Lake Nakuru National Park when they are not even an indigenous species?

The Taita Hills are a case in point. In an area of about 400km², only a few square kilometres of original forest remain, but these forests contain three endemic birds, an endemic snake, three endemic lizards, including a spectacular blade-horned chameleon, an endemic caecilian, an endemic frog, three endemic butterflies and 10 endemic plants. These organisms are not found anywhere else in the world. Yet no part of the hills is a national park, and the forest reserves there are not rigidly protected. They are disappearing. There are a lot of people in the Taita Hills and they need firewood. They get it locally. No one can deny the right of the people there to heat their homes and cook their food; but it would be a pity if, as they continue doing so, several unique animals and plants disappear from the face of the earth for ever.

Big Podocarpus partially logged, near Nyeri. (Stephen Spawls)

Depository of the Nation's riches: the original National Museum. (Stephen Spawls)

The new museum providing priceless stimulation to Kenya's young people. (Stephen Spawls)

The lack of knowledge is a problem. A lot of Kenya has never been surveyed by competent biologists. The only paper that makes any attempt to describe a snake fauna in a discrete area of Kenya was written by Arthur Loveridge more than 60 years ago. Large areas of the forest of the south-western Mau have never been visited by an ornithologist. Only one major expedition has made any attempt to find out what reptiles and amphibians occur in northern Kenya. Nobody knows how many species of fly occur within Kenya. In the past 10 years a group of Czech herpetologists have found three new endemic chameleon species, one on Mt Kulal and two on Mt Nyiro. The state of knowledge varies wildly; apart from some good atlasing work done on birds, there is very little else. Biodiversity surveys are needed, led by Kenyan institutions. Decent faunal and floral checklists for most of Kenya's protected areas do not exist. For a country that has had a functional museum and wildlife departments for the best part of a century, this is an unfortunate state of affairs. But, on the other hand, there are those who say that pure science is a waste of time; better to concentrate on the commercial aspect, get visitors who bring in money. One thing is for certain, huge amounts of money have been made available to scientists, mostly from the West, to do 'wildlife research' in Africa, but a couple of stories may be illuminating as to the way that research has been done. In the 1990s a huge research effort was aimed at the rehabilitation of Mkomazi Game Reserve, in Tanzania. Surveys of the flora and fauna were carried out. But when the organisers looked for a herpetologist to survey the reptiles, instead of inviting Tanzanians (there are several expert Tanzanian herpetologists) or involving the University of Dar es Salaam, they invited two South Africans, one of whose herpetological expertise was limited to DNA studies on lizards. The 'experts' collected a Boomslang, mis-identified it as an Olympic Snake *Dromophis lineatus* and described it as 'the easternmost record for that species'. The team carrying out an environmental impact study on the heavy mineral sands at Kwale in 2011 initially ignored the expert herpetologists at the National Museum in Nairobi and invited instead a young South African, who described himself as 'an expert on all vertebrates', to do the survey of the herpetofauna, although he had no record of publications on East African reptiles. The Serengeti warden Myles

The end of a perfect day for visitors to the Maasai Mara. (Laura Spawls)

Turner once suggested that the big sums of money spent on research achieved nothing tangible, and would have been better spent on anti-poaching and education. John Owen, director of Tanzania's national parks, established the Serengeti Research Institute, but his private opinion, according to Ian Parker, was that little would come from it … but it would attract funding to the park, and that would be to the greater good.

PROBLEMS, CONFLICTS AND POLITICS

If you take the view that the science is important, then there is a problem with the lack of information, but it's not the only one. Consider Kenya's premier tourist destination, the Maasai Mara. This is a game reserve; the funds taken at the gate all used to go to the Narok County Council. Gate funds received by KWS are used to support not only the park where they are received, but all the other parks as well, so if as a foreign visitor you pay 60 dollars for a day, your money goes to protect biodiversity all over Kenya. But that was not the case in the Maasai Mara. The reserve is the most popular in Kenya. It was declared one of the 'seven new wonders of the world', it is exquisitely beautiful and has abundant wildlife – you are hardly ever out of sight of game. Visitors never get bored here sitting in a vehicle waiting for something to see. The park receives over 300,000 foreign visitors a year, and they usually stay at least two days, paying 80 US dollars a day. That's an income of 48 million dollars. There are only a few hundred staff in the park, the wages bill is less than 2 million dollars per year. And yet, a bizarre situation has arisen in recent years where those who administer the park were appealing for foreign funds 'to protect the Maasai Mara'. It led to some angry letters in magazines and newspapers, asking why an organisation that was taking in tens of millions of dollars that were seemingly unaccounted for was appealing for funds. Where had the money gone? When Richard Leakey was head of KWS, he attempted to get the Mara taken on as a national park (as happened with Amboseli) and this brought him into conflict with the Maasai and their leaders. Leakey didn't succeed. The situation changed somewhat after 2000, when a not-for-profit body, the Mara Conservancy, was formed to manage the western third of the Mara. This philanthropic body uses gate funds to protect and maintain the park (in 10 years they caught over 1,600 poachers and the roads are superb), and supports various good works in the local area, as well as sending funds to the recently established Transmara District Council. The Mara still has its problems; many feel there are too many lodges and tented camps. *The Daily Nation* carried a story in March 2010 that suggested that there were some 108 'tourist establishments' in the Mara and only 29 of them were legal. There has been debate about actual tourist numbers in the park in the past; at the high season a pride of Lions or a Leopard up a tree can attract 50 or 60 vehicles, clustered around the animal. It has been suggested that the park is being ruined by this, that the Cheetah cannot hunt because they are disturbed by vehicles (they always hunt by day, which is when visitors are around, the other carnivores can kill at night), that the ground is being torn up and there are tracks everywhere. A recent report that gained much media attention said that '70 per cent of the Mara's wildlife had disappeared' and 'buffalo were practically non-existent', but it was based on some highly contentious figures. The debate continues, but it has been going on for over 20 years. At present it cannot be said that the Maasai Mara has been 'ruined'. Visitors there can still see the big five and the migration; in July 2011 the plains below the beautifully sited Serena Lodge were solid with wildebeest. The Mara is still an unbeatable wildlife experience … for now.

In 2005, a curious problem arose at Amboseli National Park. Amboseli became a park in 1974; before that it was a game reserve. In 2005 government suggested that Amboseli would be given back to the Maasai, and control would pass from KWS to Kajiado County Council. This had political implications. The plan was challenged in court; de-gazetting a park is a lengthy laborious business. At present KWS are still running the park, but it leads to an important point. Government are in control; if they are not keen on conservation, then protected areas can be at risk. The old question rears its head, are the government the servants of the people or the people the servants of the government? In tough times, conservation and the protection of our environment can sometimes be seen as a luxury. In Britain, in 2011, the Chancellor of the Exchequer, George Osborne, stated that he would not allow

Conservation

Not such a good sighting: the guide may be tempted to disturb the sleeping Leopard to please the clients. (Laura Spawls)

Backbone of the Kenya Wildlife Sevice: Francis Karanja explains how Black Rhino browse. (Stephen Spawls)

European Union rules and red tape on things like habitats to put ridiculous costs on British business. It was a chilling suggestion, that even in a prosperous Western country, wild places and protected areas might not be safe if they got in the way of private enterprise and the business of making money.

Conservation isn't just in the hands of the government. There are close on 50 private reserves scattered through the southern half of Kenya. Across the Laikipia Plateau on the lovely high savanna land stretching from the northern slopes of Mt Kenya west to the Rift Valley, a number of privately owned sanctuaries have been established. Collectively known as the Laikipia Wildlife Forum, the association includes places like the Lewa Wildlife Conservancy, Ol Jogi Rhino Reserve, Sweetwaters Game Reserve, Mpala Ranch and Ol Ari Nyiro Ranch. Originally cattle and sheep ranches and largely privately owned, these areas have, in the last 30 years, changed their emphasis from farming to tourism. Now the collective contains several hundred Black Rhino (10 per cent of Kenya's surviving 500-odd Black Rhinos are now protected on Ol Ari Nyiro, near Rumuruti in north-central Kenya), over 250 Lions and 7,000 elephants. In addition, a sanctuary for chimpanzees (not found naturally in Kenya) has been established and the ranches also house a number of White Rhino. A range of lodges, tented camps and homestay accommodation is present in the area. The sanctuaries are, naturally, vigorously protected by their owners; who spend their own money in fencing their land and keeping poachers out. Local schoolchildren are taken on tours of the ranches, to see wildlife. The conservancy has recently produced a lovely book, detailing the charms of the area. Thus a model network of protected areas exists, large and viable, under benevolent owners, at virtually no cost to the state, stocked with important high-profile mammals, as well as all the other small beautiful

Above: Death by starvation, Nairobi National Park. (Stephen Spawls)

Left: When wildlife meets traffic, wildlife gives way, dead wild dog near Kijabe. (Stephen Spawls)

flora and fauna. Visitors come, bringing foreign currency; genuine ecotourism is taking place here. Visitors can walk, and drive at night, which you can't do in a national park. Highly qualified guides can show you the whole range of flora and fauna, and the sanctuaries provide employment to a great many local people who take pride in their work. Who could possibly object to such a seemingly beneficial state of affairs?

The answer, of course – and here we don't intend to single out the Laikipia conservancies, they are just a convenient example; this could apply to any protected area – are four groups of people: those who have been injured or killed or whose crops and lands are being damaged by the wildlife of the sanctuary, the desperately poor who have no land of their own and would like to grow crops or graze their stock there, those who think that the land belongs to them, and finally those who think that if they could get hold of the land, they could make a profit.

So here we have a microcosm of Kenya's conservation problems. Some simply superb sanctuaries, but – as always – never free of difficulties. The problem with wildlife injuring or killing people has always been present, it isn't confined to private sanctuaries, and the main culprits are elephants (although hyenas and Lions can also be culpable). Elephants were not originally present on Laikipia in the times of the cattle ranches, now there are lots of them. They can devastate a farm, they sometimes kill people. On the outskirts of Tsavo, the Mara, around Amboseli, near the Shimba Hills, the problem repeats itself. A five-tonne animal can wreck livelihoods. Some financial compensation is available for injury or death – a rather paltry amount (it's about £250 if someone is killed) – but not for damage to crops or possessions, although this may change in the near future. To the poor, elephants are just a dangerous nuisance. In parts of Kenya, there is great resentment towards wildlife and, by association, those who protect it. A recent incident highlights the feelings of the poor towards elephants. An elephant became trapped in a muddy pool, and some of the Laikipia ranchers and wildlife officials arrived to try to rescue it. A crowd gathered, but they were not on the side of the elephant. They drove the putative rescuers away, and killed the animal. Human–wildlife conflict is a challenging problem. One solution that can work is to put up a fence. In the Aberdares, the superb 400km fence funded by the Rhino Ark project keeps farmers and wildlife apart. It benefits everybody. The fence around the northern side of Nairobi National Park does the same,

The overcrowded wilderness, lion meets visitors in the Maasai Mara. (Stephen Spawls)

but suggestions that the southern side be fenced are controversial, as plains game and attendant carnivores migrate out in the wet season. Civilisation is actually creeping around the park, and the migration corridor is narrow. The park's Lions have been killed when they get among stock in the south. Alistair Graham suggested in *The Gardeners of Eden* that it might not be a coincidence that the park entrance was opposite Lang'ata cemetery, as both were graveyards. Not yet, fortunately.

The desperately poor are present on Laikipia, as they are everywhere in Kenya. To them, the wilderness represents a chance to eke out a living, be it something as humble as grazing their stock, cutting some wood or fitting a snare on a game trail to catch a duiker. After all, who owns wild animals? And should that animal be something like a rhino, the chance of a decent amount of money is present. The poor see acres of grassland and ask why they shouldn't be allowed to take their cattle and goats there to graze. There are those who say that domestic stock represent Kenya's biggest conservation headache; but there are many pastoralists in the country, who love their cattle, sheep, goats and camels. And to the stock owner, those animals are wealth; the rural Kenyan who sees his stock dying from lack of fodder or water is in the same position as the city dweller who sees the bank that holds his or her money heading for a crash. And yet, cash in the bank doesn't – usually – do any harm. Overstocking does – especially if the domestic animals are taken into protected areas.

The Laikipia plateau is also the site of a land dispute. Originally the territory of the Laikipiak Maasai (there is more on this in Chapter 3), the clan were persuaded to move out of the area in the early twentieth century, by underhand methods. After the First World War plots were allocated to European settlers under the soldier-settler scheme. Descendants of those Maasai who were moved feel that the lease is now up and they are entitled to have their land back and the ranchers should be compensated by the former colonial government, Britain. Demonstrations were held in Nairobi. The present government feels, for obvious reasons, that it should not be bound by laws and treaties made by the colonials, while Britain feels this is a matter purely for Kenya, as a sovereign state. Hence there is an impasse. Attempts have been made to simply occupy ranch land and the response of government has been, naturally, that law and order must be maintained and property rights respected.

Charcoal sellers, Kibwezi, they see the wilderness as a chance to make a living. (Stephen Spawls)

From their point of view, there is no other approach. Kenya is a stable land, the rule of law applies. Foreign investment and visitors are important to Kenya. Any sign of a Zimbabwe-style solution, with disregard for rights of ownership, and the country's chance of attracting foreign investment goes down the drain. Government sent the security forces to remove the occupiers, and there were violent incidents; people were hurt, even killed. The more radical elements saw this as a betrayal of ethnicity, their elected government violently suppressing them in favour of rich white landowners. It was portrayed in parts of the media as a black/white issue.

There is another problem with private sanctuaries, too. The owners may decide they no longer want to maintain the wildlife, and switch back to farming, or sell the farm, or bequeath it to others. Such ranches have been broken up in the selling into many smaller plots. With a national park, one can hope that the land is reasonably safe, even if the animals are not, but with private land, it is another matter.

There is no easy solution, and these problems extend to any protected area in Kenya. The country has a huge population, which is expanding. People feel, rightly, that they are entitled to a bit of land, often with a few domestic animals on it. The poor would like to utilise wildlife. At present, they can be jailed for killing the smallest of antelope. And a nucleus of poor people is a tough issue for government. Those with no possessions and nothing to lose can be easily be inspired to take radical action. But it should be remembered that people who are not adversely affected by wildlife sometimes do take pleasure in it. A paper published on attitudes towards wildlife in Laikipia by Michelle Gadd (2005) indicated that some people did enjoy simply seeing the animals, and appreciated that they brought tourists, money and employment.

Of course, it isn't just the poor who want land in Kenya. Land-grabbing by the rich is a country-wide problem. Kenya's Nobel Laureate Wangari Maathai led the fight against the developers who tried to steal land in the KaruraForest; her autobiography *Unbowed* is a magnificent story, underlain by the theme of botanical conservation. Richard Leakey had to struggle to prevent land grabbers seizing a portion of Nairobi National Park, on the pretext of building a pipeline. But what is heartening is that there has been a raising of consciousness about the problem,

Home-grown tourism, a massive force for good, Nairobi National Park. (Stephen Spawls)

and the free press are active in exposing it. A further development is that large parts of Africa are now being leased by foreign companies and governments, often to grow food for their own people; this seems odd seeing that the food is often being grown in a country with its own food shortages. The poor stockman eking out a living from his goats in the dry land east of Kitui would no doubt be bemused to learn that rich foreigners are hiring large areas of his land to grow food for their own compatriots … or worse, to make fuel. Land in the Tana Delta has been leased to a British biofuels company and to a Middle Eastern firm to grow food for the Emirates. A massive area of Mozambique has recently been leased to Brazil, to grow food for China.

Some conservation problems in neighbouring countries may be significant for Kenya. In Tanzania and the Sudan wealthy foreigners have taken over game-rich areas and massacred game there with impunity; a wild area near Loliondo in northern Tanzania is leased to an Emirates firm so that Gulf Sheikhs and millionaires can hunt there. With increased public accountability it seems unlikely that it could happen in Kenya, but in 2009 Colonel Gaddafi's son, Al Saadi, was found in the Maasai Mara with a hunting rifle, with which he intended to shoot game. Elephant and rhino poaching continues, as the price of those commodities and demand creeps upward. It is not just elephants and rhino that are threatened. In large areas of the Far East, particularly China, the trade in bits of wild animals is fuelled by the rapidly increasing prosperity of the country and stimulated by bizarre beliefs, such as that eating the meat of an owl is good for the eyes, or eating a turtle will make you live longer. It has a certain voodoo logic. Demand for turtles to eat in China has literally exterminated the wild turtle populations of some neighbouring countries. Tiger body parts are also in high demand and the fear is that with the disappearance of the tiger, people in the Far East will decide that Lion body parts can also be used. Rare animals are also wanted for the pet trade. Foreign enthusiasts are keen to get their hands on Kenya's little endemic snakes, like the Mt Kenya Bush Viper. There is an increasing demand by rich aristocrats in the Middle East for Cheetahs, and those magnificent birds of prey, the falcons, for falconry. The impression you get is that there are countries out there that see parts of Africa simply as an unutilised resource, ripe for exploitation for the benefits of non-Africans.

THE WAY FORWARD

There is a principle in geology called uniformitarianism, a magnificently unwieldy word with a simple meaning: the present is the key to the past. In other words, the physical changes that we see happening now will give us clues as to what happened in the past. A similar idea, but looking forward, might be usefully applied to conservation, that the present is the key to the future. By this we mean that the present situation regarding conservation, the natural world and biodiversity in highly developed countries such at the United States and Britain, will give us clues as to what will take place in the future in developing countries like Kenya.

In many ways this is a depressing vision. Consider Britain. The megafauna, such as the bear, the wolf, the lynx, the boar and the beaver, have disappeared. Such animals, especially the carnivores, need space, and it is no longer available. There are a number of national parks in UK, it costs nothing to enter them but most have people living inside them, and they are relatively small. They are heavily exploited; on some days you must join a queue to walk up a magnificent mountain or hillside. Finding a place where you are out of sight of humanity and its constructions, so easy in places like Tsavo and other wild areas in Kenya even nowadays, is very difficult in Britain. When you finally leave the city (and it can take an hour or more to get out), most of what you see is heavily managed farmland. Most large British mammal species have gone, and in some areas introduced species are common and have wiped out indigenous wildlife.

Or consider India, nominally a developing country. India is a country with a lot of people. It has a few national parks, but they are mostly small and poaching is a problem in many. Ranthambore National Park, in Rajasthan is an excellent place to see tigers, but visitor demand is so great that the park authorities have divided up the park into a number of zones and have to limit the number of vehicles entering. No private vehicles are permitted in. If you manage to get in, you cannot drive your own vehicle, and your driver is limited to a set route. No such restrictions apply in any of Kenya's parks … yet.

What is the future for Kenya's wild places? It is worth quoting one of Reg Moreau's prescient comments, that 'by the time Africans are ready to become amateurs of field biology … [amateurs in this sense retains its original meaning, those who love the topic, rather than the more modern debased meaning of 'non-professional') … most of them will have to scrabble about in the ruins of their flora and fauna, as everybody else in a 'developed' country must do, and instead of studying the grand designs of natural biomes they will have to finick around with the impoverished and lopsided remnants which are the by-products of man's multiple incontinencies'. So what is going to happen eventually in Kenya? A small number of well-fenced but smallish national parks, a few of which have elephants? Most of the land covered by agriculture, with some wilderness areas set aside? Should people be allowed to live in conservation areas, and maybe even utilise the flora and fauna, as is done in other African countries, or should such protected areas be totally sacrosanct – so-called 'fortress conservation'? In Kenya, the population is concentrated in the fertile areas. Maybe it's time to prioritise a few crucial areas. And should Kenya be thinking of the benefits of biodiversity, the untapped potential for food, medicines, and useful products, or should Kenya be concentrating on of tourism and the benefits of a steady flow of visitors into Kenya, bringing their money and spending it within the country. Or both? What of the dry land of the north, which is agriculturally unrewarding? The problem there is, if it has no permanent water and very little rain, it might be unattractive to people, but it's equally unattractive to wildlife. And if you organise water for wildlife, stock owners may feel that they can also benefit.

It's not all doom and gloom. Kenya is in an unusual and favourable position. A lot of African countries have destroyed much of their wildlife, particularly the big animals, which are what the majority of visitors come to see. Perhaps we should prioritise conservation in a limited number of crucial states. There are countries within Africa that no visitor in their right mind will be visiting, now or for a long time in the future. Is it worth spending conservation resources on their elephants, when places like Tsavo and the Mara represent a much better prospect for saving Africa's megafauna? Kenya still has big parks and lots of wildlife. There are many people out there who want to see Kenya succeed in protecting its flora, fauna and wild places and are willing to help without demanding too much in return. The world is full of increasingly affluent people, who are willing to spend their money on the unique experience

The perfect setting, Voi Safari Lodge. (Stephen Spawls)

that is an African safari. (Interestingly, there is a debate here over whether the present pricing system for Kenyan sanctuaries is making the assumption that such visitors only want to come once, spend no more than one or two days in each reserve, and thus will pay very high prices for their single trip, or whether it might be worth having a decreased rate for longer stays or repeat visits, thus encouraging visitors to come back.) There are only seven or eight countries who fit the safari bill. Kenya is one, and it offers the most varied experience. No other African country has stability, an abundance of truly wild game and big sanctuaries, semi-desert, savanna and forest, a coral sea and snow-capped mountains. And yet, Kenya has been overtaken in tourist numbers by South Africa. In some ways, this is odd. South Africa has only two large parks comparable in size to Kenya's big parks, but a mass of tiny parks and private game sanctuaries have sprung up there. Most of them are stocked, with wildlife brought in from elsewhere; they are in effect just big 'safari parks', often with absurdly luxurious and hideously expensive lodges. And yet they are highly popular; it seems that the public are not worried about authenticity, they just want pampering in a sunny place with some wildlife thrown in. Of course, some game has been translocated and introduced in Kenya; many of Nairobi Park's rhino are translocated, although not all, and they did occur there naturally. But there are visitors who value authenticity and who come to see genuinely wild places and their genuinely wild inhabitants and for these people Kenya has much to offer. Consider Tsavo. Stand on Ngulia Hill and look eastwards, and virtually all the land you can see lies within the national parks. This is the way the wilderness should be, huge, wild and authentic, somewhere you can immerse yourself, somewhere with enough space to support big predators. What lodge elsewhere can beat the view out of your window in Amboseli, with the vastness of Kilimanjaro behind? Where else in Africa, in a short safari, can you see animals of the mountain forests, the savanna and the semi-desert? Nowhere. There is space for all, and enough unvisited areas to really develop ecotourism.

A big new wildlife bill is to be considered by Kenya's parliament. It is much longer than this chapter, and it may have major significance for conservation in Kenya. Although sport hunting is not being considered for re-introduction – this is too much of a political hot potato – the bill suggests that some consumptive trade in

Young people support an environmental protest against the damming of the Omo River. (Stephen Spawls)

wildlife might be permitted. It is banned at present. This may be the way forward – sustainable harvesting: wild land and wildlife to pay for itself; the people to benefit. It has worked elsewhere in Africa. But a little caution is needed. In the 1970s and 1980s in Zimbabwe, a consumptive programme to bring the benefits of wildlife to rural citizens was developed under the CAMPFIRE scheme (the acronym stands for Communal Areas Management Programme for Indigenous Resources). Local landowners were allowed to utilise wildlife on their communal or individually owned lands. The utilisation took the form of sport hunting, capture and sale of game animals, hunting for meat or game viewing; the monies earned were to be disbursed to local communities. The programme is relentlessly presented by its proponents as a success, and in fact it did bring money into communal lands. But by and large it didn't changes attitudes among rural dwellers towards wildlife. And the problems over land ownership in Zimbabwe have reduced the initiative to a shambles in many areas, particularly in the south of the country, where many regions are insecure.

Other welcome proposals in Kenya's new wildlife bill include compensation for damage by wildlife, the recognition and regulation of privately protected areas, incentives for wildlife conservation and the protection of nationally important water towers (i.e. montane forests). This has to be good if it brings benefits to the ordinary citizens of Kenya, and makes them feel they have a stake in the country's natural resources.

Some interesting small-scale conservation initiatives have taken place in Kenya – little things that make things better for the environment, a good example being the monkey bridge set up at Diani, to stop the Colobus Monkeys getting killed by vehicles. Other important initiatives for the future might involve putting more wilderness land in the hands of local cooperatives and letting more ordinary people participate in profit-making from conservation; gate fees and hotel fees might be disbursed to the local community, to build clinics, schools and wells. Schoolchildren need to be involved in conservation projects. If the local inhabitants don't feel they have a stake, they will despise the enterprise. The future is uncertain. The crucial thing is uplift, keeping a hold on Kenya's wonderful resources and heightening awareness while gradually bringing everyone out of poverty. The long road stretches ahead.

SUMMARY TIMELINE
Africa's first recorded conservation area, Menagesha Suba National Park in Ethiopia, was proclaimed a protected area by the Emperor Zar'a Ya'qob in 1450.
In 1646, Dutch settlers at the Cape in South Africa introduced Africa's first written game laws, including restricting hunting to two months of the year.
Kenya's first protected area, the Kenia Game Reserve, was created in 1897.
In 1901, Arthur Blayney Percival was appointed the first 'Ranger for Game Preservation' of the British East Africa Protectorate, as Kenya was then known, at a salary of £250 per year.
In 1902 a 'Conservator of Forests' was appointed, and by 1908 a number of major forest reserves had been established.
In 1906, an independent Game Department was formed, and in 1907 Colonel J. H. Patterson, the man who shot the man-eating lions of Tsavo, was placed in charge; his title was 'The Game Warden'.
In 1909 Teddy Roosevelt, former president of the United States, undertook a major hunting safari in Kenya.
In 1945, Sir Philip Mitchell, Governor of Kenya, proclaimed the National Parks Ordinance; the parks were to be overseen by a board of trustees.
Nairobi National Park, Kenya's first national park, came into being on 24 December 1946. The entrance fee was one shilling. Tsavo National Park was proclaimed in 1948, Mount Kenya in 1949 and the Aberdares in 1950. It now costs 60 US dollars or more for a foreign visitor to enter Tsavo for a day.
The first purpose-built lodge to be constructed within a national park, Kilaguni (meaning 'young rhino' in KiKamba) was completed in 1962, and was followed later that year by Keekorok Lodge in the Maasai Mara (Keekorok means 'black trees' in Maa, the Maasai language). Treetops Hotel, built in 1932 in the Aberdares, predates these, but the park was created after the hotel. There are now over 60 game lodges and tented camps in the Mara area alone.
In 1976 the National Parks and Kenya Game Department merged.
In the early part of the twentieth century Kenya's wildlife-orientated visitors came mostly to hunt big game, but following the building of lodges within national parks in the 1960s, game-watching tourism took off.
In 1977 Kenya banned all hunting.
Poaching, especially of elephants and rhino, became a major problem in the 1970s and 1980s. In the 1960s there were estimated to be well over 7,000 rhino in Kenya. In November 1977 there were 180 Black Rhino in the Maasai Mara Reserve but by October 1980 not a single rhino was left alive in the Mara.
In 1989 the management of conservation areas in Kenya was taken over by a new government body, Kenya Wildlife Service; its first director was the palaeontologist Richard Leakey.
On 18 July 1989, President Daniel arap Moi set fire to a pile of more than 2,000 elephant tusks in Nairobi National Park, to indicate Kenya's contempt for those who dealt in and wore ivory. Over 900 million people around the world saw the fire on television.
In 2005, over one million tourists visited Kenya, bringing the country an income of nearly 400 million US dollars or 250 million pounds.
In 2006, the American TV network ABC announced that the Maasai Mara had been declared one of the 'Seven New Wonders of the World'.

THE LITERATURE

There is an abundant, informative and excellent literature on the history and development of wildlife conservation, hunting and wildlife in Kenya. Many of the classic works are not only nicely opinionated; many are of great value as well. We have mentioned some in earlier chapters. Kenya has attracted and nurtured many prominent personalities in the wildlife field. While reading these, it is of importance to be aware of the zeitgeist. Hunting, violent suppression of those perceived to be poachers, and fortress conservation dominated by white Westerners are sensitive issues nowadays, but in many cases they served their purpose during the last century in making people aware of Kenya's magnificent wildlife and wild places, and the need for their protection.

A thoughtful essay on game protection in East Africa forms the final chapter in Edward North Buxton's book *Two African Trips* published by E. N. Stanford. Buxton reported on the status of the sanctuaries of central and eastern Africa, and offered sensible suggestions as to their maintenance. The interesting thing was that his book was published in 1902. Many early hunters visited Kenya and mused on conservation; good books from this time include Arthur Neumann's 1898 *Elephant Hunting in East Equatorial Africa* and Captain Chauncey Hugh Stigand's 1913 work *Hunting the Elephant in Africa*. Brian Herne's 1999 book *White Hunters* is a comprehensive and superb summary of the rise and fall of professional hunting in East Africa. It is effectively a history book, meticulous and thoughtful, yet enlivened by riveting stories.

A lovely book on East African conservation was Bernard Grzimek's *Serengeti Shall Not Die* (originally in German, published in translation by Hamilton in 1960). Although based in Tanzania, this book describes not only the adventures of Grzimek and his son Michael (who was killed in a plane crash, as is recorded at the end of the book) as they document the Serengeti wildlife, but is full of pertinent observations on conservation and Africa's new politics; it captures the spirit of the time. Also fascinating is Noel Simon's *Between the Sunlight and the Thunder*, published by Collins in 1962. Simon was the man who pushed for the creation of the northern Serengeti, to link with the Maasai Mara and compensate for the loss of the Ngorongoro area. He was instrumental in the creation of Lake Nakuru National Park and the Mau Forest Reserve. For good measure he was also a co-founder in 1955 of the Kenya Wildlife Society, later to become the East African Wildlife Society. This is an influential body which continues to this day and whose journal (originally *Africana*, now *SWARA*) is a major mouthpiece for Kenyan conservation. Although *Between the Sunlight and the Thunder* is slightly Eurocentric, it was a near-complete summary of the conservation situation in Kenya in the 1960s and retains its value as a historical treatise. A controversial book on Kenya's wildlife and conservation at this time, published in 1965, was the photographer Peter Hill Beard's *The End of the Game* (Hamlyn 1965). Many found it fascinating, many hated it. Its six chapters cover an eclectic range of Kenyan wildlife topics and it has an inordinate number of striking and bizarre pictures, many of dead animals, particularly elephants. Beard himself is equally controversial; he was arrested and jailed in Kenya for catching and tying up a poacher using his own wire snare. He once famously said that people would rather have air conditioning than conservation. Whatever your views on the man, this is a book worth having.

The making of Tsavo, up to 1973, is covered in Dame Daphne Sheldrick's book *The Tsavo Story* (published by Harper Collins). Although anecdotal, rather than scientific, it conveys well what it was like to be a warden in those important days of the 1950s and 1960s, when the groundwork for Kenya's parks was being laid. Equally fascinating and covering the development of both the mountain parks and Tsavo is *Elephants at Sundown* (W.H. Allen 1978), a biography of Bill Woodley and his family by Dennis Holman (who has written a string of good books about Kenya, as mentioned in earlier chapters). Kenya owes a lot to the Sheldricks and Woodleys, and the family tradition continues: Bill's sons Bongo and Danny have both been wardens. A fine book of personal adventure, which combines the story of a walk from the top of Kilimanjaro across Tsavo to the sea with a potted history of conservation in Kenya is *The Shadow of Kilimanjaro* (not to be confused with Ernest Hemingway's short story *The Snows of Kilimanjaro*), written by the adventurer Rick Ridgeway and published by Henry Holt in 1999. This is a bewitching read: Ridgeway walks with Bongo and Danny Woodley and the mountaineer Iain Allan, and along the way we meet characters like the charismatic Stephen Gichangi, who was Woodley's successor in Tsavo,

Small-scale conservation in action – the monkey bridge at Tiwi. (Stephen Spawls)

Joyce Poole, the elephant lady, the then KWS director Johan Western and the hard-bitten, six-foot-seven Samburu ranger Mohamed, whose last name we never learn.

Iain and Oria Douglas-Hamilton have published two books on elephants. The first, *Among the Elephants* (Collins 1975), is largely a fascinating account of their research among those big pachyderms at Lake Manyara National Park in Tanzania. The second, *Battle for the Elephants* (Viking 1992), paints a much bigger picture; it is the story of the gallant couple's fight, initially almost single-handedly, later with help from conservation organisations, to save the elephants of sub-Saharan Africa. The battle rages from West Africa and the Sahel to the Sudan, East and southern Africa. This is a magnificent story by a family with a mission, and earnestly manages to cover both sides.

Frequently in the Douglas-Hamilton's saga we meet a man called Ian Parker, and is has to be said that Parker is one of the most interesting men in African conservation. Equally respected or despised; as George Bush might have said, you are either for him or against him; Parker himself has publicly stated that he hates conservationists. Parker came to Kenya from Malawi in 1939, and after working on a farm was in the Kenya Regiment, he went on patrol in the mountain forests after the Mau-Mau. He joined the Kenya Game Department and then, with Alistair Graham, started a company called Wildlife Services; their business was game control and monitoring. They culled elephants. Subsequently, he was involved with the ivory trade, before finally emigrating to Australia. Parker has written and edited a number of books and scientific papers, two of which are of major importance to anyone wanting to understand the vicissitudes of wildlife management in Kenya. The first, *An Impossible Dream*, (2001) edited by Parker and the ex-warden Stan Bleazard, is ostensibly a collection of anecdotes by Kenya's 'colonial game wardens'. In reality, it is far more than that; it is almost a history of the game department, with stories of wild adventure intermixed with sober reflections on the development and changes of a dynamic and important arm of conservation; this is what it was really like. The second book of importance is Parker's semi-autobiographical *What I Tell You Three Times is True*, (2004) which is a line from Lewis Carroll's nonsense poem 'The hunting of the snark'. Originally over 300,000 words long, shortened to 200,000, overindulgent and rambling in places, it remains gripping and powerful. It is a mixture of autobiography, history, politics, adventure and philosophy, with a running theme of ivory and its trade. Parker has clashed with the highest of politicians and the most dangerous of bandits; he has strong opinions on much to do with conservation, and fires randomly at those whom he perceives to have wronged him. He calls CITES an orgy of silliness and says conservationists have a magnificent capacity to ignore evidence. This is the elephant business, the end of imperialism and 40 years of African independence told by a man at the cutting edge, as he saw it. Few have seen further, or understood it better. Both books were published by Librario, in 2001 and 2004 respectively, and are available on demand. They are essential reading for any student of the ivory trade and conservation in East Africa.

Kenya has had an unusual history. Consider, for example, Botswana, another African country colonised by Britain. In Botswana there were few white settlers, a small population under a vaguely benevolent administration, no violent conflict between the Batswana and the British, a smooth handover of power and subsequent calm prosperity. Kenya, conversely, had white settlers who took the best land and there was violent conflict between some of Kenya's people and the settlers and government forces; since independence there has been a lot of meddling by outsiders. Consequently it has been possible – for those who want to see it that way – to portray conservation issues in Kenya (and East Africa) as a continuing black against white or a 'foreigners versus natives' issue (we use the term natives here in its finest sense, to mean the original inhabitants of a land, rather than the pejorative term meaning some sort of uncivilised race). A string of books explore this theme. The first was probably Alistair Graham's thought-provoking *The Gardeners of Eden* (George Allen 1973), a blistering attack on the colonials who ran the East African parks. Graham devotes entire chapters to blasting Captain Charles Pitman and Archie Ritchie, chief game wardens in Uganda and Kenya respectively, rubbishes those who worry excessively over the disappearance of wildlife and praises poachers. The books opening sentence sets the scene, where Graham suggests that love of wild animals is a manifestation of intense hatred! It is a book that must be read by those who would protect Africa's wildlife. Other books on this theme include Jonathan Adams and Thomas McShane's *The Myth of Wild Africa* (1992) and Raymond Bonner's *At the Hand of Man* (Knopf 1993). Dan Brockington's book *Fortress Conservation* is an unswerving attack on the conservationists who were involved in the rehabilitation of Mkomazi Game Reserve, and their treatment of the pastoral people who grazed their stock there. It was published in 2002 by the International Africa Institute. Well worth reading on this theme is Edward Steinhart's *Black Poachers, White Hunters*, published in 2006 by James Currey in Oxford. Steinhart's approach is that of a Marxist; he is on the side of the ordinary people and against the rich foreigner, but he is also a rigorous historian and this book is meticulously researched, with a decent attempt to be even-handed, and it covers the Kenyan wildlife world of the twentieth century. Finally, a seminal paper on Kenya's wildlife, the related economics and the problems of externally funded NGOs is Mike Norton-Griffiths' paper 'How many wildebeest do you need?', published in *World Economics* in 2007.

There are lots more; it's a huge field that we can't cover comprehensively here. We mentioned earlier the plethora of journals that deal with conservation; in Kenya the *Journal of East African Natural History*, the *African Journal of Ecology* and *SWARA* are the main journals. The list above is a personal selection.

References and Bibliography

Adams, J. S. and McShane, T. O. 1992. *The myth of wild Africa*. W W Norton.
Adamson, G. 1968. *Bwana Game*. Collins & Harvill.
Adamson, J. 1960. *Born Free*. Harvill Press.
Ajayi, J. F. A. and Crowder, M. 1985 (3rd Edition) *History of West Africa. Vols I and II*. Longman.
Alpern, S. 2005. Did they or didn't they invent it; iron in sub-Saharan Africa. *History in Africa* Vol 32 pp 41–94
Anderson, D. 2005. *Histories of the Hanged*. Weidenfeld and Nicolson.
Archer, G. & Godman, E. M. 1937–61. *The Birds of British Somaliland and the Gulf of Aden*, 4 vols. Gurney and Jackson.
Ardrey, R. 1961. *African Genesis*. William Collins.

Backhurst, G. 1986. Check-list of the birds of Kenya. East African Natural History Society (Ornithological Sub-Committee)
Balirwa, J, Chapman, C, Chapman, L., Cox, I., Geheb, K, Kaufman, L, Lowe-Mcconnel, R., Seehausen, O.,Wanink, J. H., Welcomme, R and Witte, F. 2003. Biodiversity and Fishery Sustainability in the Lake Victoria Basin: An Unexpected Marriage? Vol. 53 No. 8. *BioScience* 703–715
Barnes, S. 2004. *How to Be a Bad Birdwatcher*. Short Books.
Battiscombe, E, 1936. *Trees and Shrubs of Kenya Colony*. Government Printer; Nairobi.
Beadle, L. C. 1974. *The Inland Waters of Tropical Africa*. Longman.
Beard, P. H. 1965. *The End of the Game*. Hamlyn.
Beentje, H.J. 1988. *Atlas of the rare trees of Kenya*. Utafiti 1(3): 71–125
Bell-Cross, G and Minshull, J. 1988. *The Fishes of Zimbabwe*. National Museums and Monuments of Zimbabwe.
Beentje, Henk. 1994. *Kenya Trees, Shrubs and Lianas*. National Museums of Kenya.
Behrensmeyer, A. K., Todd, N. E., Potts, R. and McBrinn, G. 1997. Late Pliocene Faunal Turnover in the Turkana Basin, Kenya and Ethiopia. *Science* 278; pp 1589–1594
Bennun, L.A., Dranzoa, C., & Pomeroy, D. (1996). The forest birds of Kenya and Uganda. *Journal of East African Natural History* 85: 23–48
Bennun, L. & Njoroge, P. 1999. *Important Bird Areas of Kenya*. East African Natural History Society.
Benuzzi, F. 1953. *No Picnic on Mount Kenya*. E. P. Dutton and Company.

Berger, L. & Hilton-Barber, B. 2002. *The Official Field Guide to the Cradle of Humankind*. Struik.
Blundell, M. 1987. *Collins Photo Guide to the Wild Flowers of East Africa*. Collins.
Bock, K. 1978. *A Guide to the Common Reef Fishes of the Western Indian Ocean and Kenya Coast*. Macmillan.
Bock, K. 1996. Checklist of the reef fishes of Diani and Galu, Kenya. *Journal of the East African Natural History Society* 85: pp 5–21.
Bonner, R. 1993. *At the hand of man*. Knopf.
Boswell, P. G. H. 1935. Human remains from Kanam and Kanjera, Kenya Colony. *Nature* 135: 371
Boulenger, G. A. 1909: Catalogue of the fresh-water fishes of Africa in the British Museum (Natural History). Trustees of the British Museum; London. 1: i–xii + 1–373
Boulenger, G. A. 1909: Catalogue of the fresh-water fishes of Africa in the British Museum (Natural History). Trustees of the British Museum London. 2: i–xii + 1–529
Boulenger, G. A. 1915: Catalogue of the fresh-water fishes of Africa in the British Museum (Natural History). Trustees of the British Museum London. 3: i–xii + 1–526
Boulenger, G. A. 1916: Catalogue of the fresh-water fishes of Africa in the British Museum (Natural History). Trustees of the British Museum London. 4: i–xxviii + 1–392
Bowker, R and Bowker, M. 1979. Abundance and distributions of anurans in a Kenya pond. *Copeia* 278–85.
Branch, Bill. 2005. *A photographic guide to snakes and other reptiles and amphibians of East Africa*. Struik.
Branch, Bill. 2008. *Tortoises, Terrapins and Turtles of Africa*. Struik.
Brett Young, F. 1917. *Marching on Tanga*. Collins.
Britton, P. L. (ed.). 1980. *Birds of East Africa; their habitat, status and distribution*. East African Natural History Society.
Broadley, D. G. and Howell, K. 1991. A checklist of the reptiles of Tanzania with synoptic keys. *Syntarsus* No. 1, 1991, pp 1–70.
Brockington, D. 2002. *Fortress Conservation*. The International Africa Institute.
Brown, L. & Amadon, D. 1968. *Eagles, Falcons and Hawks of the World*. Country Life Publishers.
Brown, L. 1970. *African Birds of Prey*. Collins
Brown, L. 1975. *East African Coasts and Reefs*. East African Publishing House.

Brown, L. 1979. *Encounters with Nature*. Oxford University Press.
Brown, L. 1989. *British Birds of Prey*. Collins New Naturalist.
Brown, L., Urban, E. & Newman, K. 1982. *The Birds of Africa, vol. 1*. Academic Press.
Brown, M. 1989. *Where Giants Trod*. Quiller Press.
Bryson, Bill 2003. *A Short History of Nearly everything*. Doubleday.
Bussman, R. 2002. Islands in the desert. *Journal of the East African Natural History Society* 91: pp 27–79
Buxton, E.N. 1902. *Two African Trips*. E. N. Stanford.

Campbell, N. & Reece, J. 2008. *Biology*. Pearson.
Carcasson, R. 1964. A preliminary survey of the zoogeography of African butterflies. *East African Wildlife Journal* 2: 122–57.
Carson, R. 1962. *Silent Spring*. Houghton Mifflin.
Channing, A. 2001. *Amphibians of Central and Southern Africa*. Comstock Publishing.
Channing, A. & Howell, K. 2006. *Amphibians of East Africa*. Cornell University Press.
Chenevix Trench, C. 1993. *Men Who Ruled Kenya: The Kenya Administration, 1892–1963*. Tauris.
Cheptumo, M., Madsen, Duff-Mackay, A., Hebrard, J and Rotich, D. 1986. Survey of the reptiles and amphibians in Kora national Reserve. In: Coe, M. and M. Collins (eds.). An Ecological inventory of the Kora national reserve, Kenya. Royal Geographical Society, London
Churchill, W. 1908. *My African Journey*. Chartwell.
Clausnitzer, Viola. An updated checklist of the dragonflies (Odonata) of Kakamega forest, Kenya. *Journal of East African Natural History* (2005) Volume: 94, 2, pp 239–246
Cohen, D. 1968. The River-Lake Nilotes from the Fifteenth to the Nineteenth Century. In Ogot, B. (ed.), *Zamani: A Survey of East African History*. Longman.
Cole, S. 1954 (rev. 1965). *The Prehistory of East Africa*. Penguin.
Cole, S. 1975. *Leakey's Luck*. Collins.
Collins, J. & Crump, M. 2009. *Extinction in Our Times: Global Amphibian Decline*. Oxford University Press.
Copley, H. 1941. A short account of the freshwater fishes of Kenya. *Journal of the East Africa and Uganda Natural History Society* 16(1) pp 1–24.
Copley, H. 1958. *Common Freshwater Fishes of East Africa*. Witherby, London.
Cowie, G. 1961. *Fly Vulture*. Harrap.
Cullen, A and Hemphill, P. 1971. *Crash Strike*. East African Publishing House.

Dale, I. & Greenway, J. P. 1961. *Kenya Trees and Shrubs*. Buchanan's Estates.
Darlington, P. J. 1957 *Zoogeography*. Museum of Comparative Zoology, Harvard.
Dart R. A. 1925. *Australopithecus africanus*: The man-ape of South Africa. *Nature* 115: 195–9.
Darwin, C. 1839. *The Voyage of the Beagle* (full title; 'Journal of researches into the geology and natural history of the various countries visited during the voyage of *H.M.S. Beagle* round the world'). London: John Murray.
Davis, W.M. 1899. The Geographical Cycle. *The Geographical Journal*, Vol. 14, No. 5, pp. 481–504
Dawkins, R. 1986 *The Blind Watchmaker*. Longman.
Dawkins, R. 2004. *The Ancestor's Tale*. Houghton Mifflin.
DeMenocal, P. 2004. African climate change and faunal evolution during the Pliocene-Pleistocene. *Earth and Planetary Science* Letters 220 pp 3–24.
Diamond, J. 1997. *Guns, Germs and Steel*. Jonathan Cape.
Dobiey, M. and Vogel, G. 2007. *Venomous Snakes of Africa*. Chimaira.
Douglas-Hamilton, I. & Douglas-Hamilton, O. 1975. *Among the Elephants*. Collins.
Douglas-Hamilton, I. & Douglas-Hamilton, O. 1992. *Battle for the Elephants*. Viking.
Drewes, R. C. 1984. A phylogenetic analysis of the Hyperoliidae treefrogs of Africa, Madagascar and the Seychelles'. Occasional papers of the California Academy of Sciences 139. Pp 1–70.
Drewes, R. C. 1997. Hot Spot for Frogs. *SWARA* 20:2. Pp 27–30.
Drewes, R. C. and Altig, R. 1996. Anuran egg predation and heterocannibalism in a breeding community of East African frogs. *Tropical Zoology* 9; pp 333–347.
Drewes, R.C. and Perrett, J. L. 2000. A new species of giant montane Phrynobatrachus (Anura; Ranidae) fro the central Mountains of Kenya. Proceedings of the California Academy of Sciences 52: pp 55–64.
Dudley, C. 1996. How many species of insect does Malawi have? *Nyala* 19; pp 13–16.
Duellman, W. and Trueb, L. 1986. *Biology of Amphibians*. The Johns Hopkins University Press (reprinted 1994).
Duff-MacKay, A and Schiotz, A. 1971. 'A new *Hyperolius* from Kenya', *Journal of the East African Natural History Society*, 19 (128) pp 1–3.
Duff-MacKay, A. 1980. Conservation Status Report number 1: Amphibia. National Museum of Kenya; pp 1–44.
Dupre, Gerard. *Annotated Bibliography on African Scorpions*. Available at: scorpionshttp://afras.ufs.ac.za/dl/userfiles/documents/Dupre%20unpubl%20African%20Scorpions%20Bibliography.pdf

Edwards, D. C. 1940. Vegetation Map of Kenya with Particular Reference to Grassland Types. *Journal of Ecology* Vol. 28, No. 2 (Aug., 1940), pp. 377–385
Eisner, T. 2003. *For Love of Insects*. Belknap Press.
Eliot, C. 1905. *The East Africa Protectorate*. Edward Arnold.

Fanshawe, J.H. (1994) Birding Arabuko-Sokoke Forest and Kenya's northern coast. ABC Bulletin 1(2) pp 70–89.
Fedders, A. & Salvadori, C. 1981. *Peoples and Cultures of Kenya*. Rex Collings.
Fichtl, R and Admasu Adi. 1994. *Honeybee Flora of Ethiopia*. Deutscher Entwicklungsdienst.

Finch, B. 2006. Mad about frogs. 29(3); *SWARA* pp 20–23.

Finlay. B., Thomas, J, McGavin, G., Fenchel, T and Clarke, R. T. 2006. Self-similar patterns of nature; insect diversity at local to global scales. Proceedings of the Royal Society (Biology) 273(1596); pp 1935–1941.

Finch, B, Pearson, D.J. and Lewis, A. 1989. Birds of Samburu, Buffalo Springs and Shaba National Reserves. Nairobi; Friends of Conservation.

Fischer,E and Hinkel, H. 1992. *Natur Ruandas/La Nature du Rwanda*. Ministerium des Innern und fur Sport: Mainz.

Fortey, R. 2008. *Dry Store Room Number 1*. Harper Press.

Frost, D. R. et al. 2006. *The Amphibian Tree of Life*. Bulletin of the American Museum of Natural History No. 297. Available online at http://digitallibrary.amnh.org/dspace/handle/2246/5781.

Frost, D.R. et. al. 2008. Is the amphibian tree of life really fatally flawed? *Cladistics* 24(3); pp 385–395.

Gadd, M. 2005. Conservation outside of parks: attitudes of local people in Laikipia, Kenya. *Environmental Conservation* 32: 50–63.

Gardner, J. C. M. 1957. An annotated list of East African Forest Insects. EAFRO Forestry Technical Note No. 7.

Gee, H. 2000. *Deep Time*. Fourth Estate.

Gibbons, A. 2006. *The first humans; the Race to Discover Our Earliest Ancestors*. Doubleday.

Gillson, L. 2004. Testing Non-equilibrium Theories in Savannas: 1400 years of Vegetation Change in Tsavo National Park, Kenya. *Ecological Complexity* 1; pp 281–298.

Gitonga, E and Pickford, M. 1995. *Richard E Leakey, Master of Deceit*. White Elephant Publishers, Nairobi.

Goldacre, B. 2008. *Bad Science*. Fourth Estate.

Goldschmidt, T. 1996. *Darwin's Dreampond: Drama in Lake Victoria*. MIT Press.

Gordon, A and Harrison, N. 2010. Observations of mixed-species birds flocks at Kichwa Tembo Camp, Kenya. *Ostrich* 81(3) pp 259–264.

Goudie, A. 2005. The drainage of Africa since the Cretaceous. *Geomorphology* 67: 437–56.

Grafe, T.U.and K.E. Linsenmair, K. E. 1989. Protogynous Sex Change in the Reed Frog: *Hyperolius viridiflavus*. Copeia 1989(4) pp 1024–1029.

Graham, A and Beard, P. 1974. *Eyelids of Morning*. New York Graphic Society.

Graham, A. 1973. *The Gardeners of Eden*. George Allen.

Graham, M. & Worthington, E. B. 1929. *The Victoria Nyanza and Its Fisheries*. Crown Agents for the Colonies.

Greenberg, J. 1963. *The Languages of Africa*. Republished 1966 with additions, Indiana University Press.

Greenwood, H. 1966. *The fishes of Uganda*. Uganda Society.

Greenwood, H. 1974. The cichlid fishes of Lake Victoria, East Africa: the biology and evolution of a species flock. *Bulletin of the British Museum of Natural History; Zoological Supplement* 6; pp 1–134.

Gregory, J. W. 1896. *The Great Rift Valley*. John Murray.

Grzimek, B. 1960. *Serengeti Shall Not Die*. Hamish Hamilton.

Hakansson, N. T. 2004. The Human Ecology of World Systems in East Africa; The Impact of the Ivory Trade. *Human Ecology* Vol 32; pp561–591.

Hamilton, A. C. 1981. The Quaternary history of African forests; its relevance to conservation. *African Journal of Ecology*. 19; pp 1–6.

Hamilton, A. C. 1982. *The Environmental History of East Africa*. Academic Press.

Harris, J. & White, T. 1979. Evolution of the Plio-Pleistocene African Suidae. T*ransactions of the America Philosophical Society*. 69: pp 1–128.

Harper, E, Measey, J., Patrick, D. A., Menegon, M. and Vonesh, J. 2010. *Field Guide to the Amphibians of the Eastern Arc Mountains and Coastal Forests of Tanzania and Kenya*. Camerapix

Harvey, B. 1997. Annotated checklist of the birds of Nairobi, including Nairobi National Park. British Council, Nairobi.

Hattle, Jack. 1972. *Wayward Winds*. Longman Zimbabwe.

Heaton, T. 1989. *In Teleki's Footsteps*. Macmillan.

Hebrard, J, Maloiy, G. M. O. and Alliangana, D, 1992. Notes on the habitat and diet of *Afrocaecilia taitana*. *Journal of Herpetology* 26: pp 513–525.

Hedges, N. G. 1983. *Reptiles and Amphibians of East Africa*. Kenya Literature Bureau.

Herne, B. 1999. *White Hunters*. Henry Holt and Company.

Heuvelmans, B. 1995. *On the Track of Unknown Animals*. Routledge.

Hillaby, J. 1964. *Journey to the Jade Sea*. Constable.

Hilton-Barber, B and Berger, L. 2002. *The Official Field Guide to the Cradle of Humankind*. Struik Publishers, Cape Town

Hobley, C. W. 1929. *Kenya, from Chartered Company to Crown Colony*. Witherby.

Hoffman, R. L. 1993. Biogeography of East African montane forest millipedes. Chapter 6 in Lovett and Wasser 1993.

Holman, D. 1967. *The Elephant People*. John Murray.

Holman, D. 1978. *Elephants at Sundown*. W. H. Allen.

Holmes, A. 1913. *The Age of the Earth*. Harper & Brothers.

Holmes, A. 1944. *Principles of Physical Geology*. Thomas Nelson.

Hornby, C. 2012. *Kenya; A History Since Independence*. Tauris.

House, A. 1993. *The Great Safari: Lives of George and Joy Adamson*. HarperCollins.

Hunter, J. 1952. *Hunter*. Hamish Hamilton.

Huxley, E. 1983. Foreword to reprint of R. Meinertzhagen, *Kenya Diary*. Eland Books.

Huxley, E. 1985. *Out in the Midday Sun*. Chatto and Windus.

Ibrahim, K. M. and Kabuye. C. H. S. 1988. An illustrated Manual of Kenya grasses. United Nations FAO (Rome).

Ingrouille, M. 1992. *Diversity and Evolution of Land Plants*. Chapman & Hall.

Ionides, C. J. P. 1965. *A Hunter's Story*. W. H. Allen.

Ionides, C. J. P. 1966. *Mambas and Man-eaters*. Holt, Rinehart & Winston.

Jackson, F. 1930. *Early Days in East Africa*. Arnold.

Jackson, F. 1938. *The Birds of Kenya Colony and the Uganda Protectorate*. Gurney and Jackson.

Jacobs, B. 2004. Palaeobotanical studies from Tropical Africa: relevance to the evolution of forest, woodland and savannah biomes. *Philosophical Transactions of the Royal Society of London* 359; pp 1573–1583.

Johanson, D. and Edey, M. 1981. *Lucy*. Simon and Schuster.

Johnson, O. 1941. *Four Years in Paradise*. Hutchinson.

Johnston, H. 1897. *British Central Africa*. Methuen.

Jones, S. 1993. *The Language of the Genes*. HarperCollins.

Kairo, J. G., Dahdouh-Guebas, F, Gwada, P. O., Ochieng, C and Koedam, N. 2002. Regeneration Status of Mangrove Forests in Mida Creek, Kenya: A Compromised or Secured Future? *Ambio* 31 pp 562–568.

Kalb, Jon. 2001. *Adventures in the bone trade*. Copernicus Publishers.

Kenyatta, J. 1938. *Facing Mt Kenya*. Secker and Warburg.

Kerbis Peterhans, J. & Gnoske, T. 2001. The science of 'man-eating' among Lions *Panthera leo* with a reconstruction of the natural history of the man-eaters of Tsavo. *Journal of the East African Natural History Society* 90; pp 1–40.

Kfir, R., Overholt, W. A., Khan, Z. R. and Polaszek, A. 2002. Biology and management of economically important Lepidopteran cereal stem borers in Africa. *Annual Review of Entomology* 47; pp 701–731.

Kielgast, J, Rodder, D., Veith, M and Lotters, S. 2010. Widespread occurrence of the amphibian chytrid fungus in Kenya. *Animal Conservation* 13 pp 36–43.

Kiilu, J. Dean, P. B. & Trump, E. C. 1983. Biotic Communities of Kenya. Wildlife Planning Unit; Ministry of Tourism and Wildlife, Kenya.

King, L. C. 1978. The Geomorphology of Central and Southern Africa. In Werger, M. J. A. (ed.), *The Biogeography and Ecology of Southern Africa*. Dr W. Junk.

Kinloch, B. 1972. *The Shamba Raiders*. Collins and Harvill Press.

Köhler, J., Bwong, B., Schick, S., Veith, M. and Lötters, S. 2006. A new species of arboreal *Leptopelis* from the forests of western Kenya. *Herpetological Journal* 16, pp 183–9.

Lack, P.C., Leuthold, W and Smeenck, C. 1980. Checklist of the birds of Tsavo East National Park, Kenya. *Journal of the East African Natural History Society* and National Museum. 170; pp 1–25

Lamb, David. 1984. *The Africans*. The Bodley Head.

Lambiris, A. 1989. The Frogs of Zimbabwe. Monografie Di Museo Regionale Discienze Naturali; *Torino* 10.

Lane, M. 1963. *Life with Ionides*. Viking Press.

Lang, J.C. 1973. Coral Reef project papers in memory of Dr Thomas F Goreau. 11. Interspecific aggression by scleractinian corals: why the race is not only to the swift. *Bulletin of Marine Science* 23: 260–279

Largen, M and Spawls, S. 2010. *The Amphibians and Reptiles of Ethiopia and Eritrea*. Chimaira.

Larsen, T. 1996. *Butterflies of Kenya*. Oxford University Press.

Le Carré, J. 2000. *The Constant Gardener*. Hodder and Stoughton.

Le Pelley, R. 1959. *Agricultural Insects of East Africa*. East African High Commission, Nairobi.

Leakey, L. 1927. Stone Age man in Kenya Colony. *Nature* 120: 85–86.

Leakey, L. 1935. *The Stone Age Races of Kenya*. Oxford University Press.

Leakey, L. 1937. *White African*. Hodder and Stoughton.

Leakey, L. 1974. *By the Evidence*. Harcourt Brace Jovanovich.

Leakey, M. 1984. *Disclosing the Past*. Doubleday

Leakey, M (ed). 2003. *'Lothagam', the dawn of humanity in Eastern Africa*. Columbia University Press.

Leakey, M and Harris, J. 1987. *Laetoli; a Pliocene Site in Northern Tanzania*. Clarendon Press.

Leakey, M and L. 1950. *Excavations at Njoro River Cave*. Clarendon Press.

Leakey, L. S., Tobias, P. V. & Napier, J. R. 1964. A new species of the genus Homo from Olduvai Gorge. *Nature* 202: 7–9.

Leakey, R. 1983. *One life*. Michael Joseph.

Leakey, R. & Morell, V. 2001. *Wildlife Wars: My Battle to Save Kenya's Elephants*. Macmillan.

Lewis, A and Pomeroy, D. 1989. *A Bird atlas of Kenya*. Rotterdam; Balkema.

Lieske, E and Myers, R. 1994. *Coral Reef Fishes. Indo-Pacific and Caribbean*. Collins.

Loveridge, A. 1944. *Many Happy Days I've Squandered*. Harper & Brothers.

Loveridge, A. 1957 Checklist of the Reptiles and Amphibians of East Africa. *Bulletin of the Museum of Comparative Zoology at Harvard*, vol. 117, no. 2.

Lovett, J. & Wasser, S. 1993. *Biogeography and Ecology of the Rain Forests of Eastern Africa*. Cambridge University Press.

Maathai, Wangari. 2008. *Unbowed*. Arrow.

MacDonald, D. 2001. *The New Encyclopaedia of Mammals*. Oxford University Press.

MacKay, A. & MacKay, J. 1985. *Poisonous Snakes of Eastern Africa*. Ines-May Publicity, Nairobi.

Mackinder, H. 1991. *The First Ascent of Mount Kenya*. Ohio University Press.

Maclean, G. L. 1985. *Roberts' Birds of South Africa* (5th Edition). Trustees of the John Voelcker Bird Book Fund.

Maclean, G. L. 1990. *Ornithology for Africa*. University of Natal Press.
Mackworth-Praed, C. W. and Grant, C.H.B. 1955. *Birds of Eastern and North-eastern Africa. Volume 1 and Volume 2*. Longmans, Green and Co.
Mani, M. 1968. *Ecology and Biogeography of High Altitude Insects*. Springer.
Marnham, Patrick. 1980. *Fantastic Invasion*. Jonathan Cape.
Marshall, F. 2002. Cattle before crops: the origins and spread of food production in Africa. *Journal of World Prehistory* 16: 99–143.
Martins, D and Johnson, S.D. 2009. Distance and quality of natural habitat influence hawkmoth pollination of cultivated papaya. *International Journal of Tropical Insect Science* 29(03) pp 114–123.
Martins, D.J., 2009. Differences in Odonata abundance and diversity in pesticide-fished, traditionally-fished and protected areas in Lake Victoria, Eastern Africa (Anisoptera) *Odonatologica* 38(3) pp 203–292.
Mathews, T. 2010. *The Woodpecker Calls on the Right*. David Lovatt Smith.
Mayr, E. 1950 *Taxonomic categories in fossil hominids*. Cold Spring Harbour Symposium on Quantitative Biology. 15: 109–118.
Mazrui, A. 1986. *The Africans, A Triple Heritage* (Guild Publishing 1986)
McClanahan, T. 1996. *Oceanic Ecosystems and Pelagic Fisheries*. Chapter 3 in McLanahan and Young 1996.
McClanahan, T and Obura, D. 1996. *Coral Reefs and Nearshore Fisheries*. Chapter 4 in McClanahan and Young 1996.
McClanahan, T and Young, T. P. 1996. *East African Ecosystems and their conservation*. Oxford University Press.
McKie, R. 2000. *Ape Man*. BBC Books.
Meinertzhagen, R. 1957. *Kenya Diary.* Oliver and Boyd.
Miller, C. 1976. *The Lunatic Express*. Ballantine.
Miller, C. 1974. *Battle for the Bundu*. MacMillan.
Miller, S. & Rogo, L. 2001. Challenges and opportunities in understanding and utilisation of African insect diversity. *Cimbebasia*, 17: 197–218.
Moore, Ray. 1982. *Where to Watch Birds in Kenya*. TransAfrica; Nairobi
Moreau, R. & Hall, B. P. 1970. *An Atlas of Speciation in African Passerine Birds*. British Museum.
Moreau, R. 1966. *The Bird Faunas of Africa and Its Islands*. Academic Press
Moreau, R. 1972. *The Palaearctic-African Bird Migration Systems*. Academic Press
Morel, M-Y. 1980. The co-existence of seven species of doves in a semi-arid tropical savanna of northern Senegal. In Johnson, D.N. (ed) Proceedings of the fourth Pan-African Ornithological Congress. pp 283–290. Southern African Ornithological Society.
Morrell, Virginia. 1995. *Ancestral Passions*. Simon and Schuster.
Mowat, F. 1987. *Woman in the Mists*. Warner Books.

Munyekenye, F, Mwangi, E.M. and Gichuki, N. 2008. Bird species richness and abundance in different forest types at Kakamega Forest, Western Kenya. *Ostrich* 79: pp 37–42.
Murton, M. K. & Wright, E. N. 1968. *The Problems of Birds as Pests*. Proceedings of a symposium, London, 1967. Academic Press for the Institute of Biology.
Musila, S., Syingi, A. and Sajita, N. 2010. Conservation Status of the avifauna at the North Nandi IBA. www.africanbirdclub.org/club/documents / NORTHNANDIFORESTSURVEYREPORT-final.pdf
Myers, T. 1822. *New and Comprehensive system of Modern Geography*. Sherwood, Neely and Jones.
National Atlas of Kenya. 1962. Survey of Kenya.

Neumann, A. 1898. *Elephant Hunting in East Equatorial Africa*. Rowland Ward.
Newman, J. L. 1995. *The Peopling of Africa*. Yale University Press.
Newman, K. 1983. *Newman's Birds of Southern Africa*. MacMillan South Africa.
Newmark, W. (1987) A Land-bridge Island Perspective on Mammalian Extinctions in Western North American Parks. *Nature* 325; 430–432.
Ngugi wa Thiong'o 1987. *Devil on the Cross*. Heninemann.
Norton-Griffiths, M. 2007. How many wildebeest do you need? *World Economics* 8(2) pp 41–64.
Nyamweru, C. 1980. *Rifts and Volcanoes*. Nelson Africa.

Obama, B. 1995. *Dreams from My Father: A Story of Race and Inheritance*. Times Books.
Obura, D. 1995. Environmental stress and life history strategies, a case study of corals and river sediment from Malindi, Kenya. PhD thesis, University of Miami.
Ogot, B. (ed.). 1968. *Zamani: A Survey of East African History*. Longman.
Ogot, Bethwell A. 2003. *My Footprints on the Sands of Time*. Anyange Press.
Ojany, F and Ogendo, R. 1973. *Kenya: A Study in Physical and Human Geography.* Longman Kenya.
Oliver, R and Fage, J. 1962. *A Short History of Africa* (Penguin)

Packenham, T. 1991. *The Scramble for Africa*. Weidenfeld and Nicolson.
Parker, H.W. 1942. The Lizards of British Somaliland. Bulletin of the Museum of Comparative Zoology (Harvard) Volume XCI (1)
Parker, H. W. 1949. *The Snakes of Somaliland and the Socotra Islands.* E. J. Brill.
Parker, I. 2004. *What I tell you three times is true*. Librario.
Parker, I and Bleazard, S. (Eds.) 2001. *An Impossible Dream*. Librario.
Passmore, N. I. and Carruthers, V. 1979. *South African Frogs*. Witwatersrand University Press (reprinted 1995).
Patterson, J. H. 1907. *The Man-Eaters of Tsavo*. MacMillan.

Patterson, J. H. 1925. The Man-Eating Lions of Tsavo. Field Museum of Natural History; zoology leaflet 7.
Percival, A. B. 1928. *A Game Ranger on Safari.* Nisbet & Co.
Phelps, T. 2010. *Old World Vipers.* Chimaira.
Phillipson, D. 2005. *African Archaeology* (3rd Edition). Cambridge University Press.
Picker, M., Griffiths, C. & Weaving, A. 2004. *Field Guide to the Insects of South Africa.* Struik.
Pickersgill, M. 2007. *Frog Search.* Chimaira.
Pickford, M. 1997. *Louis S B Leakey, Beyond the Evidence.* Janus Publishing.
Pinker, S. 2002. *The Blank Slate.* Viking Penguin.
Pitman, C. R. S. 1931. *A Game Warden among his charges.* Nisbet and Co.
Pitman, C. R. S. 1942. *A game warden takes stock.* Nisbet and Co.
Pitman, C. R. S. 1938. *A Guide to the Snakes of Uganda.* Uganda Society (1974 reprint published by Wheldon and Wesley)
Pounds, J. A. & Masters, K. L. 2009. Amphibian mystery misread. *Nature* 462: 38–9.
Pratt, D. J., Greenway, P. J., and Gwynne, M. D. (1966). A classification of East African rangeland with an appendix on terminology. *Journal of Applied Ecology* 3; pp 369–382.
Pratt, D. J. & Gwynne, M. D. (eds). 1977. *Rangeland Management and Ecology in East Africa.* Hodder and Stoughton.
Preston, Richard. 1994. *The Hot Zone.* Anchor.

Reader, John. 1981. *Missing Links.* Collins.
Redman, N, Stevenson, T and Fanshawe, J. 2009. *Birds of the Horn of Africa.* Christopher Helm.
Richmond, M. (ed) 1997. A guide to the seashore of eastern African and the western Indian Ocean Islands. SIDA (Swedish International Development Cooperation Agency).
Ridgeway, R. 1999. *The Shadow of Kilimanjaro.* Henry Holt.
Rieseberg, L.H., Wood, T. E. and Baack, E. J. 2006. The nature of plant species. *Nature* 440: 524–527.
Robertshaw, P. 1990. *A History of African Archaeology* (James Currey)
Ruark, R. 1967. *Use enough Gun.* Hamish Hamilton.

Salt, G. 1954. A contribution to the ecology of upper Kilimanjaro. *Journal of Ecology* 42: 375–423.
Schabel, H. G. 2006. *Forest Entomology in East Africa.* Springer, Dordrecht
Schick, S., Veith, M. and Lotters, S. 2005. Distribution patterns of amphibians from the Kakmega Forest, Kenya. *African Journal of Herpetology* 54(2) pp 185–190.
Schifter, H and Cunningham-van Someren, G.R. 1998. *The Avifauna of the North Nandi Forest, Kenya.* Annals Naturhistorisches Museum Wien, 100: pp 425–479
Schiotz, A. 1975. *The treefrogs of Eastern Africa.* Steenstrupia; Copenhagen.
Schiotz, A. 1999. *Treefrogs of Africa.* Chimaira
Shipton, E. 1943. *Upon that mountain.* Hodder and Stoughton.
Schluter, T. 1997. *Geology of East Africa.* Gebruder Borntraeger.
Schmidt, K. P. and Noble, G. K. 1998 (reprint) Contributions to the Herpetology of the Belgian Congo. Society for the Study of Reptiles and Amphibians.
Seegers, L, De Vos, L & Okeyo, D. O. 2003. Annotated checklist of the freshwater fishes of Kenya. *Journal of East African Natural History* 92: pp 11–47.
Semlitsch, R. (Ed) 2003. *Amphibian Conservation.* Smithsonian Books.
Sheldrick, D. 1973. *The Tsavo Story.* Collins.
Shell Oil. No date. *Shell Guide to East African Birds.* Shell, Kenya.
Sheuyange, A., Oba, G. and Weladji, R. B. 2005. Effects of anthropogenic fire history on savanna vegetation in northeastern Namibia. *Journal of Environmental Management* 75: pp 189–198.
Simon, N. 1962. *Between the sunlight and the thunder.* Collins.
Sinclair, I and Ryan, P. 2003. *Birds of Africa South of the Sahara.* Struik.
Skinner, H. 1974. *Snakes and Us.* Kenya Literature Bureau.
Smith, A. 1849. *Illustrations of the Zoology of South Africa, Consisting Chiefly of Figures and Descriptions of the Objects of Natural History Collected During an Expedition Into the Interior of South Africa, in the Years 1834, 1835, and 1836.* Smith, Elder and Co.
Smith, J. L. B. 1949. *The Sea Fishes of Southern Africa.* Central News Agency (South Africa).
Snoeks, J and Akinyi-Odhiambo, E. 2003. Luc De Vos; in Memoriam. *Journal of East African Natural History* 92: pp 3–9.
Spawls, S. 1978. A checklist of the snakes of Kenya. *Journal of the East African Natural History Society*, Vol. 31; pp 1–18.
Spawls, S and Branch, B. 1995. *The Dangerous Snakes of Africa.* Blandford Press.
Spawls, S., Howell, K, Drewes, R. C.and Ashe, J. 2008. *A Field Guide to the Reptiles of East Africa.* A and C Black.
Spawls, S., Howell, K and Drewes, R. C. 2006. *Pocket Guide to the Reptiles of East Africa.* A and C Black
Spawls, S and Rotich, D. 1997. An annotated checklist of the lizards of Kenya. *Journal of the East African Natural History Society*, 86 pp 61–83.
Stager, J. C., Day, J. J. and Santini, S. 2004. Comment on 'Origin of the Superflock of Cichlid Fishes from Lake Victoria, East Africa'. *Science* Vol. 304 p 963b.
Steinhart, E. 2006. *Black Poachers, White Hunters.* James Currey.
Stevenson, T. & Fanshawe, J. 2002. *Field Guide to the Birds of East Africa.* T. & A. D. Poyser.
Stewart, K. 2001. The freshwater fish of Neogene Africa. *Fish and Fisheries* 2: 177–230.

Stewart, M. M. 1967. *Amphibians of Malawi*. University of New York.
Stigand, C. H. 1913. *Hunting the Elephant in Africa*. MacMillan.
Storer, T. I. & Usinger, R.L. 1965. *General Zoology*. McGraw-Hill.
Sutton, J. 1990. *A Thousand Years of East Africa*. British Institute in Eastern Africa.

Tandy, M., Tandy, J., Keith, R. and Duff-MacKay, A. 1976. A new species of *Bufo* from Africa's dry savannas', Pierce-Sellard series; Texas Memorial Museum, vol. 24, pp 1–20.
Thomson, J. 1885. *Through Masai Land: A journey of exploration among the snowclad volcanic mountains and strange tribes of Eastern Equatorial Africa.* Sampson, Low, Marston, Searle and Rivington.
Tilbury, C. 2010. *Chameleons of Africa, an Atlas*. Chimaira.
Travis, W. 1967. *The Voice of the Turtle*. Allen & Unwin.
Trench, C. C. 1993. *Men who Ruled Kenya*. The Radcliffe Press.

Van Perlo, B. 1995. *Birds of Eastern Africa*. Collins.
Van Someren, R.A.L. & Van Someren, V.G.L. 1911. *Studies of birdlife in Uganda*. John Bale, Sons & Danielsson Ltd
Van Someren, V. G. L. 1956. Days with Birds. *Fieldiana* (Chicago Natural History Museum) Vol 38 Publication 798.
Van Someren, V. D. 1958. *A Bird Watcher in Kenya*. Oliver and Boyd.
Verheyen, E, Salzburger, W, Snoeks, J and Meyer, A. 2003. Origin of the Superflock of Cichlid Fishes from Lake Victoria, East Africa. *Science* Vol. 300 no. 5617 pp 325–329.
Verheyen, E, Salzburger, W, Snoeks, J and Meyer, A. 2004. Response to comment on 'Origin of the Superflock of Cichlid Fishes from Lake Victoria, East Africa'. *Science* Vol. 304 p 963c.

Voelker, G., Outlaw, R and Bowie, R.C.K. 2010. Pliocene forest dynamics as a primary driver of African bird speciation. *Global Ecology and Biogeography* 19, pp 111–121

Walker, A and Shipman, Pat. 1996. *The Wisdom of Bones*. Weidenfeld and Nicolson.
Wasonga, V, Afework Bekele, Lotters, S. and Balakrishnan, M. 2007. Amphibian abundance and diversity in Meru National Park. *African Journal of Ecology* 45, pp 55–61.
Wass, P.1995. *Kenya's Indigenous Forests: Status, Management and Conservation*. IUCN (International Union for Conservation of Nature).
Werger, M. J. A. (ed) 1978. *The biogeography and ecology of Southern Africa.* W. Junk.
White, F. 1983. *Vegetation Map of Africa*. UNESCO.
White, Tim. 2002. 'Rocking the cradle; review of 'The Official field guide to the cradle of humankind' by Brett Hilton Barber and Lee Berger. *South African Journal of Science* 98 (515–517).
Wiens, J. J. 2007. Review of 'The amphibian tree of life' by Frost et al. *Quarterly Review of Biology* 82: 55–56
Williams, J. G. 1963. *A field guide to the birds of East Africa*. Collins.
Williams, J. G. 1967. *Field Guide to the National Parks of East Africa*. (6th Impression 1978). Collins.
Williams, J. G. and Arlott, N. 1980. *A field guide to the birds of East and Central Africa*. Collins.
Wilson, D. E. and Mittermeier, R. A. (Eds.), 2009. *Handbook of the Mammals of the World.* Lynx Edicions
Wolpoff, M. 1971. 'Competitive exclusion among Lower Pleistocene hominids; the single species hypothesis'. *Man*, Vol 6 # 4 pp. 601–614.
Wykes, A. 1960. *Snake Man: The Story of C. J. P. Ionides*. Hamish Hamilton.

Zimmerman, D., Turner, D. & Pearson, D. 1996. *Birds of Kenya and Northern Tanzania*. Christopher Helm.
Zug, G., Vitt, L and Caldwell, J. 2001. *Herpetology*. Academic Press.

Index

Page numbers in *italic* refer to illustration captions.

Aardvark 178
Aard-wolf 192
Aberdares 25, 29, 30, 31, 35, 36, 66, 86, 90, 101, 102, 111, 113, *117*, 119, 125, 130, 131, 144, *145*, 148, 150, 157, 167, 168, 183, 187, 188, 196, 203, 204, 209, 211, 226, 240, 257, *257*, 258, 262, 271, 275, 277, 298, 300, 308, 311, 312, 319, 324, 360, 369, 402, 406, 407, 410, 420, 427
Abuodha, Pamela 371
Acacia 111, 114, 152, 154, 156
 erioloba 147
 Yellow-barked 146, 148, *232*
Achauer, Ulrich 28
Achyranthes 347
Adams, Jonathan 134, 430
Adamson, George 34, 106, 107, 159, 202, 203, 204, 209, 283, 404, 408, 413
Adamson, Joy 113, 153, 159, 170, 211, 281, 408, 410
Adder, Puff 38, 50, 184, 255, 256, 257, 271, *272*, 273, 276, 283, 284, 285, 285, 322
 Velvety-green Night 285
Adi, Admasu 170
Aepyceros melampus 56
African Queen *348*
African Violet 153, 166, 167
Afrixalus 297, 298, 299, *299*, 300
Agama 258
 Finch's 260
 Rüppell's 277
Agnew, Andrew 169
agricultural industry 121-2, 136
Agrotherium 176
Ahlquist, Jon 218
Aiello, Leslie 61
Ajayi, J.F.A. 95
Akalat, East Coast 157
Albertine Rift 24-5, 30, 32, 100, 103, 125, 129, 168, 313
Alcolapia alcalicus 114
Allan, Iain 116, 117, 134, 428
Allen, David J. 169
Alliangana, Daniel 314

aloes 153, 161, 164, 167
Alpern, Stanley 67, 82
Alpine-meadow Lizard 258
Altig, Ronald 314
altitude 25, 31, 50, 73, 80, 98, 99, 102, 106, 107, 111, 114, 119, 126, 130, 131, 141, 144, 150, 152, 159, 169, 187, 228, 239-40, 248, 258, 260, 262, 273, 295, 298, 300, 302, 304, 324, 342, 350, 352
Amadon, Dean 245
Amaras, John Offula 331
Ambatch 107
Amboseli *126*, 197, 211, 234, 327, 402, 406, 410, *414*, 418, 420, 425
Ambrose, Stanley 74, 80
Amegilla 166
Amiet, Jean-Louis 304
Amietophrynus 304, *305*, 308
amphibians 291-314, 417
Anchovy 380
Anderson, Alec 331
Anderson, David 96
Anderson, J.T. 366
Andrew, Lieutenant 107
Angelfish, Ear-spot *390*
 Emperor *383*, 384
Anomalure, Lord Derby's 191
Ant 339, 346, 348, 355, 366
 Safari 221, 286, 342, 350, 360-1, *361*
antbirds 224
Antelope 172, 175, 176, 182, 211
 Roan 84, 178, 191, 208
Anthropopithecus erectus see *Homo erectus*
Apalis, Black-collared 240
 Buff-throated 240
 Taita 230
aphids 364
Aplocheilichthys antinorii 113
Arabuko Sokoke Forest 157, 159, 167, 226, 239, 240, *261*, 277, 296, 298, 311, 313, 314, 341, 342, 399, 402
Archaeobelodon 175
archaeology 65-8

Archaeopteryx 218
Archer, Geoffrey 241, 243, *244*
Ardipithecus 58
 kadabba 57
 ramidus 57
Ardrey, Robert 184
Arewale 402
Argentavis magnificens 214
Arlott, Norman 234, 252
Army Worm 350, 359
Arthroleptides dutoiti 304
Arthroleptis 302
arthropods 338-67
Arthrospira 108
Ashe, James 245, 269, 279, 281, 285, 288
Ashe, Sanda 265, 281, *281*, 285
Assassin Bug *347*
Astley Maberly, Charles 210-11
Athi River 98, 100, *101*, 102, 160, 208, 283, 311, 318, 319, 322, 324, 336, 337
Aurochs 76
Australopithecus 58, 72
 afarensis 53, 54, 57, 71
 africanus 45, 48, 49, 62
 anamensis 57
 boisei 51, 56
 garhi 57
 sediba 57
Australosomus 317
Awach Valley 74
Awash River/Valley 53, 56, 71, 72, 74
Axelrod, D.I. 171

Babbler, Hinde's 229, 239
Baboon 58, 180, 182, 207, 211, 404
 Gelada 59
 Yellow 178, 192
Backhurst, Graeme 252
Badger, Honey 191, 206, 356
Baez, Ana Maria 295
Bagine, Richard 367
Baker, Samuel 205
Balakrishnan, Mundanthra 314
Balirwa, John 333
Balanites Tree *237*
Bally, Peter 153, 159, 161, 170, 279, 408
Bamboo 159, *159*

Baobab 136, 138, 142, 153, 154, 163
Barb, Pangani 320, *321*
 Ripon Falls *321*
Barbet, Grey-throated 240
Bark Snake *276*
Barnes, Simon 240
Barthelme, John 77
Bartlett, Des 49
Bass, Black 112, 325
Bat 172, 178, 182, 183, 192, 195, 197
 Egyptian Fruit *190*, 207
 Kenyacola Butterfly 196
 Schlieffen's Twilight 191
 Straw-coloured Fruit 191
Bat Hawk 238
Bateleur *238*, 239, 241, 251
Bates, George 243
Batis, Sokoke 230
Batrachochytrium dendrobatidis 308
Battersby, James 280
Battersby's Green Snake 288, *289*
Battiscombe, Edward 161
Baxendale, Nevil 106
beaches 371, *375*
beachflies 350, 355
Beadle, L.C. 133
Beard, Peter Hill 200, 290, 428
Beaton, Ken 406
Beautiful Sand Snake 277, 280
bedbugs 350, 358
Bee 246, 345, 346, 348, 355-7, 364
 Honeybee 355-7
 Mopane 160
Bee-eater 225
 Blue-headed 238
 Cinnamon-chested 240
 European 241
 Northern Carmine 213
 White-fronted *227*
Beelzebufo 295
Beentje, Henk 150, 170, 171
Beese, Gerhard 169
beetles 338, 339, 341, *341*, 342, 344, 345, *346*, 348, 350, 360, 361, 362, 364, 366
Behera, Swadhin 134
Behrensmeyer, Kay 52, 175

438

Bekele, Afework 314
Bell, W.D. 404
Bell-Cross, Graham 320
Bennun, Leon 226, 234, 237, 238, 239, 253
Benuzzi, Felice 117, 134
Berger, Lee 41
Bhatt, Nayan 118
bichirs 316, 317, 318
biodiversity hotspots 168-9
Bio-Ken Research Centre 269, 281, *281*
biopiracy 108-9
bipedalism 43, 44, 48, 53, 54, 56, 58, 59-60
birds 212-54
Birnie, Ann 169
Bisandi 402
Black, Alan 405
Black Fly 349
Black Widow 357
Blanding's Tree Snake 34, 267
Bleazard, Stan 106, 429
Bluegill 325
Blundell, Michael 159, 170
Boa, Kenya Sand 263, *263*, 274
Bock, Ken 384, 387, 396, 397
Boettger, Oskar 277
Boise, Charles 49
Boivin, Louis 159
Bojer, Wenceslas 159
Bongo 167, 191, *191*, 194, 208, 280
Boni 402
Bonner, Raymond 430
Boomslang 242, 264-5, *265*, *267*, 281, 285, 288, 417
Boreosomus 317
Borrow, Chris 348
Boswell, Percy 48
botanical exploration 159-62
Boulenger, George 277, 327, 335-6
Boulengerula taitanus 299, 309
Bowdler Sharpe, Richard 242
Bowie, Rauri 232
Bowker, M. 8, 314
Bowker, R. 8, 314
Brachiosaurus brancai 20
Brachyodus 175
brain size 44, 52, 53, 58, 60-1, *61*
Branch, Bill 288, 312
Brett Young, Francis 115
Breyer-Brandwik, Maria 170
Breuil, Henri 74
Breviceps 293
Bridges, Campbell 32
Brindled Cat 187
Britton, Peter 229, 252
Broadbill 242, 243
Broadley, Don 288, 290
Brockington, Dan 430
Bromhead's Site 46, 78
Broom, Robert 19, 45, 48

Brown, Leslie 115, 233, 245-6, 252, 384, 397
Brown, Monty 96, 119
Brown House Snake 267
brownbuls 253
Bruce, Henry 35
Bucephalus 264
Buckle, Colin 133
Buffalo 35, 101, 178, 180, 182, 191, 198, *199*, 201, 202, 204, 208, 211, 401, 404, 418
Buffalo Springs 192, 237, 327, 402
bugs 344, 346, *341*, *347*
Bulbul 213, 225, 240
Bullfrog, African 304, 307
  Edible 304, 307
  Groove-crowned 307
buntings 224
Burrowing Asp 283, 284, 285
Burton, Richard 94, 102, 103
Bush Pig 175, 176, 404
bushbabies 182
Bushbuck 180, *189*, 191, 354
Bush-shrike, Lüdher's 240
Bussman, Rainier 158, 162, 165
Bustard 213, 219, 225
  Kori 214, *244*
Buthus 358
butterflies 44, 220, 244, 339, *340*, *342*, *343*, 344, 345, 346, *348*, 348, 349, 363, 364, 366, 367
Butterflyfish, Racoon 384
Buttonquail 213
  Kurrichane 239
Buxton, Edward North 428
Buzzard 216
  Augur 195, 251, *251*
  Grasshopper 223
  Mountain 242
Bwong, Beryl 282, 314

Cabanis, Jean Louis 241
Caecilian 291, 292, 294, 298-9, *309*
  Mud-dwelling 299
  Ngaia Forest 415
Caldwell, Janalee 288, 311
Callulina 306, 308
Camaroptera, Grey-backed 238, 240
Camellia sinensis 136
camels 86, *87*
Campbell, Bob 169
Campbell, Neil 176, 345
CAMPFIRE 426
Camphor Tree 166
Canary, Atlantic 247
  Brimstone *247*
Candle Bush 164
Caplan, Pat 96
Caracal 191
Carcasson, Robert 336, 349, 367

Carex 136
Carp 325
Carpet Viper 50, 106, 271, 273, 274, 283, 284, *284*
Carruthers, Vincent 313
Carson, Rachel 353
cassowaries 249
catchment basins 100-1, *101*, 319
Catfish 318, 324, 332, 333
  Butter *323*
  Marbled Mountain 242
  Sharptooth 322, *323*
  Somali Giant 337
  Sudan 331
Cavendish, Henry 107
Cedar, Pencil 138
centipedes 344, 350
Central Island 402
Chameleon 35, 36, 159, 160, 255, 260, 265, 267, 273, 287-8, 417
  Blade-horned *274*, 415
  Flap-necked 257
  Jackson's 242, *260*, *287*, 288
  Kenya Pygmy *261*
  Mount Kenya Side-striped 278
  Side-striped 275
  Slender 275
  Von Hoehnel's 258
Chania River 319
Chanler, William Astor 36, 94
Channing, Alan 300, 312, 313, 314
Charaxes 244
Chat, Moorland 228
  White-browed Robin 240
Cheetah 158, 167, 173, 176, 180, 191, 194, 202, 211, 408, *411*, 418, 423
Cheptumo, Michael 314
Cherangani Hills 100, 150, 186, 226, 240, 310, 319, 324
Cheruiyot, Agnes *137*
Chesters, Katherine 139
Chiffchaff 221
Chigger 342, 350, 358
Chimpanzee *60*, 71, 178, 191, 209, 408, 419
Chlamydophila psittaci 249
Chloris 157
Cholmondeley, Hugh 110, 159, 161, 248
Chu, Peter 134
Chuah-Petiot, Min 169
Churchill, Tana 336
Churchill, Winston 144, 405
Churamiti maridididi 291
chytrid fungus 308, 310, 311, 314
Chyulu Hills 16, 29, 31, 82, 92, 167, 226, 277, 402, 408
cichlids 317, 318, 320, 324, 326-33
Cinchona bark 164

Cisticola 253
  Aberdare 226, 229
  Chubb's 240
  Singing 240
  Tana River 229
Citril, African 240
Civet 191
cladistics 57, 293
Clam, Giant 387, *389*
Clapperton, Hugh 241
Clarke, Cyril 366
Claudius, Hendrik 241
Clausnitzer, Viola 348
Clements, Frederic 149
Clifton, Mike 367
climate 124-33, 140
Clooney, George 352
Coachman *386*
coast 92, 131, 371
Coates Palgrave, Keith 170
Cobra 174, 245
  Ashe's Spitting 269, *269*, 283, 285
  Banded Water 330
  Black-necked Spitting 269, 283, 284
  Egyptian 269, 281
  Forest *268*, 269, 273, 285
  Gold's Tree 269
  Mozambique Spitting 256
  Red Spitting 106, 269, 275, 284
Cocker, Mark 251
cockroaches 339, 345, 350, 360
Coe, Malcolm 152, 170, 314, 348
Coelacanth 316, 337
Cohen, David 85
Cole, Cheryl 352
Cole, Galbraith 111
Cole, Sonia 39, 45, 46, 47, 48, 63, 76, 78
Coles, Billy 408
Collins, Billy 408
Collins, James 308, 310
Colobus 426
  Black-and-white *159*, 178, 192
  Tana River Red 172, 196
colonial history of Kenya 94-5
Combes, Simon 204
Commiphora 136, 152, *152*, 411
communication 44, 61-2, 73
  see also languages
conservation 166-8, 276-7, 310, 394-6, 398-430
continental drift 14-15, 21
Cook, I. 133
Cooke, Basil 52
Copley, Hugh 316, 334, 335
Coppens, Yves 53, 54
coral rag 22, *22*, 371
coral reefs 133, 368, 372-6, 395
  species diversity 383-90
corals 374, 389, 390
Corbett, Jim 199, 200

Coryndon, Robert 38, 278
Coryndon Museum *see* National Museum of Kenya
Cottar, Charles 405
coucals 213
coursers 241
Cowie, Mervyn 406, 410
Coypu 112, 178
Crab 325, 344, 379
  Grapsid Shore *372*
  Robber 372
Cracraft, Joel 218
Crane 322
  Grey Crowned 222
Crayfish 393
  Louisiana Red 112
Crematogaster 360
Crescent Island 77, 111, 112, 113, 205, 239
Cricket 184, 346
  Giant Burrowing 341
Crocodile 71, 84, 102, 106, 163, 255, 256, 282, 288, 290
  Nile 107, 257, 262, 283, *283*
  Saltwater 214, 283
  Slender-snouted 258
Cropper, Margaret 50
crops 80, 136, 168, 250, 341
Croton 142
Crow, House 246
  New Caledonian 71
Crowder, M. 95
Crump, Martha 308, 310
Cuckoo 219, 223
  Klaas's 241
Cullen, Anthony 397
Cunninghame, Richard 405
Curtis, Garniss 53
Cutler, William 20-1
cyanobacteria 105, *105*, 106, 108, 110, 138
cycads 138, *139*, 159, 160, 166, 167

Dactyloctenium 157
Daddy Christmas *340*
Dale, Ivan 159, 170
dams 106, 310, 318, 324, 403, *426*
Dandelot, Pierre 210
Dark Blue Pansy *342*
Darlington, Philip 318
Dart, Raymond 45, 59, 62
Darwin, Charles 40, 44, 59, 149, 241, 366, 368, 372, 384
Dasymys 177
Date, Desert 163
Daua River 86, 98, 101, 319
Davis, William Morris 97, 120
Dawkins, Richard 41, 52, 59, 328
Dawson, Alistair 201
Day, Michael 49

De Vos, Luc 319, 336-7
De Witte, Gaston-François 313
Dean, P.B. 150
Deathstalker 358
decline in amphibians 308, 310-11, 312
Delamere, Lord *see* Cholmondeley, Hugh
Dembea Stone Lapper *329*
DeMenocal, Peter B. 134, 140, 176
Denhardt, Clemens 23
Denis, Armand 408
Denis, Michaela 408
Dent, R.E. 187, 334
DeSaix, Frank 280
Desert Rose 154, *155*
deserts 128, *129*, 141
dew point 128, *129*, 130
Dharani, Najma 169
Diadema 377
Diamond, Jared 65, 94, 96
Diamond, Tony 252
Diani/Chale Marine National Park 394, 395, 402
diatomite 110, *111*
Dida-Galgalu *129*
Dik-dik 233
  Kirk's 192, 194, *194*
Dikkop, Water *218*
Diller, Helmut 210
Dingoneck 104
dinosaurs 20, 22, 41, 44, 155, 216
diseases
  human 205-8, 249, 350, 352-5, 359, 366
  livestock 77, 86, 206, 208, 359, 367
dispersal, human 68-9
Dobiey, Maik 288
Dobzhansky, Theodosius 366
Dodori 402
Dolphin 178, 380
Donaldson-Smith, Arthur 161, 242, 277
Dorst, Jean 210
Douglas-Hamilton family 403
  Iain 204, 408, 410, 429
  Oria 204, 429
  Saba 204, 210
Dove 56, 234
  Red-eyed 238, 240
Dragon Tree 140
Dragonfly 345, 348-9, *349*, 363, 364
  Violet Dropwing *363*
Dranzoa, Christine 237
Drewes, Bob 288, 302, 312, 313, 314
Drongo, Velvet-mantled 220
drought 132, 142, 146, 148, 205, 411, *412*
Drummond, Henry 34
Dryosaurus lettowvorbecki 20

Du Toit, C.A. 304
Dubois, Eugene 44
ducks 212, 213, 219
Dudley, Cornel 346
Duellman, William 311
Duff-MacKay, Alex 178, 209, 210, 282, 312, 314
Dugong 178, 376
Duiker, Aders's 172, 192
  Grey 180, *190*, 191
  Yellow-backed 280
Dung Beetle 341, *341*, 345, 348, 350
Dupre, Gerard 346
Dyson, W.S. 106, 107

Eagle 209, 213, 219, 241, 245
  Ayre's Hawk 238
  Black-chested Snake 226
  Crowned 184, 233, 238, 249
  Fish 108, 111, 377
  Long crested 238
  Martial *232*, 233
  Steppe 250
  Verraux's 253
Eagle-Owl, Cape 242
East Turkana 53
Eburru 23, 30, 46, 71, 74, 109, 111, 112
Echinocloa colona 250
Edey, Maitland 63
edible insects 341, 341
Edwards, D.C. 150
Eel 220, 336, 377
  African Mottled 324
  European 325
  Moray 390, *392*
Egg-eater, Montane 278
  Rufous 275
Eggleton, Paul 362
Egret, Cattle 322
  Great White 219, *373*
Ehret, Christopher 77, 84
Eisner, Thomas 342
El Nino 132, 368, 375
Eland 84, 180, 192, 208
Eldoret 98, 131, 177, 273, 302
Eldredge, Niles 52
Elephant, African 18, 54, 96, 106, 119, 123, 136, 142, *144*, 157, *158*, 170, 172, 174, 175, 176, 178, 186, *186*, 187, *187*, 191, 198-9, 202, 203-4, 206, *207*, 208, 209, 210, 211, 250, 356, 382, 400, 403, 404, *405*, 408, 410, 411, 412, *412*, 413, *413*, *414*, 419, 420, 423, 427, 429, 430
Elephant Shrew 177, 197
  Golden-rumped 196
  Zanj 192
Eleutherodactylus 300
Elgon, Mount 29, 30, 31, 85, 106, 119, 122, 150, 204,
207, *207*, 226, 277, 318, 304, 324, 349, 402
Eliot, Charles 110, 113, 114, 404
Elton, Charles 184
Eltonian pyramid 184, *185*, 233
Emin Pasha *see* Schnitzer, Eduard
Encephalartos tegulaneus 138
endemism 104, 113, 115, 135, 150, 153, 154, 160, 162, 166, 168, 171, 183, 192, *195*, 195-7, 212, 226, 228-32, 234, 239, *255*, *257*, 258, 260, 271, *271*, 273-7, 298, 299, 300, *302*, 304, *305*, 308, *309*, 318, 319, 320, 322-3, 324, 349, 350, 380, 415, 417, 423
Entebbe Botanical Garden 164, *164*
environmental indicators 292, 310, 314, 375
Eozygodon 175
Eragrostis tef 152
Ericson, Per 218
erosion 16, *22*, *28*, 97-121, *123*, 123-4, 379
Estes, Richard D. 211
ethnobotany 162-5
euphorbias 148, 161, *162*
Euthecodon 262
evolution 300, 326-33
  hominin 42-62, 83
Ewaso Ng'iro River 98
exploitation of marine species 390-4

Fabre, Jean-Henri 366
Fage, John 95
Fanshawe, John 229, 234, 239, 253
faulting 21, 22, 25-6, *33*
Fedders, Andrew 91, 96
Fennessy, Rena 252
Ferguson, Niall 66, 95
ferns, tree 138, *139*
Festuca 136
Fever-tree 146
Fichtl, Reichard 170
Fig, Strangler 167
Finch, Brian 237, *237*, 246, 253, 260, 314
Finlay, Bland 345
Firefinch 160
  Black-bellied 226
  Red-billed 248
Firmin, Arthur 117
Fiscal, Common 240
Fischer, Eberhard 170
Fischer, Gustav 35, 113, 114, 160, 241, 242, 277
fishes, freshwater 315-37
fishing
  freshwater 76, *103*, *333*, 334

marine 392, 393, 397
Fitch, Frank 52
Fjeldsa, John 218
Flamingo 108, 110, 111, *219*
   Greater 219
   Lesser 107, 115, 218
fleas 339, 342, 350, 358
flies 339, 340, 344, 346, 348, 349, *351*, 353-5, 366, 417
flight 213-14, 216
floods *101*, 107, 122, *123*, 132, 133, 148, 324
Flowerpot Snake 153
Flufftail 251, 286
   Buff-spotted 238
Flycatcher 213, 224, 235, 253
   African Paradise 240
   White-eyed Slaty *239*
food chains/webs 184, *184*, 339, 341, 363, 369, 370, 376, 380
food production 76-80
footprints 54, 58, 63
foramen magnum 45, 58, *59*, 61
Forbes-Watson, Alec 244
Ford, Harrison 360, 399
forests 129, 142-4, *145*, 150-1, 158-9, 168, 225, 415, 426
Fort Ternan 41, 71, 140, 156
Fortey, Richard 8, 336
Fossey, Dian 169, 281, 409, 410
fossils and origins 16, 18-22, 29, 38
   amphibians 292, 294-5
   arthropods 339-45, 359, 362
   birds 214, 216-19
   freshwater fishes 316-18, 331
   hominins 40-62
   mammals 173-6
   marine species 368
   plants 138-41
   reptiles 256-8
Fox, Bat-eared 112, 192, 206
Francolin 84, 160, 251
   Jackson's 242, *243*
freshwater habitat 315-37
Friis, Iib 171
Frog 160, 277, 288, 291, 292, 294-302
   Anchieta's Rocket *293*
   Argus Reed 293, *294*, *295*, 307
   Balfour's Reed 308
   Big-eared 308
   Blunt-faced Snout-Burrower *297*
   Bocage's Burrowing Tree 302, 307
   Common Clawed 307
   Common Reed 300, *301*
   David Sheldrick's Reed 300, 308, 314
   De Witte's River 308, 312
   Du Toit's Torrent 308
   Foam-nest 296, *297*, 298, 314

   Forest White-lipped 307
   Guinea Snout-burrower 307
   Keller's Foam-nest 308
   Kinangop Puddle 308
   Krefft's Warty *305*, 308
   Maasai Reed 300, *307*
   Mackay's Tree 308
   Mountain Reed 300, *302*
   Ornate Tree 307
   Osorio's Leaf-folding 307
   Red-banded Rubber 306, *306*, 307
   Red-legged Running 302
   Sharp-nosed Reed 300, 307
   Sharp-nosed Rocket 304
   Shimba Hills Reed 308, *310*
   Silver-bladdered Reed 308
   Southern Foam-nest 307
   Spiny Leaf-folding 298, *299*, 300, 314
   Tinker Reed 307
   West African Goliath 304
Frost, Darrel 302, 311, 312
Fry, C. Hilary 252
Fuchs, Vivian 46, 106

Gadd, Michelle 422
Gadner, J.C.M. 346
Galago, Demidoff's 191
   Somali 192
   Zanzibar 192
Galana River 20, 21, 30, 79, 98, 100, 102, 178, 229, 283, 300, 318, 319, 322, 324, 327, 354
Gallmann, Kuki 152
Gama, Vasco da 93
Gamble's Cave 41, 46, 74, 78
Gandar Dower, Kenneth 188
Gardner, J.C.M. 364, 366
Garissa 98, 107, 125, 149, 283
Garlake, Peter 67
Gaudet, John 113
Gazelle, Grant's 180, *180*, 196
   Thomson's 35, 180, 192
Gecko 258, 260, 288, 289
   Forest Dwarf *259*
   Somali–Maasai Clawed 277
   Yellow-headed Dwarf *289*
Gede 22, 92, *93*, 149, 196, *267*, 269, *270*, 355, 356
Gee, Henry 57
geese 212, 219
Genet, 182, 221
   Blotched 191
geology 11-39, 371
geothermal activity 15, *15*, 107, 108, 111, 113, 114
Gerenuk 168, 188, 192
Gerhart, John 252
Gibbons, Anne 62
Gichangi, Stephen 428
Gilgil 26, 46, 211
Gillett, Jan 153, 161, 171
Gillson, Lindsey 186

Giraffatitan brancai 20
Giraffe 54, 137, 178, 180, 182, 191, 198, 208, 209, 211
   Maasai 172
   Rothschild's 148, *198*
Gitonga, Eustace 64
glaciation *117*, *118*, 118-19, 133
Gleason, Henry 149
Gnoske, Tom 200, 201
Goatfish, Dash-and-dot 376
Godman, Eva 243, *244*
Goldacre, Ben 165
Golden Cat 191
Goldschmidt, Tijs 328
Goliath Beetle *346*
Gong Rock 300, 307
Gonolek, Black-headed *229*
   Papyrus 226
Goodall, Jane 52, 71, 408
Goodall, Vanne 52
Gordon, Alasdair 240
Gorilla 209, 210
   Lowland 176
   Mountain 125, 176, 211, 280, 408, 410
Goshawk, Pale Chanting 241
Goudie, Andrew 317
Gould, Stephen Jay 52
Grafe, Thomas 313
Graham, Alistair 290, 406, 421, 429, 430
Graham, Michael 334
Grandison, Alice 296
Grant, Claude 252, 253
Grant, James 94, 160
grasshoppers 334, 341, 344, 345, 346, 359
grasslands 148, 155-6, *158*
Gray, Tom 53
Great Zimbabwe 66-7, *67*
grebes 219
Green Keel-bellied Lizard 260, *261*, 275
Greenberg, Joseph 83, 84
Greenbul 253
   Cabanis' 240
   Eastern Mountain 238
   Yellow-whiskered 240
Greenway, Peter James 8, 150, 159, 161, 170
Greenwood, Humphry 327, 328, 330, 336
Gregory, John Walter 11, 24, 25, 26, 30, 36-7, 38, 41, 107, 118, 161, 204
Gregory Rift 24, *24*, 25, 26, 30, 36, 100, 319
Griffiths, Charlie 346
Grogan, Ewart 291, 324
Groundsel, Giant 160
Grouper, Giant 387
Grove, A.T. 133
Groves, A.W. 26
Grzimek, Bernard 249, 354, 428

Grzimek, Michael 249, 428
Guereza *see* Colobus, Black-and-white
Gueron, Moshe 358
Guineafowl 54, 225, 233
   Crested 232
   Helmeted *229*, 232
   Vulturine 232
gulls 219
Günther, Albert 327
Guppy 325
Gwyne, M.D. 122, 133, 148, 150, 170

Hackett, Shannon 218
Hadado *115*
Hadley, George 126
Hagenia 144
Haggard, Jack 376
Haile Sellassie 50
Haines, Richard 169, 171
Hakansson, Thomas 186, 403
Haldane, J.B.S. 345
Hall, B.P. 222
Hallucigenia 345
Halstead, Bev 57
Halternorth, Theodore 210
Hamerkop 226, *231*
Hamilton, A.C. 107, 131, 134, 141, 171, 230
Hamilton, Bill 208
Hamilton, Patrick 411
Hanley, Gerald 119
Hannington, James 108
Haplochromine 327
Hardinge, Arthur *230*, 242
Hare 54, 176, 178, *179*, 180, 182, 192
Harpachne 157
Harper, Elizabeth 312
Harragin, Walter 406
Harris, John 175
Harris, Wyn 117
Harrison, Nancy 240
Hartebeest 84, 201, 202
   Jackson's *196*
   Lelwel 106
Hartlaub, Karel 241
Harvey, Bill 236
Hastenrath, Stefan 118, 133, 134
Hattle, Jack 134
Hatton, Denis Finch 404
Heaton, Tom 106
Hebrard, James 314
Hedberg, Olov 162
hedgehogs 178
Hedges, Norman 288-9, 312
Heinzelin de Braucourt, Jean de 76
Heliocopris dilloni 345
Heller, Edmund 196, 229
Hell's Gate 15, *15*, 26, *27*, 35, 38, 111, 113, 184, 186, 277, 286, 402
helmetshrikes 225

441

Hemidactylus 275
Hemiscus 296, *297*, 306
Hemphill, Patrick 397
Hennig, Willi 57
Herne, Brian 428
Heron, Black-crowned Night 217, 219
  Goliath 214
  Striated 219
herpetology, history of 277-82
Herring 380
Heuglin, Martin von 241
Heuglin, Theodor von 241
Heuvelmans, Bernard 104, 188
Hewitt, C. Gordon 338
Hickes, G.W. 114
Hildebrandt, Johan 160, 241, 242, 277, 319, 365
Hildebrandt's Elephant-snout Fish 319
Hildebrandtia 160
Hill, Andrew 54
Hill, Clement 104
Hillaby, John 106
Hilton-Barber, Brett 41
Hinkel, Harald 170
Hippopotamus 76, 178, 182, 191, 198, 202, 204-5, *205*
Hirola 172, 178, *195*, 196
Hobby, African 242
Hobley, C.W. 30, 37-8, 41, 104, 113, 208
Hoffman, Richard 348, 350
Hog, Giant Forest 175, 188, *189*, 191, 242
  Red River 176
Hollis, Claud 161
Holman, Dennis 88, 280, 428
Holmes, Arthur 14
Holt, Henry 428
Homa Bay 31, 47, 85, *94*, 99, 226, *229*, *231*
Homa Mountain 47, *47*, 104
hominins 42-64
Homo antecessor 57, 69
  erectus 44, 50, 55, *55*, 56, 57, 61, 63, 68, 69, 73, 74
  ergaster 55, 57
  gautengensis 57
  habilis 50, 52, 61
  heidelbergensis 68, 69
  kanamensis 44
  neanderthalis 44, 61, 62, 68-9
  sapiens 42, 43, 50, 65, 68-9, 82
Honeyguide, Greater 246, 357
Hook, Raymond 188, 202
Hook-nosed Snake 275
Hopcraft, Rick 204
Hopwood, Arthur 46, 47
Hordern, B. 133
Horkel, A. 8
Hornbill 21, 218, 225
  Black-and-white-casqued 226
  Jackson's 242
  Southern Ground 226
  Trumpeter 238
  Von der Decken's 226
Hornby, Charles 96
Howell, Kim 288, 290, 300, 310, 311, 312, 314
humans in Kenya 65-96
  and arthropods 350-64
  and birds 246-51
  and mammals 197-209
  and reptiles 282-8
humidity 130, 131, 168, 238
hummingbirds 224
Hunter, C.S. 103
Hunter, H.C.V. 196
Hunter, John 198, 203, 204, 361, 404
Hussein, Ali Adan *87*
Huxley, Elspeth 111, 242, 404
Huxley, Julian 404
Huxley, Thomas 44
Hyena 106, 182, 206, 248, 420
  Brown 114, 188, 202
  Spotted 180, 188, 191, 198, 202
  Striped 188, 194, 202
Hyperolius 299, 300, *301*, *302*, *307*, 308, *310*, 314
Hyrax 175, 176, 178, 182, 206, 286
  Bush *175*
  Southern Tree 192
Hyrax Hill 46, 48, 77, 78, *78*, *79*, 86, *87*

Ibis, Sacred 241, *373*
Ibrahim, K.M. 171
Iha, Jackson 267, *267*, 280, 282
Imboma, Titus 246
Impala 180, 192, 208, 228
Impala National Park 402
Impatiens 153
Impregnochelys pachytectis 258
independence, Kenyan 95
Indian Ocean Dipole 132-3
Ingrouille, Martin 135
introduced species 112, 113, 138, 166, 324-5, *325*, 332, 336, 396
Ionides, C.J.P. 198-9, 244, 280
Iron Age 80-2
Irwin, Steve 377
Isemonger, Richard 256
Isiolo 98, 149, 180, 208
ivory trade 93, 186, 208, 403-4, 411-13, *414*, 427, 429, 430

Jacana 213, *215*
Jackal 180, 182, 206
  Black-backed 192, 196
  Side-striped 191
Jackson, Frederick 37, 119, 212, 242, 243, 252, 349, 366, 376, 404
Jackson's Centipede-eating Snake 242

Jackson's Tree Snake 242
Jacobs, Bonnie 156
Jameson, James Sligo 241
Jarigole Pillar complex 79
jellyfish 380, 389, 390
Jenkins, Mark 410
Jenkins, Peter 410
Jex-Blake, Arthur 170
Jex-Blake, Muriel 170
Jigger *see* Chigger
Jijabi *420*
Jimba Caves 22, 355
Johanson, Donald 53, 54, 63, 64
Johnny's Child 49-50
Johnson, Harry 277
Johnson, Martin 405, *407*
Johnson, Osa 405, *407*
Johnson, Tom 104, 107, 134, 326
Johnston, Harry 160, 241, 242, 363, 365
Jones, Steve 83, 230
Juba River 101, 160, 196
Jud, Bill 405
junipers 142

Kabochi, Paul 202
Kabuye, Christine 140, 161, *164*, 171
Kadondi, Tom 89, *90*
Kafu River 104
Kairo, James Gitundu 379
Kakamega 17, 33, 34, 74, 98, 125, 129, 136, 141, *142*, 144, *144*, 148, 152, 167, 168, 171, *190*, 226, 237, 238, 240, 269, *272*, 277, 280, 298, 310, 311, 314, 348, 350, 402
Kalb, Jon 62
Kanam 41, 47
Kanam Man 44, 64
Kantai, Parselelo 96
Kapiti Plain 30, 36, 120, 186
Karanja, Francis *419*
Kariandusi 46, 72, *72*, 73, *73*, 110, *111*
Kariandusi River 110
Kariuki, Chege Wa 246
Karomia gigas 151
Kasigau, Mount 229
Kassina 299
  Red-legged 307
  Senegal 302, *303*, 307
  Somali 302, 308
Kassner, Theodore 161
Katonga River 104
Kattwinkel, Wilhelm 41, 44
kayas 163, 166
Keay, R.W.J. 150
Keith, Arthur 50
Keith, R. 314
Keith, Stuart 252
Kelvin, Lord *see* Thomson, William

Kenya, Mount 21, 29, 30, 31, 36, 39, 66, 86, 90, 91, 94, 98, 99, 101, 102, 115-19, *119*, 122, 125, 131, 144, 148, 150, 156, 160, 166, 167, 168, 178, 187, 188, 195, 196, 197, 204, 209, 211, 226, 229, 239, 240, 271, 275, 277, 304, 308, 312, 319, 324, 334, 339, 348, 354, 360, 402, 404, 427
Kenya Wildlife Service 50, 109, 248, 313, 393, 410, 413, 414, 418
Kenyan Bombardier Beetle 338
Kenyapithecus wickeri 71
Kenyasaurus mariakaniensis 20
Kenyatta, Jomo 48, 90, 95, 116, 398, 401
Kericho 85, 98, 129, 136, 188
Kerio River/Valley 30, *31*, 32, 79, 86, *99*, 100, *101*, 106, 306, 319
Kerito, Parmois Ole 204
Kersten, Otto 365
Kestrel, Fox 226
Kfir, Rami 359
Khoisan hunter *69*
Kiandongoro Gate *117*
Kibish River 74
Kibwezi *16*, *19*, 30, 32, 92, 136, 269, 289, 298, *422*
Kielgast, J. 314
Kihansi Falls 310-11
Kiilu, J. 150
Kikuyu 90-1
Kilimanjaro, Mount 21, 29, 30, 31, 35, 36, 82, 92, 98, 102, 115, 118, 122, 148, 160, 186, 287, 319, 350, 352, 403, 425, 428
Killifish 320, 322
  Boji Plain 415
Kilombe, Mount 72
Kimathi, Dedan 95
Kimeu, Kamoya 51-2, 54-5, *60*, 63
Kinangop plateau 30, 72, 73, 74, 273
Kindaruma 98, 102
Kindoruma plateau
King, Lester 120, 122, 133
King Baboon Spider 357
Kingdon, Jonathan 59, 177, 197, 210
Kingfish 393
Kingfisher, African Pygmy 233
  Blyth's 243
  Malachite 213, 233, *234*
Kinloch, Bruce 330
Kinnear, George 406
Kinsey, Alfred 366
Kioko, Joe 410, 414
Kipipiri 30, 113, 178
Kipng'etich, Julius 414

442

Kipsaramon 41
Kisite Island 394, 395, 402
Kisumu 17, 74, 95, 98, 121, 122, 129, *346*, *349*, 402
Kite, Black 223, *223*, 240
   Black-shouldered 216
   Yellow-billed 213
Kitui 30, 160, 280, *286*, 423
Kitum Cave 207, *207*
Kiunga 394, 402
Klipspringer 192
Kob 192, *192*
Kohler, Jorn 314
Koitobos River 304
Kokwaro, John 162, 170
Kolb, George 116
Komodo Dragon 258
Kongoni 180
Koobi Fora 41, 50, 51, 52, 53, 55, 62, 72, 77, 258
Kora 171, 314, 348, 402, 408
Koru 41
Krapf, Johann Ludwig 21, 94, 115
Kudu 202, 208
   Greater 108, 175
   Lesser *209*
Kulal, Mount 158, 161, 230, 277, 417
Kwale 21, 34, 82, 417

Labeo 320, 324, 331
Lack, Peter 8, 237
Laetoli 41, 54, 58, 62, 63
lagoons 372-8
Laikipia 32, 35, 38, 86, 156, 208, 211, 319, 354, 402, 419, 421
Lak Bor River 98
Lak Dera River 98, 101
Lake Albert 24, 103, 104, 243, 328, 331
Lake Amboseli 102, 115, *219*
Lake Baringo 23, 35, 36, 39, 56, 73, 100, 102, 107, 212, 218, 237, 280, 283, 319, *333*, 334
Lake Bogoria 23, 24, 30, 35, 100, 102, 108, 111, 182, 319, 402
Lake Chala 102, 115, 283, 319, 325
Lake Chew Bahir 36, 102, 105
Lake Edward 24, 76, 89, 103, 328
Lake Elmenteita 23, 78, 85, 100, 102, 109, 110-11, 159, 204, 237, 319
Lake Fly *340*
Lake George 104, 105, 328
Lake Jipe 102, 115, 319
Lake Kamnarok 402
Lake Kanyaboli 325
Lake Kivu 24, 89, 328, 329
Lake Kyoga 85, 104, 330

Lake Logipi 100, 102, 107
Lake Magadi 24, 30, 32, 100, 101, 102, 110, 111, 114-15, 150, 319, 327
Lake Malawi 24, 160, 326, 327, 335, 336
Lake Manyara 204, 408, 429
Lake Naivasha 15, 23, 25, 30, 35, 71, 74, 77, 100, 102, 107, 108, *108*, 111-14, 136, 148, 149, 178, 205, 212, 239, 240, 269, 273, *305*, 319, 325, 334, 354
Lake Nakuru 24, 30, 76, 78, 100, *108*, 109-110, 111, 148, 152, 171, 182, 236, 237, 273, 319, 325, 402, 403, 415, 428
Lake Natron 24, 50, 98, 100, 101, 114, 246
Lake Ol Bolossat 102
Lake Oloiden 111
Lake Paradise 405, *407*
Lake Rudolf *see* Lake Turkana
Lake Rutundu 102
Lake Stephanie *see* Lake Chew Bahir
Lake Suess 26
Lake Tana 249
Lake Tanganyika 24, 102, 103, 326, 327, 328, 330, 408
Lake Turkana 22, 23, 25, 26, 30, 32, 36, 41, 42, 43, 50, 51, 52, 54, 57, 63, 66, 74, 76, 77, 78, 79, 86, 88, 96, 95, 100, 103, 105-7, 108, 158, 161, 175, 176, 253, 258, 262, 277, 283, 316, 317, 318, 319, 320, 324, 325, 331, *332*, 334, 335, 354, 404, 415
Lake Victoria 35, 38, 42, 47, 48, 63, 66, 73-4, 78, 80, 82, 89, 91, 95, 98, 99, *100*, *101*, 102-5, *105*, 126, 129, 134, 136, 150, *163*, 176, 197, 226, *234*, 258, 311, 315, 318, 319, 324, 325, 326-33, 334, 335, 336, 341, 350, 353, 354
Lake Zwai 86
Lamb, David 63
Lambiris, Angelo 313
Lammergeier 209, 236
Lampeye, Naivasha 113
Lamu 21, 22, 33, 88, 92, 98, 125, 150, 161, 277, 369, 370, 371, 376, 378, 379, 380, 382, 394, 396
land disputes 421-2
landscape
   formation 11-32
   processes 97-121
Lane, Margaret 280
Lang, Judith 374

languages 77, 83-9
Lankester, Edwin Ray 160
Lanner 249
Lantana 166
Lapparentophis 258
Largen, Malcolm 288
Lark 225
   Masked 168
   Rufous-backed *228*
   Williams's 229
Larsen, Torben 348, 349, 364, 400
Laurent, Raymond 313
Le Pelley, Richard 346, 364, 366
Le Vaillant, Francois 241
Leakey family 62
   Jonathan 46, 49, 50, 273, 280, 281, 282, 304
   Louis 20, 21, 37, 38, 39, 45-52, 63, 64, 71, 161, 218, 244, 279-80, 356, 408
   Mary 48-9, 52, 54, 63-4, 73, 75, 295
   Meave 22, 51, 55, 57
   Richard 41, 46, 50-1, 52, 55, 56, 63, 64, 367, 413-14, 418, 422, 427
Leakeyornis aethiopicus 218
Lebrun, Jean-Paul 135, 150
legends and superstitions 90, 104, 115, *147*, 163, 187-8, *245*, 246, 251, 258, 260, 267, 269, 271, 286, 287-8, 292, 302, 313, 360, 377
legislation, wildlife 425-6, 427
Leishman, William 354
Leishmania 354
Lembirdan 203
Lemudong'o 41, 258
Leopard 75, 182, 184, 191, 194, 198, 199, 201-2, 206, 209, 211, 220, 246, 279-80, *401*, 401, 404, 408, 418, *419*
Leptopelis 298, 302, 308
Leuthold, Walter 8, 237
Lewis, Adrian 237, 252
Lewis Glacier 36, 117, *117*, 118
lice 350
lichen *146*
Lichtenstein, Martin 241
Lieske, Ewald 396
Lily, Pyjama 136
Limuru 136, 258, 262, 298, 308
Lind, E.M. 170
Linnaeus, Carl 241, 291, 292
Linsenmair, Karl 313
Lion 75, 84, 106, 113, 170, 172, 180, *183*, 184, 191, 196, 198, *201*, 208, 209, 210, 211, 246, 248, 401, 404, 406, 408, 410, 418, 419, 420, *421*, 423
   man-eaters 95, 186, 199-201, 403, 410
lionfish 387

literature 38-9, 62-4, 133-4, 169-71, 209-11, 252-4, 288-90, 311-14, 335-7, 364-7, 396-7, 428-30
livestock 76-8, 156-8, 197
   diseases 77, 86, 206, 208, 359, 367
Livingstone, David 93, 94, 103, 327, 403
Lobelia deckeni 160
lobsters 389
Locke, Eric 267
Locust 117, 342, 346, 359-60
   Desert 341, 350, *351*, 359
   Green Milkweed 359, *366*
   Migratory 341, 359
   Red 341, 359
Lodwar 33, 45, 98, 124, 131, 357
Loe, Ian 364
Longclaw, Pangani *231*
   Sharpe's 228, 229
Longonot, Mount 23, *23*, 29, 30, 32, 35, 36, 111, 113, 402
Loring Brace, C. 56
Losai 402
Lothagam 41, 50, 76, 258
Lötschert, Wilhelm 169
Lotters, Stefan 314
louries 225
lovebirds 35
Loveridge, Arthur 278-9, 280, 288, 290, 304, 313, 417
Lovett, Jon 150, 170, 171
Lucky Bean Tree 146, 163
Lugard's Falls *17*, 250
Luke, Quentin 135, 150, 151-2, 162, *167*
Lukwara 104
Lumi River 115
Lungfish *235*, 317, 327, 330, 332, *333*
   Marbled 316, 331
Lye, Kare 169, 171
Lynch, Mark 80

Maasai Mara 77, 100, 142, 148, *158*, 172, 180, *181*, 183, 184, 187, 207, 211, *233*, 236, 237, 240, 249, *259*, 271, 277, 349, 354, 401, 401, 402, 403, 406, 408, 410, *411*, *417*, 418, 420, *421*, 423, 427, 428
Maathai, Wangari 399, 422
Maboko Island 41, 176, 317
Mabruk, Fundi 37
Macalders Mine *33*, 34, *34*
MacArthur, G.C. 404
MacDonald, David 176, 177, 210
Machakos 92, 117, 204, 206, 248
Macheru, Humphrey 280

443

MacInnes, Donald 46
MacKay, Alex 289, 313, 314
MacKay, Joy 289
Mackinder, Halford John 116, 118, 133
Mackworth-Praed, Cyril Winthrop 252, 253
MacLarnon, Anne 61
Maclean, Gordon 225, 235, 253
Makalia Burial Site 46
Malaki, Philista 246
Malewa River 74, 100, 113
Malindi 92, 94, 98, 226, 282, 285, 316, 318, 369, 371, 373, 374, 390, 394, 402, 403
Malka Mari 168, 402, 415
Maloiy, Geoffrey 314
Malombe, Itambo 162
Malonza, Patrick 159, *282*, 282, 314
Mamba, Black 56, 220, 221, 269, 270, *270*, 271, *281*, 285, 286, 287, 363
  Green 56, 91, 242, 267, 269, 270, *270*, 275, 280, 285
  Jameson's 269, 270
mammals 172-211
Mangabey, Tana 196
mangroves 378-80, 396
Mani, Mahadeva 350
Manjonga 44
Mann, Clive 252
Manson, Patrick 352
maps
  protected areas 402
  relief 98
  vegetation zones 143
Mara River 98, 100, 104, 122, 201, 203, 204
Marais, Johan 288
Maralal 38, 86, 318, 360, 402
Marchant, Rob 134
Margulis, Lynn 328
marine habitat 368-71
  deep water 380-3
  shallow water 371-80, 383-90
  species 368, 390-7
Markhamia 163
Marlin 382
Marnham, Patrick 63
Marsabit 88, 98, *129*, 161, 208, 211, 277, 402, 406
Marsabit, Mount 21, 29, 31, 78, 86, *128*, 130, 149, 158, 237, 405, *407*
Marshall, Fiona 77, 80
Martin, Esmond Bradley 403
Martin, W.R.H. 106, 107
martins 221
Martins, Dino 166, 334, 341, 365, *366*
mass extinctions 18, 22
Masters, Karen 310
Mastodonsaurus 294
Matekwa, Luka 269, 280

Mathews, Lloyd 36
Mathews, Rick 322
Mathews, Terry 88, 204, 205
Mathews Range 18, 36, 86, 149, 158, 159, 196, 203
Mau 26, 74, 77, 100, 111, 113, 144, *145*, 150, 166, 188, 226, 277, 280, 308, 312, 318, 300, 319, 417, 428
Mayr, Ernst 55
Mazrui, Ali 68, 95
Mbagathi River 180
Mbamba Swamp *235*
McCarthy, Michael 222
McClanahan, Tim 133, 170, 392, 396, 397
McDonald, P.G. 153, 171
McDougal, Charlie 199
McGahan, H. 133
McKenna, Virginia 408
McKie, Robin 62
McMillan, Northrup 404
McShane, Thomas 430
Mearns, Edgar 161, 229
Mears, Ray 197
Measey, John 312
medicine, traditional 164-5
Meinertzhagen, Richard 180, *189*, 204, 211, 242-3, 287, 403
Mellars, Paul 96
Menegon, Michele 312
Menengai 29, 30, 109
Mengistu, General 400
Mereroni River 110
Merille River 88
Mertens, Robert 267, 278
Meru 168, 171, 178, 187, 298, 299, *309*, 311, 325, 334, 349, 402, 410, 411
Meru, Mount 29
Mesquite Bush 166
Meyer, Axel 328
Mida Creek 378, 379, *379*, 384, 387, 394
Migori 38, *39*
migration 100, 172, *173*, 220-4, 418
  humans 68-9, 82-92
Milankovitch, Milutin 132
Milimu, Job *144*
Millennium Man 56, 58
Miller, Charles 200
Miller, Jack 52
Miller, Scott 346, 365
Miller, Stanley 14
Millet 77, 80, 136
millipedes 338, 344, 348, 350
minerals *12*, 18, 32-4, 114
Minshull, John 320
Miraa Tree 136, 169
Miskell, John 236, 302
Mitchell, Philip 406, 427
Mitre Shell 79
Mittermeier, R.A. 176

Mk'iara, Ephraim 117
Mohamed (ranger) 429
Moi, Daniel arap 413, 427
Mokoyeti Gorge *123*
Mole-rat 196
  Aberdare 192
  Naked 173, *174*, 192, 211
  Splendid 197
Mole Snake 112, 228, 257, 274
moles 178
Molo River 107
Mombasa 21, 22, 35, 36, 37, 93, 95, 98, 125, 131, 159, 160, 242, 302, 366, 369, 370, 372, 394, 402
Mombasa Island 382, 394
Money Cowrie 79, 390
Mongoose 182, 221
  Banded *181*, 191, 196, 211
  Dwarf 194
  Marsh 191
  Slender 191
Monitor 260
  Nile 191, 258, *259*, 262
  Savanna 262
monkey bridges 426, *429*
Monkey Chair 151
Monod, Theodore 150
Moore, Ray 253
Moorish Idol 384, *386*
Moreau, Reginald 221-2, 225, 238, 254, 314, 424
Morel, Marie-Yvonne 234
Morell, Virginia 63
Morgan, Dave 251
Morrison, M.E.S. 170
Mosquito Fish 325, 353
mosquitoes 198, 206, 308, 339, 342, 346, 350, 352, *352*, 353, 364
Moss, Cynthia 211
mosses and liverworts 152
moths 339, 341, 346, 359, 364, 367
mountain formation 17, 29-32
Mouse, Striped Grass 184, *185*, 341
Mousebird 218, 225
  Speckled 240
Moyer, Dave 310
Mpunguti Islands 394, 395, 402
Msuya, Charles 314
Mtito Andei 30, 169, 199, 269, 404
Muchai, Vincent 314
Mudanda Rock 121, *121*
Mudskipper 379
Muguga 161, 208, 282, 367
Mukolwe, Evans 414
Mukusu Tree 136, 141
Mukutan River 107
Mumbi, Cassian 134
Munyao, Kisoi 117
Munyekenye, Fred 238
Murchison, Roderick 34

Murchison Falls 318, 330
Muriuki, Geoffrey 90
Mus 177
Musila, Simon 239, 240
Mutisya, Mutui 280
Mutubio Gate 30, 148
Mwachala, Geoffrey 161
Mwadeghu, John *166*
Mwea 402
Mwingi 277, 402
Myers, Norman 168
Myers, Robert 396
Myers, Thomas 94
Mzima Springs 205, *259*, 320, *321*, 324

Nabokov, Vladimir 366
Nairobi 29, 30, 50, 52, 90, 95, 98, 102, 124, 125, 130, 142, 144, 152, 161, 168, 170, 201, *213*, 226, 235, 236, 258, 260, *260*, 262, 300, 302, 304, 319, 337, 353, 356, 360, 364, 366, 404, 405, 408, 421
Nairobi Eye 350, 360, 361
Nairobi National Park 29, *123*, 180, 182, 184, *189*, *201*, 202, *217*, 234, 236, 298, 402, 398, 401, 403, 406, 407, 408, 410, 413, *414*, 420, 422, *423*, 425, 427
Nairobi River 311
Nairobi Snake Park 168, 245, 267, 279, 280-1, 282, 313
Nakuru 46, 86, 98, 110, 302
Namoratunga 79, 80
Nandi *137*, 226, 236, 239, 240
Nandi Bear 115, 187, 188
Nandi Flame Tree 138, 142
Nanyuki 98, 116, 122, 126, 239, 257, 285
Napier, Evelyn 159, 161
Napier, John 49, 50, 59, 60
Nares, Peter 282
Nariokotome 41, 50, 54, 72
Narok 29, 98, 101, 258
Nasalot 402
National Museum of Kenya 45, 46, 48, 49, *59*, 64, 150, 159, 178, 209, 236, 254, 278, 279, 280, 313, 335, 337, 367, *416*, 417
national parks 183, 277, 402
  see also under named parks
NDere Island 402
Ndithia, Hendry 246
Ndoto Mountains 142, 158
Neanderthal Man 44, 61, 62, 68-9
Needle Grass 148
Neem Tree 163
Nelson, Charles 79
Nelson, Gareth 57
Nephila 357, *357*

Neumann, Arthur 88, 94, 106, 404, 428
Newman, James L. 65, 83, 91, 96
Newman, Ken 235, 245
Newmark, William 184
Newton, Len 153, 162
Ngaia Forest 152, *167*, 168, 299, 308, *309*
Ngangao Forest *166*, 299, *309*
Ngeneo, Bernard 50, 51
Ng'ethe, Sam 410
Ngoe, Hemedi 199
Ngomeni 280
Ngong Hills 29, 30, 35, 106, 180, 401
Ngulia Hills 18, *28*, 223, 425
Ng'weno, Fleur 240
Nicholson, Brian 199
Nightjar 136, 161, 223, 246, 251
   Long-tailed 236, *245*
Nile River 100, 102, 104
Njoro *239*
Njoro River Cave 46, 48, 62, 78
Njoroge, Peter 226, 234, 237, 238, 239, 246, 253
Noad, Tim 169
Nobel, Alfred 110
Noble, Gladwyn Kingsley 288
Noor, Queen of Jordan 399
Northey, Edward 279
Norton-Griffiths, Mike 430
Nothobranchius bojiensis 322
Nutcracker Man 49
Nyambeni Hills 29, 31, 90, 91, 130, 152, 196, 269, 271, 277, 300, 308
Nyamweru, Celia 31, 38-9, 117
Nyerere, Julius 330
Nyeri 90, 98, 199, *415*
Nyingi, Wanja Dorothy 337
Nyiro, Mount 152, 158, 159, 162, 417
Nzoia River 100, 324

Oak, Meru 144
   Silver 142
Oakley, Kenneth 44
Obama, Barack 400
Obura, David 374, 396, 397
Octopus 387, 389, *393*
Oddie, Bill 254
Odhiambo, Elizabeth 337
Odhiambo, Nicholas 282
Odhiambo, Thomas 366
Ogendo, Reuben 102, 133
Ogot, Bethwell 64, 66, 95, 367
O'Hara (road engineer) 200
Ohingas 89-90
Ojany, Francis 99, 102, 107, 125, 133
Okapi 160, 183, 194
Okeyo, Daniel 319, 337
Ol Donyo Sabuk 402

Olduvai Gorge 41, 43, 44-5, 46-7, 48, 50, 63, 71, 72, 75, 262, 295
Olindo, Perez 410
Olive 154
Oliver, Roland 91, 95
Ololokwe, Mount 138, 139, 196
Olorgesailie 30, 37, 38, 41, 43, 48, *51*, 72, 73, 198, 239
Olympic Snake 417
Omo Delta/River/Valley 36, 50, 52, 72, 76, 106, 176, 258, *426*
online resources 171, 234, 250, 293, 302, 311, 312, 313, 333, 336, 337, 357, 365, 367
Onyango, Joel *90*
Onyango-Abuje, John 64
Opabinia 345
Opondo, Paul 334
orbital forcing 132-3
Orb-web Weaver Spider *357*
orchids 167, 169, 170, 244
Oribi 191, *193*
Ormsby-Gore, William 406
ornithology, history of 241-6
Orrorin tugenensis 56, 64
Oryx 194, 280
   Beisa 192
Osprey 219
Ostrich 214, *217*, 225, 242, 246, 247-8, 249
Otter, Clawless 196, 325
   Spot-necked *178*
Oulton, Tom 404
Outlaw, Robert 232
ovenbirds 224
Owen, John 418
Owl 200, 286, 423
   Barn 219
   Grass 241
   Marsh 241-2
   Pel's Fishing 238, 241
   Scops 226, 251
   Spotted Eagle 251
   Wood 238, 242
Owlet, African Barred 242
Oxpecker 242
   Red-billed *230*

Pachycrocuta 198
Painted Lady 349
Pakenham, Thomas 96
palaeanthropology 40, 42-64
Palm 169
   Borassus 170
   Date *137*
Pangolin 178
   Giant 63, 241
Panicum maximum 148
papyrus 112, 148
Parabuthus 358
Paranthropus 49, 57
   boisei 49, 50

   robustus 48
Pare Mountains 82, 92, 115
Parker, H.W. 288, 334
Parker, Ian 88, 106, 242, 403, 404, 418, 429
Parrot 213, 219, 249
   Grey 226
Parrotfish, Bumphead 387
   Seagrass 376
Parthenium hysterophorus 166
Passmore, Neville 313
Patas Monkey 58, 192
Pate Island 94, 394
Patrick, David A. 312
Patterson, John Henry 199-200, 403, 427
Paulinia acuminata 112
Peake, F.G.G. 366
Pearson, David 234, 237, 253
Pel, Hendrik Severinus 241
Pelican 245
   Great White 110, 111, 214
   Pink-backed 239
Pemba Island 22, 251
Penck, Walther 120
Peninj River 50
Perch, Nile 104, 107, 317, 318, 330-3
Percival, Arthur Blayney 198, 202, 208, 209, 211, 260, 263, 324, 404, 427
Percival, Philip 209, 248, 405
Peregrine Falcon 219, 249
Periwinkle, Madagascar 164
Perret, John-Luc 313
Pesi Valley 187
Peterhans, Julian Kerbis 200, 201
Peters, Karl 37, 277
Peters, Wilhelm 277
Peters' Sand Lizard 277
Petropedetes dutoiti 304
Phelps, Tony 288
Phillipson, David 71, 72, 73, 82, 96
Philothamnus 267
Phlyctimantis 299
Phrynobatrachus 304
Piapiac 226
Picasso Bug *341*
Picker, Mike 346, 364
Pickersgill, Martin 296, 312
Pickford, Martin 39, 56, 58, 64
Pigeon 217, 242, 249
   African Green 238
pigs 175, 178, 182
Pilchard 380
Piltdown Man 44, 45
Pinchot, Gifford 398
Pinhey, Elliot 364
Pinker, Steven 43
Pipistrelle, Mount Warges 192
Pipit 221
   Golden 226

Plain-backed *228*
Pitman, Charles 280, 289-90, 330, 430
Pitta, African 238
plankton 380
plant trade 167
plants 135-71
   see also vegetation zones
Plasmodium 352, 353
Plover, Blacksmith 253
   Ringed 220, 222
Plowright, Walter 208
poaching 186, 395, 404, 406, 410-13, 423, 427, 428
Podocarpus 107, 138, 139, 141, 144, *415*
Pohl, W. 8
Poison Arrow Tree 162, 359
pollen 140-1
pollination 166, 339, 341, 355
pollution 112-13, 308, 310, 334, 353, 375, 392, 396, *400*
Pomeroy, Derek 237, 252
Poole, Joyce 204, 211, 411, 429
Poppy, Opium 164
Porcupine Fish *394*
porcupines 180, 404
Porites 387
Portal, Gerald 94
pottery 76, 78, 79, *79*, 82, 89, 90, 92
Potto 178
Pounds, Alan 310
poverty 400-1, 403, 421
Poynton, John 311
Pratt, D.J. 122, 133, 148, 150, 170
Praying Mantis *365*
pregnancy tests 307
Preston, Florence 95
Preston, Richard 207
Preston, Ronald 95
Prinia, River 234
Pritt, Denis 48
Proconsul 48, *49*, 176
Prodeinotherium hobleyi 38
protected areas 394-5, 415-18
Prys-Jones, Robert 243
Pterodon 175
Ptychadena 295, 304
Puffback, Black-backed 240
pufferfish 318
Pundamilia 330
Pupfish, Devil's Hole 333
Pyle, Robert Michael 9
Python, Reticulated 270-1
   Rock 112, 191, 263, 286, *286*

Quechon, Gerard 82
Quelea 235
   Red-billed *236*, 249-50
Quetzalcoatlus northropi 214
Ragati River 324
Rahole 402

445

railways 95, 113, 114, 138, 186, 199, 200, 209, 334, 403
rainfall 8, *22*, 38, 50, 54, 58, 80, 89, 99, 106, 113, 118, 119, 122, 123, 124-30, 131, 132, 133, 136, 138, 141, 142, 144, 146, 148, 149, 150, 151, 155-6, 157, 158-9, 168, 169, 187, 197, 225, 258, 262, 275, 292, 296, 298, 322, 324, 334, 350, 353, 380
Ramisi River 325
Rasa, Anna 211
Rat 196, 206
  Black 178, 219
  Brown 178
  Crested 174
  Giant Pouched 191, 206
  Mill 191
  Shaggy Swamp 177
Raven, P. H. 171
Ray, Blue-spotted Lagoon 377
Reader, John 62
Reagan, Ronald 55
Reagan, Tate 336
Reck, Hans 20, 44, 46, 47
Redman, Nigel 253
Redspotted Beaked Snake 267
Reece, Jane 176, 345
Reichenow, Martin 241
Remora remora 282
reptiles 255-90, 417
Rhesus blood groups 366
Rhinoceros 54, 176, 180, 182, *182*, 198, 202, 206, 209, 210, 400, 404, 410, 423, 425
  Black 38, 172, 175, 186, 191, 204, 211, 412, *419*, 419, 427
  White 176, *177*, 178, 191, 280, 411, 415, 419
Rhipsalis 148
Rhodes, Cecil 35, 66
Richmond, Matthew 396
Rider Haggard, Henry 242
Ridgeway, Rick 428
rift valleys 23-9, 30, 32, 36, 37, 38, 100, 103, 125, 129, 168, 313, 319
Riftobuthus 348
Rimoi 402
Rinkhals 304
Ritchie, Archibald 204, 404, 405, 430
Ritchie, Mark 367
Ritson, P. 133
Riva, Domenico 159
Robber 320
  Dwarf Lake Turkana 324
Robbins, Lawrence H. 76, 96
Robertshaw, Peter 77, 96
Robin 232
  White-browed Scrub 239
  White-starred 228

rock paintings *81*, 355
rocks 12-13, 15-20, 21, 27, *39*, 41, 42, 138, 372
Rodder, D. 314
Rodwell, Edward 174
Rogo, Lucie 346, 365
Roller 223
  Lilac-breasted 213, *214*
Roosevelt, Teddy 112, 161, 196, 229, 405, 427
Root, Alan 54, 205, 363, 408
Root, Joan 205, 363, 408
Ross, Ronald 352
Rotich, Damaris 282, 290
Rotich, Nehemiah 414
rove beetles 361
royal family, British 95, 102, 405, 407
Ruark, Robert 184, 198, 408
Rudolf, Crown Prince 36
Ruff 220
Rufiji Delta 92, 201
Rufous Beaked Snake *278*
Ruiru River 319
Ruma 184, *196*, *198*, *237*, 354, 402
Rumuruti 30, 152, 419
Rüppell, Eduard 277
Rushby, George 200, 201
Rusinga Island 38, 41, 48, *49*, *103*, 139, 176, 218, *218*, 258, 317, *335*, 245
Ruspoli, Eugenio 159, 161
Ruwa, Renison K. 397
Rwenzoris 25, 29, 118, 133, 177
Ryall, Charles 200
Ryan, Peter 230, 253

Sabaki River 30, 98, 100, 102, 229, 253, 283, 319, 320, 324, 373
Sable Antelope *203*
Sagalla, Mount 299, 308
Sahelanthropus tchadensis 56, 60-1
Sailfish 382
Saiwa Swamp 402
Sajita, Nickson 240
Salamander, Chinese Giant 214
salinity
  lakes 36, 104, 105, 108, 109, *109*, 110, 114, 246
  soil 123
Salt, George 350
Salvadori, Cynthia 91, 96
Salvinia molesta 112, 166
Salzburger, Walter 328
Samburu 101, 107, 182, 183, 192, 202, 203, 232, 237, 238, 284, 402
Samoilys, Melita 397
Sandai River 108
Sanderling 220
sandflies 342, 350, 354, *355*, 364

Sanimu 54
Sansevieria 44
Saola 176
Sardine 324
  Athi 336
  Lake Victoria 332, *335*
Saseeta 203
Sattler, Bernhard 20
Sausage Tree *147*
Savalli, Udo 238
savanna 119-21, 146, 155, 156
Savanna Vine Snake *266*, 275
Saw-wing, Black 240
Scaphiostreptus 344, *344*
Scarlet Tip *343*
Schabel, Hans 364, 365
Schaller, George 200, 211
Scheffler, Georg 161
Schick, Susanne 314
Schifter, Herbert 239
Schiotz, Arne 312, 314
Schluter, Thomas 38, 39, 42, 106, 107
Schmidt, Karl Patterson 288
Schnitzer, Eduard 35, 241, 365
Schouw, J.F. 150
Sclater, William 242
Scorpion 106, 338, 342, 344, 346, 348, 350, 358, 364
  Black-tailed *339*
Scorpionfish 384, *387*, 390
Scott, Francis 80
Scott, Jonathon 211
Scott-Eliot, G.F. 161
sea 369-71
sea anemones 389, 390
sea skaters 339
Sea Snake, Yellow-bellied 377
sea spiders 344
Sea Turtle 220, 255, 257, 282, 378
  Leatherback 256
Sea Urchin 377, 378, 390, *391*
seagrass 373, 376, 396
Secretary Bird 184, 218, 225, 242
Seedeater, Streaky 213, 240
Seegers, Lothar 319, 337
Seif, Mohammed 116
Selempo, Edwin 254
Selenops 357, *357*
Semliki River 25, 160
Semlitsch, Raymond 312
Senut, Brigitte 56, 58
Serengeti 77, 157, 187, 200, 204, 208, 211, 249, 287, 300, *307*, 354, 417-18, 428
Sergeant, Indo-Pacific *385*
Sergoit, Mount 177, 188
Sertich, Joseph 22
Serval 182, 184, 191, 341
Sewu, Sharp-spined 331
Seyabei River 101
Shaba 101, 192, 402, 408
Shackleton, Robert 38

Shariffu, Juma Gitau bin 48
Shark 316, 390
  Blacktip Reef 387, *388*
  Bull 319, 382
  Oceanic Whitetip 397
  Tiger 383
  Whale 214, 382
  Zebra *382*
Shebane 106
Sheldrick, Daphne 211, *405*, 428
Sheldrick, David 314, 406, 410, 411
Sheldrick Falls *120*, 354
shells 79, 389-90, 396
Sheuyange, Asser 156
Shimba Hills 20, 34, *120*, *127*, *130*, 135, 151, 152, 167, 168, 171, 178, *179*, *203*, *206*, 226, 262, 277, 299, 310, 351, 354, 399, 402, 420
Shimo La Tewa 394, 395
Shimoni 22, 99, 149, 369, 373, 394
Shimoni Tree *276*
Shine, Rick 270
Schiotz, Arne 300
Shipman, Pat 63, 70, 158, 198
Shipton, Eric 117, 133
Shoebill 234, *235*
Shrew 172, 173, 175, 176, 183, 196, 197
  Giant Otter 178
  White-toothed *193*, 210
Shrike, Magpie 253
  Northern White-crowned 160
Siafu *see* Ant, Safari
Sibiloi 277, 402, 415
Sibley, Charles 218
Sickle Bush 153, *153*
Silverbird *237*
Simiyu, Stella 171
Simon, Noel 428
Sinclair, Ian 230, 253
single-species hypothesis 55-6
sirikwa holes 86, *87*
Sitatunga 192
Sivatherium 175
Sjostedt, Yngve 365-6
Skaapsteker, Kenyan Striped 112
Skimmer, Black-tailed *349*
skin colours 83
Skink 257, 258
  Percival's Legless 260
  Peters' Writhing 277
  Red-flanked 274
  Striped 260, 266
  Western Serpentiform 260
Skinner, Hugh 289
skulls 43, 44, 45, 48-56, 58, *59*, 61, 62, 63, 68, 73, 74
slave trade 93, 201, 403
slope development *120*
Slug-eater, Common 257
Small Copper 349

Smeenk, Chris 8, 237
Smith, Andrew 241, 264, 277
Smith, Anthony 37
Smith, D.F. 334
Smith, J.L.B. 316, 397
Smithers, Reay 210
Snake Bush *149*
snakebites 165, 246, 264, 265, 273, 283-5
snakes 84, 153, 159, 160, 191, 221, 239, 242, 255, 256, 257, 262-73, 278, 287, 288, 304, 308, 330, 335, 399, 417, 423
snipe 213
Snoeks, Jos 328
snowfinches 243
Sodom Apple 154, *165*
soils *37*, 104, 121-4, 142, 326
Solomon, John 47
Sondu River 100, 102
Sorghum 77, 80
South Island 106, 402
South Kitui 402
South Turkana 157, 402
Soyinka, Wole 68
Soysambu Estate 110-11, 204
Sparrman, Anders 241
Sparrow 224, 225
  Grey-headed 234
Sparrowhawk, Eurasian 221
  Rufous-breasted 242
Spawls, Stephen 288, 290, 312
Speckled Sand Snake 106, 168, 275
Speke, John Hanning 85, 94, 102, 103, 160
spiders 338, 342, 344, 345, 348, 350, *357*, 357, 364
Spinage, Clive 211
Spirulina platensis minor 110
Spodoptera exempta 359
Spoerry, Anne 358
Spoonbill *108*
Sporobolus 157
Spotted Bush Snake *279*
Spotted Lion 187-8, 334
Spurfowl, Red-necked 226
Squeaker 317, 320
  Common 296
  Lake Victoria *322*
  Tana 336
Squirrel 178, 182, 206
  Red-bellied Coast 192, *206*
Ssemmanda, Immaculate 104
Stager, Curt 326, 328
stalk borers 350, 359
Stamp, L. Dudley 132
Stanley, Henry Morton 94
Starfish 377
  Crown-of-thorns 390
Starling 160
  Greater Blue-eared *224*
  Stuhlmann's 240
  Violetbacked 223

Steenbok 180
Steinhart, Edward 430
Stephanie, Princess 36
Sternfeld, Richard 277-8
Stevenson, Terry 229, 234, 246, 253
Stewart, James 281
Stewart, Joyce 169
Stewart, Kathlyn 317
Stewart, Margaret 313
Steyn, Peter 249
stick insects 344
Stigand, Chauncey Hugh 428
stingrays 377, *378*, 390
Stinkwood, Red 154, 164, 168
stints 241
Stoat 221
Stone Age 69-75
stonefish 371, 387, 218
Stoneman, John 330
Storer, T.I. 345
Stork, Abdim's 223
  African Openbill *215*
  Marabou 213, 218, 249, 250
  Saddle-billed *249*
  White 221, 222
Striped Bark Snake 277
Strum, Shirley 211
Stuart, Chris 210
Stuart, Tilda 210
Stuhlmann, Franz Ludwig 365
Suam River 98, 106
Suckermouth 320
Van Someren's 336
Sudex Peak 119
Suess, Eduard 26
Suguta River/Valley 25, 32, 100, 107, 319, 324
Sunbird 225, 244
  Bronze 160, 240
  Collared 240
  Golden-winged 226
  Malachite 226
  Red-chested 226
  Scarlet-tufted Malachite 160
  Tacazze 226
  Variable 240
sunis 233
Surgeonfish 390
  Powderblue 384, *385*
Sutton, John 83, 86, 96
Swainson, William 241
Swallow 221
  Barn 222
  White-headed Saw-wing 34
Swallowtail, African Mocker 366
Swamphen, Purple 112
swifts 244
Swilla, Imani 312
Swordfish 382
Syingi, Alex 240
Tachyoryctes 197
Taita Hills 18, 92, 130, 138, *139*, 149, 166, *166*, 226,

229, 230, *274*, 277, 287, 298, 299, *305*, 306, 308, *309*, 314, 353, 415
Tana Delta/River 12, 23, 35, 84, 88, 90, 92, 98, 100, *101*, 101-2, 122, 132, 152, 178, 196, 208, 226, 230, 277, 283, 299, 311, 318, 319, 322, 324, 334, 336, 337, 354, 369, 373, 376, 378, 380, 390, 396, 402, 403, 423
Tanacetum cinerariifolium 353
tanagers 224
Tandy, J. 314
Tandy, M. 314
Tanganyika Train 344, *344*
Tansley, Arthur 149
Tarantula Hawk *347*
Tarlton, Alan 286, 404
Tarlton, Leslie 248
Tattersall, Ian 57
Taung Child 45
Taylor, Royjan *281*
Tchagra, Brown-crowned 226
Teleki, Count 32, 36, 94, 106, 116
Teleki's volcano 32, 36
Temminck, Coenraad 241
temperatures
  air 106, 124, 125, 130-1, 144, 168, 369
  lakes 104, 107, 114, 327
  sea 131, 369, 370
  soil 148
Tendaguru Hill 20, 38, 44
termite hills *124*, *263*, *278*
termites 156, 339, 342, 348, 350, *362*, 362-3
Tern, Gull-billed 219
Terrapin, Helmeted 258
  Serrated Hinged 217
tetras 320
Thackeray, John 39
Tharaka 195, 262, 287, 355
Thelotornis 265
Theropithecus oswaldi 198
Thesiger, Wilfred 119, 360, 410
Thika 30, 98, 136, 366
Thimlich Ohinga *89*, *90*
Thiong'o, Ngugi wa 412
Thomas, Oldfield 242
Thomson, Joseph 35, 94, 110, 113, 160
Thomson, T. 160
Thomson, William 14
Thorburn, Archibald 243, *244*
Thrush, Abyssinian Ground 226
  Olive 160, *227*, 229-30
  Taita 229
Thua River 102
ticks 156, 339, 342, 344, 350, 359, 364
tides 370-1

Tigania *91*
Tiger Fish 107, 318
Tiger Snake *266*
Tilapia 96, 104, 201, 317, 334
  Athi River 112, 325
  Blue-Spotted 112
  Graham's 331, 334
  Korogwe 325
  Lake Chala 114
  Lake Magadi 110, 114, 325, 334
  Mozambique 325
  Nile 332, *332*
  Redbelly 112, 325
  Redbreast *325*
  Sabaki *317*
  Three-spot 325
  Victoria 331
Tilbury, Colin 288
Tilman, Bill 31, 117
Tinkerbird, Yellow-rumped 240
Tishkoff, Sarah 96
Tit, Mouse-coloured Penduline 214
Tiva River 102, 319
Tiwi *429*
Toad 291, 292, 302
  Bunty's Dwarf 296, 307
  Desert 302, 308, 314
  Garman's *305*
  Golden 308
  Kerinyaga 308
  Kihansi Spray 310-11
  Kisolo 304, 308
  Lake Turkana 304, 308
  Leopard 304, 308
  Nairobi 304, 308
  Steindachner's 308
Tobias, Philip 50
toolmaking 44, 46, 47, 48, 51, 58, 61, 68, 69-76, *78*, 90
top predators 184, 186-7
Topi 184
Tornier, Gustav 277
Tortoise 255, 273
  Leopard 228, 275
  Pancake 258, 275, 277
tortoiseshell combs *395*
Toth, Nicholas 64
Toumai 56, 60-1
tourism 111, 172, 184, 207, 212, 247, 277, 392, 396, 399, 400, 401, 405-8, 410, 418, 419, *421*, 424, 427
Travers, Bill 408
Travis, William 378, 389
Treetops Hotel 407, *409*, 427
Trench, Charles Chenevix 110, 119
Trent, Michael 187
Trewavas, Ethelwynn 327, 328, 336
Triggerfish, Clown *391*
  Picasso 384, *384*
Trinil 44

447

Trogon, Bar-tailed 228
  Narina's 241
Trout 324
Trueb, Linda 311
Trump, E.C. 150, 169
Trypanosoma spp.
Tsavo 16, 81, 95, 99, 123, 186, 197, 201, 203, 220, 226, 238, 262, 277, 300, 284, 350, 404, 406, 410, 411, 420, 425, 427
  East *17*, 88, 79, 102, *162*, 178, *181*, *186*, 237, 240, 250, 402, 412, *413*
  West *37*, 102, 115, 121, 205, 207, 223, 226, *249*, 277, 402, 408
Tsavo River 98, 102, 198, 403
Tsetse Fly 77, 85, 209, 342, 350, *351*, 353-4, 364
Tugen Hills 29, 56, 107, 139, 140, 156, 317, 318
Tumbu Fly 350, 355
Tuna 382
Tunai, Peter Ole *73*
Turaco 35, 218, 225
  Black-billed 238
  Great Blue 34, 141, 226
Turkana Boy 55, *60*, 61
Turkwell River 98, 100, 106, 319, 324, 354
Turner, Don 113, 234, 246, 252, 253
Turner, Myles 204, 417-18
Turtle 255, 258, 423
  Green 376, *377*
  Nile Soft-shelled 107, 258
Tyrannosaurus rex 22

Uasin Gishu plateau 30, 31, 35, 86
Uaso Narok River 319
Uaso Nyiro River 100, *101*, 101, 114, 318, 319, 322, 324, 354
Ukambani 92, 120, 160, 186, 188, 276, 277, 349, 353, 404
Umbrella Thorn 141
Ungwana Bay 122, 373, 394
Unicornfish *381*
Urban, Emil K. 245, 252
Usambara Mountains 161, 221
Usinger, R.L. 345

Van Couvering, John 139
Van Perlo, Ber 252
Van der Post, Laurens 364
Van Someren family
  Gurner (Chum) 239, 244
  Robert 243
  Vernon 243, 336
  Victor 243-4, 279

Van der Stel, Simon 241
Vangueriopis shimbaensis 168
Vavilov, Nikolai 168
vegetation zones 142-59, 192, 226, 274-5, 307-8, 349
Veith, Michael 314
venomous animals 173, 174, 264-73, 281-6, 306, *306*, *339*, 355, 357, 371, 377, 389-90
Venus Flytrap 137
Verdcourt, Bernard 161, 169
Verheyen, Eric 328, 329
Verreaux, Jules 241
Verrill, Hyatt 287
Vervet Monkey 182, 207
Vieillot, Louis 241
Viper, Gaboon 271, 273, 280
  Kenya Horned 271, 273, *274*, 275, 276, 334
  Kenya Montane 125, 257, *257*, 271, 275
  Mount Kenya Bush 271, *271*, 275, 277, 280, 415, 423
  Rhinoceros-horned 34, 141, 271, *272*, 273, 274, 308
  see also Carpet Viper
Vipingo 136, *373*, 395
Vitt, Laurie 288, 311
Voelker, Gary 232
Vogel, Edouard 241
Vogel, Gernot 288
Voi 30, 32, 98, 136, 186, 187, 199, 200, 258, 262, 263, 277, 298, 299, 350, 411, 412, *425*
volcanic activity 23, 26-32, 117
Von der Decken, Karl Klaus von der 160, 241, 242, 365
Von Hohnel, Ludwig 36, 94, 106
Von Lettow Vorbeck, Paul 20, 95, 198, 243
Vonesh, James R. 312
Vulture 242, 248
  Egyptian 71
  Hooded 233, 238
  Rüppell's Griffon *233*
  Turkey 247
  White-backed 233, *233*
  White-headed 249

Wabuyele, Emily 153
Wagtail, African Pied 226
Wahlberg, Johan August 241
Wahoo 380
Wajir 21, 66, *87*, 98, *115*, *124*, 136, 148, 202, 262, 298, 300, 354, 358
Wakefield, Thomas 242
Walker, Alan 52, 55, 56, 57, *60*, 61, 63, 70
Walker, Jane Brotherton 359

Wallace, Alfred Russel 44, 384
Wamba 106, 196, 204, 206
Wambuguh, Oscar 250
warblers 221, *221*, 222, 224
Warges, Mount 196
Warthog 175, 180, *181*, 191, 208, 404
  Desert 192
Warui, Charles 357
Waseges River 108
Wasonga, Victor 282, 314
Wasp 346, 355
  Paper 289, 350, *351*
Wass, Peter 171
Wasser, Samuel 150, 170
Watamu 56, 258, 262, 265, 269, 280, 281, *281*, 269, *372*, 374, *374*, 387, *388*, 394, 402
Water Hyacinth *108*, 112, 166
Water Lily 112, *163*
water towers 144, 426
waterbuck 180, 196
waterfalls *31*, 34, 35, 101, *120*, 310-11, 324
Watson, Robert 325
Watt, John 170
Watteville, Vivienne de 117
Wattle-eye, Jameson's 241
waxbills 225
Wayland, E. J. 26, 29, 38, 71, 73, 326
weather 104-5, 107, 124-33, 129
weathering 16, 97-121
Weaver 219, 224, 225, 244
  Baglafecht 240
  Clarke's 226, 229
  Golden 226
  Grey-capped Social 239
  Jackson's Golden-backed 242
  Vieillot's Black *225*
  White-headed Buffalo 226
  Yellow-mantled 220
Weaving, Alan 346
Webb, Cecil 281
Webi Shebelli River 277
Weldon, Che 307, 308
Werger, M.J.A. 133, 171
Western, David 54, 414
Western, Johan 429
Whale 178, 394
  Blue 172, 213, 380
  Killer 380
  Sperm 380
Wheatear, Northern 222
Wheeler, Peter 61
White, Frank 150, 171, 225, 274, 307, 349
White, Tim 41, 53, 54, 56-7, 64, 175
white-eyes 224, 230
Whitehead, Peter J.P. 336, 243

Whitehouse, Commander 103
Whittet, Robert 404
whydahs 225
Widowbird, Jackson's 242
  Whitewinged 239
Wiens, John 311, 312
Wild Dog 180, 191, 198, 207, *420*
Wildcat 206
Wildebeest 75, 100, 156, 172, 180, 184, 192, 208, 211, 220, 418
wildlife trade 246, 247, 248, 276, 341, 392, 397, 423
Willey's Kopje 46
Williams, John G. 229, 234, 236, 237, 238, 239, 240, 244, 252
Williamson, John Thoburn 32
Wilson, D.E. 176
Wilson, E.O. 366, 399
Winam Gulf 17, 24, 32, 64, 74, 85, 89, 103, 104, 354
winds 105, 126-8, 370
Witchweed 166
Wolpoff, Milford 56
Wood-hoopoe, Forest 220
  Green 253
Woodley, Bill 88, 406, 428
Woodley, Bongo 428
Woodley, Danny 428
woodpeckers 251
Woosnam, Richard Bowen 404
Worm Lizard 255, 258, 262
  Voi Wedge-snouted *256*
Worm Snake *264*
  Drewes' 264
Worthington, E.B. 334
Wrasse, Humphead 387, *388*
Wykes, Alan 280

Xenopus 295, 306-7

Yala River/Valley 74, 82, 100
Yamagata, Toshio 134
Yatta Plateau 15, *17*, 30
Young, Truman 133, 170, 397

Zaphiro, Dennis 204
Zebra 113, 178, 180, 182, 184, 198, 201, 206, 400
  Burchell's 192, 220
  Grevy's 168, 175, 192
Zimkuss, Breda 312
Zimmerman, Dale 230, 234, 235, 246, 253
Zinderen Bakker, E.M. van 171
Zinjanthropus boisei 49
zoogeography 188-97, 225-32, 273-5, 307-8, 318-25, 345-50
Zorilla 191, 196
Zug, George 288, 311